Control and Coordination in Hierarchical Systems

Wiley IIASA International Series on Applied Systems Analysis

9 International Series on
Applied Systems Analysis

Control and Coordination in Hierarchical Systems

W. Findeisen
*Institute of Automatic Control,
Technical University, Warsaw*

F. N. Bailey
University of Minnesota

M. Brdyś
K. Malinowski
P. Tatjewski
A. Woźniak
*Institute of Automatic Control,
Technical University, Warsaw*

A Wiley—Interscience Publication
International Institute for Applied Systems Analysis

JOHN WILEY & SONS
Chichester – New York – Brisbane – Toronto

British Library Cataloguing in Publication Data:

Control and coordination in hierarchical systems.—
   (International series on applied systems analysis).
   1. Control theory
   I. Findeisen   II. Series
   003   QA402.3   79-41726
   ISBN 0 471 27742 8

Typeset in Northern Ireland at The Universities Press (Belfast) Ltd.
and printed and bound in Great Britain at the Pitman Press, Bath.

Pamięci dr Stanisława Kurcyusza, znakomitego naukowca i
niezapomnianego kolegi

To the memory of Dr Stanisław Kurcyusz, a brilliant scientist
and a colleague whom we will never forget

Preface

The foundations of a theory of hierarchical systems were first laid down in 1970 in the classic work of M. D. Mesarović, D. Macko, and Y. Takahara, *Theory of Hierarchical, Multilevel Systems*, which provided a set of theoretical coordination principles. While this study was the impetus for much further work, both theoretical and practical, it did not focus on practical aspects. This volume treats the aspects of the theory of hierarchical systems that will be useful for applications.

Its authors are members of a team that has worked on the subject for about ten years. Fredric N. Bailey is at the University of Minnesota, and the other authors are with the Institute of Automatic Control of the Technical University of Warsaw. There has been close cooperation between the two universities in various areas of control science for several years. The authors have also had the privilege of close contact with many other places where important work is being done on hierarchical systems. Although the book incorporates important results obtained elsewhere, most of it consists of the results of research done in Warsaw and Minneapolis.

The book as a whole presents the views and achievements of a team rather than of individual authors, except perhaps for Chapter 5, which is mainly the work of Professor Bailey.

Part of the research on which this book is based was supported by NSF Grant GF-37298, which is gratefully acknowledged.

The International Institute for Applied Systems Analysis has agreed to include this volume in the International Series on Applied Systems Analysis. This decision, as well as the encouragement and assistance offered to the authors, are also acknowledged with appreciation.

The work of IIASA's editor Robert Lande made a very important contribution to seeing this book from manuscript into its published form; the authors acknowledge this contribution with special appreciation.

Contents

Introduction

The purpose of this book is to present the theory of control and coordination in hierarchical systems—that is, in systems where the decision-making responsibility has been divided. Since it aims to present theory that will be useful for applications, it not only encompasses the basic, general, and consequently somewhat abstract principles of coordination, but also considers such practical features as differences between models and the reality they describe, constraints, possible use of feedback information, and time horizons.

The control of complex systems may be and very often actually is structured hierarchically for several reasons. For example:

- The decision-making capability of an individual is limited, but it can be extended by the hierarchy in a firm or organization
- Subsystems (parts of the complex system) may be far apart and have limited communication with one another
- There is a cost, delay, or distortion in transmitting information
- Subsystems make decisions autonomously using private information (e.g., in the economic system)

It is to a certain degree irrelevant whether we discuss a hierarchy of human decision makers or a similar multilevel arrangement of computerized decisions, as long as we assume that both use the same rational bases. The book is intended to be useful for both cases. In particular, the structural principles and the features of the coordination methods, e.g., constraints and sensitivity problems, apply to both human and computerized decision making. For computerized decisions, we also need numerical procedures and algorithms; wherever possible, these are discussed with recommendations as to which procedures are best suited to which situations.

Since the book deals mainly with control problems, two assumptions are essential:

- The system under control is in operation and influenced by disturbances.
- Current information about system behavior or disturbances is available and can be used to improve control decisions.

These two assumptions differ from those of studies of problems of planning and scheduling, where the only data we can use to determine a control or a policy come from an *a priori* model, and where, therefore, the accuracy of results depends heavily on and is limited by the exactness of the models. In contrast, the success of control systems, the performance obtained, depends also on the structure in which we use the feedback information.

Chapter 1 is an introduction. It explains, with illustrative examples, the multilayer and multilevel structures used in hierarchical control systems. It presents the possibility of separating the stabilization task from the optimization task, the allocation of different time horizons to adjacent layers of the hierarchy, and the partitioning of the optimization problem into coordinated subproblems. It also briefly presents the informational aspects of hierarchical systems, and describes the difficulties involved in the formulation of control problems.

Chapter 2 describes decomposition and coordination as applied to optimization problems. These concepts provide a background and a source of ideas for hierarchical control structures. The presentation in this chapter reflects the need for optimizing complex, interconnected systems and stresses, for example, equality constraints in the problem. There is also some emphasis on the dependence of the solutions on the parameters in the models. Procedures and algorithms are presented that are especially designed for applications: the accuracy of the solution may be sacrificed in order to reduce the number of iterations. Open-loop control structures follow directly from the coordination methods discussed.

Chapter 3 is devoted to control and coordination of steady-state systems, that is, to static problems. Steady-state control is discussed separately from dynamic control because in steady-state control we may use feedback information iteratively. Sections 3.2 and 3.3 present complete results for the structure in which the measured outputs of the subsystems are available only to the coordinator, i.e., only the coordinator receives feedback. In particular, the sensitivity of performance to the differences between the model and reality is shown. Section 3.4 describes the other case, where the feedback information is available to local decision-making units, which must, at the same time, respond to the coordinator. The existing theoretical results are less comprehensive in this case, but they are adequate to explain the main properties of this structure. Section 3.5 deals with another question: if we admit that the real system is time-varying, how often should the coordinator intervene? Answers to this question are available for at least some of the cases. Iterative coordination procedures are essentially applicable only in steady-state control. A dynamic version of these procedures (iterations in function space) can be produced easily and may be used in step-by-step improvements of batch processes. The methods used to control a subsystem must be feasible; that is, they must not violate real constraints that may not be reflected in the model because of its limited accuracy. The last section of Chapter 3 discusses this problem.

Chapter 4 presents a few structures and procedures of coordination that are applicable to dynamic control, with an appropriate use of feedback from the system. Possible structures based on three different principles are presented: the dynamic Lagrangian, state feedback, and conjugate variables. Dynamic price coordination (the dynamic Lagrangian approach) is discussed extensively in two sections. Section 4.4 gives practical procedures for price coordination for static systems with dynamic inventory couplings. The dynamic price coordination structure seems to be the most natural one for decision-maker hierarchies; it makes considerable use of the so-called repetitive control algorithms, whose properties are studied in section 4.5. State feedback multilevel structures are presented in section 4.6; unlike other structures, they are restricted mainly to linear systems.

Chapter 5 is devoted to problems of information in hierarchical structures. It shows some ways in which the value and the cost of information can be incorporated into the system design. In many systems the information costs, especially the cost of communication, are significant and must be considered in choosing control structures. If we decide to have restricted information flows in the system, then the coordination procedures and the performance will be affected. The last section of the chapter presents some of the results.

The theory presented in this book may assist in:

- Explaining the behavior of existing systems
- Designing the structure of a new system (for example, determining what decisions are to be made at each level, what coordination instruments are to be used)
- Implementing decision making based on the computer

The book is addressed to researchers, high-level practitioners, and to students of control science and decision theory in organizations.

1 Complex Systems and Hierarchical Control

1.1. INTRODUCTION

THE CONTROL SYSTEM

Control is a means of influencing an *object* to behave in a desired way. The object may be an economic system, a technological process (such as a chemical plant), a water-resource system, or an ecological system. Thus, control problems have been worked on in many fields, differing terminologies have arisen, and different approaches have been developed. Consequently, it is worthwhile to identify at the outset some basic notions and formulations relating to control systems, and at the same time provide a rudimentary glossary of terms.

Figure 1.1 presents a typical control system schematically, showing the *controlled system* (also called process, plant, object, environment) and the *control unit* (controller, decision unit). The controlled system has some *manipulated inputs* (denoted by vector m) which may cause a change in the *outputs* y (also called *outcomes*). The controlled system is also subject to another group of inputs, which are beyond our influence and are termed *disturbance inputs* or *disturbances z*. Disturbances should essentially be considered random variables.

The task of the control unit (which in a complex case may be an arrangement of several units) is to determine the value of m that achieves a certain goal, i.e., that meets a certain specification on the behavior of the controlled system. We refer to the values determined by the control unit as *control decisions* or simply *controls*. The control unit bases its current control decisions on *observed variables* or *observations* (*measurements*, denoted by v) that are related to the controlled system or the disturbances or both. In shaping the control decisions, the control unit—along with the

4

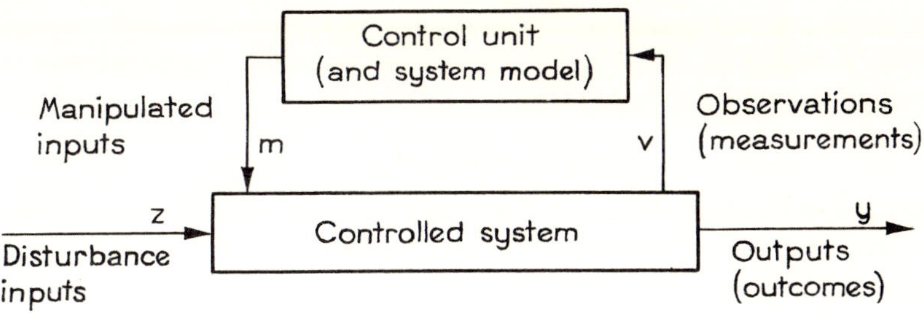

FIGURE 1.1   A schematic drawing of a control system.

observations—makes use of its understanding of the controlled system. We refer to this understanding as the *system model.*

Let us proceed to a more formal presentation of what we have said so far. We have to state first that in many controlled systems of practical interest the output y at a particular time t, that is, $y(t)$, depends not only on the inputs $u(t)$, $z(t)$ at the same instant but also on all their past values (we say that the system is *dynamic* or is a system *with memory*). We could never know the past values for all t ranging from infinity to the present, but this difficulty can be overcome when the concept of the *state* of a system is introduced. The vector of *state variables* x is such that its value at t_0, denoted by $x(t_0)$, and the inputs m and z over the interval $[t_0, t]$ determine the state $x(t)$ uniquely.

An example of a state variable is the water level in a reservoir. To know the level $x(t)$ one would have to know all the past inflows and outflows, *or* the level at a particular time t_0 and the inflows and outflows thereafter.

We can say that $x(t_0)$ "summarizes" all the past inputs.

For the state $x(t)$ we write

$$x(t) = \phi_{[t_0,t]}(x(t_0), m_{[t_0,t]}, z_{[t_0,t]}) \tag{1.1}$$

and then we can assume for the output

$$y(t) = g(x(t), m(t), z(t)), \tag{1.2}$$

by which we say that the output depends on the past inputs through the present state of the system. (We note that Eq. (1.1) would have a slightly different form for a system with time delays.)

Note that (1.1) and (1.2) combined yield

$$y(t) = \eta_{[t_0,t]}(x(t_0), m_{[t_0,t]}, z_{[t_0,t]}), \tag{1.3}$$

which is a direct way of expressing the dependence of the present output on the inputs over the interval $[t_0, t]$, but also on the initial state $x(t_0)$.

It should be noted that in some circumstances the variations of state are negligible. This happens when some of the manipulated inputs are used to enforce a *steady-state* condition. The output y may, however, still depend on the (time-varying) inputs m and z. The value of state x_s is a parameter in this dependence (compare Eq. (1.2)):

$$y(t) = g(x_s, m(t), z(t)). \qquad (1.2')$$

It should be noted that in Eq. (1.2′), $y(t)$ is related to $m(t)$ at any given time, but the dependence changes over time because of the external input $z(t)$. Equation (1.2′) describes a *static, time-varying system*.

The observation $v(t)$ may be assumed to depend on the values of $x, m,$ and z at any given time; thus,

$$v(t) = h(x(t), m(t), z(t)), \qquad (1.4)$$

it being noted that all dynamics are included in Eq. (1.1).

Figure 1.1 indicates that the control unit is generating control decisions related to observations. We can assume that this is being done according to a rule

$$m(t) = d(v_{[t_0, t]}), \qquad (1.5)$$

which will be referred to as the *control law* or *control strategy* (or the *decision rule*). Note that in Eq. (1.5), decision $m(t)$ may be based on observations made over a time interval $[t_0, t]$.

We have mentioned earlier that the control decisions aim at achieving a certain goal. It may be, for example, that we are given a *preference ordering* of the outputs,

$$y_1 < y_2 < y_3 < \dots .$$

The goal of the control unit would then be to obtain a value of y with the highest possible ordering. If y is not a single variable but a vector, preference ordering is difficult to achieve. It may, however, be done in a simple form, in which one defines control decisions that maximize (or minimize) a scalar-valued *performance index Q (pay off, utility, welfare function)*.

In many cases we are interested in optimizing the time-integrated value of the performance. In that case two equivalent formulations are being used:

$$Q = \int_{t_0}^{t_f} q(t)\, dt, \qquad Q = \frac{1}{t_f - t_0} \int_{t_0}^{t_f} q(t)\, dt,$$

where $q(t)$ is the value of the performance rate at time t, t_0 is the *initial time*, and t_f is the *final time*. The interval $t_f - t_0$ is referred to as the *control horizon*. What one chooses for the control horizon is very important in some applications.

The actual formulation of a performance index is up to the *user*, that is, it depends on what he decides is of value to him. It may be, for example, that the value of q is explicitly related to $y(t)$, $q(t) = f_0(y(t))$, which means that we attach a value to the outputs only. But it may also be $q(t) = f_0(y(t),$

$m(t), v(t))$ which means that we consider not only the outputs, but also the cost of control and the cost of observation.

We should not overstress the importance of "optimal control." It will be optimal only when we are really able to justify reducing the general goal to a scalar performance index. We may be asked, for example, to design a control arrangement where the main requirement is that the output or the state of the controlled system follow some given path (a desired trajectory). We may still require that the distance of the real trajectory from the desired one be strictly minimized, or that the follow-up be done at minimum cost, and so on, but such optimality requirements would here play a secondary role in the system design.

When the goal of the control has been formulated in one way or another, for example in the form of a performance index, we are able to discuss not only the best possible control law but also the structural aspects of the control system. For the simple case of Figure 1.1, we may be first concerned with *information structure*: what should the form of Eq. (1.4) and the arguments in $h(\cdot)$ be, i.e., what should be observed or measured in the controlled system? The preferred information structure would be such that the obtainable result of control (optimal value $\hat{Q}$ of performance) is better.

It remains to say something about treating the uncertainty or randomness involved in the operation of the control system because of the disturbances z. The value of Q depends on both the control m and disturbance z. We may, according to our needs, set the goal of the control unit to be the maximizing of the expected value of Q, or the maximizing of the worst-case value of Q:

$$\text{maximize } E[Q], \qquad \text{maximize } \min_{z} Q$$

or set up some other formulation that is appropriate from the user's point of view.

The decision problem, i.e., the problem of determining m, is referred to as *deterministic* if we assume that z has a known value or is a known function of time. If the probabilistic properties of z are known and we make an appropriate use of them, the decision problem is *stochastic*. If we use the deterministic approach, it is important to investigate how the control system behaves when the real disturbance differs from the value used in the model: we must require that the performance of the control system be relatively insensitive to this difference between the model and reality. We return to this question often.

THE COMPLEX SYSTEM

A *complex system* is an arrangement of elements in which outputs are connected with inputs, as, for example, in an industrial plant (Figure 1.2). If

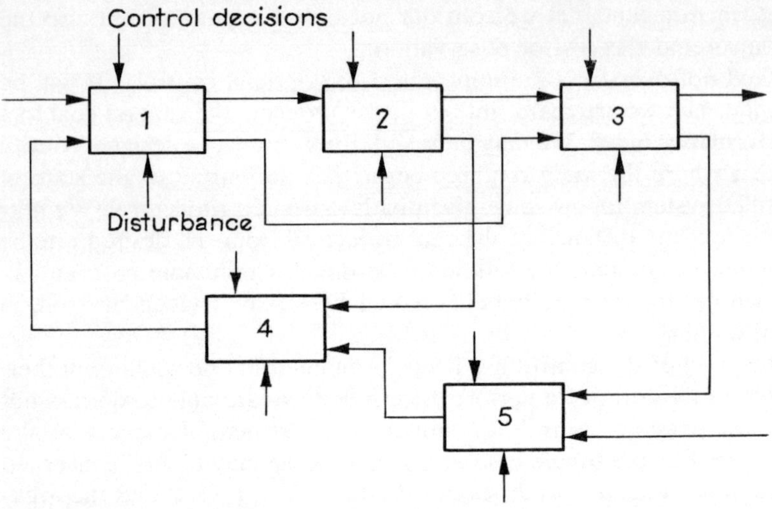

FIGURE 1.2   A schematic drawing of a complex system.

we introduce an orderly input–output interconnection matrix H, we obtain a scheme as in Figure 1.3. The matrix H represents the *structure* of the system. Each row of this matrix is associated with a single input of a subsystem. The elements in the row are zeros except where a one shows the single output that the given input is connected to.

HIERARCHICAL CONTROL CONCEPTS

We are now interested in controlling systems represented by Figure 1.3 by use of some special *hierarchical* structures. There are two fundamental and

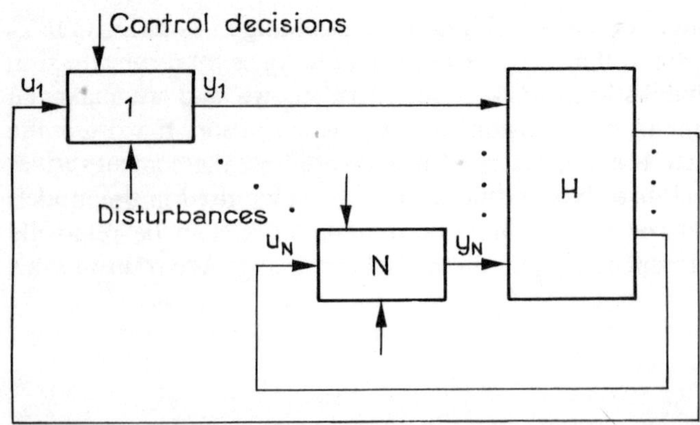

FIGURE 1.3   The complex system presented with an ordering matrix H.

by now classical ideas in hierarchical control:

- The multilayer concept (Lefkowitz 1966), where the control of an object is split into algorithms, or *layers*, each of which acts at different time intervals.
- The multilevel concept (Mesarović *et al.* 1970), where control of an interconnected, complex system is divided into local goals, local control units are introduced, and their action is coordinated by an additional supremal unit.

MULTILAYER SYSTEMS

The *multilayer concept* is best depicted by Figure 1.4, where the task of determining control m is split into:

- Follow-up control, causing *controlled variables* c to be equal to their desired values c_d.
- Optimization, or an algorithm to determine optimal values of c_d, assuming some fixed parameters β in the model of the plant or the environment or both.
- Adaptation, with the aim of setting optimal values of β.

The vector of parameters β may be interpreted more generally as determining also the structure of the algorithm performed in the optimization layer, and may be divided into several parts that are adjusted at different time intervals. Thus, we may have several adaptation layers.

The most essential features of the structure in Figure 1.4 are the interaction of the layers at different points in time and with decreasing frequency, and the use in each layer of some feedback or information from the environment. The latter linkages are shown in the figure by dotted lines.

Structures like Figure 1.4 are usually associated with controlling industrial processes, e.g., chemical reactors or furnaces, but the structure can be applied elsewhere. For example, a similar division of functions may exist in a hierarchy of decision makers, where the higher level of authority prescribes the target values (performs the optimizing control), and a lower level makes the detailed decisions necessary to achieve these targets. It performs a kind of follow-up control and it may not know the criterion of optimality by which the target values have been set.

A system like the one represented in Figure 1.4 is designed to implement control m, which cannot be strictly optimal because the actions of the higher layers are discrete and are thus unable to follow the strictly optimal continuous time pattern. It is hoped, however, that the total cost of performing all control calculations (the total effort of making all decisions) is

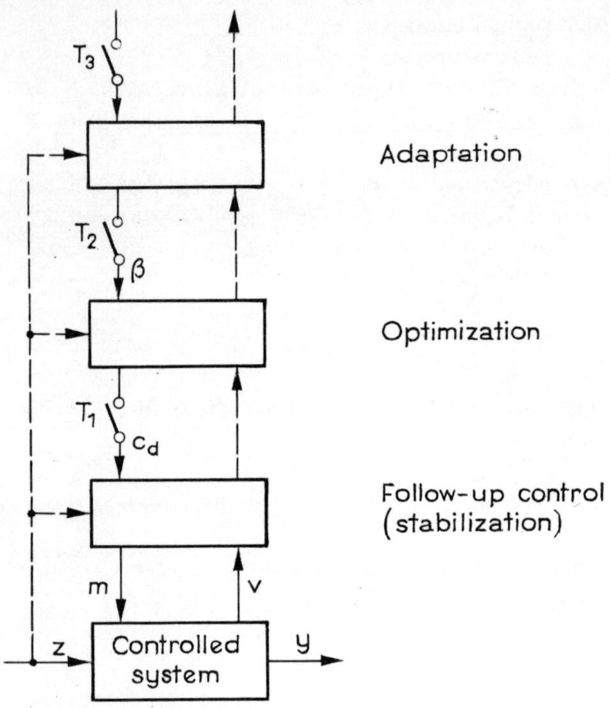

FIGURE 1.4   Multilayer control—a functional hierarchy.

less when the structure of Figure 1.4 is used than it would be without the functional division, that is, when the current control m is determined directly by an optimization algorithm. The essential problem must therefore be the tradeoff between loss of optimality and the computational and informational cost of control. A practical problem of this kind is difficult to formalize so as to permit effective theoretical solutions.

The multilayer concept can also be related to a control system where the dynamic optimization horizon has been divided, as illustrated in Figure 1.5. These two features are now essential:

- Each of the layers considers a different time horizon; the highest layer has the longest horizon.
- The *model* used at each layer, or the degree to which details of the problem are considered, is also different: the top level is the least detailed.

Control structures of the kind presented in Figure 1.5 have been most widely applied, for example, in industrial or other organizations and in production scheduling and control. These applications seem to be ahead of formal theory, which in this case—as it also was for Figure 1.4—fails to

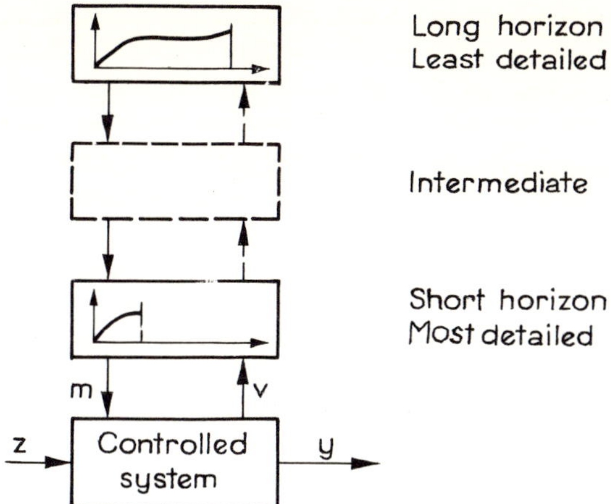

FIGURE 1.5   Multilayer system formed by separating the time horizons.

supply explicit methods to design such systems. For example, we would like to determine how many layers to form, what horizon to consider at each layer, and how simple the models may be. The answers to such questions have to be found on a case-by-case basis.

MULTILEVEL SYSTEMS

The *multilevel concept* of hierarchical control, which makes it essential to introduce local goals and appropriate coordination, has been inspired by decomposition and coordination methods developed for mathematical programming or for solving other kinds of formally specified problems. We should especially note the difference between the following approaches:

• Decomposition used to solve optimization problems, where we operate with mathematical models only and the goal is to save computational effort
• The multilevel approach to on-line control, where these features are important: the system is disturbed and the models are inadequate, reasonable measurements are available, no vital constraints can be violated, computing time is limited

Mathematical programming decomposition can be applied directly only as an open-loop control, though with model adaptation, as shown in Figure 1.6. But here in fact, any method of solving the optimization problem can be used and the results achieved will all be the same, depending on model accuracy. Nevertheless, it is highly desirable to study and develop decomposition methods in programming, even for control. The open-loop structures like the one represented in Figure 1.6 should not be dismissed, since

12

they offer the advantages of inherent stability and fast operation. Structuring the optimization algorithm as in Figure 1.6 with the multilevel approach may also yield efficient computational methods in the software and allow multicomputer hardware arrangements. We devote Chapter 2 to decomposition and coordination of optimization problems of the *nonlinear* and *dynamic* kind encountered in control applications. In this book we shall pay much more attention to the multilevel structures of control that use feedback information from the real system to improve control decisions. Figure 1.7 illustrates what we mean.

In Figure 1.7, there are *local decision units* and a *coordinator*, whose aim is to influence the local decision units to achieve the overall goal. All these units use information in the form of mathematical models of the system elements, but they may also use observations.

If we now look at the hierarchical systems as a whole (compare Figures 1.4, 1.5 and 1.7) we see that they have one feature in common: the decision making has been divided. Moreover, it has been divided in a way leading to hierarchical dependence. This means, that *there exist several decision units in the structure, but only some of them have direct access to the controlled system. The others are at a higher level of the hierarchy—they define the tasks and coordinate the lower-level units, but they do not override their decisions.*

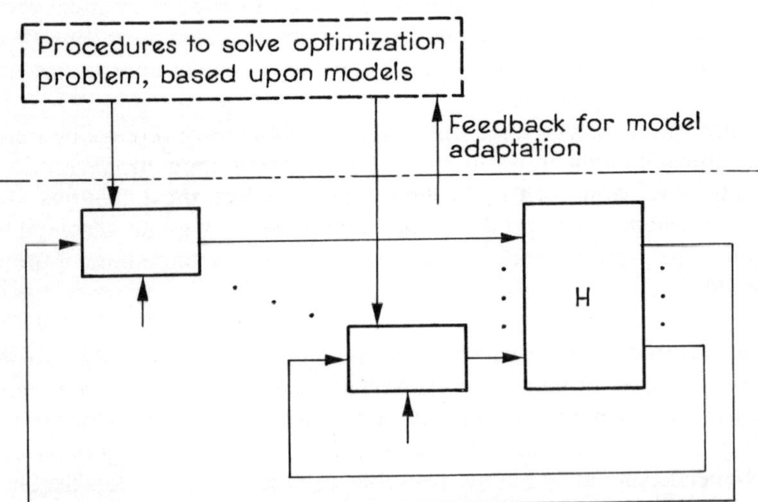

FIGURE 1.6   Open-loop control of a complex system with model adaptation.

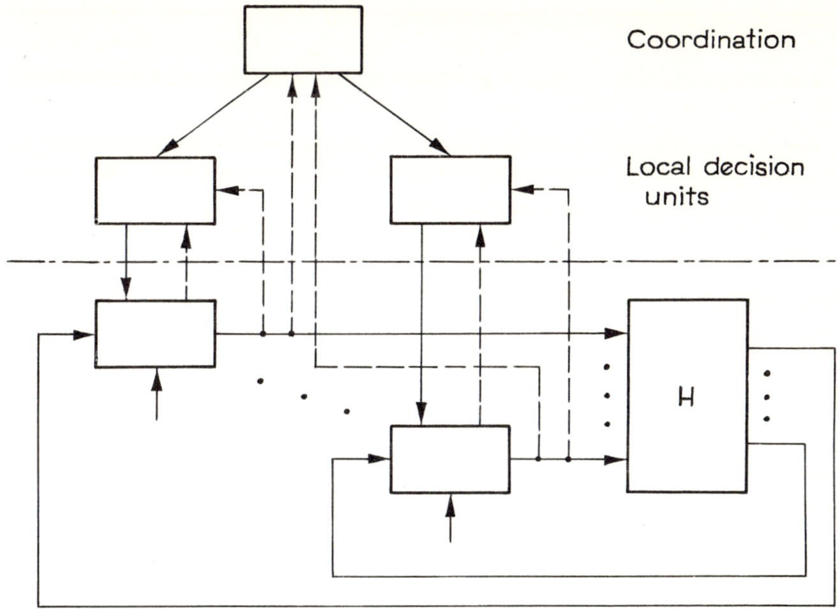

FIGURE 1.7   Multilevel control of a system. Dotted lines show possible feedback paths.

As explained in the Introduction to the book, there may be many reasons for the division of the decision making. In terms of human decision making, one person would be unable to make all the decisions required to run a complex organization. Moreover, the subsystems of a large organization often are distant from one another and the transmission of information is both expensive and subject to distortion. Parallel, decentralized decision making suggests itself as a solution and is indeed satisfactory as long as the subsystem or local goals are not in conflict. If they arc in conflict, a coordinating agent is necessary and the hierarchy depicted in Figure 1.7 results.

In industrial control applications the trend towards hierarchical control can be associated with the technology of control computers. The advent of microprocessors has made control computers so cheap and handy that they are being introduced at almost every place in the process where previously the so-called analog controllers had been used. The information processing capabilities of the microprocessors are much more than is needed to replace the analog controllers and they may easily be assigned an appropriate part of the higher-layer control functions, e.g., optimization.

Some of the other reasons for using decentralized or hierarchical rather

14

than centralized structures of control are:

- The desire to increase the overall system reliability or *robustness* so the system survives if one of the control units breaks down.
- The possibility that the system as a whole will be less sensitive to disturbance inputs if the local units can respond faster and more adequately than a more remote central decision unit.

The purpose of studying hierarchical control systems may be twofold: we may be interested in the *design* of such systems for industrial or organizational applications, or we may want to know how an existing hierarchical control system *behaves*. The second case applies to economic systems, for example. The focus of the two cases differs very much, as do the permissible simplifications and assumptions that can be made in the investigation.

For example, if we want to design a multilevel system like the one represented in Figure 1.7, we would have to deal with questions like:

- What kind of coordination instruments should the coordinator be allowed to use and how will his decisions enter into the local decision processes?
- How much feedback information should be made available to the coordinator and to the local decision units?
- What procedures (algorithms) should be used at each level in determining the coordinating decisions and the control decisions (control actions) to be applied to the real system?
- How will the whole of the structure perform when disturbances appear?
- What will be the effect of distortion of information transmitted between the levels?

In an existing system, some of the above questions were answered when the system was designed and put into operation. However, we are often interested in modifying and improving an existing system, and the same system design problems will come up again.

1.2. MULTILAYER SYSTEMS

Section 1.1 has introduced us briefly to the concept of multilayer systems. We shall now discuss their two principal varieties in more detail.

In a dynamic control problem, for control or a decision to be made and applied at the current time t, we must take into account the future behavior of the system. When the *state* of the controlled system is relevant for the control decision, we have to consider what this state is going to be in the future. In a dynamic system, future states will depend on the past decisions; compare Eq. (1.1). Within the framework of optimal control, we speak about the optimization *horizon*. As shown in section 1.1 and Figure 1.5, the optimization horizon can be divided in a hierarchical system.

Let us illustrate the operation of such a hierarchy by reference to control of a water supply system with retention reservoirs. The top layer would determine, at time zero, the optimal state trajectory of water resources up to a final time, e.g., one year later. This would require long-term planning and the model simplification mentioned in section 1.1 could consist of dropping the medium-size and small reservoirs, or lumping them into a single equivalent capacity. The model would be of low order and have only a few state variables (the contents of larger water reservoirs). We can see from this example why it is necessary to consider the future when the present decision is being made and we deal with a dynamic system: the amount of water at any time t may be used right away, or left for the next week, or for the next month, and so on. Note that the outflow rate which we command today will have an influence on the retention state at any future t.

There is a difference between control of a dynamic system and control of a static time-varying system (see Eq. 1.2′). In the latter case nothing is being accumulated or stored and the present control decision does not influence the future. An example might be supplying water to a user who has a time-varying demand, but no storage facility of any kind.

The long-horizon solution supplies the state trajectory for the first month, but this solution is not detailed enough: the states of medium-size and small reservoirs are not specified. The intermediate layer would now be acting and compute—at time zero—the more-detailed state trajectory for the month.

From this trajectory, we could derive the optimization problem for the first day of system operation. Here, in the lowest layer, a very detailed model must be considered, since we have to specify what is to be done to each reservoir. For example, we might have to specify outflow rates. We consider each reservoir in detail, but we have here the advantage of considering a short horizon.

Let us now describe this hierarchy more formally. Assume that the water system problem was

$$\text{maximize} \int_{t_0}^{t_f} q_0^1(x^1(t), m^1(t), z^1(t))\, dt$$

and that the system is described by state equation

$$\dot{x}^1(t) = f^1(x^1(t), m^1(t), z^1(t))$$

where the state $x^1(t_0)$ is given and $x^1(t_f)$ is free or specified as the required water reserve at $t = t_f$. The control input is m^1 and the time-varying disturbance is z^1. The value of q_0^1 is the rate of net profit in the water system. $q_0^1(\cdot)$ is the functional dependence of this rate on the variables x^1, m^1, z^1. Let us divide this problem among three layers. (Note that the superscripts here are not exponents.)

Top layer (*long horizon*)

$$\text{maximize} \int_{t_0}^{t_f} q_0^3(x^3(t), m^3(t), z^3(t)) \, dt,$$

where $\dot{x}^3(t) = f^3(x^3(t), m^3(t), z^3(t))$, $x^3(t_0)$ is given, and $x^3(t_f)$ is free or specified as above. Here, x^3 is the simplified (aggregated) state vector, m^3 is the simplified control vector, z^3 is the simplified or equivalent disturbance. Hence, q_0^3 is the same rate of profit as in the original formulation, but $q_0^3(\cdot)$ is a function relating it to aggregated variables x^3, m^3, z^3.

Solution of the long-horizon problem determines, among other things, state $\hat{x}^3(t_f')$, i.e., the state to be obtained at time t_f' (this could be one month in the water system example). This state is a target condition for the problem considered at the next layer down in the hierarchy.

Intermediate layer (*medium horizon*)

$$\text{maximize} \int_{t_0}^{t_f'} q_0^2(x^2(t), m^2(t), z^2(t)) \, dt$$

where $\dot{x}^2(t) = f^2(x^2(t), m^2(t), z^2(t))$, $x^2(t_0)$ is given, and $x^2(t_f')$ is given by $\hat{x}^3(t_f')$. The final state requirement cannot be introduced directly because vector x^2 has a lower dimension than x^3, according to the principle of increasing the number of details in the model as we step down the hierarchy. We must introduce a function γ^2 and require

$$\gamma^2(x^2(t_f')) = \hat{x}^3(t_f').$$

Function γ^2 is related to model simplification (aggregation of state as we go upwards) and should be determined together with those simplifications. A suitable profit function $q_0^2(\cdot)$ has to be determined as well.

Solution to the intermediate layer problem determines, among other things, the value of $\hat{x}^2(t_f'')$, i.e., the state to be obtained at $t = t_f''$ (this could be one day in the water system example).

Lowest layer (short horizon)

$$\text{maximize} \int_{t_o}^{t_f''} q_0^1(x^1(t), m^1(t), z^1(t))\, dt$$

where $\dot{x}^1(t) = f^1(x^1(t),\ m^1(t),\ z^1(t))$, $x^1(t_0)$ is given, and $x^1(t_f'')$ is given by $\gamma^1(x^1(t_f'')) = \hat{x}^2(t_f'')$.

We omit explanation of the details of this problem since they are similar to those of the previous problems. The functions $q_0^1(\cdot)$, $f^1(\cdot)$ used here are the same as in the original problem (the "full" model), but the time horizon is considerably shorter. The lowest layer solution determines the control actions $\hat{m}^1$ to be taken in the real system. See Figure 1.8 for a sketch of the three layers and their linkages.

If no model simplifications were used, the multilayer structure would make little sense. If we used the full model at the top layer, we would have determined the trajectory $\hat{x}^1$ and the control actions $\hat{m}^1$ right there, and not only for the interval (t_0, t_f'') but for the whole horizon (t_0, t_f). The lower layers would only repeat the same calculations.

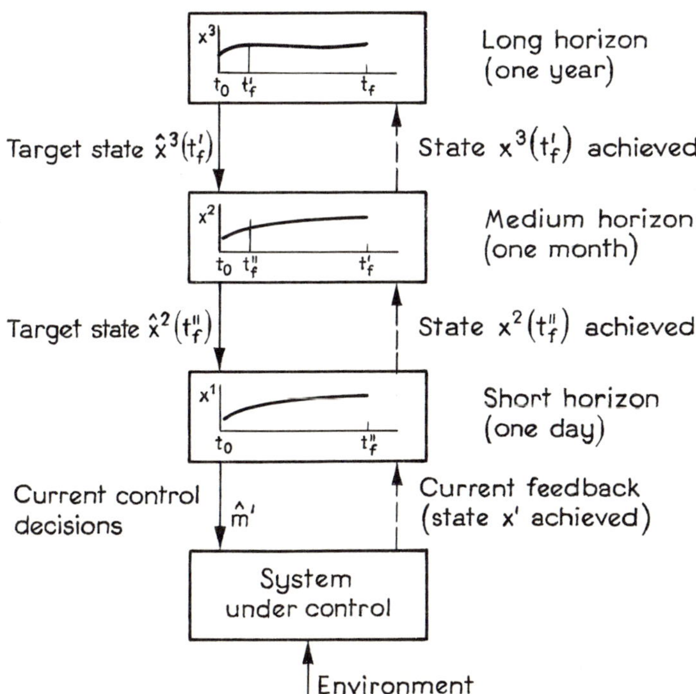

FIGURE 1.8 The multilayer concept applied to multihorizon dynamic control.

18

Let us now introduce feedback and use the system operation to improve control. We could use the actual value of $x^1(t_f'')$ as the initial condition for the intermediate layer problem. This means that at time t_f'' (one day in the example) we resolve the intermediate layer problem using as the initial condition:

$$x^2(t_f'') = \gamma^1(x^1(t_f'')).$$

After the second day, i.e., at $t = 2t_f''$, we would use

$$x^2(2t_f'') = \gamma^1(x^1(2t_f''))$$

and so on.

This way of using feedback is often referred to as *repetitive optimization* because the computational (open-loop) solution will be repeated many times in the course of the control system operation.

The same feedback principle could be used to transmit feedback information to the higher layers with a decreased repetition rate. We shall refer to this concept of feedback quite extensively in Chapter 4, which deals with dynamic coordination in multilevel systems.

Consider what would happen if we used no feedback from the system. The system would be a multilayer structure but its performance might be unnecessarily worse than if feedback were used. Note that for the dynamic optimization calculation that is performed in the control system, we need to know the initial state and a prediction of the behavior of the environment. The prediction itself calls for repetition of the optimization calculation at appropriate intervals. Not using the feedback in the form of the measured state would mean that we will be setting an assumed initial state instead of the real one in each of the consecutive calculations. The errors may accumulate. Needless to say, feedback would be redundant when the model at the lowest layer exactly describes reality, inclusive of all disturbances— but this is not likely.

An example of an existing multilayer hierarchy is shown in Figure 1.9, which is based on a state-of-the-art report on integrated control in steel industries (Lefkowitz and Cheliustkin 1976). We can see there how the time horizon gets shorter when we step down from long-range corporate planning to process control. It is also obvious that the problems considered at the top do not encompass details like what should be done with every piece of steel in the plant. At the bottom level, however, each piece must receive individual consideration, because the final action (manipulation) must be specified there.

One may ask if the model at the highest level can really be an aggregated one, and if so, how aggregated it can be. A qualitative answer is as follows: the details of the present state have little influence on the distant future, and the prediction of details in the distant future makes no sense because it cannot be reliable. Quantitative answers are possible for specific cases.

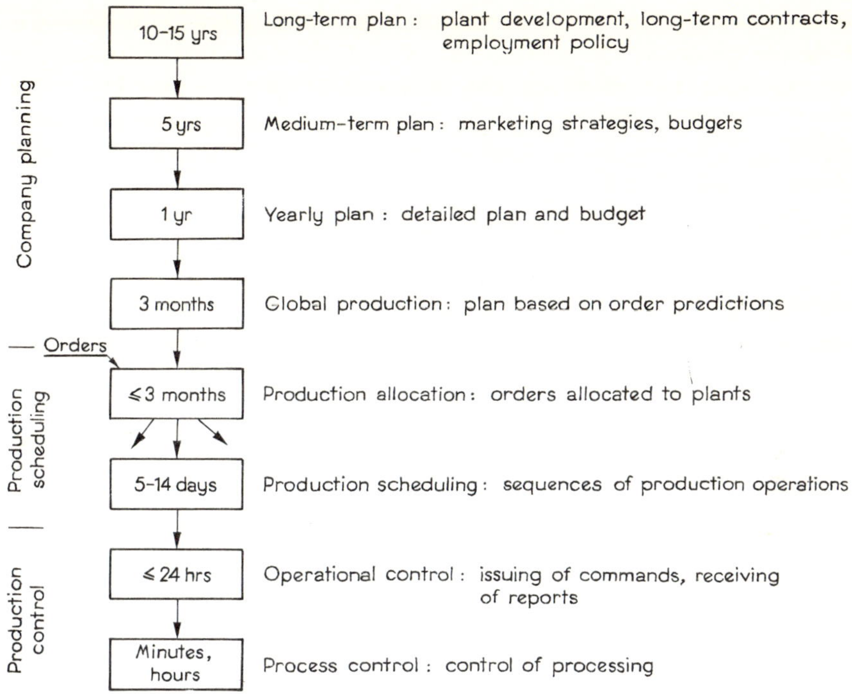

FIGURE 1.9 Hierarchy of models and time horizons in a steel company.

THE OPTIMIZATION HORIZON

Roughly speaking, we may distinguish two kinds of dynamic optimization problems:

- Problems where the time horizon is inherent in the problem itself
- Problems where the choice of time horizon has to be made by the problem solver

Examples of the first variety are a ship's cruise from harbor A to B, a spaceship flight to the moon, one batch in an oxygen steel-making converter. Examples of the second kind could be operation of an electric power system, a continuous production process, operation of a shipping company, operation of a steel-making shop.

For the problems of the second kind, it is necessary to choose an optimization horizon. We are going to show, in a rather qualitative way, how this choice may depend upon two principal factors: the dynamics of the system and the characteristics of the disturbance.

20

Assume we have first chosen a fairly long time horizon t_f and formulated a problem

$$\text{maximize } Q = \int_{t_0}^{t_f} q_0(x(t), m(t), z(t))\, dt$$

for a system described by

$$\dot{x}(t) = f(x(t), m(t), z(t))$$

where $x(t_0)$ is known and $x(t_f)$ is free. Because of the disturbance z, this is an optimization problem under uncertainty and we should speak about maximizing the expected value of Q, for example. Let us drop this approach and assume that we convert the problem into a deterministic one by taking $\bar{z}$, a predicted value of z, as if it was a known input. Assume that the problem has been solved numerically yielding state trajectory $\hat{x}$ and control $\hat{m}$ for the interval (t_0, t_f).

Figure 1.10 shows what is expected to result in terms of a predicted $\bar{z}$ and the solution $\hat{x}$. There seem to be two crucial points here. First, a predicted $\bar{z}$ will start from the known value $z(t_0)$ and always end up being either constant or periodic—in other words, the initial value $z(t_0)$ has no influence on the estimated value of the disturbance and what we get as $\bar{z}$ must be either the mean value or a function with periodic properties. Secondly, if (t_0, t_f) is large enough (say one year for an industrial plant) we expect that in a period far from $t = t_0$, the initial state $x(t_0)$ no longer has any influence on the optimal values $\hat{x}(t)$. If we are still long before $t = t_f$, the final conditions have no influence either.

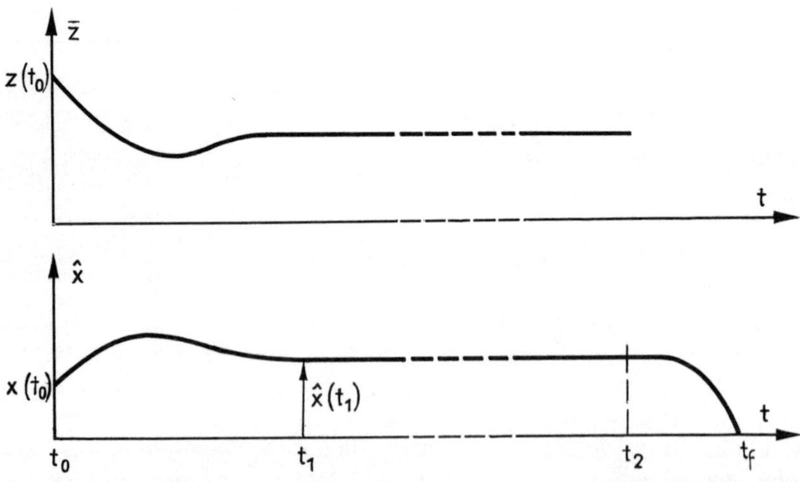

FIGURE 1.10   An optimization horizon.

Thus, what we expect is that the optimal trajectory $\hat{x}$ that is calculated at $t = t_0$ will exhibit a quasi-steady state interval (t_1, t_2) where $\hat{x}$ depends only on $\bar{z}$. But since $\bar{z}$ is either constant or periodic, $\hat{x}$ will be the same; a more thorough discussion of this can be found in Findeisen (1974).

The above qualitative consideration allows us to explain why, practically speaking, we would be allowed to consider only (t_0, t_1) as the optimization horizon for our problem. Note that if we decide to use this short horizon we must formulate the problem as one with the given final state:

$$\text{maximize } Q = \int_{t_0}^{t_1} q_0(x(t), m(t), \bar{z}(t))\, dt$$

for a system described by

$$\dot{x}(t) = f(x(t), m(t), \bar{z}(t))$$

where $x(t_0)$ is known and $x(t_1)$ is given as $\hat{x}(t_1)$ from Figure 1.10.

The solution $\hat{x}$ from this problem and the control $\hat{m}$ are correct only for a short portion of (t_0, t_1) because the real z will not follow the predicted $\bar{z}$. Thus, we have to repeat the solution after some interval δ much shorter than the horizon $t_1 - t_0$ and use the new initial values $x(t_0 + \delta)$ and $z(t_0 + \delta)$. The horizon should now reach to $t_1 + \delta$. It is relatively easy to verify our reasoning by studying a problem that would have an analytical solution, by simulation, or by just imagining how some real systems operate. The proper length of an optimization horizon can be defined very briefly as follows: The optimization horizon is long enough if it permits a proper control decision at $t = t_0$.

FUNCTIONAL MULTILAYER HIERARCHY. THE STABILIZATION AND OPTIMIZATION LAYERS

Section 1.1 explained very briefly what we intend to achieve by a functional multilayer hierarchy: a reduction in the frequency and hence in the effort of making control decisions. There are also other features of this hierarchy, such as a division of decision making and data base requirements.

Let us discuss the division of control between the first two layers: stabilization (direct control, follow-up control) and optimization; see Figure 1.4.

Assume that for a dynamic system described by

$$\dot{x}(t) = f(x(t), m(t), z(t)) \tag{1.6}$$

we have made a choice as to what variables of the plant should become the controlled (stabilized) variables; see Figure 1.11. This choice is equivalent to setting up some functions $h(\cdot)$, relating $c(t)$ to the values of plant variables

$x(t)$ and $m(t)$ at the same instant in time

$$c(t) = h(x(t), m(t)). \tag{1.7}$$

We will assume that the values of c are directly measured (observed). Functions $h(\cdot)$ would be identities $c \triangleq x$ if we chose the state vector itself as controlled variables—but this choice may be neither possible nor desirable and a more general form expressed by function $h(\cdot)$ is appropriate.

The direct control layer (Figure 1.11) will have the task of providing a follow-up of the controlled variables c with respect to their set-points (desired values) c_d; the direct control layer ensures that $c = c_d$. The optimization layer has to impose the c_d which would maximize the performance index of the controlled system (or plant in the industrial context); the optimization layer determines c_d so as to maximize Q. Note that Q has to be performance assigned to the operation of the controlled system itself, for example, the chemical reactor's yield, with no consideration yet of the controllers or of the control structure. In other words, Q is a performance measure which we should know from the user of the system.

The question is how to choose the controlled variables c, that is, how to structure the functions $h(\cdot)$. It is too easy to say that the choice should be such that there is no deterioration of the control result in the two-layer system as compared with a direct optimization. Rather, it should be

$$\max_{c_d} Q = \max_{m} Q$$

where the number on the left is plant performance achieved with the two-layer system of Figure 1.11 and the number on the right is the

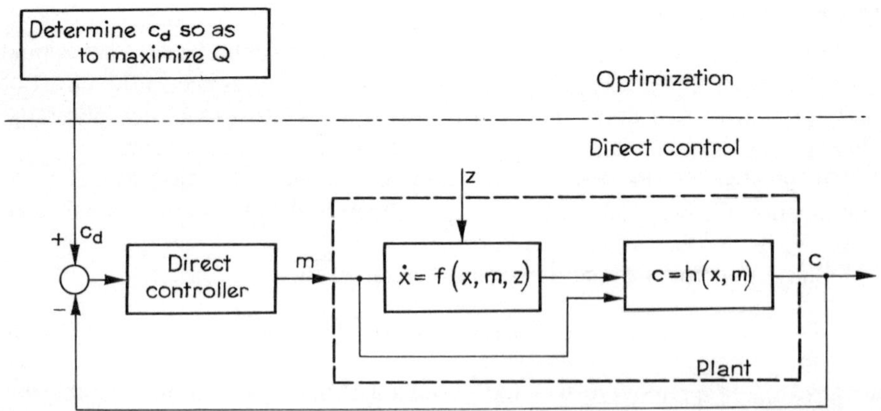

FIGURE 1.11   A two-layer system.

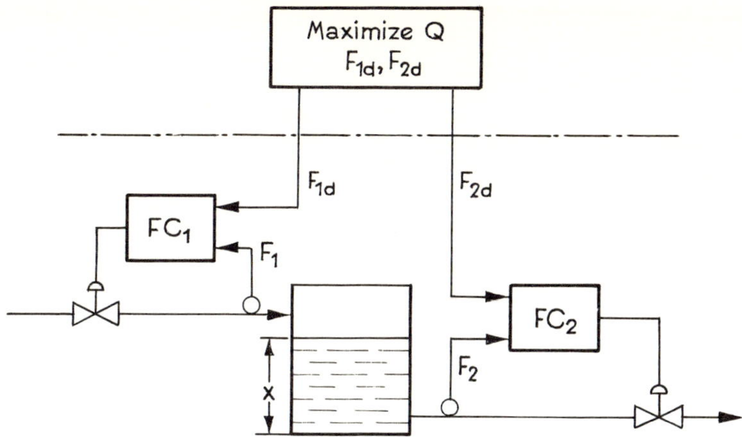

FIGURE 1.12 A poor choice of controlled variables in a two-layer system to control a tank.

maximum achievable performance of the plant itself, where the available manipulated inputs are optimized directly.

In order to get some more constructive conclusions, let us require that a setting of c_d should uniquely determine both state x and control m which will be established in the system of Figure 1.11 when a c_d is imposed. Since we are interested in getting optimal values x and m, let us demand the following property:

$$c = \hat{c}_d \rightarrow x = \hat{x}, \; m = \hat{m}.$$

A trivial solution and a wrong choice of controlled variables could be $c \triangleq m$. Imposing $m = \hat{m}$ on the plant would certainly do the job, but it is a poor choice because the state x that results from an applied m depends also on the initial condition $x(t_0)$—the optimizer that sets $\hat{c}_d$ would have to know $x(t_0)$.

A trivial example explains the pitfall. Assume we made a two-layer system to control a tank using two flow controllers as in Figure 1.12. We delegate to the optimizer the task of determining the optimal flows, F_{1d} and F_{2d}. The optimizer would have no idea about what level x will be established in the tank, unless it memorized $x(t_0)$ and all the past actions. We can see it better while thinking of a steady state: if the optimizer imposes *correct* steady-state optimal values $F_{1d} = F_{2d} = \hat{F}_d$, it still *would not* determine the steady level x in the tank.

Let us therefore require that the choice of c should free the optimizer from the necessity to know the initial condition:

$$c(t) = \hat{c}_d(t) \rightarrow x(t) = \hat{x}(t), \; m(t) = \hat{m}(t), \qquad \forall t \geq t_1 > t_0 \qquad (1.8)$$

and the implications shall hold for any $x(t_0)$.

An example of what we aim at may be best given by considering that we want a steady-state $x(t) = x =$ a constant in the system, while the system is subjected to a constant, although unknown disturbance $z(t) = z$. In that case, m and $c = c_d$ will also not be time-varying. The state equations of the plant reduce to

$$f_j(x, m, z) = 0, \qquad j = 1, \ldots, \dim x \qquad (1.9)$$

because $\dot{x}(t) \triangleq 0$, and if we add the equations that are set up by our choice of the controlled variables

$$h_i(x, m) = c_i, \qquad i = 1, \ldots, \dim c \qquad (1.10)$$

we have a set of equations (1.9) (1.10) for which we desire that x, m as the dependent variables be uniquely determined by c. But we also want (1.9) and (1.10) to be a consistent set of equations; their number should not exceed the number of dependent variables x, m, and thus we arrive at the requirement that $\dim c = \dim m$: the number of controlled variables should be equal to the number of manipulated inputs. Then, from the implicit function theorem, it is sufficient for the uniqueness of x, m that f_j, h_i are continuously differentiable, and

$$\det \begin{bmatrix} \dfrac{\partial f_j}{\partial x_k} & \dfrac{\partial f_j}{\partial m_k} \\[2ex] \dfrac{\partial h_i}{\partial x_k} & \dfrac{\partial h_i}{\partial m_k} \end{bmatrix} \neq 0. \qquad (1.11)$$

We leave it to the reader to verify that the system of Figure 1.12 is not consistent with Eq. (1.11).

We should warn the reader of a possible misinterpretation of our argument. We have shown the conditions under which steady-state x, m in the control system will be single-valued functions of c, but these functions may still contain z as a parameter. In other words, we did not say that a certain value of c will ensure the value of x, m in the plant, irrespective of the disturbance. If, for example, we are interested in ensuring the value of the state, we could choose $c \triangleq x$. But note that this may not be entirely feasible if we have too few manipulated inputs (remember that $\dim c = \dim m$).

Of course, the structure of Figure 1.11 can be used when the plant state x is time-varying. In that case, we would write, instead of (1.9) and (1.10):

$$\dot{x}_j(t) = f_j(x(t), m(t), z(t)), \qquad j = 1, \ldots, \dim x \qquad (1.9a)$$

$$h_i(x(t), m(t)) = c_i(t), \qquad i = 1, \ldots, \dim c. \qquad (1.10a)$$

The value of the state at time t, $x(t)$, will still be related to the value $c_d(t)$ that is being enforced on the system, but $\dot{x}(t)$ is also involved in the relationship. This means that in order to obtain a certain state $x(t)$, we must take into account the initial state $x(t_0)$, the disturbance input over the interval $[t_0, t]$, $z_{[t_0,t]}$, and appropriately shape the control decision $c_{d[t_0,t]}$.

If we want to ensure the value of state $x(t)$ in spite of the disturbances and without dependence on the initial state, we must investigate the *follow-up controllability*: is it possible, using the input m, to cause state x to follow a desired trajectory x_d? The follow-up controllability is a stronger requirement than controllability in the usual sense, where we check whether a state can be reached but do not insist on the trajectory by which it will be reached.

Assume that the follow-up has been achieved, that is $x(t) = x_d(t)$, $\dot{x}(t) = \dot{x}_d(t)$, $\forall t$. Then the state equations give, in the linear case,

$$\dot{x}_d(t) = Ax_d(t) + Bm(t) + Fz(t).$$

These equations have to hold for any $z(t)$, at the expense of varying or adjusting $m(t)$, which has to be done by the controller. The value of $m(t)$ that is necessary to nullify effects of a disturbance $z(t)$ in order to satisfy the follow-up condition is

$$m(t) = -B^{-1}[Ax_d(t) + Fz(t) - \dot{x}_d(t)]$$

from which it follows that there should be a dim $m = $ dim x and that B should be an invertible matrix (a case of redundant controls, dim $m >$ dim x, could be described as well).

In the more general, nonlinear case we have to write that in the follow-up condition the state equtions give

$$f_j(x_d(t), m(t), z(t)) - \dot{x}_d(t) = 0, \qquad j = 1, \ldots, \text{dim } x. \qquad (1.12)$$

We should note the meaning of Eq. (1.12). If Eq. (1.12) has to hold, we have to adjust $m(t)$ so as to offset the effect of $z(t)$. This must of course require certain properties of the functions $f_j(\cdot)$ and we also expect to have enough manipulated inputs. The requirements will be met if the set of equations (1.12) will define $m(t)$ as single-valued functions of $z(t)$. The conditions for this are that $f_j(\cdot)$ be continuously differentiable and moreover that

$$\text{rank} \left[\frac{\partial f_j}{\partial m_k} \right] = \text{dim } x. \qquad (1.13)$$

This implies that dim $m \geq$ dim x. Let us note also that the actual value $m(t)$, as required by the disturbance $z(t)$, should never lie on the boundary of the constraint set of manipulated inputs. We must always have the possibility of adjusting $m(t)$ up or down in order to offset the influence of the random disturbance. The actual value of this required reserve or margin depends of course on the range of possible disturbances.

26

Remember that Eq. (1.13) is a requirement related to controllability, that is, to the properties of the plant itself. Controllability does not give control m such that $x = x_d$; it tells us only that this control exits. If we decide to build a feedback control system as shown in Figure 1.11, we have to choose the controlled variables c in an appropriate way. For the dynamic follow-up to be ensured by the condition $c = c_d$, the choice would have to be $c \triangleq x$, that is, the state variables themselves (as opposed to $c = h(x, m)$ which was alright for the steady-state uniqueness of x).

Until now, the choice of controlled variables has been discussed from the point of view of the *uniqueness* property: how to choose c in such a way that when $c = c_d$ is enforced, some well-defined values x, m will result in the plant. We have done this for the plant described by ordinary differential equations (1.6). An extension of this consideration to distributed parameter plants with lumped manipulated inputs is possible (Findeisen 1974).

We turn now to the more spectacular aspect of choosing the controlled variables: can we choose them in a way permitting reduction or elimination of the on-line optimization effort, that is, elimination of the optimization layer in Figure 1.11, leaving only the follow-up control? To make the discussion more simple, let us consider steady-state optimization.

For a plant

$$f_j(x, m, z) = 0, \qquad j = 1, \ldots, \dim x$$

we are given the task

$$\text{maximize } Q = f_0(x, m, z)$$

subject to inequality constraints

$$g_i(x, m) \le b_i, \qquad i = 1, \ldots$$

Assume the solution is $(\hat{x}, \hat{m})$. At point $(\hat{x}, \hat{m})$ some of the inequality constraints become equalities (active constraints), and other inequalities are irrelevant. Thus at $(\hat{x}, \hat{m})$ we have a system of equations

$$\begin{aligned}
f_j(\hat{x}, \hat{m}, z) &= 0, & j &= 1, \ldots, \dim x \\
g_i(\hat{x}, \hat{m}) &= b_i, & i &= 1, \ldots, k \le \dim m
\end{aligned} \tag{1.14}$$

If it happens that $k = \dim m$ then the rule is simple: choose the controlled variables as follows

$$\begin{aligned}
h_i(\cdot) &\triangleq g_i(\cdot), & i &= 1, \ldots, \dim m \\
\hat{c}_{di} &= b_i.
\end{aligned}$$

This simply says that the controllers should keep the plant variables (x, m) at the appropriate border lines of the constraint set. Note two things:

- We have assumed that $g_i(x, m)$ and not $g_i(x, m, z)$, i.e., the disturbance, did not affect the boundaries of the constraint set.

- We have assumed that $k = \dim m$ (the number of active constraints equals the number of controls), and we also did not consider that even in this case the solution $(\hat{x}, \hat{m})$ may lie in different "corners" of the constraint set for different z.

Even under these assumptions, however, the case makes sense in many practical applications, since solutions to constrained optimization problems tend to lie on the boundaries. For example, the yield of a continuous-flow, stirred-tank chemical reactor would increase with the volume of chemicals in the tank. The volume is obviously constrained by tank capacity; therefore, the design would result in the use of a level controller and in setting the desired value of the level at the full capacity. The level controller would adjust inflow or outflow to maintain the level. No on-line optimization would be necessary.

We have mentioned already in section 1.1 that the approach we have taken, letting the *direct controller* make continual control decisions and providing for an upper level to set a rule or goal to which the direct control must conform, has more than just industrial applications. It is also clear that a rule or goal does not have to be changed as often as those decisions and hence a two-layer structure makes sense.

If the solution $(\hat{x}, \hat{m})$ fails to lie on the boundary of the constraint set, or the number of active constraints $k < \dim m$, we may still try to structure our functions $h_i(\cdot)$ in such a way as to make the optimal value $\hat{c}_d$ independent of disturbances z. The way to consider this may be as follows. We have solutions $\hat{m} = \hat{m}(z)$ and $\hat{x} = \hat{x}(z)$. We put them into the functions $h_j(\cdot)$ for $j = k+1, \ldots, \dim m$:

$$h_j(x, m) = h_j(\hat{x}(z), \hat{m}(z)), \qquad j = k+1, \ldots, \dim m$$

By an appropriate choice of $h_j(\cdot)$ we may succeed in getting

$$\frac{\partial h_j}{\partial z} = 0, \qquad j = k+1, \ldots$$

in the envisaged range of disturbances z. We turn now to a more elaborate example of building up a two-layer system.

EXAMPLE OF TWO-LAYER CONTROL

Consider the stirred-tank, continuous-flow reactor presented in Figure 1.13. Material B flows in at rate F_B and has temperature T_B, material A flows in

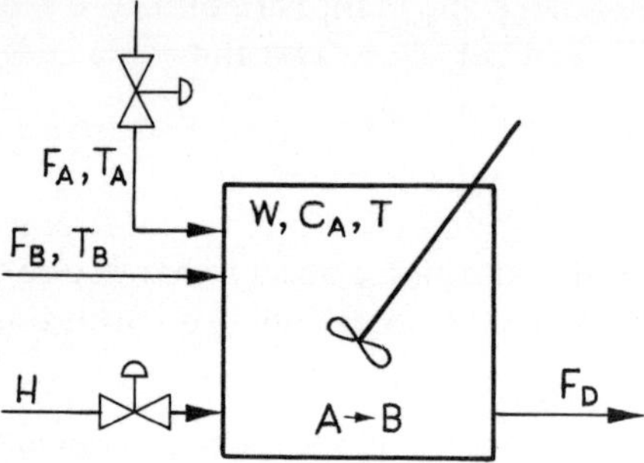

FIGURE 1.13   A stirred-tank reactor.

at rate F_A and temperature T_A. They mix, and reaction $A \rightarrow B$ takes place in the vessel, resulting in composition C_A of the product. Heat input H is needed for temperature T to be obtained in the reactor. Outflow F_D carries the mixture of A and B out of the vessel. We want a control structure that optimizes the operation of this reactor; F_A and H are the manipulated inputs.

Description of the plant

There will be three state variables and state equations:

$$\dot{W} = f_1(\cdot) = F_A + F_B - F_D$$
$$\dot{C}_A = f_2(\cdot)$$
$$\dot{T} = f_3(\cdot).$$

We drop the detailed structure of the functions $f_2(\cdot)$, $f_3(\cdot)$ because it is not important for the example.

Formulation of the optimization problem

Assume that we want to maximize production less the cost of heating:

$$\text{maximize } Q = \frac{1}{t_2 - t_1} \int_{t_1}^{t_2} [(1 - C_A)F_D - \psi(T)]\, dt$$

where $\psi(T)$ expresses the cost of reaching temperature T.

There will be inequality constraints

$$W \leq W_m, \qquad C_A \leq C_{Am}, \qquad T \leq T_m$$

and we also have to consider the state equations and initial and final conditions. If there are reasons to assume that the optimal operation of the reactor is at a steady-state, $\hat{x} = a$ constant, then the plant equations reduce to

$$f_1(\cdot) = F_A + F_B - F_D = 0$$
$$f_2(\cdot) = 0$$
$$f_3(\cdot) = 0$$

and the optimization goal is

$$\text{maximize } Q = (1 - C_A)F_D - \psi(T).$$

Solution of the optimization problem

Assume that the optimization problem has been solved and the results are (the problem has been solved for a fully detailed example):

$$\hat{W} = W_m$$
$$\hat{C}_A = C_{Am} \quad \text{if} \quad z \in Z_1; \qquad \hat{C}_A = \phi_1(z) < C_{Am} \text{ otherwise}$$
$$\hat{T} = \phi_2(z) < T_m \quad \text{if} \quad z \in Z_1; \qquad \hat{T} = T_m \text{ otherwise}$$
$$\hat{F}_A = \phi_3(z)$$
$$\hat{H} = \phi_4(z)$$

where z stands for disturbance vector (F_B, F_D, T_A, T_B) and Z_1 is a set in z-space, that is, a range of disturbance values. Note that for $z \in Z_1$ we have a simple solution $\hat{C}_A = C_{Am}$ and for $z \notin Z_1$ there is $\hat{T} = T_m$.

Examination of the solution and choice of control structure

Let us make a wrong decision and choose as controlled variables the flows F_A, H. We would then fail to get a uniquely determined steady-state volume W in the tank (a check on the matrix determinant condition (1.11) would show it) and the optimizer that sets the desired F_{Ad}, H_d would have to know disturbance vector z and functions $\phi_3(\cdot)$, $\phi_4(\cdot)$. Note that this would involve an accurate knowledge of the state equations of the plant.

Inspection of the optimization solution reveals that volume W is the best candidate for the controlled variable. The optimal $\hat{W}$ is W_m under all circumstances; no on-line optimization or knowledge of plant state equations will be required. Since we have two manipulated inputs, we shall have two controlled variables, and the second choice could be either concentration C_A or temperature T. Let us consult Figure 1.14 for a discussion of the options. We have displayed there the feasible set in the (W, C_A) plane and have shown where the optimal solution lies in the two cases, that is, when $z \in Z_1$ (point 1) and $z \notin Z_1$ (point 2). Note that the solution is in a corner of

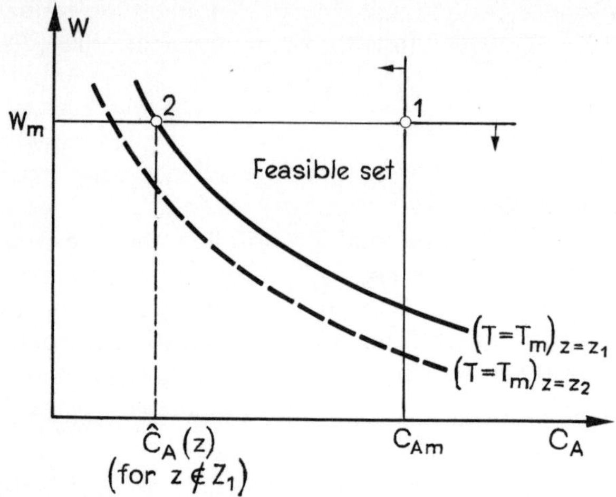

FIGURE 1.14 Position of optimal solutions 1 and 2 in the feasible set for the chemical reactor problem. Note that the curve $T = T_m$ moves in the (W, C_A) plane depending on the disturbances.

the constraint set, but unfortunately not in the same corner for all z. One may have two choices:

• Take C_A as a controlled variable and ask the optimizer to watch disturbances z and perform the following

$$C_{Ad} = C_{Am} \quad \text{for} \quad z \in Z_1$$
$$C_{Ad} = \phi_1(z) \quad \text{otherwise}$$

whereby a knowledge of the function $\phi_1(\cdot)$ is required.

• Take C_A as a controlled variable when $z \in Z_1$ and then set $C_{Ad} = C_{Am}$, whereby for $z \notin Z_1$ one would switch to T as a controlled variable and set $T_d = T_m$. In this case, the second-layer of control would perform the switching, that is, it would detect if $z \in Z_1$. This may be easier to do than to identify the function $\phi_1(\cdot)$, which was required in the first alternative.

APPLICABILITY OF STEADY-STATE OPTIMIZATION

Steady-state optimization, following the structure of Figure 1.11, is a fairly common practice. It might be worthwhile to consider when it is appropriate. If we exclude the cases where the *exact* solution for the optimal state is $\hat{x} =$ a constant, we may think of the remaining cases in the following way. Let (a) in Figure 1.15 be the optimal trajectory of the operation of a plant over optimization horizon (t_0, t_1). Assume that we control the plant by a two-layer system, have x as the controlled variable, and choose to change the desired value x_d at intervals T that are a small fraction of (t_0, t_1). Then (b) is the plot of $x_d(t)$. Note that we have thus decided to be nonoptimal because x_d should be shaped like (a), and should not be a step-wise changing function. Note also that the step values of x_d would have to be calculated from a dynamic, though discrete, optimization problem.

Now let us look at the way in which the real x will follow the step-wise changing x_d in the direct control system of Figure 1.11. In case (c) of Figure 1.15, x almost immediately follows x_d. In case (d) the dynamics are apparently slow and it cannot be assumed that x follows x_d. It is only in case (c) of

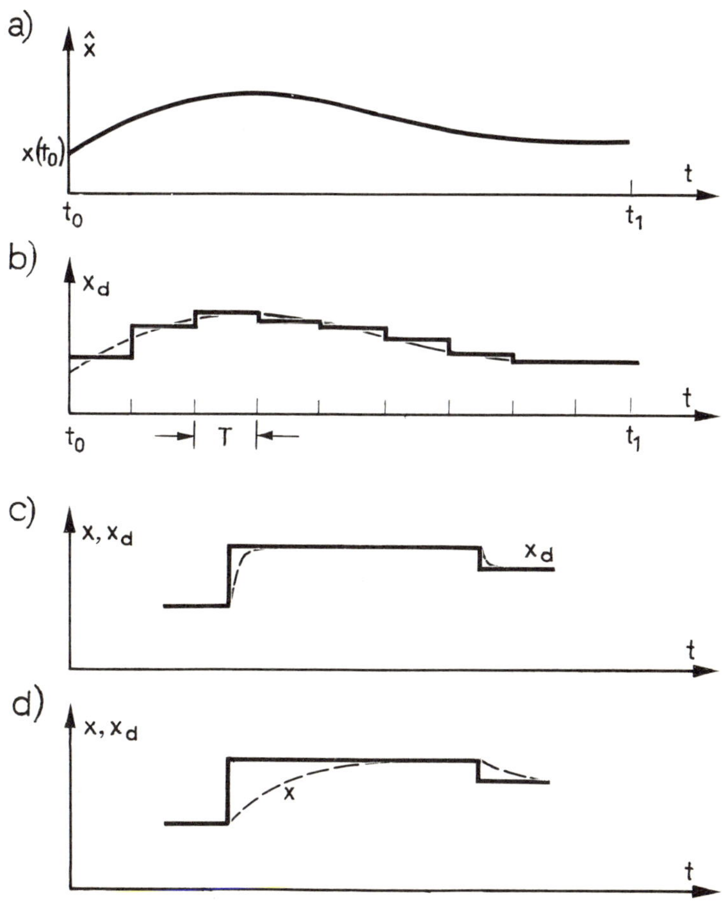

FIGURE 1.15   Trajectories of operation of a plant.

32

Figure 1.15 that we may be allowed to assume that state x is *virtually constant* over periods T, thus permitting the setting of $\dot{x}=0$ in the state equations and the calculation of the step value of x_d from a steady-state optimization problem.

When will case (c) occur? By no means are we free to choose the interval T at will. We must relate it to the optimization horizon (t_0, t_1); interval T would be a suitable fraction of this ($\frac{1}{10}$ or $\frac{1}{50}$ for example). Here is the qualitative answer to the main question: if (t_0, t_1) has resulted from slow disturbances acting on a fast system, case (c) may be applicable, that is, we may perhaps be allowed to calculate a step of x_d under the steady-state assumption. It should be noted that the actual behavior (c) or (d) results from two factors: the characteristics of the plant itself and the performance of direct controls.

The importance of the possibility of replacing the original dynamic optimization problem by an almost equivalent static optimization done in the two-layer system cannot be overemphasized. The reason is of a computational nature: dynamic problems need much more effort to be solved than static problems, and for many control tasks, for example, for a chemical plant, may be practically unsolvable in the time available. There are also many cases where the dynamic properties of the plant are not sufficiently known. On the other hand, the operation of many plants is close to steady state and the optimization of set-points done by static optimization may be quite close to the desired result.

We devote considerable space in this book to steady-state, on-line optimization structures, procedures, and algorithms (Chapter 3). We should point out that the procedures for static optimization are different in principal from those suitable for dynamic control, if feedback from the process is being used.

Let us come back to Figure 1.4 of section 1.1. We have presented there an adaptation layer and assigned to it the task of readjusting some parameters β that influence the setting of the value of c_d. Assume that this setting is done by means of a fixed function $k(\cdot)$:

$$c_d = k(\beta, z)$$

where z stands for the disturbance acting on the plant. We assume, at this point, that it is measured and thus it can enter the function $k(\cdot)$.

We may of course assume that the strictly optimal value of c_d, referred to as $\hat{c}_d(z)$, exists. With $\hat{c}_d(z)$ we would get a maximum value of performance denoted by $Q(\hat{c}_d(z))$.

Optimal values of β in the optimizer's algorithm could be found by solving the problem

$$\underset{\beta}{\text{minimize}}\ E_z \| Q(\hat{c}_d(z)) - Q(k(\beta, z)) \|. \tag{1.15}$$

We drop discussion of this formulation because we should assume that the optimizer has only restricted information about z, denoted z^* (it could, for example, be samples of z taken at some intervals). This leads to $c_d = k(\beta, z^*)$ and the parameter adjustment problem should now be

$$\underset{\beta}{\text{minimize}}\ E_{z,z^*}[Q(\hat{c}_d(z)) - Q(k(\beta, z^*))] \tag{1.16}$$

which means that the choice of β should aim at minimizing the loss of performance with respect to the best plant operation. An indirect way which may be easier to perform, but which is not equivalent, would be

$$\underset{\beta}{\text{minimize}}\ E_{z,z^*} \| \hat{c}_d(z) - k(\beta, z^*) \|. \tag{1.17}$$

Here, unlike in (1.15), we would not be able to get $\beta = \hat{\beta}$ such that $E\|\cdot\|$ is zero, since the basis for $k(\beta, \cdot)$ is z^* and not z. It means that even with the best possible parameters, the control is inferior to a fully optimal one because of the restricted information.

In formulations (1.16) and (1.17), β is adjusted once, and kept constant for a period of time; it is over this period that the expectations $E\|\cdot\|$ in (1.16) or (1.17) should be taken. In some practical adaptive systems, though, we try to obtain the values of parameters of the plant, and thus also the values of β, by some kind of on-line identification procedure. We may refer to it as on-line parameter estimation. One of the important questions is how often should the parameters β be updated; the obvious answer seems to be the more often the better. A limit case in which β are estimated continuously may be of interest. Let us consider what this limit case could supply. Note that for each z, an optimal value $\hat{\beta}(z)$ maximizing the performance exists and yields perfect control. We must assume, however, that we do not have $\hat{\beta}(z)$ but only an estimated value of it, $\tilde{\beta}(z)$. With $\tilde{\beta}(z)$, our optimizing control would be

$$c_d = k(\tilde{\beta}(z), z^*)$$

where we assume that only z^* is available as current information.

The application of this control gives a loss of optimality which amounts to

$$E_{z,z^*}[Q(\hat{c}_d(z)) - Q(k(\tilde{\beta}(z), z^*))].$$

This value could be discussed with respect to the quality of estimating β, insufficiency of disturbance information z^*, and so on. In other words, it measures the overall efficiency of adaptation.

1.3. MULTILEVEL SYSTEMS: DECOMPOSITION AND COORDINATION IN STEADY-STATE AND DYNAMIC CONTROL

In this section we shall consider the multilevel control structures shown in Figure 1.7 in more detail. We will indicate the practical difference between steady-state and dynamic control structures, which will be much more thoroughly investigated in Chapter 3 and Chapter 4, respectively. The mathematical needlework presented there may, however, conceal the basic principles and features of the structures. It is easier to make the comparisons in this section.

THE STEADY-STATE CONTROL PROBLEM

Let us first describe the complex system of Figure 1.2 and Figure 1.3 more carefully. For the subsystem i, x_i denotes the state vector, m_i the manipulated input, z_i the disturbance, u_i the input from other subsystems, and y_i the output connected to other subsystems. The subsystem state equation (compare Eq. (1.1)) will then be

$$x_i(t) = \phi_{i[t_0,t]}(x_i(t_0), m_{i[t_0,t]}, u_{i[t_0,t]}, z_{i[t_0,t]}). \tag{1.18}$$

In this section we assume that Eq. (1.18) is in the form of an ordinary differential equation

$$\dot{x}_i(t) = f_i(x_i(t), m_i(t), u_i(t), z_i(t)). \tag{1.18'}$$

The output y_i will be related to (x_i, m_i, u_i, z_i) by output equation

$$y_i(t) = g_i(x_i(t), m_i(t), u_i(t), z_i(t)). \tag{1.19}$$

Now assume that the first-layer or direct controls are added to the subsystem such that the following is enforced:

$$c_i(t) = h_i(x_i(t), m_i(t), u_i(t)) = c_{di}(t). \tag{1.20}$$

See section 1.2, where this equation is introduced. Assume that we are in steady state, $\dot{x}_i(t) = 0$, $\forall t$, $x_i(t) = x_{si} = $ a constant; the functions $h_i(\cdot)$ have been chosen properly so as to ensure uniqueness of the state x_{si} and manipulated output $m_i(t)$ in response to the imposed $c_i(t)$ and $u_i(t)$, with $z_i(t)$ as a parameter. Then Eq. (1.18') becomes

$$f_i(x_{si}, m_i(t), u_i(t), z_i(t)) = 0 \tag{1.21}$$

and Eq. (1.21) along with (1.20) provide for x_{si}, $m_i(t)$ to be functions of $c_i(t)$. Therefore Eq. (1.19) becomes the following input–output dependence:

$$y_i(t) = F_i(c_i(t), u_i(t), z_i(t)). \tag{1.22}$$

Equation (1.22) is a relation between instantaneous values. We have obtained it by assuming that the system is in steady-state, $x(t) = x_s = $ a constant.

We can consider the state to be time-varying; then Eq. (1.22) is true only under the assumption that the actual state x_i follows the desired state trajectory x_{di}. As mentioned in section 1.2, this is possible if the subsystem conforms to the follow-up controllability condition (see Eq. (1.13)) and if $h_i(\cdot)$ is chosen, for example, such that $c_i \triangleq x_i$.

In the general case of a time-varying state, we would have to put formula (1.18) for $x_i(t)$ into (1.19), making $y(t)$ dependent on the initial state $x_i(t_0)$ and the inputs m_i, u_i, z_i over interval $[t_0, t]$. The existence of an appropriate equation (1.20) allows elimination of m_i in favor of c_i and thus we obtain

$$y_i(t) = F'_{i[t_0,t]}(x_i(t_0), c_{i[t_0,t]}, u_{i[t_0,t]}, z_{i[t_0,t]}). \tag{1.22'}$$

The input–output relation in the form of Eq. (1.22′) is not very convenient for notational reasons. We may assume that the initial state is known, or we can treat $x_i(t_0)$ as part of the disturbance z_i. Additionally, if we use notation y_i, c_i, u_i, z_i to express time functions, then Eq. (1.22′) becomes

$$y_i = F_i(c_i, u_i, z_i). \tag{1.22''}$$

The difference between Eqs. (1.22) and (1.22″) is that Eq. (1.22″) denotes a mapping between time functions (i.e., it describes a dynamic system). When the subsystem is in steady-state, Eq. (1.22) will hold. Its practical meaning is that the dynamics of the subsystem are suppressed and that is why we have a static input–output relation. We will usually write Eq. (1.22) in abbreviated form, dropping the argument t and sometimes also the disturbance input:

$$y_i = F_i(c_i, u_i), \qquad i \in \overline{1, N}. \tag{1.23}$$

Note that the form of Eq. (1.23) is similar to Eq. (1.22″) and the notation does not indicate whether we describe a static or a dynamic system. This is rather convenient for considerations of a general nature. Below we are going to speak about the steady state and we consider y_i, c_i, u_i to stand for $y_i(t)$, $c_i(t)$, $u_i(t)$.

The interconnections in the system are described by

$$u_i = H_i y = \sum_{j=1}^{N} H_{ij} y_j, \quad \text{so that} \quad u = Hy \tag{1.24}$$

where H_i is part of matrix H:

$$H = \begin{bmatrix} H_1 \\ \vdots \\ H_N \end{bmatrix}.$$

36

We assume that a resource constraint is imposed on the system as a whole

$$\sum_{1}^{N} r_i(c_i, u_i) \leq r_0 \tag{1.25}$$

and also that some local constraints restricting (c_i, u_i) may exist

$$(c_i, u_i) \in CU_i, \qquad i \in \overline{1, N}. \tag{1.26}$$

We further assume that a local performance index (local objective function) is associated with the subsystem

$$Q_i(c_i, u_i), \qquad i \in \overline{1, N} \tag{1.27}$$

whereby a global system performance is also defined as

$$Q = \psi(Q_1, Q_2, \ldots, Q_N). \tag{1.28}$$

The function ψ is assumed to be strictly order-preserving (strictly monotonic). This kind of assumption will be made throughout this book. It eliminates gamelike situations among decision makers from our discussion; see, e.g., Germeyer (1976).

Note that Eqs. (1.27) and (1.28) may result from two practical cases. It might be that there were already some local decision makers and we decided to set up an overall Q to provide for some harmony in their actions. But it also might be that we had an overall Q first and we then decided to distribute the decision making among the lower-level units.

We are now ready to define the goal of the coordination level: it has to ensure that the overall constraints are preserved and the overall performance maximized. Coordination will be accomplished by influencing decision making in the local units, and not by overriding control decisions already made.

COORDINATION BY THE DIRECT METHOD

The simplest way to present direct coordination (also called primal or parametric coordination) is to assume that the coordinator prescribes the outputs y_i and demands an equality $y_i = y_{di}$. If a resource constraint (1.25) is present, the coordinator also allocates a value r_{di} to each local problem. The local decision problem becomes

$$\text{maximize } Q_i(c_i, u_i)$$

subject to

$$u_i = H_i y_d$$
$$F_i(c_i, u_i) = y_{di}$$
$$(c_i, u_i) \in CU_i$$
$$r_i(c_i, u_i) \leq r_{di}.$$

When this problem is solved, results depend on (y_d, r_{di}). Note that they depend on the whole y_d, not on y_{di} only, because we had $u_i = H_i y_d$. We denote the results as $\hat{c}_i(y_d, r_{di})$ and $Q_i(\hat{c}_i(y_d, r_{di}), H_i y_d) \triangleq \hat{Q}_i(y_d, r_{di})$.

The coordination instruments (y_d, r_d) have to be adjusted to an optimum by solving the problem

$$\underset{(y_d, r_d)}{\text{maximize}} \; Q = \psi(\hat{Q}_1(y_d, r_{d1}), \ldots, \hat{Q}_N(y_d, r_{dN}))$$

subject to

$$r_{d1} + r_{d2} + \cdots + r_{dN} \leq r_0.$$

The main difficulty of the method lies in the fact that a local problem may have no solution for some (y_d, r_d) because of the constraints; an output value may not be achievable or the allocated resources inadequate, or both. Therefore, the values (y_d, r_d) set by the coordinator must be such that the local problems have solutions

$$(y_d, r_d) \in YR.$$

The set YR cannot be easily determined because it implicitly depends on local equations and constraints. Moreover, the boundaries of set YR are affected by the system uncertainty or by the disturbances, since they are related to local constraints and to the subsystem equations. This has the implication that the coordinator would have to keep his decisions (y_d, r_d) in a "safe" region of YR such that (y_d, r_d) are feasible even in the worst case of system disturbances. Apart from the difficulty of defining the safe region, we of course realize that the worst-case approach may force this region to be very small or even empty. Before trying to find a remedy for this situation, we shall make an additional remark on the direct method.

REMARK ON THE LOCAL CONTROLS

We should note that by prescribing the outputs we also preset the inputs and hence, in the local equation we have only c_i as a free variable:

$$F_i(c_i, H_i y_d, z_i) = y_{di}.$$

Strictly speaking, we should consider the interconnected system as a whole, where we have

$$F(c, u, z) = y$$

$$u = Hy$$

and $y = y_d$, giving

$$F(c, Hy_d, z) = y_d$$

which means that a c must be available such that a system of equations

$$K(c, z) = y_d$$

38

could be satisfied by adjusting c, the control decisions, for any y_d, z in their envisaged range.

The question is: do we have an adequate number of control variables c_j, $j = 1, \ldots,$ dim c, and are they appropriately placed in the system equations?

Let us clarify the implications by using the chemical reactor in Figure 1.13 of section 1.2. The output vector y would in this case be (F_D, C_A, T) since the outflow from the reactor is characterized by flow rate (F_D), composition (uniquely expressed by C_A), and temperature (T). We have only two manipulated variables, F_A and H, and hence two controlled variables, say W and C_A. Therefore, dim $c = 2$ while dim $y = 3$. We should be unable to prescribe an arbitrary value for the output vector. Indeed, the steady-state equation $y = K(c, z)$ of the reactor inclusive of direct controls would be, in scalar notation,

$$F_D = z_1$$

$$C_A = C_{Ad}$$

$$T = K_3(W_d, C_{Ad}, z)$$

where z_1 stands for the flow rate demanded (imposed) by the receiving end of the pipe, and z for the whole vector of disturbances. By choosing W_d and C_{Ad}, we would be able to steer the output C_A and T, but not F_D. Note that our control influence on the output T is rather complicated and the actual T also depends on disturbances. Nevertheless, we can influence it by adjusting W_d, which means that we have "adequate c" for the purpose.

The question of local controls is vital for the direct method. We should, however, consider that when this hierarchical structure is applied, the number of local controls always exceeds the number of outputs that are being prescribed. Otherwise, we might doubt if it makes sense to use the structure since the coordinator could make all the c_i decisions directly.

PENALTY FUNCTIONS IN DIRECT COORDINATION

We can propose an iterative procedure to be used at the coordination level such that the feasible set YR would not have to be known. The main idea is to use penalty functions in the local problems while imposing there the coordinator's demands. If we use the penalty function for the matching of the output, the local problem will take the form:

$$\underset{c_i}{\text{maximize}} \; Q_i' = Q_i(c_i, u_i) - K_i(y_i - y_{di})$$

with the substitutions

$$u_i = H_i y_d$$

$$y_i = F_i(c_i, u_i)$$

and subject to the constraints

$$(c_i, u_i) \in CU_i$$

$$r_i(c_i, u_i) \le r_{di}.$$

As can be seen, we used the penalty function to ensure the condition $y_i = y_{di}$. The resource constraint could also be dealt with by a penalty term, if necessary. Also, the substitution $u_i = H_i y_d$ may be replaced, if needed, by the penalty term. Interaction input u_i would then become a free decision variable in the local problem. The result of using the penalty formulation is that the solution to the local problem would exist even for an infeasible y_{di}. The demand would simply not be met.

We must now have a mechanism to let the coordinator know that he is demanding something impossible. We let his optimization become:

$$\text{maximize}_{y_d, r_d} Q' = \psi[(\hat{Q}_1(y_d, r_{d1}) - K_1(\hat{y}_1 - y_{d1})), \dots, (\hat{Q}_N(y_d, r_{dN})$$

$$- K_N(\hat{y}_N - y_{dN}))]$$

where the clue is that we introduce the local performances less the penalty terms. Hence, the coordination iterations will try to adjust y_d so as to reduce the values of the penalty terms, while the local optimizers do the same on their part by influencing y_i.

It is shown in section 2.3 of Chapter 2 that when the iterations reach the limit where the penalty terms vanish, the values y_d obtained there are both feasible and strictly optimal.

A MECHANISTIC SYSTEM OR A HUMAN DECISION-MAKING HIERARCHY?

Three clarifications are in order here because the reader of the previous section may be confused about what our considerations are applicable to. First, we can obviously think of coordination used in an off-line, model-based solution of a set of local problems. This would be *decomposition and coordination in mathematical programming*. Sections 2.2 and 2.3 of Chapter 2 are appropriate here. If we apply the solution of the optimization problem, that is, the final control values $\hat{c}_i$, to a real system, the feasibility of the result with respect to the real system must be considered. The problem of generating feasible controls may arise. Nevertheless, from the viewpoint of control, this will be an open-loop control structure. Section 2.7 of Chapter 2 deals with this subject.

Second, we can consider that the coordination level acts on local decision makers who control the real system elements and try to comply with the coordinator's demands. Here we may not even know what the local decision making process is. Let us look at this situation by assuming that the coordinator works by iteration. At each step of the iterative procedure, the local decision makers "do their best" with respect to the real system outputs. If we know the algorithm that the local decision maker is using, the time-behavior of the system from one coordination step to another could be discussed. Let us only state that the behavior may be unstable because of the many separate decision makers acting on the same system. If a steady state is achieved, the coordinator may take his next step to try to improve the value of his performance function (whether in the penalty form or without it). Note that in the case where no penalty terms are used, the direct coordination can in principle be achieved in one step: the coordinator sets values (y_d, r_d) that should optimize the system to the best of his knowledge (i.e., according to the model of the system) and then the local decision makers do their job by achieving $y_i = y_{di}$ and complying with the resources constraint. It is in this case, however, that y_{di} should be feasible for the real system; otherwise the expectations of the coordinator may not be realized.

If the coordinator's demands are feasible for the real system (for instance, because he knows the constraints exactly, or he has decided to move in the "safe region" only), then the demands are feasible in every step of the iterative procedure of the direct method. Hence, the direct method is sometimes referred to as the *feasible method*. In contrast, the direct-penalty coordination is using infeasible demands in the course of the iterations. When the local decision maker is trying to comply with an infeasible demand, his output may violate the constraints related to the input of another subsystem.

We can also consider a mechanistic decision-making hierarchy of control, where we attribute certain formal algorithms to decision making at the local level. We have already mentioned that formal algorithms can be used to set up an open-loop control structure. This would only very seldom be a satisfactory and ultimate solution. The performance of the control can be improved by using feedback information; the human decision makers postulated above were using such information implicitly. Now we would have to say very explicitly what kind of feedback information is available and how it is being used in the formal algorithms. For example, we can assume that the real subsystem outputs y_{*i} are measured. Then we can consider them to be used in essentially two ways: in the local algorithm and in the coordination algorithm. The second possibility has been quite satisfactorily explored and is the subject of the major part of Chapter 3. We are able to obtain coordination algorithms that

- End in a point not violating the real system constraints (provided they are of the form $(c_i, u_i) \in CU_i$ and $y \in Y$),
- Provide for a value of overall performance that is superior to the result of open-loop control.

The first possibility, i.e., the use of measured values y_{*i} in the local level algorithms, is discussed in section 3.4 and some useful results are provided. We can also make use of y_{*i} at both levels; one of the price coordination structures of section 3.3 works in that way.

A MORE COMPREHENSIVE EXAMPLE OF STEADY-STATE OPTIMAL CONTROL

A typical area of application of steady-state optimal control is the continuous chemical processes, where the hierarchical approach was first used. Let us describe how the multilevel approach could be applied to control of an ammonia plant.

Description of the process

Figure 1.16 displays the principal processes in the plant. The first is *methane conversion*, in which H_2 comes from the methane and N_2 from atmospheric air, and steam is added to provide for stoichiometric balance. The second is *conversion* of carbon oxide, in which CO is turned into CO_2, because CO could not be removed directly. Then we have *decarbonization*, where CO_2 is removed from the gas stream. At this point there should be no CO or CO_2 present in the gas stream—what remains of them is neutralized by conversion into methane in the *methanization* part of the plant because CO and CO_2 are toxic to the catalyst used in the synthesis reactor. The *synthesis reactor* is the last essential part of the plant—here the mixture $3H_2+N_2$ reacts to make $2NH_3$ at high pressure and temperature. A cooled liquid (essentially pure ammonia) F_a leaves the plant. The characteristic feature of the ammonia synthesis process is that the synthesis reactor works with recycling, whereby its input flow consists of both the fresh gas and the recycled gas—the latter with NH_3 removed (transferred to the liquid F_a). The fresh gas, however, contains not only H_2 and N_2 but also some "inerts," i.e., components not reacting in the process. They are mainly argon from the atmospheric air and CH_4 from the methanization process used for removing the remaining CO and CO_2. Inerts are no harm, but they would cycle in the synthesis reactor loop endlessly; as new inerts continuously flow in with the fresh gas we would have a considerable increase of inerts in the loop gas,

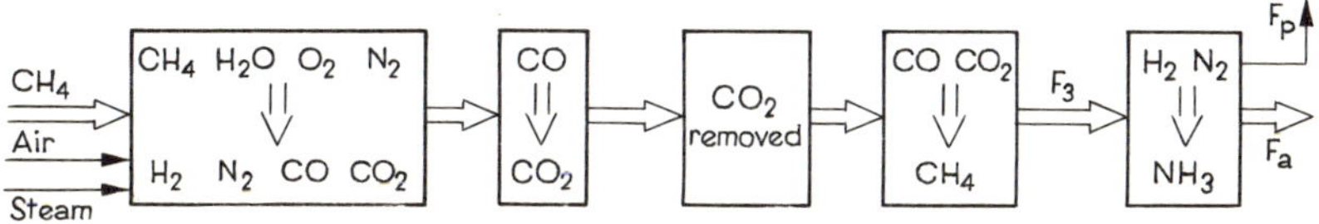

FIGURE 1.16   Principal processes in an ammonia plant.

leaving no space for the useful H_2 and N_2. Inerts have to be removed. There is, however, no practical way to remove them selectively, so the inert level is kept down by a very simple measure: part of the loop gas is blown out into the atmosphere as the so-called purge, F_p.

The optimization problem

Assume that we want to maximize the steady-state production rate Q of ammonia (in kilograms per hour). We have

$$Q = F_a - F_a \sum_j r_j \tag{A}$$

where r_j is the solubility of component j of the circulating gas in liquid ammonia. In order to get variables of other parts of the plant involved in the expression for Q, let us write two mass balance equations. The overall mass balance of the synthesis loop will be:

$$F_a + F_p = F_s \tag{B}$$

where F_s is the fresh gas inflow. The mass balance of the inerts in the synthesis loop will be:

$$F_a r_{in} + F_p y_{pi} = F_s y_{si} \tag{C}$$

where r_{in} is the solubility of the inerts in the liquid ammonia, y_{pi} is the concentration of the inerts in the purge gas, and y_{si} the same for fresh gas.

The use of (B) and (C) allows us to arrive at

$$Q = F_s \left(1 - \frac{y_{si} - r_{in}}{y_{pi} - r_{in}}\right)\left(1 - \sum_j r_j\right). \tag{D}$$

At this stage we infer from physical and chemical knowledge that r_j, r_{in} do not depend on any plant variables, and $y_{pi} > y_{si} > r_{in}$. Under these circumstances we can see that Q is maximized when F_s is maximized, y_{si} is minimized, and y_{pi} is maximized (please recall the physical meanings). We thus would have, in terms of local performance indices Q_i,

$$Q = \psi(Q_1, Q_2, Q_3) = bF_s\left(1 - \frac{y_{si} - a}{y_{pi} - a}\right)$$

where a and b are constants. Note ψ is in this case a strictly order-preserving function.

There could be three local problems: maximize F_s, minimize y_{si}, and maximize y_{pi}. Since the local problems are of course interconnected, a coordination will be needed to provide for max Q while preserving all constraints. In the actual study, it was assumed that F_s is given. It was, however, found reasonable to replace y_{si} by two local performance indices, both to be minimized:

$$Q_1 \triangleq y^1_{CH_4} + y^1_{CO}, \qquad Q_2 = y^2_{CH_4}$$

and to form three subsystems as shown in Figure 1.17. They have the performance indices Q_1, Q_2, and $Q_3 \triangleq y_{pi}$, respectively.

We denoted by $y^1_{CH_4}$ the concentration of CH_4 at the output of the first subsystem. This CH_4 directly contributes to the inert content in the gas F_s; therefore, it makes sense to minimize it right away. The same applies to CO content here, because CO will not be removed in decarbonization. The performance index Q_2 for the second subsystem is the CH_4 concentration in the fresh gas stream F_s. This CH_4 involves the result of methanization which had to be done on the CO_2. Local control can decrease this CH_4 by improving decarbonization, i.e., by decreasing the CO_2 content of the gas stream. Operation of the second subsystem is subject to the constraint that methanization is always complete, i.e., no CO_2 or CO can be left in the stream.

In the third subsystem we have to maximize $Q_3 \triangleq y_{pi}$, the concentration of inerts in the purge gas. This means of course that as little H_2 and N_2 is lost as possible, because in the balance all incoming inerts must be let out. We then have

$$F_p y_{pi} = \text{a constant.}$$

Note that we could replace the goal maximize y_{pi} by the equivalent minimize F_p.

Coordination variables and coordination method

For the nonadditive function ψ in

$$Q = \psi(Q_1,\ldots, Q_N)$$

we have to use coordination by the direct method (price coordination, described below, could not be used here). Let us look at the possible coordination variables. In principle, they should be all the subsystem outputs (or inputs). The coordinator would prescribe their values and thus separate one subproblem from another.

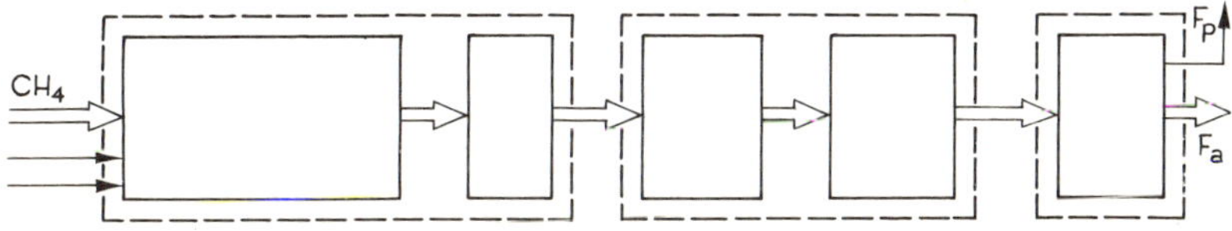

FIGURE 1.17   The ammonia plant divided into three subsystems.

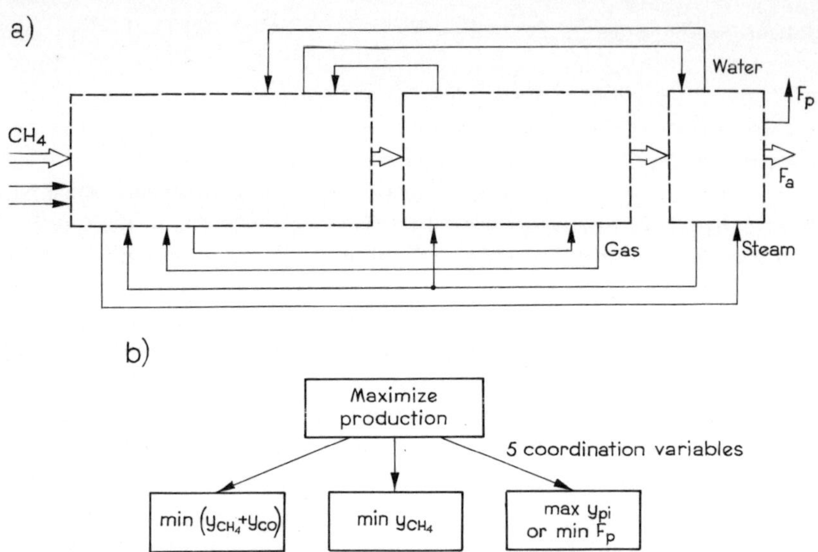

FIGURE 1.18   Subsystem interactions and the control structure.

Here a serious failure of the approach was encountered. Examination of the plant showed that there are many feed-forward and recycle links between parts of the system, not only in the main stream. This was due to the plant design where the links utilize the heat energy generated in the plant and thus make the plant self-supporting in this respect. The main links are shown in Figure 1.18. The failure of the approach consisted in the fact that to describe a cross section through all links would take about 40 variables (flow rates, temperatures, concentrations and so on); these would have to be decision variables in the coordination problem. In the actual task, however, all parts of the plant together had only 22 control variables to be adjusted (the set points of 22 different controllers). Hence we would replace a 22-variable problem by a 40-variable problem at the coordination level, in addition to having to solve the local problems. The two-level problem was more complex and expensive than the direct one.

An examination of the quantitative properties of the problem and of the actual operating experience has permitted an approximate solution. Only 5 out of 40 variables were found to be "essential" and were consequently chosen as coordination variables:

v_1—gas (CH$_4$) flow into the process
v_2—steam flow into the system
v_3—gas pressure in the gas preparation section
v_4, v_5—two principal heat steam flows

The other variables were found to be either directly related to the five, or were assumed to be constant and needing no adjustment by the coordinator, or their values were almost irrelevant for the plant optimization. Note, for example, that the coordinator would not have to prescribe the air inflow to the process. If he sets gas and steam, the amount of air is automatically dictated by the required N_2 to H_2 ratio. The ammonia process has indicated an important topic for hierarchical control studies: *subcoordination*, that is, the use of fewer coordination variables than would be required for a strict solution.

SUBCOORDINATION

Let us very briefly present the problem of subcoordination for the case of the direct coordination method. The main point is that the coordinator would prescribe the output y by using a vector v instead of y_d, where $\dim v < \dim y$. There are two principal ways of using v in coordination. One way of using v is to set up a fixed matrix R and specify for the local problems: $y_d = Rv$, that is, $y_{di} = R_i v$ for each subsystem. Note that if we knew our system accurately, we could construct an adequate matrix $R = \hat{R}$ and a value $v = \hat{v}$, obtaining $y_d = \hat{y}_d$ (the strictly optimal value), whatever the dimension of v. This makes little sense, however; the differences between the model and reality must be assumed.

Another way of using v is to set a fixed function $\gamma\,(\cdot)$ and require that the local problems comply with $\gamma(y) = v$, that is, $\gamma_i(y_i) = v_i$ for each subsystem. This makes more sense intuitively, since we are granting the subproblems their freedom except for the fulfillment of the demands specified in v. For example, we demand a total production but do not specify the individual items. However, in this case the subproblems are not entirely separated and analysis of such a system is much more difficult. The subcoordination approach is also possible in the framework of the price method, which will now be discussed.

COORDINATION BY THE PRICE METHOD

Let us recall the description of the system and of the control problem, as was given by Eqs. (1.23) to (1.27) at the beginning of this section. We have subsystem equations:

$$y_i = F_i(c_i, u_i), \qquad i \in \overline{1, N} \tag{1.23}$$

the system interconnection equation:

$$u_i = H_i y = \sum_{j=1}^{N} H_{ij} y_j, \qquad u = Hy \tag{1.24}$$

46

resource constraints:

$$\sum_{1}^{N} r_i(c_i, u_i) \leq r_0 \tag{1.25}$$

local constraints:

$$(c_i, u_i) \in CU_i, \qquad i \in \overline{1, N} \tag{1.26}$$

and local performance indices:

$$Q_i(c_i, u_i), \qquad i \in \overline{1, N}. \tag{1.27}$$

Even before we define the global performance index of the system, we can define the task of coordination, which will be to influence the local decision makers in such a way that system constraints will be preserved and the overall optimum will be achieved. Price coordination consists in letting the coordinator prescribe prices on inputs, outputs, and resources and then permitting the local decision makers to define their own choices of the values of these variables. The system is *coordinated* when the local choices satisfy the interconnection equation (1.24) and do not violate the global constraint (1.25). The prices that effect this state of the system can be termed *equilibrium prices,* since satisfaction of Eq. (1.24) means an equilibrium of the inputs and outputs.

Price coordination brings about an overall system optimum if the *global performance index* is a sum of local ones

$$Q = \sum_{i=1}^{N} Q_i. \tag{1.29}$$

It is worth remembering that the direct and penalty function coordination methods presented before allowed a more general form of the global performance index.

The discussion of price coordination that follows omits the resource constraint (1.25) and focuses on interconnections in Eq. (1.24). Suitable extensions will be mentioned. A full discussion can be found in Chapter 2, section 2.4. See also section 2.5, where the prices (Lagrangian multipliers) are used along with appropriate penalty terms, and have a broad range of applicability.

INTERACTION BALANCE METHOD (IBM)

In economic systems, price coordination has been known for a long time. Prices were called upon to equate supply and demand, that is, equate the corresponding outputs and inputs. In terms of the system description, the aim of price adjustment is to provide for satisfaction of the interconnection equation (1.24).

Let us look at this in some detail. The *local problems*, i.e., problems associated with the individual system elements, can be formulated as follows (assuming that $Q_i(c_i, u_i)$ has to be minimized):

$$\text{minimize } Q_{\text{mod}i} = Q_i(c_i, u_i) + \langle \lambda_i, u_i \rangle - \langle \mu_i, F_i(c_i, u_i) \rangle \tag{1.30}$$

subject to

$$(c_i, u_i) \in CU_i$$

with the results $\hat{c}_i(\lambda)$, $\hat{u}_i(\lambda)$, $\hat{y}_i(\lambda) = F_i(\hat{c}_i(\lambda), \hat{u}_i(\lambda))$, where λ is the price vector.

If (1.30) is related to a finite-dimensional problem (as is the case in steady-state optimization), then the scalar product

$$\langle \lambda_i, u_i \rangle \text{ means } \sum_{j=1}^{\dim u_i} \lambda_{ij} u_{ij},$$

and

$$\langle \mu_i, F_i(c_i, u_i) \rangle \text{ means } \sum_{j=1}^{\dim y_i} \mu_{ij} F_{ij}(c_i, u_i).$$

In problem (1.30), we assumed coordination by the price vector λ, composed of prices of inputs in the whole system. Hence, λ_i are prices of interaction input u_i; the prices μ_i of output y_i are defined as well by virtue of (1.24), namely

$$\mu_i = \sum_{j=1}^{N} H_{ji}^T \lambda_j.$$

It is therefore right to say that the results of (1.30) are exclusively dependent on vector λ.

The *interaction balance* or *equilibrium* prices $\hat{\lambda}$ will be defined so as to provide for

$$\hat{u}(\hat{\lambda}) - H\hat{y}(\hat{\lambda}) = 0 \tag{1.31}$$

where $\hat{y}(\lambda) = F(\hat{c}(\lambda), \hat{u}(\lambda))$.

Providing for condition (1.31) is the task of the coordinator. In classical economics this could be assigned to a procedure at the stock exchange: a person not connected with the negotiating parties would vary the price λ, watch the responses $\hat{u}(\lambda)$ and $\hat{y}(\lambda)$, and stop the procedure at $\lambda = \hat{\lambda}$.

In order to apply the price method, we need the following:

- Existence conditions for $\hat{\lambda}$, the equilibrium price
- System optimality with control $\hat{c}(\hat{\lambda})$
- Procedures to obtain $\hat{\lambda}$

These will be discussed in Chapter 2. They are based upon discussion of the Lagrangian function of the global problem. After the local minimizations (1.30) have been performed, the Lagrangian is

$$\phi(\lambda) = \sum_{i=1}^{N} Q_i(\hat{c}_i(\lambda), \hat{u}_i(\lambda)) + \langle \lambda, \hat{u}(\lambda) - HF(\hat{c}(\lambda), \hat{u}(\lambda)) \rangle$$

and we require that it has a maximum at $\lambda = \hat{\lambda}$:

$$\phi(\hat{\lambda}) = \max_{\lambda} \phi(\lambda).$$

If such a $\hat{\lambda}$ exists, its further use to determine optimal control is restricted to the case where $(\hat{c}, \hat{u})$, the solutions, are single-valued functions of λ. This requirement appears to be vital for applications. The most simple sufficient conditions are: $(\hat{c}, \hat{u})$ are single-valued if the functions $Q_i(\cdot)$ are strictly convex and the mappings $F_i(\cdot)$ are affine (linear). With $\lambda = \hat{\lambda}$, the unique solutions $\hat{c}(\hat{\lambda})$, $\hat{u}(\hat{\lambda})$ are optimal.

It may be appropriate to indicate that the requirement of uniqueness of $\hat{c}(\hat{\lambda})$, $\hat{u}(\hat{\lambda})$ in response to a change in λ has a simple interpretation: since the prices λ aim at providing a match of the outputs to the inputs of other subsystems, they should have a well-defined influence. In many real-life problems, the uniqueness of response can be predicted by physical considerations for systems that are far from linear (remember that we do not know the necessary conditions, while the sufficient ones are too severe to be of much practical use).

It is easy to provide an example in which the uniqueness of response will fail to appear. If λ is the price imposed by the coordinator on some product and $\hat{y}(\lambda)$ the optimal amount produced by a subsystem according to its own local optimization procedure, $\hat{y}(\lambda)$ will not be well-defined when the unit production cost is equal to λ. Note that there would be no local gain or local loss associated with the size of production y, which means that no value of output y will be preferred and this subsystem will give no single-valued response to price λ.

Let us now turn back to the main stream of our considerations. What procedures could be used at the coordination level in the search for $\hat{\lambda}$? It will be shown in Chapter 2 (section 2.4) that if $Q_i(\cdot)$ are continuous and $F_i(\cdot)$ are continuous, then gradient procedures for λ can be used, if we find a way to deal with the points where the $(\hat{c}, \hat{u})$ are not unique and where the gradient is not defined (subgradients can be considered there). In those regions of λ-space where the $(\hat{c}, \hat{u})$ are unique, the following formula holds for the (weak) derivative of $\phi(\lambda)$: (see section 2.4 for details)

$$\nabla \phi(\lambda) = \hat{u}(\lambda) - HF(\hat{c}(\lambda), \hat{u}(\lambda)). \tag{1.32}$$

Note that this is exactly the input–output difference, the *discoordination*, in the system, which has to be brought to zero. The second derivative, $\nabla^2\phi(\lambda)$, does not exist in the general case.

Let us mention that the interaction balance method (IBM) described so far can be applied to both static and dynamic problems. In this method, the search for λ is based on the difference, $\hat{u}(\lambda) - H\hat{y}(\lambda)$. It is, therefore, a computational concept rather than a control structure. In a system that is already in operation, the interconnection equation is satisfied all the time for any control c. We could never find out if λ is correct. We could, therefore, use the method for open-loop control only. It means that we would first compute $\hat{c}(\hat{\lambda})$ and then apply it to the real system. The control result will of course strongly depend on the accuracy of the models.

Let us now come back to the resource constraint (1.25):

$$r_1(c_1, u_1) + \cdots + r_N(c_N, u_N) \leq r_0.$$

This additive form of global constraint can be incorporated into the price coordination scheme by using an additional price vector η (the resource price) and adding to each local problem a value $\langle \eta, r_i(c_i, u_i) \rangle$, so that the local objective function becomes:

$$Q_{i\,\mathrm{mod}} = Q_i(c_i, u_i) + \langle \lambda_i, u_i \rangle - \langle \mu_i, F_i(c_i, u_i) \rangle + \langle \eta, r_i(c_i, u_i) \rangle. \tag{1.33}$$

By varying η, the coordinator would change the resource requirements of the local problems so as to satisfy the overall constraint. In mathematical programming terminology, η would be a Kuhn–Tucker multiplier. The next paragraphs will discuss some other ideas of price coordination, in which feedback from the operating system will be used to improve control.

PRICE COORDINATION IN THE STEADY STATE WITH FEEDBACK TO THE COORDINATOR (THE IBMF METHOD)

In this section we shall consider the optimization problem to be in finite-dimensional space, i.e., to be a problem of nonlinear programming. This means that we optimize the steady state in a complex system. We remember from section 1.2 that steady-state control is an appropriate technique if the optimal state trajectory of a dynamic system is slow enough to assume that the value of state vector x is at any time related to control only, the time derivative of x being negligible.

The mappings F_i, Q_i are now functions in finite-dimensional space. We have therefore the following model-based global problem:

$$\text{minimize } Q = \sum_{i=1}^{N} Q_i(c_i, u_i)$$

subject to

$$y_i = F_i(c_i, u_i), \qquad i \in \overline{1, N}$$

$$u = Hy$$

$$(c_i, u_i) \in CU_i, \qquad i \in \overline{1, N}.$$

We have dropped the resource constraint for simplicity. A solution to the model-based problem yields *model-based control* $\hat{c}$. We intend now to pay considerable attention to the difference between model and reality; let us therefore formulate the following *real problem*:

$$\text{minimize } Q = \sum_{i=1}^{N} Q_i(c_i, u_i)$$

subject to

$$y_i = F_{*i}(c_i, u_i), \qquad i \in \overline{1, N}$$

$$u = Hy$$

$$(c_i, u_i) \in CU_i, \qquad i \in \overline{1, N}.$$

We should notice that here the only difference between model and reality is assumed to exist in the subsystem equations, that is, the functions $F_{*i}(\cdot)$ are different from the functions $F_i(\cdot)$ in the model. We shall indicate below some effective ways to deal with the consequences of this difference.

It must be stressed, however, that differences also may exist in the performance function and in the constraint set. For example, if a performance function is explicitly $Q_i(c_i, u_i, y_i)$, then it will reduce to some $Q_i(c_i, u_i)$ by using the subsystem equation, but this makes it model-based. The real $Q_{*i}(c_i, u_i)$ would be different from $Q_i(c_i, u_i)$. In a similar way, the set CU_{*i} may be different from CU_i.

The solution to the real problem will be termed *real-optimal control* $\hat{c}_*$. It is not obtainable by definition since reality is not known. We can only look for a structure that would yield a control better than the purely model-based $\hat{c}$, but in principle what we will achieve is bound to be inferior to $\hat{c}_*$.

One possible structure is iterative price coordination with feedback to the coordinator. It is shown schematically in Figure 1.19. The *local problems* are exactly the same as in the open-loop interaction balance method, that is, we have for each $i \in \overline{1, N}$:

$$\text{minimize } Q_i(c_i, u_i) + \langle \lambda_i, u_i \rangle - \langle \mu_i, F_i(c_i, u_i) \rangle$$

subject to

$$(c_i, u_i) \in CU_i.$$

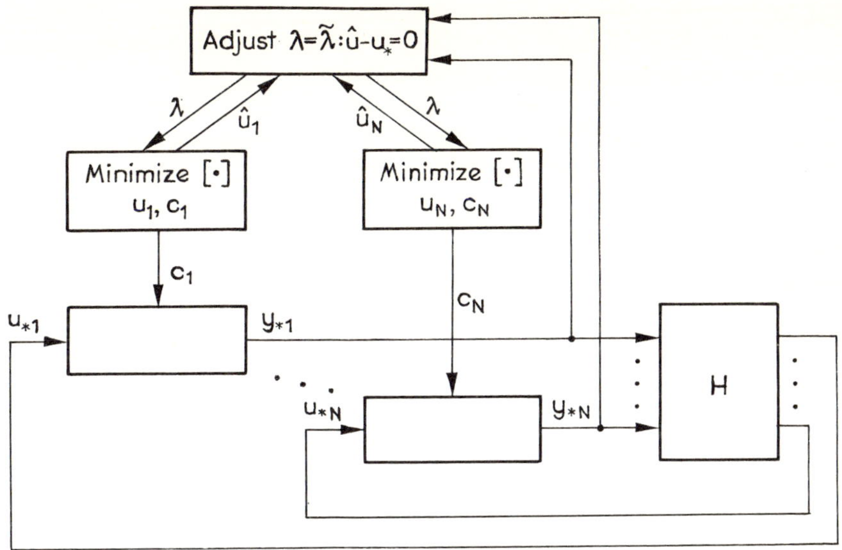

FIGURE 1.19   Iterative price coordination with feedback to the coordinator.

The controls $\hat{c}_i(\lambda)$ determined by solving this problem computationally for the current value of λ are applied to the real system, giving some u_* and y_*. The coordination concept consists of the following upper-level problem:

$$\text{find } \lambda = \tilde{\lambda} \quad \text{such that} \quad \hat{u}(\tilde{\lambda}) - u_*(\hat{c}(\tilde{\lambda})) = 0. \tag{1.34}$$

Condition (1.34) is an equality of model-based optimal input $\hat{u}(\lambda)$ and u_*, measured in the real system and the product of control $\hat{c}(\lambda)$. Providing for this equality is the basic idea of the interaction balance method with feedback (IBMF).

The properties of control based on condition (1.34) have been studied quite extensively, and are discussed in Chapter 3, section 3.3. The usual questions of the existence of $\tilde{\lambda}$, system optimality with control $\hat{c}(\tilde{\lambda})$, and procedures to obtain $\tilde{\lambda}$ have been discussed there and detailed answers have been formulated. The essence of these answers is as follows.

Solution $\tilde{\lambda}$ exists, if solution $\tilde{\lambda}$ of the open-loop interaction balance method (IBM) exists for all s-shifted systems

$$u = HF(c, u) + s$$

where $s \in S$, and S is the set of all possible values of the model-reality difference

$$HF_*(c, u) - HF(c, u) = s$$

52

with $(c, u) \in CU = CU_1 \times \ldots \times CU_N$. When the models do not differ from reality, $\hat{c}(\tilde{\lambda})$ is strictly optimal control and $\tilde{\lambda}$ equals the equilibrium prices $\tilde{\lambda}$ which would be obtained by solving the problem by the interaction balance method described above. When models differ from reality, the control based on (1.34) is in the first approximation no worse than the control based on open-loop value $\hat{\lambda}$. In the particular case where

$$F_{*i}(c_i, u_i) = F_i(c_i, u_i) + \beta_i \qquad i \in \overline{1, N}$$

that is, where the model-reality difference of the subsystems consists in a shift, the control based on condition (1.34) is strictly real-optimal. The open-loop would, of course, be much inferior in this case. See section 3.3 in Chapter 3 for some numerical results.

Control based on condition (1.34) has the very important property of keeping to the constraints in the real system. Note that this real control c_* equals the model-based $\hat{c}$ for any λ, because the result $\hat{c}(\lambda)$ is applied to the system. For $\lambda = \tilde{\lambda}$, we also have $u_* = \hat{u}$. Since the model-based solution will keep $(\hat{c}_i, \hat{u}_i) \in CU_i$, $i = \overline{1, N}$, the same will be kept in the real system, but only at $\lambda = \tilde{\lambda}$. Note that the open-loop control $\hat{c}(\hat{\lambda})$ may violate the constraints in the real system because at $\lambda = \hat{\lambda}$ we will generally have $u_* \neq \hat{u}$. The control based on $\lambda = \tilde{\lambda}$ does not violate the constraints $(c_i, u_i) \in CU_i$ if the real constraint sets equal the model ones, $CU_{*i} = CU_i$, $i \in \overline{1, N}$. Section 3.3 shows, however, a modified method (MIBMF) where the case $CU_{*i} \neq CU_i$ is also covered by appropriate use of feedback information.

As far as the procedures to find $\tilde{\lambda}$ are concerned, iterations have to be done at a rate appropriate for the real system, i.e., a rate permitting new values u_* to establish themselves after a change of λ. Unfortunately, the expression

$$R_*(\lambda) = \hat{u}(\lambda) - u_*(\hat{c}(\lambda)) \tag{1.35}$$

which has to be brought to zero is not a derivative of any function. The value of $\tilde{\lambda}$ has to be found by solving the equation $R_*(\lambda) = 0$. It should be stressed that if there are inequality constraints in the local problems, $R_*(\lambda)$ will in general be nondifferentiable. Suitable numerical methods to find $\tilde{\lambda}$ are proposed and discussed in section 3.3 of Chapter 3.

We should emphasize that the application of coordination principle (1.34) means that iterations must be done on the real system, as an on-line control. This can be performed in a steady-state optimization, but not in a dynamic one. The only exception is iterative optimization of batch or cyclic processes, in which various functions of time are applied in each consecutive batch. The main field of application of the coordination principle (1.34) is, however, steady-state optimizing control, and this is the framework in which Chapter 3 has been written.

Let us add an example to explain what on-line price coordination really means. Consider the electric power system and its customers. The amount of power that is produced is matched to the current load. How can we tell whether the price of electric energy is correct when there is no difference between demand and supply? From the on-line price adjustment proposed in this section, we can infer that the price is correct when the production-load balance of the power that has established itself in the real system (u_*) is equal to the model-based optimal value ($\hat{u}$). Any difference would be used to change the price of electricity.

DECENTRALIZED CONTROL WITH PRICE COORDINATION (FEEDBACK TO LOCAL DECISION UNITS)

The structure of Figure 1.19, although effective and superior to open-loop, model-based control, may be criticized because the information about real system u_* is made available only to the coordinator. The local units calculate their controls and their inputs u_i for each λ on the basis of models, "knowing" that reality is different. The scheme of Figure 1.19 is therefore suitable for a mechanistic control system, but does not reflect the situation in which local problems are solved by decision makers with more freedom of choice.

We can expect that the local decision maker would tend to use the real value u_{*i} in his problem, that is, he would

$$\text{minimize } Q_i(c_i, u_{*i}) + \langle \lambda_i, u_{*i} \rangle - \langle \mu_i, F_i(c_i, u_{*i}) \rangle \tag{1.36}$$

subject to

$$(c_i, u_i) \in CU_i.$$

Schematically, this is presented in Figure 1.20 as feeding u_{*i} to the corresponding local problem. Even with fixed λ, the control exercised by local decision makers on the system as a whole remains to some extent coordinated, since the value of λ will influence the control decisions. However, since u_{*i} are used locally, we may call the structure of Figure 1.20 *decentralized*.

A problem in itself is system stability, or the convergence of iterations made by local optimizers while trying to achieve their goals. It is obvious that all the iteration loops in the system are interdependent, since a u_{*i} will depend on the decisions $c = (c_1, \ldots, c_N)$ in the previous stage, that is, on the decisions of all the decision units. If the iterations converge, some steady-state values $\hat{c}(\lambda)$, $\hat{u}_*(\lambda)$ and $\hat{y}_*(\lambda)$ will be obtained for the given price vector λ. It may be predicted that if this λ would happen to be $\tilde{\lambda}$ from the previous paragraph, the result of decentralized control would also be the same as in the previous structure. This does not say that we should aim at it, since the

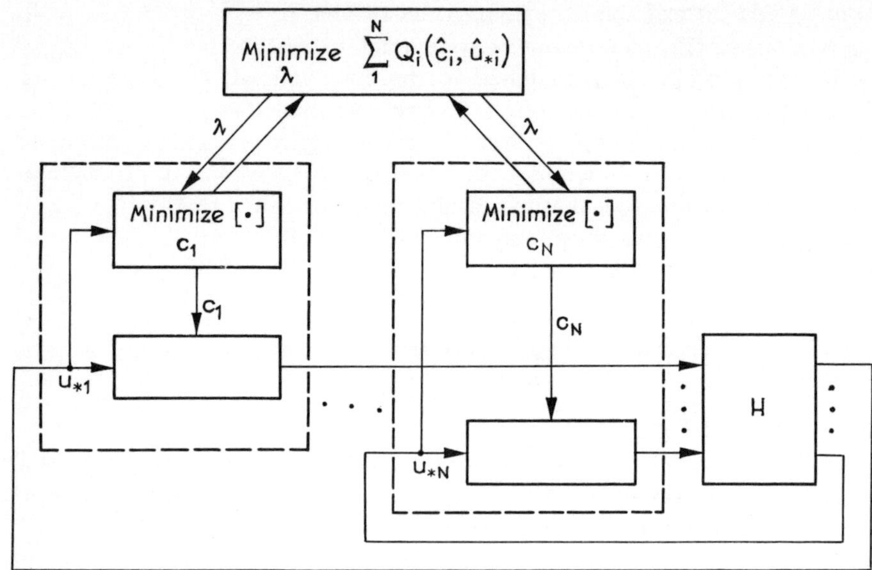

FIGURE 1.20 Decentralized control with on-line price coordination.

results obtained with $\tilde{\lambda}$ are not real-optimal and a better value of λ may exist.

We should look for some way to iterate the prices λ in the system of Figure 1.20. A possibility might be

$$\text{minimize } Q = \sum_{i=1}^{N} Q_i(\hat{c}_i(\lambda), \hat{u}_{*i}(\lambda)) \tag{1.37}$$

which simply means that we should find a price λ such that the overall result of local controls is optimized.

Two properties of the problem seem predictable. If the models are adequate, and all iterations converge, they will converge to the strict overall optimum for the system. If the models differ from reality, then the constraints $(c_i, u_i) \in CU_i$ will be secured (as in the structure in Figure 1.19), but the overall result will be suboptimal; in performing the local optimizations we continue to have an inadequate (model-based) value of the output y_i. Section 3.4 of Chapter 3 is devoted to a more thorough discussion of this control structure.

DYNAMIC MULTILEVEL CONTROL

The structures of on-line dynamic control using decomposition of the control problem differ from those applicable to the steady state. The differences lie in the use of feedback from the system in operation. In steady-state control we could use feedback in the form of measured inputs or outputs of the system elements and provide for an extremum of a current or "instantaneous" performance index, as described above in this section. In dynamic optimization, we need to consider at time t the future behavior of the system, that is, to consider an *optimization horizon.* Since the future behavior depends on both control and the initial state, we cannot determine the optimal control input unless we know the present state of the system; if we wish to have a control structure with feedback from reality, this feedback must contain information on the state $x(t)$.

There are three principal ways in which local dynamic control problems can be formulated and subsequently coordinated by an appropriate master problem. They are the following:

- *Dynamic price coordination,* in which time-varying prices on the inputs and outputs are imposed by the coordinator, who also decides on the target states to be achieved by each subsystem over the local optimization horizon
- *Structure based on the state-feedback concept,* in which the local decision making is reduced to a static (instantaneous) feedback decision rule, and the coordinator supplies signals that modify either the local decisions or the local decision rules so as to account for the performance of the system as a whole
- *Structures using conjugate variables,* in which local decision making is a kind of static (instantaneous) optimization, and the optimal dynamic policy is secured by a vector of prices on the derivatives of the state variables (the trend of the subsystem state), i.e., the vector of conjugate variables imposed on the subsystems and readjusted by the coordinator

In this section, which is of an introductory nature, we shall briefly discuss only the first of these alternatives, and focus on the dynamic features. A more thorough discussion of the dynamic price coordination structure, as well as a description of the other two possibilities, are contained in Chapter 4.

Assume that the global control problem of the interconnected system is as follows

$$\text{minimize } Q = \sum_{i=1}^{N} \int_{0}^{t_f} q_{0i}(x_i(t), m_i(t), u_i(t)) \, dt \tag{1.38}$$

subject to

$$\dot{x}_i(t) = f_i(x_i(t), m_i(t), u_i(t)), \qquad i \in \overline{1, N} \text{ (state equations)}$$

$$y_i(t) = g_i(x_i(t), m_i(t), u_i(t)), \qquad i \in \overline{1, N} \text{ (output equations)}$$

$$u(t) = Hy(t) \text{ (interconnections)}$$

56

with $x(0)$ given, and $x(t_f)$ free or specified. Consider that in solving the problem we incorporate the interconnection equation into the following Lagrangian:

$$L = \sum_{i=1}^{N} \int_0^{t_f} q_{0i}(x_i(t), m_i(t), u_i(t))\, dt + \int_0^{t_f} \langle \lambda(t), u(t) - Hy(t) \rangle\, dt$$

where $\langle \lambda(t), u(t) - Hy(t) \rangle$ means

$$\sum_{j=1}^{\dim u} \lambda_j(t)(u(t) - Hy(t))_j.$$

Assume that the solution to the global problem using this Lagrangian has been found and it has provided for

$$\hat{x}_i, \; i = 1, \ldots, N \; \text{(optimal state trajectories)}$$

$$\hat{m}_i, \; i = 1, \ldots, N \; \text{(optimal controls)}$$

$$\hat{u}_i, \; i = 1, \ldots, N \; \text{(optimal inputs)}$$

$$\hat{y}_i, \; i = 1, \ldots, N \; \text{(optimal outputs)}$$

$$\hat{\lambda} \qquad \text{(the value of the Lagrangian multipliers that provides for the solution)}$$

Note that now our Lagrangian can be split into additive parts, thus allowing the formation of local problems of the kind:

$$\text{minimize } Q_i = \int_0^{t_f} [q_{0i}(x_i(t), m_i(t), u_i(t)) + \langle \hat{\lambda}_i(t), u_i(t) \rangle$$

$$- \langle \hat{\mu}_i(t), y_i(t) \rangle]\, dt \tag{1.39}$$

where

$$y_i(t) = g_i(x_i(t), m_i(t), u_i(t))$$

and optimization is subject to

$$\dot{x}_i(t) = f_i(x_i(t), m_i(t), u_i(t))$$

with $x_i(0)$ given, $x_i(t_f)$ free or specified, as in the original problem.

In the local problem, the price vector $\hat{\lambda}_i$ is an appropriate part of $\hat{\lambda}$, and $\hat{\mu}_i$ is also given as a function of $\hat{\lambda}$ as

$$\hat{\mu}_i = \sum_{j=1}^{N} H_{ji}^T \hat{\lambda}_j.$$

Notice that we have put the optimal value of price vector $\hat{\lambda}$ into the local problems, which means that we have solved the global problem before and that the solutions of the local problems will be strictly optimal. There is little sense, however, in solving the local problems if the global problem was solved before, because the global solution would provide not only $\hat{\lambda}$ but also $\hat{x}$ and $\hat{m}$ for the whole system. To make local solution practical, let us try to shorten the local horizons and use feedback in the local problems. If we shorten the horizon from t_f to t_f', the local problem (1.39) becomes

$$\text{minimize } Q_i = \int_0^{t_f'} [q_{0i}(x_i(t), m_i(t), u_i(t)) + \langle \hat{\lambda}_i(t), u_i(t) \rangle$$
$$- \langle \hat{\mu}_i(t), y_i(t) \rangle] \, dt \qquad (1.40)$$

with $x_i(0)$ given as before, but the target state taken from the global long-horizon solution, $x_i(t_f') = \hat{x}_i(t_f')$. Here we might remind the reader of the discussion of multilayer hierarchies with the divided time horizon in section 1.2 (see Figure 1.9).

For the local problem (1.40), we must of course supply the price vectors $\hat{\lambda}_i$, $\hat{\mu}_i$. It may also be reasonable to use $\hat{u}_i$, that is, the "predicted" input value, from the global solution. The short-horizon formulation (1.40) will pay off if we have to repeat the solving of (1.40) many times in order to use feedback information; we would have to solve the global problem only once. See Figure 1.21, where the principle of the proposed control structure is presented.

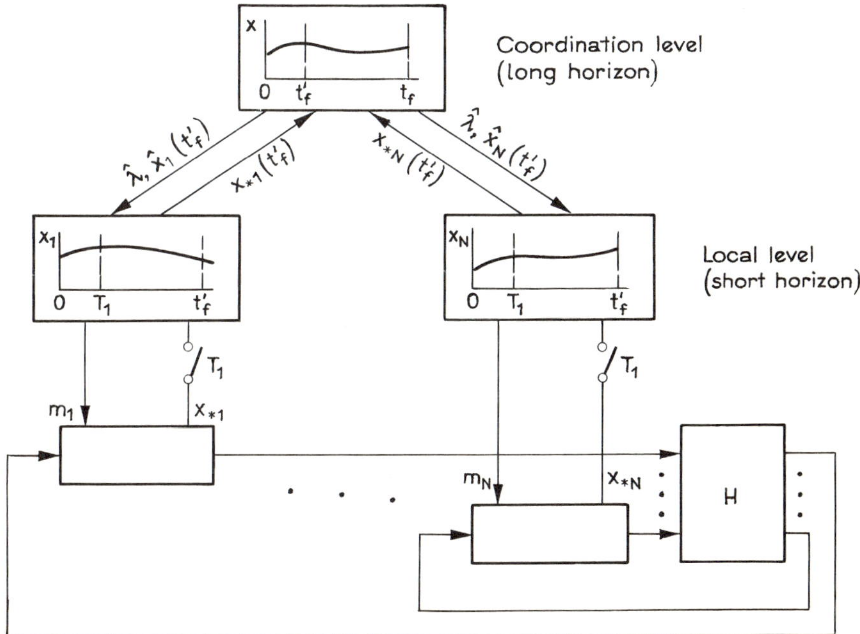

FIGURE 1.21   A structure for on-line dynamic price coordination.

Feedback at the local level consists in solving the short-horizon local problems at some intervals $T_1 < t_f'$ and using the actual value of the measured state $x_{*i}(kT_1)$ as the new initial value for each repetition of the optimization problem. This gives us something new; we now have a truly on-line control structure and can expect, in appropriate cases, to get results better than those dependent on the models alone. Note that we have now gained from both decomposition (reducing dimension) and a shortening of the horizon. The feedback algorithm just described would be referred to as a repetitive optimization scheme.

We should mention disturbances that act on the real system but were not shown explicitly in the formulations. Disturbance prediction would be used while solving (1.38) and (1.40), that is, the global and the local problems. It is indeed because of the disturbances, which in reality will differ from how they are predicted, that we are inclined to use the feedback structure of Figure 1.21.

The feedback introduced thus far cannot compensate for the errors made by the coordination level in setting the prices $\hat{\lambda}$. Another repetitive feedback can be introduced to overcome this problem, for example, by bringing to the coordinator the actual value x_{*8i} at time t_f', $2t_f'$, ... and asking that the global problem be resolved for each new initial value. This structure of control is presented in Figure 1.21. We should note that feeding back the actual values of the state achieved makes sense if the models used in computation differ from reality, for example, because of disturbances. Otherwise, the actual state is exactly equal to what the models have predicted and the feedback information is irrelevant.

In Chapter 4 we discuss the structure of Figure 1.21 and its different varieties in much more detail. In particular, we consider there the important case in which the subsystems are interconnected not only in a direct ("stiff") way, but also through storage elements (see Figure 4.1). It is believed that these kinds of interconnections may be quite common in dynamic systems, for example, where some subsystem products are stored and other subsystems take their inputs from these stores.

1.4. INFORMATION STRUCTURES IN HIERARCHICAL CONTROL

As noted earlier in the discussion of Figure 1.1, the functioning of the elementary control unit involves two separate processes: an observation process and a decision process. The observation process collects information about the controlled system and its environment. The decision process uses this information, plus additional information already on hand, to select desirable controls. The effectiveness of these two processes is judged by evaluating results using a performance index q or its integral Q. Note that the information used by the decision process is made up of two parts: the observations $v(t)$ defined in Eq. (1.4) and the prior information which we will denote by the vector p. We will refer to these two items collectively as *the information.*

The identification of the two separate processes performed by the elementary control unit leads to the identification of two separate parts of the control system design problem:

1. *The design of the decision process*
2. *The design of the observation process*

Using notation introduced in section 1.1 we see that the second part involves the choice of the information function h in Eq. (1.4) where

$$v(t) = h(x(t), m(t), z(t)) \tag{1.4}$$

and the first part involves the choice of the decision rule d in Eq. (1.5). We will modify Eq. (1.5) to show explicitly the dependence on prior data as

$$m(t) = d(v_{[t_0,t]}, p). \tag{1.5}$$

Much classical control theory simplifies the design problem by considering only the first part: the choice of an optimal control law when the information function h is given. This approach is appropriate when h is chosen intuitively from a limited range of possibilities or when h is given in the problem specifications. For example, a common assumption is that h provides perfect information about the state of the controlled system. While this is clearly simplistic, it is in many cases close enough to reality to yield useful results.

More generally, a control system design involves a choice of both h and d. Moreover, in an optimal design they cannot be chosen independently because m, and thus q, depends on h by way of d. That is, because

$$q(t) = q(y(t), m(t), v(t))$$
$$= q(y(t), d(h(x(t), m_{[t_0,t]}, z_{[t_0,t]}, p), v(t)). \tag{1.41}$$

Thus, in the general case we must evaluate alternative (h, d) pairs to find the optimal design.

As we move to the consideration of hierarchical control structures, the informational part of the design problem grows considerably more complex. In a hierarchical control system, there are several separate decision (control) units. Each decision unit makes its own independent observations v_i and has its own prior information p_i and its own manipulated variables m_i. In the initial design, we might simply maintain this partition of information and

60

have each unit choose local controls based only on its local information (v_i, p_i). This approach to the control of complex systems is commonly employed in many situations, but particularly those in which the controlled subsystems (Figure 1.2) are only weakly coupled, either by accident or by design. However, even in such apparently completely decentralized control schemes, there is usually some "tuning" (i.e., some coordinated adjustment of all the d_i) to improve performance. Thus it appears quite reasonable to expect improved performance through some sort of information sharing among control units, and we now add a third part to the control system design problem to allow such sharing.

3. *The design of an information distribution mechanism*

The fundamental questions to be considered here are what information should be sent to whom and when. As in the single decision unit case, we may be tempted to avoid the complexity of parts 2 and 3 by simply assuming at the outset an observation process and an information distribution mechanism. For example, we may try to extend the single control unit approach and simply assume that each decision unit receives all the available information (observations v_i and prior data p_i from everyone else). Unfortunately, this simplistic approach is now somewhat dangerous because its basic assumption is inconsistent with the original justification for a hierarchical control system presented in section 1.1. There we argued that the decision process should be divided precisely because of informational constraints. That is, no decision unit had both access to all the information and the capacity to process all the information. Thus we might expect to obtain rather strange results if we eliminate the information constraints by assumption. To avoid inconsistency, it appears that we must explicitly recognize the information constraints in our design procedure. Yet this is not easily done. The explicit inclusion of informational constraints in a hierarchical control system design problem is very difficult, or perhaps even impossible. At the moment, a fundamental aspect of the design problem cannot be solved and yet cannot be ignored or eliminated without considerable peril of obtaining meaningless results. For example, suppose we assume, either explicitly or implicitly, that all of the decision units can have access to all of the available information. In this case, any one of them can solve the entire decision problem and simply direct the actions of the others. All but one is then superfluous and our design reduces to a single control unit solution. Yet this type of solution was initially ruled unacceptable!

We are forced to conclude that, where possible, the informational constraints must be explicitly included in the hierarchical control system design procedure. Where this is not possible, we may be forced to design without constraints, but we must then check the results for constraint satisfaction.

However, in any case, it is clear that we cannot simply ignore the informational aspects of the design problem. Unfortunately, these informational aspects are very complex and at present only poorly understood. We examine these questions in greater detail in Chapter 5.

INFORMATION STRUCTURE

In the above paragraphs we have argued that there are two informational parts to the hierarchical control problem: the design of the observation process and the design of the information distribution mechanism. We now lump these two parts together as the design of the *information structure.*

The observation process will be represented by a set of information functions h_i where

$$v_i(t) = h_i(x(t), m(t), z(t)) \quad i \in \overline{1, N}.$$

We will represent the information distribution mechanism by defining messages γ_i and a message generation process

$$\gamma_i(t) = \phi_i(x(t), m(t), z(t), \gamma(t), p) \qquad i \in \overline{1, N}. \tag{1.42}$$

where $\gamma = (\gamma_1, \ldots, \gamma_N)$. Here γ_i is the message received by decision unit i. We now describe the two parts of the hierarchical control system design problem as:

- The choice of N decision rules d_i for a given information structure
- The design of the information structure (h, ϕ) where $h = (h_1, \ldots, h_N)$ and $\phi = (\phi_1, \ldots, \phi_N)$.

As in the single decision unit problem, the simpler problem is the first, that is, the problem of selecting an optimal $d = (d_1, \ldots, d_N)$ given h and ϕ. However, the complete design requires an optimum choice of the triple (d, h, ϕ). Moreover, the information structures considered in the design must be consistent with information constraints that originally motivated the choice of a hierarchical design. We will note three common types of constraints and indicate some possible procedures for incorporating them into the design process.

PRIVACY CONSTRAINTS

There is certain information in the individual decision units that is private and should not be revealed during the decision process. Examples of such private information might be process models or performance data that is proprietary. Here we face a problem of designing an information structure that preserves privacy and at the same time leads to satisfactory decisions.

62

This problem has received some attention in the economics literature, e.g., Hurwicz (1971). As might be expected, there are some trade-offs to be made between the degree of privacy and performance of the control system. In general, problems involving privacy constraints are beyond the scope of this book and will not be discussed in Chapter 5.

CAPACITY CONSTRAINTS

In many problems, the constraint is on information processing or computational capacity of the decision units. A typical example is the case in which the decision makers are human operators who have physiological limits to the data processing rates they can handle. Such problems have been addressed in the organization theory literature under the title of bounded rationality (e.g., see Simon (1972)). Unfortunately, the translation of capacity constraints into statements about acceptable information structures is quite difficult. One of the major problems appears to be the lack of a fundamental theory relating these two topics. Some initial results are discussed in Chapter 5.

COST CONSTRAINTS

In some problems, the major justification for a hierarchical system arises from the cost of centralizing the information or providing decision units capable of processing the centralized information, or both. For example, in a recent discussion of the design of a hierarchical computer control system for a large testing process, the designer was asked why there was a hierarchy of decision units when any one of the installed computers could handle the decision involved. The reply was that it was actually cheaper simply to buy three additional computers than it would have been to wire all the required signals from each location to a single central computer (see Aronson (1972)). As might be expected, this leads to a cost-benefit approach to the information structure design: the cost of additional information is weighed against the benefits in terms of improved performance. The major problem here is that the benefits, measured in terms of a change in the performance index, are not immediately comparable with the costs, measured in additional communication capacity. Moreover, these costs may include technological coefficients that change significantly with innovations in communications technology. The basic concepts of benefits (value) and cost of information are explored in greater detail in Chapter 5. There we develop a rudimentary technique for investigating cost-benefit trade-offs when information costs are dominated by transmission costs. These results suggest that extensions of classical information theory may provide a useful tool for the analysis of information costs in hierarchical control system design.

1.5. FORMULATION OF CONTROL PROBLEMS

In this section we present a few remarks on the way in which control problems are formulated. We are concerned, in particular, with the role of optimization, with the question of how the performance indices are defined, and with the kinds of constraints and their effect on the problem. An experienced engineer, economist, or systems analyst may find this section oversimplified, if not trivial. We hope, however, that our simple presentation will help some of the readers by indicating the important features which should not be ignored when a problem is being formulated. The importance of problem formulation could hardly be overemphasized. The formulation determines the solution that will be obtained; nothing can be more misleading or confusing than the right solution to the wrong problem.

THE TASKS OF CONTROL

Section 1.1 introduced the basic notions of control. Very briefly, the control task may be described as follows: there is a system to be controlled, it has manipulated inputs m, and we are to act on m to bring about a desired performance of the system. As pointed out in section 1.1, it would be wrong to assume that the desired behavior as seen by the system user always involves formulation of a scalar-valued performance index. In fact, there are many cases in which the user of the controlled system has no preference with regard to the performance index to be maximized or minimized by the control. The user is often more directly interested in maintaining certain properties such as system stability, in spite of varying external conditions (for example, maintaining the stability of a power system in spite of varying loads), or ensuring a satisfactory time response of the system to an external command, in spite of varying parameters (for example, ensuring the response of the airplane to the pilot's commands at various speeds and altitudes). A follow-up task attributed to the direct control layer in section 1.2 (see Figure 1.11) would be another example: we do not care very much about optimization of the follow-up as long as the difference between c and c_d is reasonably small.

In all these cases, however, a performance index could be introduced, perhaps by the designer of the control system himself, in order to reduce ambiguity, that is, to provide for a well-defined, unique solution to the problem. A follow-up controller mentioned above may have several alternative designs, none of which are able to provide for a perfect follow-up because of physical constraints, measurement errors, etc. If we decide on a certain scalar measure of the follow-up error, for example, its mean-squared value, we have a clue by which to choose the alternative design for which the mean-squared error is a minimum. We can also optimize the design, formally or by simulation, so as to obtain the least possible error.

There are also many cases of control tasks in which optimization is directly relevant, that is, in which the user of the system has a definite optimization goal in mind. The ammonia production process described briefly in section 1.3 is such an example. It may be argued that the number of control problems that are formulated as an optimization problem by the users will increase as more people learn that optimal control is both possible and practicable. This conviction is, of course, related to developments in the area of control computers, in particular to the invasion of microprocessors. We do not want to say, however, that the limitations in the use of optimal control have been only of a technical nature. It should rather be said that in some applications it may not be possible to agree on a rational performance index. Therefore, it may be more appropriate to formulate the control task in a different way. For example, in the long-range planning of a national economy, it may be more reasonable to agree on a certain desired rate of growth, or even on a growth trajectory (a *turnpike*), than to formulate an overall utility function. We will come back to this example later on. For the moment, let us conclude that an acceptable performance index is possible in those cases in which the factors that count in our judgment of the system behavior are of a technical or economic nature. Social values and qualities are less likely to be measurable and are thus more difficult to include in a performance index.

THE OPTIMIZATION OBJECTIVES

We know from the previous sections of this chapter that optimization problems can be static or dynamic. The distinction is important not only for the methods of solution, but also for the problem formulation. For static problems (remember they may be time-varying) it is enough to agree on an instantaneous, or current, performance index. For example, for the steady-state optimization of the ammonia process, it was appropriate to require that the production rate in kilograms per hour be maximized. The maximum of integrated production, over a week, a month, or any other period, would be obtained if the production rate was maximized at all instants in time.

In a dynamic problem, the optimization does not reduce to instantaneous values; because of the accumulation properties of the controlled system, the problem solution has to be defined as a control policy over time. Therefore the performance index must be appropriately related to the system behavior over the entire optimization horizon.

In most of the examples that are considered in textbooks, and this one is hardly an exception, the performance over time is judged by taking an integral:

$$Q = \int_{t_0}^{t_f} q(\ldots)\, dt$$

where q is a rate value, for example, a production rate in kilograms per hour, or a cost rate in dollars per hour.

The integral formulation is appropriate if the user of the system is really sensitive to the value of the integral and not to other aspects of the time behavior of the system. A user trying to minimize fuel consumed by a ship to reach a target harbor would be sensitive to the value of the integral, as would one who wants a desired temperature and composition of steel in an electric blast furnace at minimum energy expense.

But the value of the integral may also be inadequate. For example, in a study of a national economy we could propose that the development should follow the goal

$$\text{maximize } Q = \int_0^T \text{GNP } dt$$

where *GNP* means gross national product per year and T is the planning horizon (optimization horizon).

A solution to such a problem will inevitably say that investments in new production capacities should prevail in the first part of interval $(0, T)$ and decrease later on. The consumption will be correspondingly low (or zero) in these years, and rise considerably when time T is approached. The problem formulation was evidently not done with an awareness of what the society needs. The deficiency will not be remedied if we make T shorter or longer or even if we propose planning with a sliding horizon; the "invest now—consume later" feature will not disappear. No essential change occurs if we replace the bare GNP by some other and better justified "social utility" function. The error of the formulation lies in using an integral form of performance measure which ignores the fact that the society using the system is rather sensitive to what it has each year or each month.

A more adequate formulation of the optimization problem is needed. Several ways are possible, but all of them involve some arbitrary choice. For example, we can leave the integral form as above, but add a constraint that the consumption C should rise at a given minimum rate:

$$C(t) \geq C(0) + kt.$$

Note that the added constraint will make the solution more reasonable, but that we have to specify the coefficient k. Its value depends on a judgment of how much people would be willing to sacrifice of their immediate welfare for a better but distant future. The introduction of k improves the problem formulation but at the same time introduces subjective judgment and arbitrariness into the solution. Another popular method of getting out of trouble is to use some discounting rate for future benefits; this, however, is another arbitrary choice in the problem formulation.

The discussion of the above example should not completely discourage the use of optimization techniques and optimal control. The big advantage of optimization is that, once the problem has been formulated, one usually gets a unique solution. The message of the example is only that we should not treat the solution obtained by optimization techniques as the single recommended alternative (although the label "optimization" would suggest it). We should rather vary the problem formulation and display a range of solutions; thus the optimization procedure becomes a useful tool for the generation of alternatives.

However, if the performance index chosen is too far from what the user feels is important in the system performance, then optimization will generate inadequate solutions, when much better alternatives may exist. An instructional example may be given by the design of closed-loop control systems of the kind shown in Figure 1.11. For some time it was quite popular to design them as linear–quadratic–optimal, that is, to minimize a weighted sum of the squares of the deviations of the state variables and the squares of the manipulated inputs. If we changed the weighting coefficients, a family of designs emerged differing in time behavior (for example, in the response to a step change in the external command). Then, if we were interested in a particular feature of the step response, for example, in the minimal settling time, the most correct design would probably not be generated by the method: the quadratic performance to be minimized would be too far from what we were really after. An entirely different form of performance index, that includes the settling time explicitly, would have to be used instead.

CONSTRAINTS

To maximize or minimize the performance index of a system without any constraints would in most cases lead to economically unreasonable or physically impossible solutions. For example, the time response of a dynamic system could be made very fast but at the cost of a prohibitively large control input. To take another example, the maximum production rate of a chemical installation is constrained by the volumes of the reactors and the admissible pressures and temperatures.

The above examples concern constraints of a technological or physical nature; we may have an influence on them in the design stage of the system, for example, while developing the chemical process installation, but not in the operating stage. This means that in control problems (that is, in the field with which this book is concerned) the technological constraints are not flexible.

There is also another kind of constraint for the optimal control problems, that is, the constraints which are in a certain sense external to the physical system. For example, we are told to maximize production rate P of the ammonia process, but can we maximize at any cost? We might be given a constraint: the per unit cost of produced ammonia c should not exceed a certain value c_0. A reversed formulation of the problem suggests itself. Instead of "maximize production P, subject to cost constraint $c \leq c_0$," we may have "minimize cost c, subject to $P \geq P_0$," that is, subject to the requirement that the production rate should be no less than a limiting value, P_0.

In this example, the two elements, cost and production, are in principle commensurate, that is, both can be expressed in monetary units if a unit value γ of the product is introduced. Then the constraint becomes a well-justified part of the performance index:

$$\text{maximize } (\gamma P - C)$$

where C is the total cost of producing P. Such an incorporation is not always possible. For example, assume that an increase in the production rate of ammonia increases the rate of environmental pollution, $p(P)$. If a limit value for p is prescribed, that is, a pollution standard p_0, then the problem becomes

$$\text{maximize } (\gamma P - C), \text{ subject to } p(P) \leq p_0.$$

The pollution constraint can hardly be incorporated into the performance index of the ammonia plant because the pollution affects other users of the environment, not only the ammonia plant itself. The negative value attributed to pollution cannot be assessed from the viewpoint of the polluter. Thus, we must treat $p \leq p_0$ as an external constraint for the ammonia plant.

It is interesting to note that a simple analysis can reveal how much it costs us to keep the performance (production rate of ammonia) down because of the external (pollution) constraint. We can express it in terms of the marginal benefit with respect to the constraint. For the ammonia plant, it would be

$$\text{marginal benefit} = \frac{\partial[\gamma P - C]}{\partial p_0}.$$

This value is relatively easy to obtain if an optimization computation is being made. The marginal benefit shows how much of the optimal result is lost (or gained) per unit change in the setting of the constraint limit (p_0 in the example). In that way the marginal benefit measures the importance of the constraint (the requirement that $p \leq p_0$) in the optimization problem at hand. When the marginal benefit is low, it means that the constraint is relatively "harmless" for the system, from the point of view of the performance index that was considered. If it is high, we should perhaps reconsider the constraint setting (that is, the value p_0).

Compared with technological constraints, which may have to be observed strictly at all times, external constraints like cost, production volume, and

68

pollution effects are usually more elastic. For one thing, they are as a rule set up for average values, not for instantaneous ones. A constraint on pollution, for example, would perhaps be limiting in terms of a daily or weekly average, even if formally prescribed in a more restrictive way. The same would apply to other constraints such as cost. This means that we are somewhat free to violate the value at any instant in time. Secondly, a violation of the nonphysical constraint usually does not mean any irreversible damage to the system or to the environment. Consequently, the two kinds of constraints can be treated differently in the control problems.

It may be worthwhile to give a general warning with respect to formulating the constraints (in particular the nonphysical ones) for an optimization problem. If they are too conservative, too much "on the safe side," the results, that is, the system performance, may be worse than it needs to be. If, on the contrary, they are overly loose or even ignored, the problem solution, that is, the system operation, may have side effects that are both unexpected and undesirable.

Let us remember the multilayer hierarchy of problems with differing time horizons described in section 1.2 (see Figures 1.5, 1.8, and 1.9). The question arises whether the long-term objectives should be the same as the short-term ones. More precisely, should we have essentially the same performance index for all horizons, or should they be different. We indicated in section 1.2 that each layer in the hierarchy uses a different model in the sense that the higher you go the more simple is the model. We want to point out now that in some cases the performance index itself has to be different. This is related to the differences between the short-term and long-term effects of a policy, or to the existence of constraints that hold in the short run, but may not hold in the long run.

Let us show this by an example. The water supply available in a water system has to be divided among the users. One of the users is a farming region. When the water resource allocation is planned at the beginning of a year, the part of the overall objective function that describes the farm users is derived by taking into account the possibility of using less water but more fertilizers, or of changing the crops. In other words, the performance index for the long horizon is calculated with due consideration for possible substitutions.

A water allocation schedule for the whole year can be determined accordingly. Then the system is put into operation and the water dispatchers are told to optimize over some shorter time periods, as described in section 1.2. But the short-run characteristics of the system are different: if we are already in late spring, for example, no more crop choices are left for the farmers. They will now be much more sensitive to a lack of water than the performance index used before would indicate. Therefore, the water dispatching optimization must use a different performance function to adequately describe the farming region benefits related to water quantity.

Note that this example was not a case of a simple or complicated performance index used for various horizons. It was a case in which the short-horizon problem had a different set of constraints, that is, it was more tightly constrained, than the long-horizon problem. These local constraints have provided for a difference between the short-run and the long-run allocation problem.

The following example indicates how, at different levels of a hierarchy, static or dynamic formulations may be appropriate. Let us consider steady-state optimization for a sugar plant (Findeisen *et al.* 1970). The plant is composed of three principal parts that are referred to as sugar extraction, juice clarifying, and juice evaporation; see Figure 1.22. For a given amount of sugar beets to be processed per day, the steady-state optimization can be reduced to minimizing the cost rate of processing, inclusive of the sugar losses:

$$\text{minimize } Q(F_B, z), \text{ subject to a prescribed } F_B$$

where $Q(F_B, z)$ is in dollars per day, and F_B is the amount of beets to be processed in tons per day.

The optimization is a steady-state one because the process is a continuous one and the disturbance z, which is mainly a change in the quality of the sugar beets, varies slowly in comparison with the dynamics of the processing

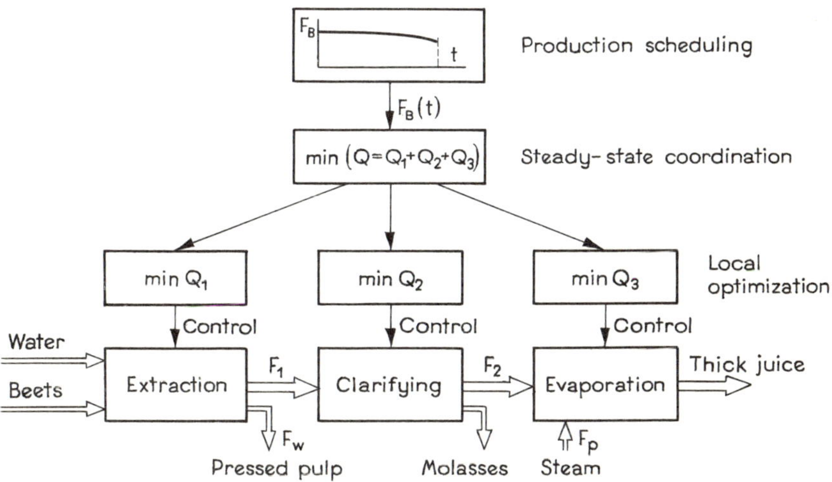

FIGURE 1.22   The sugar plant and its control structure.

70

system. However, if we ask what F_B, the optimal processing volume per day, is, a dynamic "production scheduling" problem arises. Its formulation is as follows.

There is a total amount of beet crop B assigned to be processed by a plant (note that this production allocation is another optimization problem, to be solved at a higher level of the hierarchy). We want to make sugar, that is, to process the amount B, with minimal cost and losses. The formulation is

$$\underset{F_B}{\text{minimize}} \int_{t_0}^{t_f} \hat{Q}(F_B(t), z(t))\, dt$$

subject to

$$\int_{t_0}^{t_f} F_B(t)\, dt \le B.$$

Note that we have an inequality constraint; it may not pay to process the beets at the end of the season when they are of bad quality. The constraint can also be written in the usual differential form, if the inventory variable W of the beets is introduced. We then have a state equation

$$\dot{W}(t) = -F_B(t), \quad \text{with} \quad W(t_0) = B, \qquad W(t_f) \ge 0.$$

The function $\hat{Q}(F_B(t), z(t))$ describes the optimal steady-state operation of the plant. It is nonlinearly dependent on $F_B(t)$. The quality of the sugar beets $z(t)$ depends on the weather conditions prior to time t. The beets always deteriorate over time and for that reason the solution to the dynamic problem varies over time and is not $\hat{F}_B(t) = $ a constant. It is better to lose some sugar during high-speed processing at the beginning of the season, than to store the beets too long and lose the sugar by natural processes.

Incidentally, sugar beet processing is an interesting example of optimal dynamic control where the initial time t_0 can be freely chosen. Indeed, if the estimated crops are high or the weather forecast is not very favorable for a long beet season, it pays to start processing the first beets before they are entirely ripe, that is, to move time t_0.

The steady-state performance index Q was divided into three parts, associated with the three sections of the plant that were used in the hierarchical scheme of control shown in Figure 1.22. It may be worthwhile to look at this example in some detail because of its methodological value. The optimization problem was to minimize the current rate of operation cost and sugar losses. This rate is expressed by

$$Q = \gamma_1 F_w c_{kw} + \gamma_2 (F_1 c_{k1} - F_2 c_{k2}) + (\gamma_2 - \gamma_3) F_2 \sum_{i=1}^{19} c'_{Mi} K_{Mi}$$
$$+ \gamma_4 (q_0 i_{p0})^2 + \gamma_5 F_2 (T_1 - 100)^2 + \gamma_6 q_4$$

where the following notation is used. $F_w c_{kw}$ is the loss due to sugar content in the pressed pulp, c_{kw} being the weight concentration of sugar in the pulp and F_w the pressed pulp flow rate (see Figure 1.22). $F_1 c_{k1} - F_2 c_{k2}$ is the net sugar loss in the juice-clarifying section of the plant. The term $\gamma_4 (q_0 i_{p0})^2$, where q_0 is steam flow and i_{po} its enthalpy, is the cost of steam used for evaporation. It is assumed to be squared because the cost of steam rises steeply with the load of the steam boilers. The term proportional to $(T_1 - 100)^2$ approximates the sugar losses and the losses due to deterioration of the sugar when the temperature T_1 in the first section of the evaporator is too high. The last term $\gamma_6 q_4$ is the value of steam loss due to undesirable condensation in the fourth section of the evaporator. Let us draw attention to some of the prices γ_i. In the formula, γ_1 is the price of sugar at the extraction outlet, that is, without the cost of clarifying. γ_2 is the price of sugar in the clarified thin juice, and γ_3 is the price of molasses. Hence, a coefficient $(\gamma_2 - \gamma_3)$ is associated with the term that expresses the loss due to a transformation of sugar into the molasses. This transformation is associated with the content of 19 different substances, referred to as nonsugars, in the clarified juice c'_{Mi} and with the sugar adsorption coefficients K_{Mi}.

The global performance index Q is split into local performances in such a way that their sum is Q:

$$Q_1 + Q_2 + Q_3 = Q.$$

These local indices are

$$Q_1 = \gamma_1 F_w c_{kw} + (\gamma_2 - \gamma_3) F_1 \sum_{i=1}^{19} c_{Mi} K_{Mi},$$

$$Q_2 = \gamma_2 (F_1 c_{k1} - F_2 c_{k2}) - (\gamma_2 - \gamma_3)\left[F_1 \sum_{i=1}^{19} c_{Mi} K_{Mi} - F_2 \sum_{i=1}^{19} c'_{Mi} K_{Mi} \right],$$

$$Q_3 = \gamma_4 (q_0 i_{p0})^2 + \gamma_5 F_2 (T_1 - 100)^2 + \gamma_6 q_4.$$

The performance index Q_1 deserves special attention. The loss due to sugar content in the molasses depends on the concentration of nonsugars c'_{Mi} in the clarified juice. But the presence of nonsugars results from poor operation of the extraction section. Therefore, it appeared necessary to consider these substances in the performance evaluation of the extraction; so the index Q_1 contains concentrations c_{Mi}. These concentrations are then reduced in the juice-clarifying section and the potential loss is decreased; hence, a term in Q_2 has a minus sign.

It is quite obvious that the plant could not operate properly if subjected to local optimizations only. For example, it is advantageous to use a lot of water in the extraction because it improves diffusion of sugar from the beets to the fluid. But more water means that more heat has to be used for evaporation. A compromise has to be reached by the coordinator, who acts to minimize the overall Q.

REFERENCES

Aronson, R. L. (1972). "Four computer net-makes and tests carburetor parts." *Control Engineering* 19:50–51.

Bailey, F. N. (1976). "Decision processes in organizations." in R. Saeks, ed., *Large Scale Dynamical Systems*. Point Lobos Press, Los Angeles, pp. 81–112.

Bailey, F. N., and K. Malinowski (1977). "Problems in the design of multilayer–multiechelon control structures," *Proceedings, 4th IFAC Symposium on Multivariable Technological Systems*, Fredericton, Canada, pp. 31–38.

Bernhard, P. (1976). *Commande optimale, décentralisation et jeux dynamiques*. Dunod, Paris.

Brdyś, M. (1975). "Methods of feasible control generation for complex systems." *Bull. Pol. Acad. Sci., Ser. Tech. Sci.* 23 (12):1025–1031.

Brdyś, M., and K. Malinowski (1977). "Price coordination mechanism and control of steady-state processes," *Proceedings, IFAC-IFORS-IIASA Workshop on Systems Analysis Application to Complex Programs*, Bielsko-Biala, Poland.

Brosilow, C. B., L. S. Lasdon, and J. D. Pearson (1965). "Feasible optimization methods for interconnected systems." *Proceedings, Joint Automatic Control Conference*, Troy, New York, pp. 79–84.

Chong, C. Y., and M. Athans (1975). "On the periodic coordination of linear stochastic systems." *Proceedings, 6th IFAC Congress*, Pt. IIIA, Boston, Massachusetts.

Davison, E. J. (1977). "Recent results on decentralized control of large scale multivariable systems," *Proceedings, 4th IFAC Symposium on Multivariable Technological Systems*, Fredericton, Canada, pp. 1–10.

Dirickx, I., and L. P. Jennergren (1979). *Multilevel Systems Analysis: Theory and Applications*. Wiley, London.

Donoghue, J. F., and I. Lefkowitz (1972). "Economic tradeoffs associated with a multilayer control strategy for a class of static systems," *IEEE Trans. Autom. Control*, AC-17 (1): 7–15.

Findeisen, W. (1968). "Parametric optimization by primal method in multilevel systems," *IEEE Trans. Syst. Sci. Cybern.* SSC-4 (2): 155–164.

Findeisen, W. (1974). *Wielopoziomowe uklady sterowania* [*Multilevel Control Systems*]. PWN, Warsaw. German translation: *Hierarchische Steuerungssysteme*. Verlag Technik, Berlin, 1977.

Findeisen, W. (1975). "A structure for on-line dynamic coordination," *Bull. Pol. Acad. Sci., Ser. Tech. Sci.* 23(9):81–87.

Findeisen, W. (1976a). "Lectures on hierarchical control systems." *Report*, Center for Control Sciences, University of Minnesota, Minneapolis, Minnesota.

Findeisen, W. (1976b). "Hierarchical control structures," *Proceedings, IRIA Symposium on New Trends in Systems Analysis*, Paris.

Findeisen, W. (1977). "Multilevel structures for on-line dynamic control," PP-77-6. International Institute for Applied Systems Analysis, Laxenburg, Austria; and *Ricerche di Automatica* 8 (1).

Findeisen, W. ed. (1979). *Second Workshop on Hierarchical Control*. Institute of Automatic Control, Technical University of Warsaw, Warsaw.

Findeisen, W., and I. Lefkowitz (1969). "Design and applications of multilayer control," *Proceedings, 4th IFAC Congress*, Tech. Sess. 42, Warsaw.

Findeisen, W., J. Pulaczewski, and A. Manitius (1970). "Multilevel optimization and dynamic coordination of mass flows in a beet sugar plant," *Automatica* 6(2): 581–589.

Findeisen, W., and K. Malinowski (1976). "A structure for on-line dynamic coordination," *Proceedings, IFAC Symposium on Large-Scale Systems Theory and Applications*, Udine, Italy.

Findeisen, W., M. Brdyś, K. Malinowski, P. Tatjewski, and A. Woźniak (1978). "On-line hierarchical control for steady-state systems," *IEEE Trans. Autom. Control* AC-23: 189–209.

Foord, A. G. (1974) "On-line optimization of a petrochemical complex," *Doctoral dissertation*, Cambridge University.

Germeyer, Yu. B. (1976). *Games with Non-Antagonistic Goals*. Nauka, Moscow. (in Russian).

Grateloup, G., and A. Titli (1971). "Methode de decomposition mixte et convergence d'un algorithme coordinateur de type gradient pour systèmes de commande a deux niveaux," *Bull. Acad. Pol. Sci., Ser. Tech. Sci.* 19 (5).

Gutenbaum, J. (1974). "The synthesis of direct control regulator in systems with static optimization," *Proceedings, 2nd Polish–Italian Conference on Applications of Systems Theory*, Pugnochiuso, Italy.

Haimes, Y. Y., and D. A. Wismer (1972). "A computational approach to the combined problem of optimization and parameter identification," *Automatica* 8 (3): 327–347.

Hakkala, L., and H. Blomberg (1976). "On-line coordination under uncertainty of weakly interacting dynamical systems," *Automatica* 12: 185–193.

Heescher, A., K. Reinisch, and R. Schmitt (1975). "On multilevel optimization of nonconvex static problems—application to water distribution of a river system." *Proceedings, 6th IFAC Congress*, Pt. IIIC, Boston, Massachusetts.

Himmelblau, D. M., ed. (1973). *Decomposition of Large Scale Problems*. North-Holland, Amsterdam.

Hurwicz, L. (1971). "Centralization and decentralization in economic processes." in A. Eskstem, ed., *Comparison of Economic Systems: Theoretical and Methodological Approaches*. University of California Press, Berkeley, California.

Kulikowski, R. (1970). *Control in Large-Scale Systems*. WNT, Warsaw. (in Polish).

Lasdon, L. S. (1970). *Optimization Theory for Large Systems*. MacMillan, London.

Lefkowitz, I. (1966). "Multilevel approach applied to control system design," *Trans. ASME, Series B, J. Bas. Eng.* 88 (2) 392–398.

Lefkowitz, I. (1975). "Systems control of chemical and related process systems," *Proceedings, 6th IFAC World Congress*, Pt. IID, Boston, Massachusetts.

Lefkowitz, I., and A. Cheliustkin, eds. (1976). *Integrated Systems Control in the Steel Industry*, CP-76-13, International Institute for Applied Systems Analysis, Laxenburg, Austria.

Malinowski, K. (1975). "Properties of two balance methods of coordination." *Bull. Pol. Acad. Sci., Ser. Tech. Sci.* 23 (9).

Malinowski, K. (1976). "Lectures on hierarchical optimization and control." *Report*, Center for Control Sciences, University of Minnesota, Minneapolis, Minnesota.

Malinowski, K., and A. Ruszczyński (1975). "Application of interaction balance method to real process coordination," *Control and Cybernetics* 4 (2).

Mesarović, M. D., D. Macko, and Y. Takahara (1970). *Theory of Hierarchical, Multilevel Systems*. Academic Press, New York.

Milkiewicz, F. (1977). "Multihorizon–multilevel operative production and maintenance control," *Proceedings, IFAC-IFORS-IIASA Workshop on Systems Analysis Application to Complex Programs*, Bielsko-Biala, Poland.

Pearson, J. D. (1971). "Dynamic decomposition techniques," in D. A. Wismer, ed, *Optimization Methods for Large-Scale Problems*. McGraw-Hill, New York.

Piervozvanski, A. A. (1975). *Mathematical Models in Production Planning and Control*. Nauka, Moscow. (in Russian)

Pliskin, L. G. (1975). *Optimization of Continuous Production*. Energiya, Moscow. (in Russian).

Ruszczyński, A. (1976). "Convergence conditions for the interaction balance algorithm based on an approximate mathematical model," *Control and Cybernetics* 5 (4): 29–43.

Sandell, N. R. Jr., P. Varaiya, and M. Athans (1976). "A survey of decentralized control methods for large scale systems," *Proceedings, Engineering Foundation Conference on Systems Engineering for Power: Status and Prospects*, Henniker, New Hampshire, pp. 334–352.

Šiljak, D. D. (1976). "Competitive economic systems: stability, decomposition, and aggrega-

74

tion," *IEEE Trans. Autom. Control* AC-21 (1): 149–160.

Šiljak, D. D., and M. K. Sundereshan (1976). "A multilevel optimization of large-scale dynamic systems," *IEEE Trans. Autom. Control* AC-21 (1): 79–84.

Šiljak, D. D., and M. B. Vukcević (1976). "Decentralization, stabilization and estimation of large-scale linear systems." *IEEE Trans. Autom. Control* AC-21 (3): 363–366.

Simon, H. A. (1972). "Bounded rationality in organizations," in C. B. McGuire and R. Radner, eds., *Decision and Organization.* North-Holland, Amsterdam.

Singh, M. G. (1977). *Dynamical Hierarchical Control.* North-Holland, Amsterdam.

Singh, M. G., S. A. W. Drew, and J. F. Coales (1975). "Comparisons of practical hierarchical control methods for interconnected dynamical systems," *Automatica* 11 (4): 331–350.

Singh, M. G., M. F. Hassan, and A. Titli (1976). "Multilevel feedback control for interconnected dynamical systems using the prediction principle," *IEEE Trans. Syst. Man. Cybern.* SMC-6: 233–239.

Singh, M. G., and A. Titli (1978). *Systems: Decomposition and Control.* Pergamon Press, Oxford.

Smith, N. J., and A. P. Sage (1973a). "An introduction to hierarchical systems theory," *Computers and Electrical Engineering* 1: 55–71.

Smith, N. J., and A. P. Sage (1973b). "A sequential method for system identification in hierarchical structure," *Automatica* 9: 667–688.

Szymanowski, J., M. Brdyś, and A. Ruszczyński (1976). "An algorithm for real system coordination," *Proceedings, IFAC Symposium On Large-Scale Systems Theory and Applications,* Udine, Italy.

Tamura, H. (1975). "Decentralised optimization for distributed lag models of discrete systems," *Automatica* 11: 593–602.

Tatjewski, P. (1977). "Dual methods of multilevel optimization". Bull. Pol. Acad. Sci., Ser. Tech. Sci. 25(3): 247–254.

Tatjewski, P. (1978). "A penalty function approach to coordination in multilevel optimization problems." R.A.I.R.O. Automatique/Systems Analysis and Control 12(3): 221–244.

Tatjewski, P., and A. Woźniak (1977). "Multilevel steady-state control based on direct approach". *Proceedings IFAC-IFORS-IIASA Workshop on Systems Analysis Application to Complex Programs,* Bielsko-Biala, Poland.

Titli, A. (1972). "Contribution a l'étude des structures de commande hierarchisées en vue de l'optimization de processus complexes". *Doctoral dissertation,* Université Paul Sabatier, Toulouse.

Titli, A. (1978). *Analyse et Commande des Systemes Complexes.* C.E.P.A.D.U.E.S. Editions, Toulouse, France.

Tsuji, K., and I. Lefkowitz, (1975). "On the determination of an on-demand policy for a multilayer control systems", *IEEE Trans. Autom. Control* AC-20 (4): 464–472.

Vatel', I. A., and N. N. Moiseev (1977). "On the modelling of economic mechanisms," *Ekonomika i Matematicheskie Metody,* 13(1): 16–30 (in Russian)

Wilson, I. D. (1977). "The design of hierarchical control systems by decomposition of the overall control problem," *Proceedings IFAC-IFORS-IIASA Workshop on Systems Analysis Application to Complex Programs,* Bielsko-Biala, Poland.

Wismer, D. A., ed. (1971). *Optimization Methods for Large-Scale Problems,* McGraw-Hill, New York.

Woźniak, A. (1976). "Parametric method of coordination using feedback from the real process," *Proceedings, IFAC Symposium on Large-Scale Systems Theory and Applications,* Udine, Italy.

Woźniak, A. (1977). "Decomposition of the control problem: a unified approach." *Proceedings, IFAC Workshop on Control and Management of Integrated Industrial Complexes.* Toulouse, France.

2   Coordination in Optimization Problems

2.1. PROBLEM DESCRIPTION

In Chapter 1 we introduced the reader to the complex system as a subject of control. We have also described the principal ways in which a hierarchical control structure for the complex system can be created. Chapter 2 will be devoted primarily to solutions of the system control problem based on decomposition and coordination. Various methods of coordination will be presented, and the conditions under which they are applicable as well as appropriate numerical procedures will be described. The coordination methods and procedures presented in Chapter 2 involve no feedback from the real system; they supply, therefore, purely model-based solutions. As such, they are primarily useful for solving system optimization problems in the system design stage, for example, for determining the best operating state for a process before it is put into operation. The model-based algorithms are a basis for on-line optimizing control done in an open-loop mode. We come to this topic in section 2.7. Moreover, the model-based solution can be used to great advantage as a starting point for on-line iterations involving feedback from the real system; such feedback structures are described in Chapter 3. In this section we describe the complex system and its control problem a bit more precisely and rigorously than they were described in Chapter 1.

We assume that the complex controlled system is an arrangement of N subsystems (see Figure 1.3), each of which is described by a state equation

$$(\forall t \geq t_0)\ \dot{x}_i(t) = f_i(x_i(t),\ m_i(t),\ u_i(t),\ z_i(t),\ t) \tag{2.1}$$

75

with the initial condition $x_i(t_0) = x_{0i}$, where

f_i is a mapping from $E_{X_i} \times E_{M_i} \times E_{U_i} \times E_{Z_i} \times [t_0, \infty)$ into E_{X_i}
x_i is the state time function, $x_i : [t_0, \infty) \to E_{X_i}$
m_i is the manipulated input time function, $m_i : [t_0, \infty) \to E_{M_i}$
u_i is the interaction input time function, $u_i : [t_0, \infty) \to E_{U_i}$
z_i is the disturbance input time function, $z_i : [t_0, \infty) \to E_{Z_i}$

and an output equation

$$\forall t \geq t_0 \quad y_i(t) = g_i(x_i(t), m_i(t), u_i(t), z_i(t), t) \tag{2.2}$$

where the mapping g_i is

$$g_i : E_{X_i} \times E_{M_i} \times E_{U_i} \times E_{Z_i} \times [t_0, \infty) \to E_{Y_i}$$

and y_i is the subsystem output time function, $y_i : [t_0, \infty) \to E_{Y_i}$.

Following the discussion in section 1.2, we assume that first-layer (direct) controllers are introduced so that the following is enforced

$$\forall t \geq t_0 \quad c_i(t) = h_i(x_i(t), m_i(t), u_i(t), y_i(t), t) \tag{2.3}$$

where $c_i : [t_0, \infty) \to E_{C_i}$ is a time function describing the controlled variable, and h_i is a mapping that was chosen by the first-layer control system design (see sections 1.2 and 1.3),

$$h_i : E_{X_i} \times E_{M_i} \times E_{U_i} \times E_{Y_i} \times [t_0, \infty) \to E_{C_i}.$$

The arguments in $h_i(\cdot)$ show that we assume that $c_i(t)$ may be based on measured values of the subsystem state, the manipulated inputs, the interaction inputs, and the subsystem outputs. Function $h_i(\cdot)$ may also be made time-dependent if it is explicitly programmed or related to a measured disturbance. In general, though, disturbances z are assumed to be unmeasured and do not enter into $h_i(\cdot)$.

A perfect follow-up in the direct control of each subsystem is assumed so that no distinction between c_i (the controlled variable) and c_{di} (the desired value of the controlled variable) has to be made and c_i can be considered to be the new control input of the subsystem (input from the second control layer). With proper choice of $h_i(\cdot)$ (see section 1.2) for each $t \geq t_0$, a functional dependence of $y_i(t)$ on $c_i(\cdot)$, $u_i(\cdot)$, $z_i(\cdot)$ exists. Therefore, there exists a subsystem input-output equation (compare Eq. (1.22″) in section 1.3):

$$y_i = F_i(c_1, u_i, z_i) \tag{2.4}$$

where

$y_i \in \mathcal{Y}_i \subseteq \{[t_0, \infty) \to E_{Y_i}\}$ is the subsystem output
$c_i \in \mathcal{C}_i \subseteq \{[t_0, \infty) \to E_{C_i}\}$ is the control input
$u_i \in \mathcal{U}_i \subseteq \{[t_0, \infty) \to E_{U_i}\}$ is the interaction input
$z_i \in \mathcal{Z}_i \subseteq \{[t_0, \infty) \to E_{Z_i}\}$ is the disturbance input

and

$F_i : \mathcal{C}_i \times \mathcal{U}_i \times \mathcal{Z}_i \rightarrow \mathcal{Y}_i$ is the subsystem input-output mapping.

We pointed out in section 1.3 that y_i also depends on the initial state $x_i(t_0)$, but we include it in z_i. If the subsystem explicitly depends on time, this is also included in disturbance z_i. Note that we do not yet specify any properties of the function spaces $\mathcal{Y}_i, \mathcal{C}_i, \mathcal{U}_i, \mathcal{Z}_i$; we will have to do so when solution methods of the system optimization problem are discussed.

Sometimes it may be convenient or necessary to describe the subsystem in implicit form

$$y_i = F_i^0(c_i, u_i, y_i, z_i) \tag{2.5}$$

where F_i^0 is the implicit subsystem input-output mapping from $\mathcal{C}_i \times \mathcal{U}_i \times \mathcal{Y}_i \times \mathcal{Z}_i$ into $\mathcal{Y}_i$.

INTERCONNECTIONS

The interconnections between subsystems are described by N linear coupling equations

$$u_i = H_i y, \qquad i \in \overline{1,N} \tag{2.6}$$

where $y \triangleq (y_1, \ldots, y_N) \in \mathcal{Y}_1 \times \ldots \times \mathcal{Y}_N \triangleq \mathcal{Y}$ and H_i is interconnection matrix i composed of zeros and ones. It follows that the couplings are separable:

$$u_i = \sum_{j=1}^{N} H_{ij} y_j. \tag{2.7}$$

When we denote $u \triangleq (u_1, \ldots, u_N) \in \mathcal{U}_1 \times \ldots \times \mathcal{U}_N \triangleq \mathcal{U}$ then

$$u = Hy \tag{2.8}$$

where

$$H = \begin{bmatrix} H_1 \\ \cdot \\ \cdot \\ \cdot \\ H_N \end{bmatrix}.$$

CONSTRAINTS

Next we assume that the constraints restricting subsystem inputs have the following form

$$(c_i, u_i) \in CU_i \triangleq \{(c_i, u_i) \in \mathcal{C}_i \times \mathcal{U}_i : G_i(c_i, u_i) \in S_i\} \tag{2.9}$$

where $G_i : \mathcal{C}_i \times \mathcal{U}_i \rightarrow \mathcal{S}_i$ is a constraint function, and S_i is a given subset of

78

the constraint space $\mathscr{S}_i$. Subsystem constraints may also be given in implicit form

$$(c_i, u_i, y_i) \in CUY_i \triangleq \{(c_i, u_i, y_i) \in \mathscr{C}_i \times \mathscr{U}_i \times \mathscr{Y}_i : G_i^0(c_i, u_i, y_i) \in S_i\} \quad (2.10)$$

where $G_i^0 : \mathscr{C}_i \times \mathscr{U}_i \times \mathscr{Y}_i \to \mathscr{S}_i$.

If we define $c \triangleq (c_1, \ldots, c_N) \in \mathscr{C}_1 \times \ldots \times \mathscr{C}_N \triangleq \mathscr{C}$, then the constraints (2.9) taken together for the whole system can be written as:

$$(c, u) \in CU \triangleq \{(c, u) \in \mathscr{C} \times \mathscr{U} : G(c, u) \in S\} \quad (2.11)$$

where $S \triangleq S_1 \times \ldots \times S_N$ and $G \triangleq (G_1, \ldots, G_N)$.

We may have some global system constraints in addition to Eq. (2.11):

$$(c, u) \in CU_0 \triangleq \{(c, u) \in \mathscr{C} \times \mathscr{U} : r(c, u) \in R\}, \quad (2.12)$$

where $r : \mathscr{C} \times \mathscr{U} \to \mathscr{R}$, $R \subseteq \mathscr{R}$. In the implicit case it would be

$$(c, u, y) \in CUY_0 \triangleq \{(c, u, y) \in \mathscr{C} \times \mathscr{U} \times \mathscr{Y} : r^0(c, u, y) \in R\}, \quad (2.13)$$

where $r^0 : \mathscr{C} \times \mathscr{U} \times \mathscr{Y} \to \mathscr{R}$. We may also have a constraint on the output

$$y \in Y \subseteq \mathscr{Y}, \quad (2.14)$$

which, when $Y = Y_1 \times \ldots \times Y_N$, is decomposable into

$$y_i \in Y_i \subseteq \mathscr{Y}_i \quad (2.15)$$

and then becomes a simplified version of Eq. (2.10).

For many of the decomposition–coordination solution methods, it is essential that constraint (2.12) be of a less general form, namely,

$$r(c, u) = \sum_{i=1}^{N} r_i(c_i, u_i),$$

where $\forall i \in \overline{1, N}\ r_i : \mathscr{C}_i \times \mathscr{U}_i \to \mathscr{R}$. We also require that the set R of (2.12) be defined as follows

$$R = \{r \in \mathscr{R} : r \leq r_0\},$$

where $\leq$ is appropriately defined and where $r_0 \in \mathscr{R}$. The resulting form of (2.12),

$$(c, u) \in CU_0 \triangleq \{(c, u) \in \mathscr{C} \times \mathscr{U} : \sum_{i=1}^{N} r_i(c_i, u_i) \leq r_0\} \quad (2.16)$$

will be called the resource constraint.

PERFORMANCE INDEX

We assume that a local, scalar-valued performance index (a local objective function)

$$Q_i : \mathscr{C}_i \times \mathscr{U}_i \to \mathbb{R} \quad (2.17)$$

is associated with each subsystem, and that it should be minimized. A local performance index could also be given that explicitly involves the output

$$Q_i : \mathscr{C}_i \times \mathscr{U}_i \times \mathscr{Y}_i \to \mathbb{R} \tag{2.18}$$

and also the disturbance

$$Q_i : \mathscr{C}_i \times \mathscr{U}_i \times \mathscr{Y}_i \times \mathscr{Z}_i \to \mathbb{R}. \tag{2.19}$$

Note that if the performance index is primarily given in a form involving the output, that is, in the form of (2.18), we can in principle use the mapping F_i (see Eq. (2.4)) to eliminate y_i. But then we get a Q_i that depends on the disturbance as Q_i in (2.19) does.

A global or overall performance index (a utility function) is assumed to exist as

$$Q = \psi \circ (Q_1, \ldots, Q_N) \tag{2.20}$$

where the function $\psi : \mathbb{R}^N \to \mathbb{R}$ is assumed to be strictly order preserving.

METHODS OF SOLUTION

The overall optimization problem is to find a control for each of the subsystems, such that the global performance index Q is minimized. As mentioned at the beginning of this section, we intend to use *decomposition*; therefore, throughout this chapter local optimization problems will be introduced. Each of them will deal with the local performance index Q_i. Independent local optimization does not ensure global optimality (except for trivial cases) and a supremal problem has to be introduced in order to bring about a *coordination* of the local problems.

There are various methods of coordination. Each of them uses a different way of intervening in the local problem. The formulations of the local problems differ accordingly and that is why they are not being shown here, but rather in the subsequent sections of the chapter. There will also be considerable differences in the supremal problem of the various methods, as well as in the procedures by which coordination is achieved. Various methods will have different areas of applicability, that is, they will require differing properties of the mappings and the constraint sets that describe the system and its optimization problem.

THE STATIC CASE

We should point out that the problem formulation given in this section is fairly general; in particular, it applies to both static and dynamic optimization problems.

For static optimization only, some elements of the formulation can be made more simple. The input-output equation is written as before

$$y_i = F_i(c_i, u_i, z_i) \tag{2.4'}$$

but the spaces $\mathcal{Y}_i$, $\mathcal{C}_i$, $\mathcal{U}_i$, $\mathcal{Z}_i$ to which y_i, c_i, u_i, z_i belong are finite-dimensional. Therefore, the mapping F_i or the mapping F_i^0 (see Eq. (2.5)) are functions from a finite-dimensional space into another such space. The local constraints (2.9) are now

$$(c_i, u_i) \in CU_i \triangleq \{(c_i, u_i) \in \mathcal{C}_i \times \mathcal{U}_i : G_i(c_i, u_i) \le 0\} \tag{2.9'}$$

where $G_i : \mathcal{C}_i \times \mathcal{U}_i \to \mathbb{R}^{m_i}$, that is, G_i is a function with values in m_i-dimensional real space. Similarly, we can get finite-dimensional versions of implicit local constraints (2.10), global constraints (2.11), (2.12) and (2.13), resource constraints (2.16) and of the performance indices (2.17), (2.18), or (2.19). The overall utility function has the same form, (2.20), for dynamic as well as for static systems.

HISTORICAL NOTE

The complex system described in this section consists of interconnected subsystems; therefore, its mathematical model has a specific structure, and a specific (but very general) structure of the performance index was assumed. Hence, the coordination methods that will be presented in detail in Chapter 2 can be regarded as a class of rather complex, block programming problems. The attempts to develop specific optimization methods for such problems were initiated in linear programming; many results and first reviews can be found in Arrow *et al.* (1958) and Lasdon (1970). The best-known method is the Dantzig-Wolfe decomposition. Geoffrion (1970) gives a review of some decomposition approaches both for linear and nonlinear cases. A compact review of the most important classes of block programming problems is given by Pervozvanski (1975). A problem formulation very close to that considered in this chapter was used by Mesarović *et al.* (1970), Wismer (1971), and Findeisen (1974, 1976). The coordination approaches suggested and developed in subsequent sections of this chapter (such as the fixing of interconnection variables and the use of prices) have their own history; the history can be found in the literature listed above and will be described in the following sections.

2.2. THE DIRECT METHOD

2.2.1. FORMAL PRESENTATION

In the preceding section, the system control problem was described as the minimization of the global performance index on the feasible set defined by subsystem input-output mappings (2.4), linear coupling equations, (2.6), local constraints (2.11), and global constraints (2.12, 2.14). Let us describe this problem more formally. The overall control problem (OP) is

Find control $\tilde{c} = (\tilde{c}_1, \ldots, \tilde{c}_N)$ giving output $\tilde{y} = (\tilde{y}_1, \ldots, \tilde{y}_N)$ such that

$$(\tilde{c}, \tilde{y}) = \arg \min_{CY^z} \psi \circ (Q_1(\cdot, H_1(\cdot)), \ldots, Q_N(\cdot, H_N(\cdot))) \qquad (2.21)$$

where

$$CY^z \triangleq \{(c, y) \in \mathscr{C} \times \mathscr{Y} : y \in Y \wedge (c, Hy) \in CU_0 \wedge$$
$$\forall i \in \overline{1, N} \ [(c_i, H_i y) \in CU_i \wedge y_i = F_i(c_i, H_i y, z_i)]\}.$$

The integrated problem defined as above may be too difficult to solve as a whole. If we thoroughly examine the constraint set CY^z we easily see that by fixing outputs $y_1, \ldots, y_N$ we can decompose the original problem into a set $(LP_1, \ldots, LP_N)$ of independent problems (LP_i):

For output fixed at some value, say v, find control

$$\hat{c}_i(v) = \arg \min_{C^z(v)} Q_i(\cdot, H_i v) \qquad (2.22)$$

where

$$C_i^z(v) \triangleq \{c_i \in \mathscr{C}_i : (c_i, H_i v) \in CU_i \wedge v_i = F_i(c_i, H_i v, z_i)\}.$$

It is obvious that when $v = \tilde{y}$, the control $(\hat{c}_1(\tilde{y}), \ldots, \hat{c}_N(\tilde{y}))$ obtained by solving the above set of independent problems is optimal overall. But how do we find the appropriate value of the output?

Based on the solutions of the problems LP_i, let us define the following problem (CP):

Find the value of the output

$$\hat{v} = \arg \min_{V \cap V_r} \psi \circ (Q_1(\hat{c}_1(\cdot), H_1(\cdot)), \ldots, Q_N(\hat{c}_N(\cdot), H_N(\cdot))) \qquad (2.23)$$

where

$$V \triangleq \{v \in \mathscr{Y} : v \in Y \wedge (\hat{c}(v), Hv) \in CU_0\}$$

and

$$V_r \triangleq \{v \in \mathscr{Y} : \text{for each } i \in \overline{1, N} \text{ there exists a solution } \hat{c}_i(v) \text{ of problem } LP_i\}.$$

In the problems defined above, the disturbance input z is a parameter. Therefore, the dependence of objects on z will be omitted below for convenience.

82

The following theorem shows the relationships between problems OP and CP.

THEOREM 2.1. *If mapping ψ strictly preserves order on the set*

$$\Omega \triangleq \bigcup_{v \in V} (\Omega_1(v) \times \ldots \times \Omega_N(v)) \subseteq \mathbb{R}^N$$

where

$$(\forall v \in V)(\forall i \in \overline{1, N}) \quad \Omega_i(v) \triangleq \{q_i \in \mathbb{R} : \exists c_i \in C_i(v) \; q_i = Q_i(c_i, H_1 v)\},$$

then

- *Solution of the overall problem* OP *exists iff a solution of problem* CP *exists.*
- $(\hat{c}_1(\hat{v}), \ldots, \hat{c}_N(\hat{v}))$ *is a solution of overall problem* OP *whenever* $\hat{v}$ *is a solution of problem* CP.

Proof.   First, we shall show that

$$\forall i \in \overline{1, N} \;\; \hat{c}_i(v) = \arg \min_{C_i(v)} Q_i(\cdot, H_i v) \;\Leftrightarrow$$

$$\hat{c}(v) = \arg \min_{C(v)} \psi(Q_1(\cdot, H_1 v), \ldots, Q_N(\cdot, H_N v))$$

where $C(v) \triangleq C_1(v) \times \ldots \times C_N(v)$ and $\hat{c}(v) \triangleq (\hat{c}_1(v), \ldots, \hat{c}_N(v))$. We assumed that function ψ is strictly order preserving on Ω, hence

$$\forall q \in \Omega \; q' \leq q \Leftrightarrow \forall q \in \Omega \; \psi(q') \leq \psi(q)$$

where $q' \leq q$ means that for all $i \in \overline{1, N} \; q_i' \leq q_i$. Therefore

$$\forall i \in \overline{1, N} \;\; \hat{c}_i(v) = \arg \min_{C_i(v)} Q_i(\cdot, H_i v) \;\Leftrightarrow$$

$$(\forall i \in \overline{1, N})(\forall q_i \in \Omega_i(v)) \;\; Q_i(\hat{c}_i(v), H_i v) \leq q_i \;\Leftrightarrow$$

$$\forall q \in \Omega_1(v) \times \ldots \times \Omega_N(v)$$

$$\psi(Q_1(\hat{c}_1(v), H_1 v), \ldots, Q_N(\hat{c}_N(v), H_N v)) \leq \psi(q) \;\Leftrightarrow$$

$$(\hat{c}_1(v), \ldots, \hat{c}_N(v)) = \arg \min_{C(v)} Q(\cdot, H v).$$

Now, let us notice that the definition of overall feasible set CY may be expressed as follows

$$CY = \bigcup_{v \in V} C(v) \times \{v\}.$$

Then

$$(\exists(\tilde{c}, \tilde{y}) \in CY)(\forall(c, y) \in CY) \ Q(\tilde{c}, H\tilde{y}) \leq Q(c, Hy) \Leftrightarrow$$

$$(\exists(\tilde{c}, \tilde{y}) \in CY)(\forall v \in V \cap V_r)(\forall c \in C(v)) \ Q(\tilde{c}, H\tilde{y}) \leq Q(c, Hv) \Leftrightarrow$$

$$(\exists(\tilde{c}, \tilde{y}) \in CY)(\forall v \in V \cap V_r) \ Q(\tilde{c}, H\tilde{y}) \leq \min_{C(v)} Q(\cdot, Hv) \Leftrightarrow$$

$$(\exists \tilde{y} \in V \cap V_r)(\forall v \in V \cap V_r) \ \min_{C(y)} Q(\cdot, H\tilde{y}) \leq \min_{C(v)} Q(\cdot, Hv) \Leftrightarrow$$

$$(\exists \tilde{y} \in V \cap V_r)(\forall v \in V \cap V_r) \ Q(\hat{c}(\tilde{y}), H\tilde{y}) \leq Q(\hat{c}(v), Hv). \quad \square$$

In order to have the above relationship between problems OP and CP further on, we shall assume that ψ is a strictly order-preserving function on $\Omega \subseteq \mathbb{R}$.

From Theorem 2.1 it follows that when problem CP is solved, the obtained value of output $\hat{v}$ is such that corresponding solutions $\hat{c}_1(\hat{v}), \ldots, \hat{c}_N(\hat{v})$ of problems $(\text{LP}_1, \ldots, \text{LP}_N)$ compose a solution of the overall problem OP. On the other hand, in order to solve problem CP, one needs to know how the solution to problem LP_i depends on a fixed value of output. This interdependence suggests a sequential mode of solving both problems with control units arranged on two levels. That is, first an estimate v of optimal output $\hat{v}$ is made on the upper level. Next, on the lower level the problems $(\text{LP}_1, \ldots, \text{LP}_N)$ are solved independently for values $\hat{c}_i(v)$, $i \in \overline{1, N}$. These values are in turn transmitted to the upper level, where an algorithm for solving problem CP produces a better estimate of the outputs. The process is then repeated. Thus, the effort to find the solution of the overall problem OP is divided between the two levels: the lower (first, infimal) level, where the set $(\text{LP}_1, \ldots, \text{LP}_N)$ of independent problems is solved by the so-called

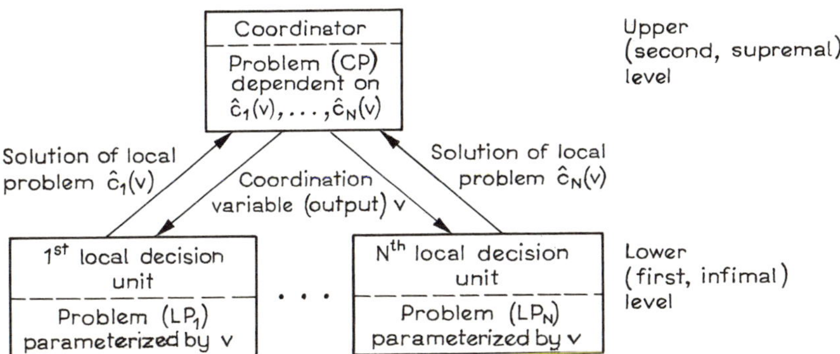

FIGURE 2.1   Two-level solution of the overall problem using the direct method of coordination.

local decision units, and the upper (second, supremal) level where the so-called coordinator solves the problem CP.

This method of producing a decomposed, multilevel form of the control problem has been termed the direct method of coordination (DM) (Findeisen 1976). The label *coordination* comes from the fact that the upper level coordinates the independent activity of the local decision units. Therefore, the supremal level decision unit is called the coordinator and the variables (parameters) imposed in the local problems are called coordination variables (instruments). The term *direct* comes from the use of variables that were in the original problem formulation as coordination instruments.

The other name—feasible method (Brosilow *et al.* 1965)—arises from the fact that throughout the iterative calculation, all intermediate values of controls, outputs, and interactions are feasible or allowed. The other names—parametric method (Findeisen 1968) and model (image) coordination method (Schoeffler 1971, Mesarović *et al.* 1970)—arise from the fact that decomposition is made possible by adding new equality constraints $y_i = v_i$ and $u_i = H_i v$ (parameterization) to the mathematical model of the subsystem.

2.2.2.   EXISTENCE OF AN OVERALL PROBLEM SOLUTION

Before investigating the properties of the direct methods we will show when a solution of the overall problem exists. Let us assume that

There exists a set $C^K \subseteq \mathscr{C}$ such that for each c in C^K there exists exactly one output y in the interconnected system, i.e., there exists a mapping

$K : C^K \to \mathscr{Y}$ such that

$$\forall c \in C^K \ [y = K(c) \Leftrightarrow \forall i \in \overline{1, N} \ \ y_i = F_i(c_i, H_i y)].$$

It is obvious that in formulating the overall problem, we assume tacitly that for some controls, output in the interconnected system exists; the assumption requires only the uniqueness of the output in addition. We assume furthermore that the sets $\mathscr{C}_i$, $\mathscr{U}_i$ and $\mathscr{Y}_i$ are Hausdorff topological spaces.

THEOREM 2.2.   *If the controlled system, constraint sets, and overall performance function are such that*

1.   *The above assumption is fulfilled and mapping K is continuous*
2.   *The functions ψ and Q_i, $i \in \overline{1, N}$, are continuous*
3.   *The sets CU_i, $i \in \overline{1, N}$ are compact*
4.   *The sets Y and CU_0 are closed*

then the overall optimal control $\tilde{c}$ exists.

Proof. It is easy to prove that

$$\tilde{c} = \arg\min_{\tilde{C}} \psi \circ (Q_1(\cdot, H_1 K(\cdot)), \ldots, Q_N(\cdot, H_N K(\cdot)))$$

where

$$\tilde{C} \triangleq \{c \in \mathscr{C} : K(c) \in Y \wedge (c, HK(c)) \in CU_0 \wedge \forall i \in \overline{1, N} \ (c_i, H_i K(c)) \in CU_i\}.$$

From the Weierstrass theorem, it follows that $\tilde{c}$ exists whenever the minimized function is continuous and the feasible set is compact. Under assumptions 1 and 2, the function $\psi \circ (Q_1(\cdot, H_1 K(\cdot)), \ldots, Q_N(\cdot, H_N K(\cdot)))$ is continuous. Similarly, the set

$$\tilde{C} = K^{-1}(Y) \cap D^{-1}(CU_0) \cap \bigcap_{i=1}^{N} D_i^{-1}(CU_i) \subseteq$$

$$\pi_{\mathscr{C}}(\{(c, u) \in \mathscr{C} \times \mathscr{U} : \forall i \in \overline{1, N} \ (c_i, u_i) \in CU_i\}) \triangleq C^0$$

where

$$c \mapsto D(c) = (c, HK(c))$$
$$c \mapsto D_i(c) = (c_i, H_i K(c))$$

is compact because the mappings $H_i K$ are continuous and the set C^0 is compact. $\square$

The most difficult assumption in the above theorem to verify in an application is assumption 1. Given that the sets $\mathscr{C}$ and $\mathscr{Y}$ are metric linear spaces, we can verify fulfillment of assumption 1 by the following theorem (Opoytsev 1976).

THEOREM (*Opoytsev*). *Let us denote $F \triangleq (F_1, \ldots, F_N)$. If mapping F is such that*

1. *There exists (c, y) in $\mathscr{C} \times \mathscr{Y}$ such that $y = F(c, Hy)$,*
2. *For each (c, y) satisfying the equation $y = F(c, Hy)$ there exist neighborhoods $\mathcal{O}_c$ and $\mathcal{O}_y$ such that there exists a continuous mapping $k : \mathcal{O}_c \to \mathcal{O}_y$ such that*

$$\forall \bar{c} \in \mathcal{O}_c \ [k(\bar{c}) = F(\bar{c}, Hk(\bar{c})) \wedge \forall y \neq k(\bar{c}) \ y \neq F(\bar{c}, Hy)],$$

3. *The set $\{y \in \mathscr{Y} : \exists c \in C \ y = F(c, y)\}$ is compact whenever the set $C \subseteq \mathscr{C}$ is compact,*

then assumption 1 of Theorem 2.2 is fulfilled on $\mathscr{C}$.

2.2.3. PROPERTIES OF LOCAL AND COORDINATOR PROBLEMS

For fixed values of the coordination instruments, the local problems LP_i are standard problems of nonlinear programming defined in terms of the original overall control problem OP. Consequently, they are well defined and the methods of investigation of their properties are the same as in mathematical programming. But the coordinator problem CP is completely different because in the definition we used objects that were not in the original problem formulation. They were the solvability set:

$$V_r = \{v \in \mathcal{Y} : (\forall i \in \overline{1, N})(\exists \hat{c}_i(v))\},$$

and a function to be minimized that is a composition of the solutions of the local problems:

$$V_r \ni v \mapsto \hat{Q}(v) = \psi(Q_1(\hat{c}_1(v), H_1 v), \ldots, Q_N(\hat{c}_N(v), H_N v)) \in \mathbb{R}. \quad (2.24)$$

Both of the above definitions are nonconstructive, because the solution of infimal problems $\hat{c}_i(\cdot)$ and $Q_i(\hat{c}_i(\cdot), H_i(\cdot))$ are used in them.

2.2.4. PROPERTIES OF THE SET V_0

We first consider the properties of the solvability set. Its definition is extremely nonconstructive. Therefore, to remove at least some of the difficulties, conditions ensuring the existence of solutions to the local problems for nonempty local feasible sets are frequently assumed. These assumptions lead to the replacement of the solvability set V_r with the so-called set V_0, where

$$V_0 \triangleq \{v \in \mathcal{Y} : \forall i \in \overline{1, N} \ C_i(v) \neq \varnothing\}.$$

Let us write the standard assumptions in the form of a lemma whose proof, based on the generalized Weierstrass theorem (e.g., Céa 1971), is obvious.

LEMMA 2.1. *If $\mathcal{C}_i$, $i \in \overline{1, N}$, are reflexive Banach spaces and for all $v \in V_0$ and $i \in \overline{1, N}$:*

- *Local performance function $Q_i(\cdot, H_i v)$ is weakly lower semicontinuous*
- *The set $C_i(v)$ is bounded and weakly closed or*

$$\lim_{\substack{\|c\| \to \infty \\ c \in C_i(v)}} Q_i(c, H_i v) = +\infty,$$

then

$$V_r = V_0.$$

We will assume that $\mathcal{C}$, $\mathcal{U}$, and $\mathcal{Y}$ are reflexive Banach spaces and that the original control problem OP is such that $V_r = V_0$. Hence, we will deal only with set V_0.

LEMMA 2.2. *The set $V \cap V_0$ is a projection of the set CY on axis $\mathcal{Y}$. If the set CY is compact (convex) then the set $V \cap V_0$ is compact (convex).*

Proof.

$$V \cap V_0 = \{v \in V : C(v) \neq \varnothing\} = \{v \in V : \exists c \in C(v)\} = \pi_{\mathcal{Y}}(CY).$$

Now, compactness (convexity) of the set $V \cap V_0$ follows from the fact that the projection is a continuous linear mapping. $\square$

THEOREM 2.3 (*corollary*).   *If* $\mathscr{C} \times \mathscr{Y}$ *is a finite-dimensional space and CY is a polyhedron, then* $V \cap V_0$ *is a polyhedron and consequently it is described by a finite set of affine equations and inequalities. The method of determining the set* $V \cap V_0$ *in this case has been considered by Grabowski (1969).*

In general, the problem of defining inequalities and equations describing the set V_0 is unsolved. The solution is known in two cases: when the original problem OP has appropriate convexity properties and when we have a particular partial description of the set V_0. However, before we present the solutions, we will state the conditions ensuring the nonemptiness of the interior of the set V_0.

LEMMA 2.3.   *If for all* $i \in \overline{1, N}$

1.   int $CU_i \neq \varnothing$,
2.   *The mapping*

$$(c_i, u_i) \mapsto f_i(c_i, u_i) = (u_i, F_i(c_i, u_i))$$

is open i.e., converts open sets on open sets, and

3. $(\exists \bar{c}, \bar{v})(\forall i \in \overline{1, N})[(\bar{c}_i, H_i \bar{v}) \in \text{int } CU_i \wedge \bar{v}_i = F_i(\bar{c}_i, H_i \bar{v})]$

then the set V_0 *has a nonempty interior.*

Proof.   The definition of the set V_0 can be rewritten in the following form

$$v \in V_0 \Leftrightarrow \forall i \in \overline{1, N} \ C_i(v) \neq \varnothing \Leftrightarrow$$
$$(\forall i \in \overline{1, N})(\exists c_i, u_i) \ [u_i = H_i v \wedge F_i(c_i, u_i) = v_i \wedge (c_1, u_i) \in CU_i] \Leftrightarrow$$
$$(\forall i \in \overline{1, N})(\exists u_i) \ [(u_i, v_i) \in f_i(CU_i) \wedge u_i = H_i v].$$

Define for every $i \in \overline{1, N}$ the set

$$V_i^H \triangleq \{v \in \mathscr{Y} : \exists u_i \ [(u_i, v_i) \in f_i(CU_i) \wedge u_i = H_i v]\}.$$

From the above chain of equivalences it follows that the definition of the set V_0 can now be rewritten in the following form

$$V_0 = \bigcap_{i=1}^{N} V_i^H.$$

Let j be a fixed number in $\overline{1, N}$. By assumptions 1 and 2, the set

$$f_j(\text{int } CU_j) \triangleq UV_j' \subseteq \mathscr{U}_j \times \mathscr{Y}_j$$

88

is open. Let us denote

$$UV_j \triangleq \{(u_j, v) \in \mathcal{U}_j \times \mathcal{Y} : (u_j, v_j) \in UV_j'\}.$$

Now, from the definition of set V_j^H we have

$$\pi_{\mathcal{Y}}(UV_j \cap \{(u_j, v) : u_j = H_j v\}) \triangleq \Omega_j \subseteq V_j^H.$$

By assumption 3, the point $(\bar{u}_j, \bar{v})$, where $\bar{u}_j \triangleq H_j \bar{v}$, belongs to the set UV_j. This set is open because the set UV_j' is open; hence, there exists $\varepsilon > 0$, such that

$$(1) \qquad \mathcal{B}(\bar{u}_j, \varepsilon) \times \mathcal{B}(\bar{v}, \varepsilon) \subseteq UV_j.$$

The continuity of mapping H_j implies that

$$(2) \qquad (\exists \delta(\varepsilon) > 0)(\forall v \in \mathcal{B}(\bar{v}, \delta(\varepsilon))) \ H_j v \in \mathcal{B}(\bar{u}_j, \varepsilon).$$

Let us denote $\eta \triangleq \min(\varepsilon, \delta(\varepsilon))$. We shall show that

$$\mathcal{B}(\bar{v}, \eta) \subseteq \Omega_j \subseteq V_j^H.$$

Suppose, on the contrary, that there exists $\tilde{v} \in \mathcal{B}(\bar{v}, \eta)$ such that $\tilde{v} \notin \Omega_j$, i.e.,

$$(\exists \tilde{v} \in \mathcal{B}(\bar{v}, \eta))(\forall u_j) \ [(u_j, \tilde{v}) \notin UV_j \vee u_j \neq H_j \tilde{v}].$$

Because $\mathcal{B}(\bar{u}_j, \varepsilon) \times \mathcal{B}(\bar{v}, \eta) \subseteq UV_j$, the above predicate means that

$$(\exists \tilde{v} \in \mathcal{B}(\bar{v}, \eta))(\forall u_j \in \mathcal{B}(\bar{u}_j, \varepsilon)) \ u_j \neq H_j \tilde{v}.$$

Since $\mathcal{B}(\bar{v}, \eta) \subseteq \mathcal{B}(\bar{v}, \delta(\varepsilon))$, we have

$$\exists \tilde{v} \in \mathcal{B}(\bar{v}, \delta(\varepsilon)) \ H_j \tilde{v} \notin \mathcal{B}(\bar{u}_j, \varepsilon),$$

which contradicts 2 above. Therefore $\mathcal{B}(\bar{v}, \eta) \subseteq V_j^H$.

By assumption 3, $\bar{v} \in V_0 = \cap_{j=1}^{N} V_j^H$, hence the above inclusion means that int $V_0 \neq \varnothing$. $\square$

Remark. By the Banach open mapping theorem, the function $f_i(\cdot, \cdot) = (\cdot, A_i(\cdot, \cdot) + y_i^0)$ where $A_i : \mathcal{C}_i \times \mathcal{U}_i \to \mathcal{Y}_i$ is a continuous linear operator, and $y_i^0 \in \mathcal{Y}_i$, is open whenever A_i is a mapping onto $\mathcal{Y}_i$. The sufficient condition for the openness of mapping f_i in the nonlinear case will be given in Chapter 3, section 3.2 (Lemma 3.1). $\square$

We now consider the problem of finding the constructive description of the set V_0 in the convex case. First we state the conditions of convexity of this set. Recall the definitions of the set CY (cf. Eqs. (2.9), (2.11), and (2.14)) and the set $C_i(y)$:

$$CY = \{(c, y) \in \mathcal{C} \times \mathcal{Y} : y \in Y \wedge r(c, Hy) \in R$$
$$\wedge \ \forall i \in \overline{1, N} \ [G_i(c_i, H_i y) \in S_i \wedge y_i = F_i(c_i, H_i y)]\},$$
$$C_i(y) = \{c_i \in \mathcal{C}_i : G_i(c_i, H_i y) \in S_i \wedge y_i = F_i(c_i, H_i y)\}.$$

LEMMA 2.4.   *If for all $i \in \overline{1, N}$*

1. S_i *is a convex cone in linear-topological space $\mathscr{S}_i$,*
2.   *The mapping G_i is concave with respect to S_i, i.e.,*

$$(\forall x', x'' \in \mathscr{C}_i \times \mathscr{U}_i)(\forall 0 \le \alpha \le 1)$$
$$G_i(\alpha x' + (1 - \alpha)x'') - \alpha G_i(x') - (1 - \alpha)G_i(x'') \in S_i,$$

3. $\qquad\qquad (\forall c_i, u_i) F_i(c_i, u_i) = A_i(c_i, u_i) + y_i^0$

where A_i is a linear operator and $y_i^0 \in \mathscr{Y}_i$,

then the set V_0 is convex.

Proof.   Let $v', v'' \in V_0$, hence

$$(\forall i \in \overline{1, N})(\exists c_i') \quad (G_i(c_1', H_i v') \in S_i \wedge v_i' = A_i(c_i', H_i v') + y_i^0)$$
$$(\exists c_i'') \quad (G_i(c_i'', H_i v'') \in S_i \wedge v_i'' = A_i(c_i'', H_i v'') + y_i^0).$$

G_i is concave, so for all $\alpha \in [0, 1]$

$$G_i(\alpha c_i' + (1 - \alpha)c_i'', H_i(\alpha v' + (1 - \alpha)v'')) - \alpha G_i(c_i', H_i v') - (1 - \alpha)G_i(c_i'', H_i v'') \in S_i.$$

The convexity of S_i implies that $s', s'' \in S_i \Rightarrow s' + s'' \in S_i$, therefore

$$(1) \qquad\qquad G_i(\alpha_i c_i' + (1 - \alpha)c_i'', H_i(\alpha v' + (1 - \alpha)v'')) \in S_i.$$

Since operator A_i, is linear, we have

$$(2) \qquad \begin{aligned} \alpha v' + (1 - \alpha)v'' &= \alpha A_i(c_i', H_i v') + (1 - \alpha)A_i(c_i'', H_i v'') + y_i^0 \\ &= A_i(\alpha c_i' + (1 - \alpha)c_i'', H_i(\alpha v' + (1 - \alpha)v'')) + y_i^0. \end{aligned}$$

Formulae (1) and (2) show that

$$\alpha c_i' + (1 - \alpha)c_i'' \in C_i(\alpha v' + (1 - \alpha)v''),$$

therefore $\forall i \in \overline{1, N}$ $C_i(\alpha v' + (1 - \alpha)v'') \ne \varnothing$ and consequently

$$\alpha v' + (1 - \alpha)v'' \in V_0. \qquad \square$$

THEOREM 2.4.   *If for all $i \in \overline{1, N}$:*

1. S_i *is a proper, convex, and closed cone in Banach space $\mathscr{S}_i$,*
2.   *The mapping G_i is continuous, concave with respect to S_i and such that*

$$\forall y \quad \{c_i : G_i(c_i, H_i y) \in S_i\} \subseteq C_i^0$$

where C_i^0 is convex and compact,

3. $(\forall c_i, u_i)$ $F_i(c_i, u_i) = A_i(c_i, u_i) + y_i^0$, *where A_i is a continuous linear operator, then*

$$V_0 = \{v : \forall i \in \overline{1, N} \ \min_{p_i \in P_i} \max_{c_i \in C_i^0} \langle p_i, \bar{G}_i(c_i, v) \rangle \ge 0\}$$

where

$$P_i \triangleq \{p_i \in S_i^* \times \mathcal{Y}_i^* : \|p_i\| \leq 1\} \ (S_i^* \text{ is a dual (polar) cone of } S_i) \text{ and}$$

$$(\forall c_i, v) \ \bar{G}_i(c_i, v) \triangleq (G_i(c_i, H_i v), A_i(c_i, H_i v) + y_i^0 - v_i).$$

(In the above, $\langle \cdot, \cdot \rangle$ denotes a duality relation between the given Banach space and its dual space; in Hilbert spaces it denotes a scalar product.)

Proof. Let us observe that in the notation introduced in the assumptions of the theorem, the definition of the set $C_i(y)$ can be rewritten as follows

$$\forall y \ C_i(y) = \{c_i : c_i \in C_i^0 \wedge \bar{G}_i(c_i, y) \in S_i \times \{0\}\}.$$

We shall first prove that

$$\bar{v} \in V_0 \ \Rightarrow \ \forall i \in \overline{1, N} \ \min_{p_i \in P_i} \max_{c_i \in C^0} \langle p_i, \bar{G}_i(c_i, \bar{v}) \rangle \geq 0.$$

Let $\bar{v}$ lie in V_0. Hence, for each $i \in \overline{1, N}$, there exists $\bar{c}_i$ in C_i^0 such that $\bar{G}_i(\bar{c}_i, \bar{v}) \in S_i \times \{0\}$. So, from the definition of the dual cone, it follows that for each p_i in $S_i^* \times \mathcal{Y}_i^*$

$$\langle p_i, \bar{G}_i(\bar{c}_i, \bar{v}) \rangle \geq 0.$$

To prove the converse implication, let us observe that by assumptions 2 and 3 the mapping $\bar{G}$ is concave with respect to $S_i \times \{0\}$. Then it can be shown that for each $i \in \overline{1, N}$ there exists $\bar{c}_i \in C_i^0$ such that $\bar{G}_i(\bar{c}_i, \bar{v}) \in S_i \times \{0\}$, which means that $\bar{v}$ lies in V_0. $\square$

Theorem 2.4 gives a complicated but constructive description of set V_0 in the convex case: in order to verify the membership of an element v in set V_0 it is sufficient to solve the collection $\text{AP1}_1, \ldots, \text{AP1}_N$ of the following auxiliary, linear–concave, min–max problems AP1_i:

$$\text{find } \alpha_i(v) \triangleq \min_{p_i \in P_i} \max_{c_i \in C_i^0} \langle p_i, \bar{G}_i(c_i, v) \rangle.$$

We must next identify the signs of the obtained values $\alpha_1(v), \ldots, \alpha_N(v)$. If for all $i \in \overline{1, N}: \alpha_i(v) \geq 0$, then $v \in V_0$; otherwise $v \notin V_0$.

We shall now present the constructive representation of the set V_0 if we have partial knowledge of its description.

We recall that the local feasible set $C_i(v)$ is defined as follows

$$C_i(v) = \{c_i \in \mathcal{C}_i : (c_i, H_i v) \in CU_i \wedge v_i = F_i(c_i, H_i v)\}.$$

This definition can be rewritten in the following form

$$C_i(v) = \pi_{\mathcal{C}i}(CU_i \cap F_i^{-1}(\{v_i\}) \cap (\mathcal{C}_i \times \{H_i v\})) \tag{2.25}$$

where

$$F_i^{-1}(\{v_i\}) \triangleq \{(c_i, u_i) \in \mathcal{C}_i \times \mathcal{U}_i : v_i = F_i(c_i, u_i)\}$$

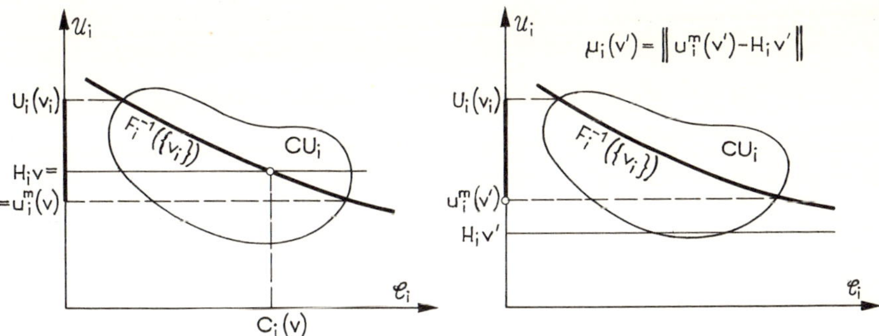

FIGURE 2.2   The location of sets CU_i, $F_i^{-1}(\{v_i\})$, $C_i(v)$, and $U_i(v_i)$.

and $\pi_{\mathscr{C}i}$ is a projection from $\mathscr{C}_i \times \mathscr{U}_i$ onto $\mathscr{C}_i$. Now, the condition defining the set V_0

$$v \in V_0 \Leftrightarrow (\forall i \in \overline{1, N})(\exists c_i \in C_i(v)),$$

can be stated as follows

$$v \in V_0 \;\Leftrightarrow\; (\forall i \in \overline{1, N})(\exists c_i)\; (c_i, H_i v) \in CU_i \cap F_i^{-1}(\{v_i\}).$$

Consequently

$$V_0 \subseteq \mathop{\times}_{i=1}^{N} \{y_i \in \mathscr{Y}_i : CU_i \cap F_i^{-1}(\{y_i\}) \neq \varnothing\} = \mathop{\times}_{i=1}^{N} F_i(CU_i) \triangleq V_F,$$

where $F_i(CU_i)$ is the image of the set CU_i, i.e.,

$$F_i(CU_i) \triangleq \{y_i \in \mathscr{Y}_i : \exists (c_i, u_i) \in CU_i \;\; y_i = F_i(c_i, u_i)\}.$$

The location of sets CU_i, $F_i^{-1}(\{v_i\})$, $C_i(v)$, and $U_i(v_i)$ is shown in Figure 2.2.

It follows from the above formulae that for v lying in V_F the set

$$U_i(v_i) \triangleq \pi_{\mathscr{U}i}(CU_i \cap F_i^{-1}(\{v_i\})) \tag{2.26}$$

is nonempty and

$$\forall i \in \overline{1, N} \;\; H_i v \in U_i(v_i) \;\Leftrightarrow\; v \in V_0.$$

Therefore we obtain

$$V_0 = \{v \in \mathscr{Y} : v \in V_F \wedge \forall i \in \overline{1, N} \;\; H_i v \in U_i(v_i)\}. \tag{2.27}$$

Now, let $\mu_i : V_F \to \mathbb{R}$ be the function defining the distance between the point $H_i v$ and the set $U_i(v_i)$, i.e.,

$$v \mapsto \mu_i(v) = \inf_{u_i \in U_i(v_i)} \|u_i - H_i v\|. \tag{2.28}$$

It is easy to see that using this function, definition (2.27) can be rewritten as follows

$$V_0 = \{v \in \mathscr{Y} : v \in V_F \wedge \forall i \in \overline{1, N} \;\; \mu_i(v) = 0\}$$

whenever the sets $U_i(v_i)$ are closed. To summarize our consideration, we can state the following theorem.

THEOREM 2.5. *If for all* $i \in \overline{1, N}$, *(a) the set* CU_i *is compact, and (b) the mapping* F_i *is continuous, then*

$$V_0 = \{v : v \in V_F \wedge \forall i \in \overline{1, N} \ \mu_i(v) = 0\}. \tag{2.29}$$

Proof. To complete the proof we must show that for all $i \in \overline{1, N}$ and v_i in $F_i(CU_i)$, the sets $U_i(v_i)$ are closed; this follows directly from definition (2.26) and the assumptions of the theorem. $\square$

It is obvious that the representation of the set V_0 given by the distances μ_i can be used to verify the membership of an element v in the set V_0 only when we know the set V_F resulting from all local constraints and subsystem input-output mappings. In engineering language, we would have to know for each subsystem the range of output variation corresponding to the feasible variation (i.e., belonging to the set CU_i) of subsystem inputs (controls and interaction). For many industrial processes, we can identify or at least estimate this range. When we know the set V_F, we can verify that a point v fulfills the remaining constraints describing the set V_0 by solving the collection of problems AP2$_i$:

$$\text{find } \mu_i(v) = \min_{u_i \in U_i(v_i)} \|u_i - H_i v\|$$

and checking if $\mu_i(v)$ equals zero for all $i \in \overline{1, N}$. In the following lemma, the representation of the set V_0 given by the distances μ_i will be used to prove the compactness of this set.

LEMMA 2.5. *If for all* $i \in \overline{1, N}$, *(a) the set* CU_i *is compact, and (b) the mapping* F_i *is continuous and open on* CU_i, *then (1) for each* $i \in \overline{1, N}$ *mapping* μ_i *is continuous, and (2) the set* V_0 *is compact.*

Proof. Let us introduce the following mapping

$$F_i(CU_i) \ni y_i \mapsto g_i(y_i) = (F_i | CU_i)^{-1}(\{y_i\}) \in C\mathscr{P}(CU_i),$$

where $(F_i | CU_i)^{-1}(Y^0) \triangleq \{(c_i, u_i) \in CU_i : F_i(c_i, u_i) \in Y^0 \subseteq \mathscr{Y}\}$.

(For any topological space $X : C\mathscr{P}(X)$ denotes the family of all compact, nonempty subsets of the set X.) Now, the definition (2.26) of the set $U_i(v_i)$ may be rewritten as follows

$$U_i(v_i) = \pi_{\mathscr{U}_i}(g_i(v_i))$$

and it is obvious that this set is compact. In accordance with definition (2.28)

$$\forall v \in V_F \mu_i(v) = \min_{u_i \in \Pi_{\mathscr{U}_i}(g_i(v_i))} \|u_i - H_i v\|,$$

where

$$C\mathscr{P}(CU_i) \ni X \mapsto \Pi_{\mathscr{U}_i}(X) = \pi_{\mathscr{U}_i}(X) \in C\mathscr{P}(U_i^0),$$
$$U_i^0 \triangleq \pi_{\mathscr{U}_i}(CU_i).$$

The norm and mapping H_i are continuous, hence by the Debreu theorem (see Berge 1963, Theorems 1 and 2 of section 3 in Chapter 6), functional μ_i is continuous whenever mapping

(1) $$\Pi_{\mathscr{U}_i} \circ g_i \circ \Pi_i \,|\, V_F : V_F \to C\mathscr{P}(U_i^0),$$

where

$$\mathscr{Y} \ni (v_1, \ldots, v_i, \ldots, v_N) \mapsto \Pi_i(v) = v_i \in \mathscr{Y}_i,$$

is continuous with exponential (Vietoris) topology in $C\mathscr{P}(U_i^0)$ (cf. Berge 1963, Chapter 6; and Kuratowski 1966, Chapter 1, section 17).

The composition (1) above is continuous whenever the mappings of which it is composed are continuous. The continuity of $\Pi_i \,|\, V_F$ and $\Pi_{\mathscr{U}_i}$ is obvious, thus only the proof of the continuity of mapping g_i remains to be shown.

We assumed that the mapping $F_i \,|\, CU_i$ is open. Hence, from Kuratowski (1966, Theorem 2 of section 17, III) it follows that the mapping g_i is continuous, from which it follows that the mapping μ_i is continuous.

Let us rewrite the definition (2.29) of the set V_0 in the following form

$$V_0 = V_F \cap \mu_1^{-1}(\{0\}) \cap \ldots \cap \mu_N^{-1}(\{0\}).$$

It is easy to see that the set V_F is compact. The continuity of μ_i implies that the set $\mu_i^{-1}(\{0\})$ is closed, hence the set V_0 is compact. $\quad\square$

2.2.5. THE PROPERTIES OF FUNCTION $\hat{Q}$ TO BE MINIMIZED IN THE COORDINATOR PROBLEM

Now that we know the properties and methods of description of the set V_0, when suitable assumptions about the elements making up the original problem OP have been made, we may investigate the properties of function

$$v \mapsto \hat{Q}(v) = \psi(Q_1(\hat{c}_1(v), H_1 v), \ldots, Q_N(\hat{c}_N(v), H_N v))$$

that will be minimized in the coordinator problem CP. We start with the simplest question—the convexity of $\hat{Q}$.

LEMMA 2.6.   *If for all $i \in \overline{1, N}$, (a) the set CU_i is bounded, closed, and convex, (b) $\forall c_i, u_i \;\; F_i(c_i, u_i) = A_i(c_i, u_i) + y_i^0$, where A_i is a continuous linear operator, and $y_i^0 \in \mathscr{Y}_i$, and (c) function $Q(\cdot, H(\cdot)) \triangleq \psi \circ (Q_1(\cdot, H_1(\cdot)), \ldots, Q_N(\cdot, H_N))$) is lower-semicontinuous and convex, then function $\hat{Q}$ is convex.*

Proof.   It is easy to see that for each v and $i \in \overline{1, N}$, the set $C_i(v)$ is bounded and weakly closed; hence, owing to Lemma 2.1, the function $\hat{Q}$ is defined

94

on the set V_0. The convexity of the set

$$CV \triangleq \{(c, v) \in \mathscr{C} \times \mathscr{Y} : v \in V_0 \wedge \forall i \in \overline{1, N} \ [(c_i, H_i v) \in CU_i \wedge v_i = A_i(c_i, H_i v) + y_i^0]\}$$

and the set V_0 follows from the fact that for all i in $\overline{1, N}$ and α in $[0, 1]$

$$(\alpha c_i' + (1-\alpha)c_i'', \alpha H_i v' + (1-\alpha)H_i v'') \in CU_i$$

and

$$A_i(\alpha c_i' + (1-\alpha)c_i'', H_i(\alpha v' + (1-\alpha)v'')) + y_i^0 = \alpha A_i(c_i', H_i v')$$
$$+ (1-\alpha)A_i(c_i'', H_i v'') + y_i^0 = \alpha v_i' + (1-\alpha)v_i''$$

whenever (c', v') and (c'', v'') belong to CV.

Now, from the convexity of $Q(\cdot, H(\cdot))$ and the convexity of CV we have

$$(\forall 0 \leq \alpha \leq 1)(\forall (c', v'), (c'', v'') \in CV)$$

$$Q(\alpha c' + (1-\alpha)c'', H(\alpha v' + (1-\alpha)v'')) \leq \alpha Q(c', Hv') + (1-\alpha)Q(c'', Hv'').$$

This implies that for fixed α, v', and v''

(1)
$$\inf_{(c', c'') \in C(v') \times C(v'')} Q(\alpha c' + (1-\alpha)c'', Hv^\alpha)$$

$$\leq \alpha \min_{C(v')} Q(\cdot, Hv') + (1-\alpha) \min_{C(V'')} Q(\cdot, Hv'') = \alpha \hat{Q}(v') + (1-\alpha)\hat{Q}(v'')$$

where

$$C(v) = \{c : \forall i \in \overline{1, N} \ [(c_i, H_i v) \in CU_i \wedge v_i = A_i(c_i, H_i v) + y_i^0]\}$$

and

$$v^\alpha \triangleq \alpha v' + (1-\alpha)v''.$$

Let us rewrite the left side of the above inequality in the following form

(2)
$$\inf_{(c', c'') \in C(v') \times C(v'')} Q(\alpha c' + (1-\alpha)c'', Hv^\alpha) = \inf_{\alpha C(v') + (1-\alpha)C(v'')} Q(\cdot, Hv^\alpha).$$

We know that

$$(\forall c' \in C(v'))(\forall c'' \in C(v'')) \ \alpha c' + (1-\alpha)c'' \in C(v^\alpha)$$

$[C(v^\alpha) \neq \varnothing$ because the convexity of V_0 implies that $v^\alpha \in V_0]$, so

$$\alpha C(v') + (1-\alpha)C(v'') \subseteq C(v^\alpha).$$

Therefore from (1) and (2) we have

$$\alpha \hat{Q}(v') + (1-\alpha)\hat{Q}(v'') \geq \min_{C(v^\alpha)} Q(\cdot, Hv^\alpha) = \hat{Q}(\alpha v' + (1-\alpha)v'')$$

which proves the lemma. $\square$

From the above lemma and properties of convex functions we have the following result.

Theorem 2.6 (corollary). *If the assumptions of Lemma 2.6 hold, then (1) function $\hat{Q}$ is continuous on int V_0, and (2) for all $v \in$ int V_0, there exists nonempty subdifferential $\partial\hat{Q}(v)$ of $\hat{Q}$ at v. If in addition $\mathcal{Y}$ is finite-dimensional, then (3) for almost all v in int V_0, function $\hat{Q}$ is Gâteaux differentiable.*

Proof. The assumptions of Lemma 2.6 imply that the convex function $\hat{Q}$ takes on finite values on the set V_0, so in int V_0 it is bounded. The application of the theorem of convex analysis (see Ekeland and Temam 1976, Propositions 2.5. and 5.2; and Rockafellar 1970, Theorem 25.4) completes the proof. $\square$

Corollary 2.6 gives a very important but nonconstructive result—the subdifferentiability of $\hat{Q}$ on int V_0. The next theorem will have the constructive character. It will show how to compute the subgradient of $\hat{Q}$ for each point $v \in$ int V_0.

Theorem 2.7. *If for all $i \in \overline{1, N}$*

1. *S_i is a proper, convex, closed cone with nonempty interior,*
2. *The mapping G_i is continuous, Fréchet differentiable, concave with respect to S_i (cf. Lemma 2.4) and such that the set $CU_i = G_i^{-1}(S_i)$ is bounded,*
3. *$\forall c_i, u_i \;\; F_i(c_i, u_i) = A_i(c_i, u_i) + y_i^0$,*
where A_i is a continuous linear operator on $\mathcal{Y}_i$,
4. *There exists a pair $(\bar{c}, \bar{v})$ such that*

$$\forall i \in \overline{1, N} \;\; [G_i(\bar{c}_i, H_i\bar{v}) \in \text{int } S_i \wedge A_i(\bar{c}_i, H_i\bar{v}) + y_i^0 - \bar{v}_i = \theta],$$

5. *The function $Q(\cdot, H(\cdot))$, cf. Lemma 2.6, is continuous, convex, and Gâteaux differentiable,*

then subdifferential $\partial\hat{Q}(v)$ of $\hat{Q}$ at v for all v in V_0 has the form

$$\partial\hat{Q}(v) = \{g \in \mathcal{Y}^* : (\exists\hat{c} = (\hat{c}_1, \ldots, \hat{c}_N), \; p_1, \ldots, p_N, \; l_1, \ldots, l_N)(\forall i \in \overline{1, N})$$

$$\times [\hat{c}_i = \hat{c}_i(v) \wedge (p_i, l_i) \in P_i(v, \hat{c}) \wedge g = H^*\nabla_u Q(\hat{c}, Hv)$$

$$+ \sum_{i=1}^{N} [H_i^* G_{iu}'^*(\hat{c}_i, H_iv)p_i + (A_{iu}'(\hat{c}_i, H_iv) \circ H_i - I_{\mathcal{Y}i})^* l_i]]\},$$

where $\forall i \in \overline{1, N}$

$$P_i(v, \hat{c}) \triangleq \{(p_i, l_i) \in \mathcal{S}_i^* \times \mathcal{Y}_i^* : p_i \in -S_i^* \wedge \langle p_i, G_i(\hat{c}_i, H_iv)\rangle = 0$$

$$\wedge \nabla_{ci} Q(\hat{c}, Hv) + G_{ic}'^*(\hat{c}_i, H_iv)p_i + A_{ic}'^*(\hat{c}_i, H_iv)l_i = 0\}.$$

96

Proof. Let us define the following set and function

$$CV = \{(c, v) \in \mathscr{C} \times \mathscr{Y} : \forall i \in \overline{1, N} \ [G_i(c_i, H_i v) \in S_i \wedge v_i = A_i(c_i, H_i v) + y_i^0]\}$$

$$= \{(c, v) : c \in C(v)\},$$

$$\forall v \in V_0 \ \mathscr{C} \times \mathscr{Y} \times \mathscr{Y}^* \ni (c, y, \lambda) \mapsto L_v(c, y, \lambda) = Q(c, Hy) + \langle \lambda, v - y \rangle \in \mathbb{R}.$$

From Lemma 2.6 and assumptions 1, 2, 3, and 5 above, it follows that function $\hat{Q}$ is defined on the set V_0, i.e.,

$$(\forall v \in V_0)(\exists \hat{c}(v)) \ \hat{c}(v) = \arg \min_{C(v)} Q(\cdot, Hv).$$

We shall show that each $g \in \mathscr{Y}^*$ satisfying the inequality

$$(1) \qquad \forall (c, y) \in CV \ L_v(c, y, g) \geq L_v(\hat{c}(v), v, g)$$

is a subgradient of $\hat{Q}$ at v, i.e.,

$$\forall w \in V_0 \ \hat{Q}(w) - \hat{Q}(v) \geq \langle g, w - v \rangle.$$

The converse will also be proved.

Let g in $\mathscr{Y}^*$ be such that for each (c, y) in CV we have

$$L_v(c, y, g) \geq L_v(\hat{c}(v), v, g),$$

in other words

$$\forall (c, y) \in CV \ Q(c, Hy) + \langle g, v - y \rangle \geq Q(\hat{c}(v), Hv) + \langle g, v - v \rangle = \hat{Q}(v).$$

Since for each w in V_0 point $(\hat{c}(w), w)$ belongs to CV, the above implies that

$$\forall w \in V_0 \ \hat{Q}(w) = Q(\hat{c}(w), Hw) \geq \hat{Q}(v) - \langle g, v - w \rangle$$

which proves the first part of our assertion.

To prove the converse, we let g in $\mathscr{Y}^*$ be any subgradient of $\hat{Q}$ at v, and let $(c, y) \in CV$. Since $c \in C(y)$, we have

$$L_v(\hat{c}(v), v, g) = \hat{Q}(v) \leq \hat{Q}(y) - \langle g, y - v \rangle \leq Q(c, Hy) + \langle g, v - y \rangle = L_v(c, y, g).$$

Point (c, y) in CV was chosen arbitrarily, so the above inequality implies that subgradient g together with $(\hat{c}(v), v)$ satisfies (1).

Thus, we have shown that for each v in V_0

$$\partial \hat{Q}(v) = \{g \in \mathscr{Y}^* : (\forall (c, y) \in CV) \ L_v(c, y, g) \geq L_v(\hat{c}(v), v, g)\}$$

$$= \{g \in \mathscr{Y}^* : L_v(\hat{c}(v), v, g) = \min_{CV} L_v(\cdot, \cdot, g)\}$$

$$= \{g \in \mathscr{Y}^* : (\hat{c}(v), v) = \arg \min_{(c,y) \in CV} [Q(c, Hy) + \langle g, v - y \rangle]\}.$$

From assumptions 1 and 4 it follows that the set CV fulfills the generalized Slater regularity condition (see Gol'shtein 1971, p. 89). We infer from assumptions 2, 3, and 5, that there exists $\arg \min_{CV} L_v(\cdot, \cdot, g)$ for each

(v, g) in $\mathcal{Y} \times \mathcal{Y}^*$. Now, combining Theorem 2.1 of Chapter 3, section 2.2, and Theorem 4.3 of Chapter 4, section 4.4 in Gol'shtein (1971), we obtain

$$(\hat{c}(v), v) = \arg \min_{(c,y) \in CV} [Q(c, Hy) + \langle g, v - y \rangle] \Leftrightarrow$$

$$(\forall i \in \overline{1, N})(\exists (p_i, l_i) \in (-S_i^*) \times \mathcal{Y}_i^*) \left[(\nabla_c Q(\hat{c}(v), Hv), H^* \nabla_u Q(\hat{c}(v), Hv)) \right.$$

$$- (0, g) + \sum_{j=1}^{N} [(G_{jc}'^*(\hat{c}_j(v), H_j v), H_j^* G_{ju}'^*(\hat{c}_j(v), Hv))p_j$$

$$+ (A_{jc}'^*(\hat{c}_j(v), H_j v), (A_{ju}'(\hat{c}_j(v), H_j v) \circ H_j - I_{\mathcal{Y}_j})^*)l_j] = 0$$

$$\left. \wedge \forall j \in \overline{1, N} \ \langle p_j, G_j(\hat{c}_j(v), H_j v) \rangle = 0 \right]$$

from which, after algebraic transformations, the theorem is proved. $\square$

Theorem 2.7 and the properties of convex functions suggest the following.

THEOREM 2.8 (*corollary*). *The function* $\hat{Q}$ *is Gâteaux differentiable at* $v \in V_0$ *whenever* (a) *the assumptions of Theorem 2.7 hold, and* (b) *for all* $i \in \overline{1, N}$, *the solution* $\hat{c}_i(v)$ *of local problem* (LP$_i$) *is unique and the set* $P_i(v, \hat{c}_i(v))$ *contains a single element.* $\square$

One can infer from the above corollary that in the convex case the nondifferentiability of $\hat{Q}$ in int V_0 is due to the nonuniqueness of the solution and Lagrange multipliers in the local problems. The sufficient conditions for the uniqueness of the solutions of local problems, and multipliers are, e.g., the strict convexity of $Q_i(\cdot, u_i)$ for all u_i and the linear independence of gradients of constraints.

The assumptions of Theorem 2.7 ensure the convexity of V_0 (Lemma 2.4), and the convexity, continuity, and subdifferentiability of $\hat{Q}$ (Lemma 2.6 and Theorem 2.6). The representation of subdifferential $\partial \hat{Q}(v)$ is also obtained. The "price" that is paid for these strong results is the assumption that the input-output equations of the subsystems are affine. In the general case we can only prove that $\hat{Q}$ is continuous.

THEOREM 2.9. *If for all* $i \in \overline{1, N}$

1. *The set* CU_i *is compact,*
2. *The mapping* F_i *is continuous and such that the mapping*

$$(c_i, u_i) \mapsto f_i(c_i, u_i') = (u_i, F_i(c_i, u_i))$$

is open on CU_i,
3. *The functions* Q_i *and* ψ *are continuous,*

then $\hat{Q}$ *is continuous on* V_0.

98

Proof. From assumption 1 and the continuity of F_i, we infer (cf. Eq. (2.25)) that

$$\forall v \in V_0 \; C_i(v) \in C\mathscr{P}(\mathscr{C}_i),$$

hence the continuity of $Q_i(\cdot, H_i v)$ implies that

$$V_r = V_0.$$

The definition (2.24) of the function $\hat{Q}$ can be rewritten as follows

$$\hat{Q} = \psi \circ (\hat{Q}_1, \ldots, \hat{Q}_N),$$

where

$$\forall i \in \overline{1, N} \;\; V_0 \ni v \mapsto \hat{Q}_i(v) = \min_{C_i(v)} Q_i(\cdot, H_i v) \in \mathbb{R}.$$

So, $\hat{Q}$ is continuous whenever ψ and $\hat{Q}_i$ are continuous. The function ψ is continuous by assumption 3. From the continuity of H_i and Q_i, it follows from the Debreu theorem (see Berge 1963, Theorems 1 and 2 of Chapter VI, section 3) that function $\hat{Q}_i$ is continuous whenever mapping $C_i(\cdot): V_0 \to C\mathscr{P}(\mathscr{C}_i)$ is continuous with exponential topology introduced in $C\mathscr{P}(\mathscr{C}_i)$. The proof of this continuity follows.

It is easy to observe that the definition of mapping $C_i(\cdot)$ may be rewritten as (cf. Eq. (2.25))

$$v \mapsto C_i(v) = \pi_{\mathscr{C}_i}(CU_i \cap f_i^{-1}(\{(H_i v, \Pi_i(v))\}))$$

where

$$\mathscr{Y} \ni (v_1, \ldots, v_i, \ldots, v_N) \mapsto \Pi_i(v) = v_i \in \mathscr{Y}_i.$$

For convenience we employ the following notation

$$f_i(CU_i) \ni (u_i, y_i) \mapsto \tilde{f}_i(u_i, y_i) = (f_i \mid CU_i)^{-1}(\{(u_i, y_i)\}) \in C\mathscr{P}(CU_i).$$

Thus, mapping $C_i(\cdot)$ is the following composition

(1) $$C_i(\cdot) = \Pi_{\mathscr{C}_i} \circ \tilde{f}_i \circ (H_i, \Pi_i \mid V_0)$$

where

$$C\mathscr{P}(CU_i) \ni CU \mapsto \Pi_{\mathscr{C}_i}(CU) = \pi_{\mathscr{C}_i}(CU) \in C\mathscr{P}(\mathscr{C}_i).$$

Assume that in the sets $C\mathscr{P}(CU_i)$ and $C\mathscr{P}(\mathscr{C}_i)$, exponential topology is introduced. The projections Π_i and $\Pi_{\mathscr{C}_i}$ are continuous; the mapping H_i is continuous by definition. The continuity of mapping $\tilde{f}_i$ remains to be proved. We assumed that the mapping f_i is open on CU_i. Hence, from Theorem 2 of Chapter 1, section 17, III in Kuratowski (1966), it follows that the mapping $\tilde{f}_i$ is continuous, so composition (1) above is continuous. $\square$

The crucial assumption in Theorem 2.9 is the requirement that the mappings $(c_i, u_i) \mapsto (u_i, F_i(c_i, u_i))$ are open on CU_i. Unfortunately, an example (Woźniak 1976) can be constructed that shows that when this assumption is dropped, function $\hat{Q}$ is not even lower-semicontinuous. In section 3.2.2. we shall discuss this question extensively (cf. Lemma 3.1).

THEOREM 2.10 (*corollary*).   *If* (a) *the assumptions of Theorem 2.9 hold, and* (b) *the set Y is closed and the set CU_0 equals $\mathscr{C} \times \mathscr{U}$, then the solution of the original overall problem* OP, *definition* (2.21), *exists.*

Proof.   By Lemma 2.5 and Theorem 2.9 the set

$$V \cap V_r = Y \cap V_0$$

is compact and function $\hat{Q}$ is continuous. Therefore, by the Weierstrass theorem a solution of the coordinator problem CP exists. The theorem then follows from Theorem 2.1. $\square$

2.2.6.   THE OVERALL PROBLEM WITH A GLOBAL RESOURCE CONSTRAINT

If the global system constraint

$$(c, u) \in CU_0 \tag{2.30}$$

has the form of a resource constraint

$$(c, u) \in CU_0 \;\Leftrightarrow\; \sum_{i=1}^{N} r_i(c_i, u_i) \leq r_0$$

where

$$\forall i \in \overline{1, N} \;\; r_i : \mathscr{C}_i \times \mathscr{U}_i \to \mathscr{R},$$

[$\mathscr{R}$ is a linear space ordered by relation $\leq$, and $r_0 \in \mathscr{R}$ (see Eq. 2.16)], then the original overall problem OP may be equivalently transformed into the following form in which the global constraint is not present.

The overall problem with resource constraint (OPr) is

Find control $\tilde{c} = (\tilde{c}_1, \ldots, \tilde{c}_N)$ giving output $\tilde{y} = (\tilde{y}_1, \ldots, \tilde{y}_N)$ and distribution of resource $\tilde{r}_d = (\tilde{r}_{d1}, \ldots, \tilde{r}_{dN})$ such that

$$(\tilde{c}, \tilde{y}, \tilde{r}_d) = \arg \min_{(c, y, r_d) \in CYR} (Q_1(c_1, H_1 y), \ldots, Q_N(c_N, H_N y))$$

where

$$CYR \triangleq \{(c, y, r_d) \in \mathscr{C} \times \mathscr{Y} \times \mathscr{R}^N : y \in Y \wedge \sum_{i=1}^{N} r_{di} \leq r_0$$

$$\wedge \forall i \in \overline{1, N} \;\; [(c_i, H_i y) \in CU_i \wedge y_i = F_i(c_i, H_i y) \wedge r_{di} = r_i(c_i, H_i y)]\}.$$

When we denote

$$\mathscr{Y} \times \mathscr{R}^N \triangleq \mathscr{W} \ni w \triangleq (y, r_d)$$

100

and

$$\mathscr{C}_i \times \mathscr{U}_i \ni (c_i, u_i) \mapsto \tilde{F}_i(c_i, u_i) = (F_i(c_i, u_i), r_i(c_i, u_i)) \in \mathscr{Y}_i \times \mathscr{R} \triangleq \mathscr{W}_i$$

$$\mathscr{W} \ni (y, r_d) \mapsto \tilde{H}_i(y, r_d) = H_i y \in \mathscr{U}_i, \qquad i \in \overline{1, N},$$

$$YR \triangleq \{(y, r_d) \in \mathscr{W} : y \in Y \wedge \sum_{i=1}^{N} r_{di} \leq r_0\}$$

then the definition of constraint set CYR will have the same form as the definition of the set CY:

$$CYR = \{(c, w) \in \mathscr{C} \times \mathscr{W} : w \in YR \wedge \forall i \in \overline{1, N} \ (c_i, \tilde{H}_i w) \in CU_i \wedge w_i = \tilde{F}_i(c_i, \tilde{H}_i w)]\}.$$

Therefore, the investigation of the properties of problem (OPr) may be carried out in the same manner as for overall problem OP; it is, however, more complicated since there are more local equality constraints and more coordination variables and constraints.

2.2.7   METHODS OF SOLVING LOCAL PROBLEMS AND COORDINATOR PROBLEMS

With the background provided by the above study, we can discuss briefly the choosing of methods of solving the local problem LP$_i$ and the coordinator problem CP. We recall that for fixed values of the coordination variable, the local problem LP$_i$ is a standard problem of infinite-dimensional nonlinear programming. Hence, the choice of numerical procedure depends on properties like the differentiability of the performance function, the linearity of subsystem input-output mappings, and the convexity of subsystem constraints.

The coordinator problem CP has a special structure. We do not know the analytical form of the minimized function or in most cases of the feasible set because they are defined in terms of the solutions of the local problems. It is the cost that we must pay for dividing the effort in finding the solution of the overall problem OP between units arranged on two levels. In such a situation, the choice of the method of generating an optimal coordination variable, the so-called coordinator strategy, is a rather hard problem.

2.2.8.   COORDINATOR STRATEGY

In the simplest case considered in sections 2.2.4 and 2.2.5, when local constraint sets CU_i are convex, subsystem input-output mappings are affine, the global performance function is convex (i.e., in the case of convexity), and we know the analytical description of the set V_0, the coordinator strategy can be chosen on the basis of Theorem 2.7 as a subgradient minimization method. The problems and difficulties that arise in this approach are discussed by Malinowski and Szymanowski (1976).

In the second case, we assumed that we had a partial description of the set V_0, i.e., we know the set V_F, and we can construct the functions μ_i (see Eq. (2.28)) giving a representation of the set V_0. As we recall

$$\forall i \in \overline{1, N} \mu_i(v) \begin{cases} = 0 & \text{for } v \text{ in } V_0 \\ > 0 & \text{for } v \text{ in } V_F \backslash V_0. \end{cases}$$

So, the sum of these functions multiplied by a positive constant can be used as an exterior penalty function related to the set V_0. Let us describe this approach more carefully.

First, we have to consider the problem of continuously extending function $\hat{Q}$, primarily defined on set V_0, on set V_F. In order to solve this problem, for each $i \in \overline{1, N}$, we define the following modified local problem ($m\mathrm{LP}_i$)

For given v in V_F, find $\hat{c}_i^m(v)$ and $u_i^m(v)$, such that

$$u_i^m(v) = \arg \min_{u_i \in U_i(v_i)} \|u_i - H_i v\|$$

$$\hat{c}_i^m(v) = \arg \min_{C_i^m(v)} Q_i(\cdot, H_i v) \tag{2.31}$$

where

$$U_i(v_i) = \{u_i \in \mathscr{U}_i : \exists c \ [(c, u_i) \in CU_i \wedge v_i = F_i(c, u_i)]\} \quad \text{(cf. Eq. (2.26))}$$

$$C_i^m(v) \triangleq \{c_i \in \mathscr{C}_i : (c_i, u_i^m(v)) \in CU_i \wedge v_i = F_i(c_i, u_i^m(v))\}.$$

It is easy to observe (see Eq. (2.27)) that from the above definitions it follows that for each $i \in \overline{1, N}$

$$\mu_i(v) = \|u_i^m(v) - H_i v\| \quad \text{and} \quad \forall v \in V_0 \ u_i(v) = H_i v,$$

consequently

$$\forall v \in V_0 \qquad C_i^m(v) = C_i(v).$$

Now we can define the following function

$$v \mapsto \hat{Q}^m(v) = \psi(Q_1(\hat{c}_1^m(v), H_1 v), \ldots, Q_N(\hat{c}_N^m(v), H_N v)).$$

The proof of the following lemma is obvious.

LEMMA 2.7. *If for all $i \in \overline{1, N}$: (a) the set CU_i is compact and mapping F_i is continuous, and (b) for all v function $Q_i(\cdot, H_i v)$ is continuous, then (1) for each $v \in V_F$ and $i \in \overline{1, N}$ there exists solution $(\hat{c}_i^m(v), u_i^m(v))$ of problem $m\mathrm{LP}_i$, and (2) the domain of function $\hat{Q}^m$ is the set V_F and $\hat{Q}^m \mid V_0 = \hat{Q}$.* $\square$

Therefore, we obtain the needed extension of function $\hat{Q}$. The following theorem gives the conditions of continuity of function $\hat{Q}^m$.

THEOREM 2.11. *If we assume that (a) the assumptions of Theorem 2.9 hold, (b) the mapping F_i is open on CU_i, and (c) for each $v \in V_F$ and $i \in \overline{1, N}$ there* *exists no more than one*

$$u_i^m(v) = \arg \min_{u_i \in U_i(v_i)} \|u_i - H_i v\|,$$

then the function $\hat{Q}^m$ is continuous.

Proof.   As in the proof of Theorem 2.9 from the definition of function $\hat{Q}^m$, we have

$$\hat{Q}^m = \psi \circ (\hat{Q}_1^m, \ldots, \hat{Q}_N^m),$$

where

$$\forall i \in \overline{1, N} \ \ V_F \ni v \mapsto \hat{Q}_i^m(v) = \min_{C_i^m(v)} Q_i(\cdot, H_i v) \in \mathbb{R}.$$

Therefore, $\hat{Q}^m$ is continuous whenever the mapping $C_i^m(\cdot): V_F \to C\mathscr{P}(\mathscr{C}_i)$ is continuous for each $i \in \overline{1, N}$ with exponential topology introduced in $C\mathscr{P}(\mathscr{C}_i)$.

Let us rewrite the definition of $C_i^m(\cdot)$ in the following form (cf. the proof of Theorem 2.9)

$$v \mapsto C_i^m(v) = \pi_{\mathscr{C}_i}(CU_i \cap f_i^{-1}(\{(u_i^m(v), \Pi_i(v))\}))$$
$$= (\Pi_{\mathscr{C}_i} \circ \tilde{f}_i \circ (u_i^m(\cdot), \Pi_i \,|\, V_F))(v)$$

where

$$f_i(CU_i) \ni (u_i, y_i) \mapsto \tilde{f}_i(u_i, y_i) = (f_i \,|\, CU_i)^{-1}(\{(u_i, y_i)\}) \in C\mathscr{P}(CU_i).$$

An argument similar to that used in the proof of Theorem 2.9 shows that $C_i^m(\cdot)$ is continuous whenever the mapping

$$V_F \ni v \mapsto u_i^m(v) = \arg \min_{u_i \in U_i(v_i)} \|u_i - H_i v\| \in U_i^0 \triangleq \Pi_{\mathscr{U}_i}(CU_i)$$

is continuous. The proof of this continuity follows.

Let us define for every $i \in \overline{1, N}$ the following mapping

$$V_F \ni v \mapsto W_i(v) = \{u_i \in U_i^0 : \mu_i(v) - \|u_i - H_i v\| \geq 0\} \in C\mathscr{P}(U_i^0).$$

We assumed that the mapping F_i is open on CU_i. Therefore, by Lemma 2.5, the function μ_i is continuous. Norm and mapping H_i are continuous as well. So, the set

$$\{(v, u_i) \in V_F \times U_i^0 : u_i \in W_i(v)\} = \{(v, u_i) : \mu_i(v) - \|u_i - H_i v\| \geq 0\}$$

is closed. The set U_i^0 is compact; hence, by Theorem 4 of section 43, I in Kuratowski (1968), the mapping W_i is upper semicontinuous with exponential topology introduced in $C\mathscr{P}(U_i^0)$.

Let us observe now that for each v in V_F

$$U_i(v_i) \cap W_i(v) = \{u_i \in U_i^0 : u_i \in U_i(v_i) \wedge \|u_i - H_i v\| \leq \mu_i(v)\}$$

$$= \text{Arg} \min_{u_i \in U_i(v_i)} \|u_i - H_i v\| \triangleq \hat{U}_i^m(v) \in C\mathscr{P}(U_i^0).$$

In the proof of Lemma 2.5 we showed that the openness of the mapping F_i on CU_i and the compactness of the set CU_i imply the continuity of mapping $V_F \ni v \mapsto U_i(v_i) \in C\mathscr{P}(U_i^0)$. Hence, by Theorem 1 of section 18, V in Kuratowski (1966), the mapping $v \mapsto \hat{U}_i^m(v) = U_i(v_i) \cap W_i(v)$ is upper semicontinuous.

Because our assumptions imply that for each v in V_F the set $\hat{U}_i^m(v)$ contains exactly one element $\hat{U}_i^m(v) = \{u_i^m(v)\}$, the upper semicontinuity of mapping $\hat{U}_i^m$ means that mapping $v \mapsto u_i^m(v)$ is continuous. $\square$

We are now in a position to approach the choosing of the coordinator strategy. Let $\hat{Q}_{pty} : V_F \times \mathbb{R}_+ \to \mathbb{R}$ be defined as follows

$$(v, \rho) \mapsto \hat{Q}_{pty}(v, \rho) = \hat{Q}^m(v) + \rho \sum_{i=1}^{N} \mu_i(v).$$

THEOREM 2.12. *If we assume that* (a) *the assumptions of Theorem* 2.9 *hold,* (b) *the mapping F_i is open on CU_i,* (c) *for each $v \in V_F$ and $i \in \overline{1, N}$ there exists no more than one $u_i^m(v)$,* (d) *the set Y is closed and the set CU_0 equals $\mathscr{C} \times \mathscr{U}$, and* (e) *the sequence $\{\rho^n\}_{n=1}^{\infty}$ of positive real numbers tends to infinity, then* (1) *the sequence*

$$\{\arg \min_{Y \cap V_F} \hat{Q}_{pty}(\cdot, \rho^n)\}_{n=1}^{\infty} \triangleq \{\hat{v}^n\}_{n=1}^{\infty}$$

has an accumulation point, and (2) *any accumulation point of this sequence is the solution $\hat{v}$ of the coordinator problem* CP. *If, in addition, coordinator problem* CP *has no more than one solution, then*

$$\lim_{n \to \infty} \hat{v}^n = \hat{v}.$$

Proof. The proof consists of a direct application of the main theorem for the exterior penalty function method, e.g., Theorem 35 of section 4.1 in Polak (1971). $\square$

If a given system control problem satisfies the assumptions of Theorem 2.12, one can quite readily propose a coordinator strategy. It ought to consist in minimization of function $\hat{Q}_{pty}(\cdot, \rho)$ on the set $Y \cap V_F$ with increasing value of the penalty coefficient ρ. Figure 2.3 shows a diagram of information flows in this case. The main difficulty in this approach is connected with the possibility that the function $\hat{Q}_{pty}$ may be nondifferentiable at some points. We recall that in the case of convexity, $\hat{Q}_{pty} | \text{int } V_0$ has a subgradient that can be used in the minimization algorithm. In the nonconvex, finite-dimensional case, the direct search methods can be proposed as the minimization algorithm. The detailed discussion of this approach can be found in Woźniak (1975).

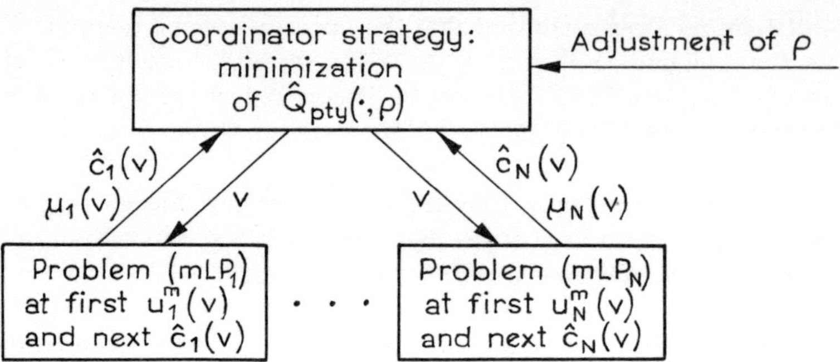

FIGURE 2.3   Information flows within the chosen coordinator strategy.

2.3.   THE PENALTY FUNCTION METHOD

2.3.1.   INTRODUCTORY PRESENTATION

The direct method has its advantages and drawbacks. Troubles connected with set V_0 and the impossibility of applying the method when there are fewer controls than outputs in some subsystems (stiff subsystems) create the most significant drawbacks. These drawbacks can be significantly decreased or eliminated when penalty functions are used (Pearson 1971, Findeisen 1974). Findeisen also noticed the significant role that local constraints could play in applications of the direct method.

The use of penalty functions as the coordination mechanism will be called the *penalty function method* (PFM); it will be introduced below for the overall system optimization problem:

Find control $\hat{c}$ giving output $\hat{y}$, such that

$$(\hat{c}, \hat{y}) = \arg\min \psi(Q_1(c_1, H_1 y), \ldots, Q_N(c_N, H_N y)) \tag{2.32}$$

subject to

$$y_i = F_i(c_i, H_i y, z_i), \qquad (c_i, H_i y) \in CU_i, \qquad i \in \overline{1, N}, \qquad y \in Y.$$

All considerations of this section can be easily extended to the more general case, with various types of overall system constraints (2.13). This will be discussed later in this section.

The disturbances z_i, $i \in \overline{1, N}$, are assumed to be constant throughout the optimization and will be omitted for convenience.

Note that the drawbacks of the direct method are caused by the requirement to fulfill strictly the output-structure equations

$$v_i = F_i(c_i, H_i v), \qquad i \in \overline{1, N} \tag{2.33}$$

after each minimization of the local decision problems. In PFM we eliminate this requirement and try to fulfill Eqs. (2.33) using penalty functions, which are introduced *locally* regardless of the form (possibly nonadditive) of the utility function ψ. Thus, local decision problem i (LP$_{\text{pty}i}$) is

For given coordination variable $v \in \mathcal{Y}$ and penalty coefficient $\rho_i \in \mathbb{R}_+$ find control

$$\hat{c}_i(v_i, \rho_i) = \arg \min_{C_i(v)} Q_{\text{pty}i}(\cdot, v, \rho_i) \tag{2.34}$$

where

$$\forall c_i \in \mathscr{C}_i \qquad Q_{\text{pty}i}(c_i, v, \rho_i) \triangleq Q_i(c_i, H_i v) + \rho_i \|v_i - F_i(c_i, H_i v)\|^2,$$

$$C_i(v) \triangleq \{c_i \in \mathscr{C}_i : (c_i, H_i v) \in CU_i\}.$$

The above quadratic penalty functions are used in this section. Other types of penalty functions are also possible. We continue by defining

$$\hat{Q}_{\text{pty}i}(v, \rho_i) \triangleq Q_{\text{pty}i}(\hat{c}_i(v, \rho_i), v, \rho_i), \qquad i \in \overline{1, N},$$

$$\hat{Q}_{\text{pty}}(v, \rho) \triangleq \psi(\hat{Q}_{\text{pty}1}(v, \rho_1), \ldots, \hat{Q}_{\text{pty}N}(v, \rho_N)), \rho \triangleq (\rho_1, \ldots, \rho_N) \in \mathbb{R}_+^N.$$

Coordination variables v in PFM are the same kind as those in the direct method, and the *coordinator problem* (CP$_{\text{pty}}$) takes the form

Find the value $\hat{\rho}$ (called the coordinating value) of $\rho \in \mathbb{R}_+^N$ and coordination variable

$$\hat{v}(\hat{\rho}) = \arg \min_{Y \cap V_0'} \hat{Q}_{\text{pty}}(v, \hat{\rho}), \tag{2.35}$$

where $V_0' = \{v \in \mathcal{Y} : \forall i \in \overline{1, N} \ C_i(v) \neq \varnothing\}$, such that $(\hat{c}(\hat{v}(\hat{\rho}), \hat{\rho}), \hat{v}(\hat{\rho}))$ is a satisfactory approximation to the solution of the initial problem (2.32).

If ρ is fixed, then local problems LP$_{\text{pty}i}$ together with the problem of finding $\hat{v}(\rho)$ create the two-level optimization problem that will be called the *parametric problem* (with parameter ρ) below. The parametric problem can be regarded as a two-level optimization problem of the direct-method type, but with an extended performance index when compared with the two-level problem of the direct method from the previous section, and no output equations. Thus, it is better-suited to two-level optimization: there is no problem with "stiff" subsystems and the only troubles may be caused by the dependence of the constraint sets of the local problems LP$_{\text{pty}i}$ on the coordination variables v. Hence, troubles with the set V_0' may occur, but it is *generally much easier to handle* V_0' than the set V_0 (see Eq. 2.27). Moreover, for the important class of applications with separable local constraint sets, i.e.,

$$CU_i = C_i \times U_i, \qquad i \in \overline{1, N}, \tag{2.36}$$

the set V_0' reduces to the precisely known set

$$V_L = \{v \in \mathcal{Y} : Hv \in \bar{U} = \overset{N}{\underset{i=1}{\times}} U_i\}.$$

In this case, the constraint sets in local problems $\mathrm{LP}_{\mathrm{pty}i}$ are $C_i(v) = C_i$—they are not dependent on the coordination variables—and the coordinator constraint set $Y \cap V_L$ for the variables v is well defined.

However, one should realize that all the advantages of this approach are obtained at a cost: the coordinating values of the penalty coefficients ρ must be effectively chosen. To ensure that the parametric problem can be solved using two-level optimization, the values of the parameters ρ should be constrained to the set

$$D_{\mathrm{pty}} = \{\rho \in \mathbb{R}_+^N : \forall v \in V \quad (\hat{Q}_{\mathrm{pty}1}(v, \rho_1), \ldots, \hat{Q}_{\mathrm{pty}N}(v, \rho_N)) \in \Omega\}, \quad (2.37)$$

where $V = Y \cap V_0'$ and $\Omega \subseteq \mathbb{R}^N$ is the set on which function ψ strictly preserves order. (Precisely speaking, $(\hat{Q}_{\mathrm{pty}1}(v, \rho_1), \ldots, \hat{Q}_{\mathrm{pty}N}(v, \rho_N))$ should belong to Ω for all values of v that are sent from coordinator $\mathrm{CP}_{\mathrm{pty}}$ to local decision problems $\mathrm{LP}_{\mathrm{pty}i}$. The choice $v \in V$ in Eq. (2.37) is proper when the coordinator strategy does not use infeasible values of v.) The above requirement follows from the standard requirements of the direct method (see section 2.2.) and assures that two-level minimization of the parametric problem is equivalent to minimization of the overall penalized performance index

$$Q_{\mathrm{pty}}(c, v, \rho) \triangleq \psi(Q_{\mathrm{pty}1}(c_1, v, \rho_1), \ldots, Q_{\mathrm{pty}N}(c_N, v, \rho_N)) \quad (2.38)$$

on the set

$$CV \triangleq \underset{v \in V}{\bigcup} (\overset{N}{\underset{i=1}{\times}} C_i(v) \times \{v\}) = \{(c, v) \in \mathcal{C} \times \mathcal{Y} : (c, Hv) \in \overset{N}{\underset{i=1}{\times}} CU_i, \, v \in V\}.$$

The penalty function method will be analyzed within the framework of duality theory; the analysis will use the methodology introduced by Rockafellar (1974). To consider the method as a dual one, let us define *the primal problem* (P)

$$\text{minimize } \underset{D_{\mathrm{pty}}}{\sup} \, Q_{\mathrm{pty}}(c, v, \cdot)$$

$$\text{subject to } (c, v) \in CV, \quad (2.39)$$

and *the dual problem* (D)

$$\text{maximize } \underset{CV}{\inf} \, Q_{\mathrm{pty}}(\cdot, \cdot, \rho)$$

$$\text{subject to } \rho \in D_{\mathrm{pty}}. \quad (2.40)$$

It can be easily shown that relation

$$\sup (D) \leq \inf (P) \leq \mu$$

always holds, where $\sup (D)$ and $\inf (P)$ denote solutions (i.e., the performance index values) of the dual and primal problems, and μ is the solution of the initial system optimization problem (2.32). Thus, the coordinating values of ρ are simply values solving the dual problem—hence, the method can be regarded as a dual one and penalty coefficients ρ as dual variables.

To complete the introductory presentation of PFM, we observe that the overall penalized performance index (2.38) used in it differs from the function commonly used in penalty techniques

$$\psi(Q_1(c_1, H_1 v), \ldots, Q_N(c_N, H_N v)) + \sum_{i=1}^{N} \rho_i \|v_i - F_i(c_i, H_i v)\|^2$$

which in general (i.e., when utility function ψ is not necessarily additive) is nondecomposable.

2.3.2. APPLICABILITY CONDITIONS

One can see intuitively that the larger the values of the penalty coefficients, the closer the solutions of the parametric problem will be to the minimum of system optimization problem (2.32) that we are trying to find. The theory confirms this supposition as long as some conditions are satisfied; these conditions will be formulated in this section.

Let us assume at the beginning that the following rather moderate requirements concerning utility function ψ and the set Ω on which ψ strictly preserves order are satisfied:

$$\overline{\lim_{n \to \infty}} \|a^n\| = \infty \ \Rightarrow \ \overline{\lim_{n \to \infty}} \psi(a^n) > \psi(a), \tag{2.41}$$

where a is any element from Ω, and $\{a^n\}$ any sequence contained in Ω and bounded from below;

$$(\forall a \in \Omega)(\forall a' \in \mathbb{R}^N) a' \geq a \Rightarrow a' \in \Omega. \tag{2.42}$$

Conditions (2.41) and (2.42) are assumed to be satisfied in this section. They define a very broad class of nonadditive performance indices, including the most frequent cases as additive, multiplicative, or a mixture of the two. Woźniak (1973) gives some examples of functions that strictly preserve order. Recall that all spaces considered here are assumed to be Hilbert spaces, though the main results are true for more general spaces.

The first question to be answered concerning the applicability of PFM is whether the dual admissible set D_{pty} is nonempty and sufficiently "large." That and even more is ensured by a typical condition of boundedness from below, which we now introduce.

108

THEOREM 2.13. *If condition (2.42) is satisfied and*

$$(\exists \bar{\rho} \in \mathbb{R}_+^N)(\exists \bar{a} \in \Omega)(\forall (c, v) \in CV) \qquad \bar{Q}_{\text{pty}}(c, v, \bar{\rho}) \geq \bar{a}, \qquad (2.43)$$

where

$$\forall (c, v) \in \mathscr{C} \times \mathscr{Y} \qquad \bar{Q}_{\text{pty}}(c, v, \rho) \triangleq (Q_{\text{pty}1}(c_1, v, \rho_1), \ldots, Q_{\text{pty}N}(c_N, v, \rho_N)),$$

then

(1)
$$[\bar{\rho}, +\infty) \in D_{\text{pty}},$$

(2)
$$\lim_{\rho \to +\infty} d(\rho) = \sup (D),$$

where $d(\rho) = \inf_{CV} Q_{\text{pty}}(\cdot, \cdot, \rho)$ is called the dual function and $\rho \to +\infty$ means that all components $\rho_i \in \mathbb{R}_+$ of $\rho \in \mathbb{R}_+^N$, $i \in \overline{1, N}$, tend to plus infinity.

Proof. (1) It follows directly from (2.43) that $\bar{\rho} \in D_{\text{pty}}$. Take any $\rho \geq \bar{\rho}$ and denote $\rho_i = \bar{\rho}_i + \Delta\rho_i$, $i \in \overline{1, N}$. For every $v \in V$

$$\hat{Q}_{\text{pty}i}(v, \rho_i) = \inf_{C_i(v)} [Q_i(\cdot, H_i v) + \bar{\rho}_i \|v_i - F_i(\cdot, H_i v)\|^2$$

$$+ \Delta\rho_i \|v_i - F_i(\cdot, H_i v)\|^2] \geq \inf_{C_i(v)} [Q_i(\cdot, H_i v) + \bar{\rho}_i \|v_i - F_i(\cdot, H_i v)\|^2] = \hat{Q}_{\text{pty}i}(v, \bar{\rho}_i),$$

because any norm is non-negative. Hence, it follows from condition (2.42) and definition (2.37) that $\rho \in D_{\text{pty}}$, which implies that $[\bar{\rho}, +\infty) \in D_{\text{pty}}$.

(2) According to (2.42) it follows from definition (2.37) that

$$\rho \in D_{\text{pty}} \implies \forall (c, v) \in CV \ \bar{Q}_{\text{pty}}(c, v, \rho) \in \Omega.$$

Arguing analogously to the proof of (1) we get

$$\rho_1 \geq \rho_2 \implies \forall (c, v) \in CV \ \bar{Q}_{\text{pty}}(c, v, \rho_1) \geq \bar{Q}_{\text{pty}}(c, v, \rho_2).$$

Hence, the dual function is nondecreasing on D_{pty}, which implies relation (2). $\square$

Theorem 2.13 shows that large penalty coefficients (dual variables) are admissible (part 1), and that they should be coordinating if the duality relation $\sup (D) = \inf (P) = \mu$ is to be satisfied (part 2). Conditions ensuring this relation will now be presented.

LEMMA 2.8. *If we assume that*

1. *Conditions (2.41), (2.42), and (2.43), with $\bar{a} \in \text{int } \Omega$, are satisfied,*
2. *Functions Q_i, $i \in \overline{1, N}$ are (weakly) lower semicontinuous on (weakly) compact sets $CV_i = \bigcup_{v \in V}(C_i(v) \times \{H_i v\})$,*

3. *Mappings F_i are (weakly) continuous on CV_i, $i \in \overline{1, N}$,*

4. *Function ψ is lower semicontinuous on $\bar{Q}^H(CV)$ and order preserving on Ω^0, where:*

$$\forall (c, v) \in \mathscr{C} \times \mathscr{Y} \quad \bar{Q}^H(c, v) \triangleq \bar{Q}(c, Hv) \triangleq (Q_1(c_1, H_1v), \ldots, Q_N(c_N, H_Nv)),$$

$$\Omega^0 = \{a \in \mathbb{R}^N : \exists (c, v) \in CV \quad a \geq \bar{Q}^H(c, v)\},$$

then

$$\sup(D) = \mu.$$

Outline of the proof. The proof is based on the following partial results A, B, and C. First, we introduce some definitions:

$$\forall t \triangleq (t_1, \ldots, t_N) \in \mathscr{Y} \qquad X_t \triangleq \{(c, v) \in CV : v_i - F_i(c_i, H_iv) = t_i, i \in \overline{1, N}\},$$

$$T \triangleq \{t \in \mathscr{Y} : X_t \neq \varnothing\},$$

$$\forall t \in T \qquad \hat{Q}(t) \triangleq \inf_{X_t} \psi \bar{Q}^H(\cdot, \cdot).$$

The function $\hat{Q}: T \to \mathbb{R}$ is usually called the perturbation function; see Rockafellar (1974). Notice that the optimal value μ of problem (2.32) is equal to $\hat{Q}(0)$.

A. If assumptions (2.41), (2.42), and (2.43) with $\bar{a} \in \operatorname{int} \Omega$ are satisfied, then

$$\sup(D) \geq \varliminf_{t \to \theta} \hat{Q}(t).$$

The arduous proof of the above inequality can be found in Tatjewski (1976, 1978).

B. If the function $\psi \circ \bar{Q}^H$ is (weakly) lower semicontinuous on CV, mappings F_i are (weakly) continuous on CV_i, $i \in \overline{1, N}$, the set CV is (weakly) closed and

(1) $\quad (\exists \varepsilon > 0)(\exists \delta_0 > 0)(\forall \delta \in [0, \delta_0]) \quad (C_\delta = \{(c, v) \in CV :$

$$\psi(\bar{Q}^H(c, v)) \leq \varepsilon \wedge \|v - F(c, Hv)\| \leq \delta\} \quad \text{is nonempty and bounded}),$$

then

$$\varliminf_{t \to \theta} \hat{Q}(t) \geq \hat{Q}(0) = \mu.$$

The above result is based on an inf-stability theorem given in Wierzbicki and Kurcyusz (1977).

C. If the functions Q_i are (weakly) lower semicontinuous on CV_i, $i \in \overline{1, N}$, the function ψ is lower semicontinuous on $\bar{Q}^H(CV)$ and order preserving on Ω^0, then the function $\psi \circ \bar{Q}^H$ is (weakly) lower semicontinuous on CV.

Using the above results, proof of the lemma can be easily derived: The inequality $\sup(D) \leq \mu$ always holds and $\sup(D) \geq \underline{\lim}_{t \to 0} \hat{Q}(t)$ from A, so the proof amounts to establishing that the assumptions of B are satisfied. The set CV is (weakly) compact since all sets CV_i are (weakly) compact, hence, it is (weakly) closed and for every $\delta_0 > 0$ and $\delta \in [0, \delta_0]$ the sets $X_\delta \triangleq \{(c, v) \in CV : \|v - F(c, Hv)\| \leq \delta\}$ are bounded. Choosing any $\varepsilon > \mu$, we conclude that assumption (1) is fulfilled, and by virtue of C all other assumptions of B are also true. $\square$

Remark. In the most typical cases, the sets Ω are open; hence, assumption (2.43) alone ensures that $\bar{a} \in \operatorname{int} \Omega = \Omega$. $\square$

The results obtained in Theorem 2.13 and Lemma 2.8 make it now possible to formulate applicability conditions of PFM.

THEOREM 2.14 (*applicability conditions of* PFM). *Denote by* $\{\rho^n\}$ *any sequence of penalty coefficients tending to* $+\infty$ *and by* $\{\alpha^n\}$ *any sequence of positive scalars converging to zero. If we assume that* (a) *the assumptions of Lemma 2.8 are satisfied, and* (b) *function* ψ *is lower semicontinuous on* Ω, *then for sufficiently large n there exist points* (c^n, v^n) *satisfying*

$$Q_{\text{pty}}(c^n, v^n, \rho^n) \leq \inf_{CV} Q_{\text{pty}}(\cdot, \cdot, \rho^n) + \alpha^n = d(\rho^n) + \alpha^n$$

and

$$\lim_{n \to \infty} \rho_i^n \|v_i^n - F_i(c_i^n, H_i v^n)\|^2 = 0, \qquad i \in \overline{1, N},$$

$$\lim_{n \to \infty} Q_{\text{pty}}(c^n, v^n, \rho^n) = \lim_{n \to \infty} \psi(\bar{Q}(c^n, Hv^n)) = \mu,$$

and any (weakly) convergent sub-sequence of the sequence $\{(c^n, v^n)\}$ *(weakly) converges to the solution of problem* (2.32).

Proof. By Theorem 2.13 $[\bar{\rho}, +\infty) \subseteq D_{\text{pty}}$. Thus, we can assume, for sufficiently large n, that $\rho^n \in D_{\text{pty}} + \varepsilon$, $\varepsilon \in \mathbb{R}_+^N$. We have $\psi(\bar{a}) > -\infty$ since $\bar{a} \in \operatorname{int} \Omega$ and ψ strictly preserves order on Ω. The above results together with assumption (2.43) imply that for sufficiently large n there exist finite lower bounds of the function $Q_{\text{pty}}(\cdot, \cdot, \rho^n)$, and hence points (c^n, v^n).

Since for every $(c, v) \in CV$ the function $Q_{\text{pty}}(c, v, \cdot)$ is nondecreasing on $[\bar{\rho}, +\infty)$, for sufficiently large n

$$d(\rho^n) + \alpha^n \geq Q_{\text{pty}}(c^n, v^n, \rho^n) \geq Q_{\text{pty}}(c^n, v^n, \rho^n - \varepsilon) \geq d(\rho^n - \varepsilon).$$

By Theorem 2.13

$$\lim_{n \to \infty} (d(\rho^n) + \alpha^n) = \lim_{n \to \infty} d(\rho^n - \varepsilon) = \sup(D),$$

and therefore by Lemma 2.8

(1)
$$\lim_{n\to\infty} Q_{\mathrm{pty}}(c^n, v^n, \rho^n) = \mu.$$

We now demonstrate that $\overline{\lim}_{n\to\infty} \|v_i^n - F_i(c_i^n, H_i v^n)\| = 0$ for every $i \in \overline{1, N}$, since in the opposite case

$$\exists j \in \overline{1, N} \quad \overline{\lim_{n\to\infty}} \|v_j^n - F_j(c_j^n, H_j v^n)\| > 0,$$

which implies that

$$\overline{\lim_{n\to\infty}} \rho_j^n \|v_j^n - F_j(c_j^n, H_j v^n)\|^2 = +\infty.$$

The sequence $\{\bar{Q}_{\mathrm{pty}}(c^n, v^n, \rho^n)\}$ is contained in the set Ω and is bounded from below; hence, by assumption (2.41)

$$\lim_{n\to\infty} Q_{\mathrm{pty}}(c^n, v^n, \rho^n) > \psi(\bar{Q}(\tilde{c}, H\tilde{v})) \geq \mu,$$

where $(\tilde{c}, \tilde{v})$ is any point belonging to CV and satisfying the output equations (2.33), i.e., $(\tilde{c}, \tilde{v}) \in CV \cap CV^F$, $CV^F \triangleq \{(c, v) \in \mathscr{C} \times \mathscr{Y}: v = F(c, Hv)\}$. This strict inequality contradicts (1).

The set CV is (weakly) compact and therefore the sequence $\{(c^n, v^n)\}$ consists of its (weakly) convergent sub-sequences. Denote any of them by $\{(c^{n'}, v^{n'})\}$, and its (weak) limit by $(\bar{c}, \bar{v}) \in CV$. Since the output mappings F_i are (weakly) continuous and we have shown above that $\|v_i^n - F_i(c_i^n, H_i v^n)\| \to 0$, $i \in \overline{1, N}$, we know that $(\bar{c}, \bar{v}) \in CV \cap CV^F$.

It is not possible for any $j \in \overline{1, N}$ that $\lim_{n'\to\infty} \rho_j^{n'} \|v_j^{n'} - F_j(c_j^{n'}, H_j v^{n'})\|^2 = +\infty$, since it leads to contradiction with (1), as we have shown above. Let us examine, therefore, if it is possible that

$$\exists j \in \overline{1, N} \qquad \overline{\lim_{n'\to\infty}} \rho_j^{n'} \|v_j^{n'} - F_j(c_j^{n'}, H_j v^{n'})\|^2 = \delta_j > 0,$$

and assume, without loss of generality, that

$$\overline{\lim_{n'\to\infty}} \rho_j^{n'} \|v_j^{n'} - F_j(c_j^{n'}, H_j v^{n'})\|^2 = \lim_{n'\to\infty} \rho_j^{n'} \|v_j^{n'} - F_j(c_j^{n'}, H_j v^{n'})\|^2.$$

Denote by $\{(c^{n_1}, v^{n_1})\}$ a sub-sequence of $\{(c^{n'}, v^{n'})\}$ such that

$$\alpha_1 = \underline{\lim_{n'\to\infty}} Q_1(c_1^{n'}, H_1 v^{n'}) = \lim_{n_1\to\infty} Q_1(c_1^{n_1}, H_1 v^{n_1}),$$

by $\{(c^{n_2}, v^{n_2})\}$ a sub-sequence of $\{(c^{n_1}, v^{n_1})\}$ such that

$$\alpha_2 = \underline{\lim_{n_1\to\infty}} Q_2(c_2^{n_1}, H_2 v^{n_1}) = \lim_{n_2\to\infty} Q_2(c_2^{n_2}, H_2 v^{n_2}),$$

112

and so on up to sub-sequence $\{(c^{n_N}, v^{n_N})\}$. $(\bar{c}, \bar{v}) \in CV \cap CV^F$ implies, by assumption (2.43), that $\bar{Q}(\bar{c}, H\bar{v}) \in \Omega$. Since the functions Q_i are (weakly) lower semicontinuous $\alpha = (\alpha_1, \ldots, \alpha_N) \geq \bar{Q}(\bar{c}, H\bar{v})$ and (2.42) implies that $\alpha \in \Omega$. Taking into account that ψ is lower semicontinuous and strictly preserves order on Ω and that penalty functions are non-negative, we get

$$(2) \quad \mu = \lim_{n_N \to \infty} Q_{\text{pty}}(c^{n_N}, v^{n_N}, \rho^{n_N}) \geq \psi(\alpha_1, \ldots, \alpha_{j-1}, \alpha_j + \delta_j, \alpha_{j+1}, \ldots, \alpha_N)$$

$$> \psi(\bar{Q}(\bar{c}, H\bar{v})) \geq \mu,$$

which is, of course, impossible. Hence, for any $j \in \overline{1, N}$

$$(3) \qquad \lim_{n \to \infty} \rho_j^n \|v_j^n - F_j(c_j^n, H_j v^n)\|^2 = 0.$$

Equality (2) implies therefore that $\psi(\bar{Q}(\bar{c}, H\bar{v})) = \mu$. Thus, any (weakly) convergent sub-sequence of $\{(c^n, v^n)\}$ (weakly) converges to the solution of problem (2.32) since $(\bar{c}, \bar{v}) \in CV \cap CV^F$.

We need only show that $\lim_{n \to \infty} \psi(\bar{Q}(c^n, Hv^n)) = \mu$ to complete the proof. It follows from assumption (2.43) with $\bar{a} \in \text{int } \Omega$ that

$$\exists \beta > 0 \qquad \bar{a}_\beta = (\bar{a}_1 - \beta, \ldots, \bar{a}_N - \beta) \in \Omega.$$

By virtue of Eq. (3), for sufficiently large n the values of penalty functions are not greater than β. Thus, for such n, $\bar{Q}(c^n, Hv^n) \in \Omega$, and $Q_{\text{pty}}(c^n, v^n, \rho^n) \geq \psi(\bar{Q}(c^n, Hv^n))$. Using now the lower semicontinuity of $\psi \circ \bar{Q}^H$ (see the proof of Lemma 2.8) and Eq. (1), we get finally

$$\mu = \lim_{n \to \infty} Q_{\text{pty}}(c^n, v^n, \rho^n) \geq \lim_{n \to \infty} \psi(\bar{Q}^H(c^n, v^n)) \geq \psi(\bar{Q}(\bar{c}, H\bar{v})) = \mu,$$

which completes the proof. $\square$

Remark. If in Lemma 2.8 above (and hence also in Theorem 2.14) lower semicontinuity of Q_i and ψ is replaced by continuity, then the requirement for order-preserving on Ω^0 is superfluous because it was used only to ensure lower semicontinuity of $\psi \circ \bar{Q}^H$ (see the proof of Lemma 2.8, part C). This remark is important especially in finite-dimensional spaces, when continuity is a natural assumption. $\square$

It is always the case that $CV_i \subseteq CU_i$, because the sets CV_i consist of such points $(c_i, u_i) \in CU_i$ that $u_i = H_i v$ for some $v \in V$, and hence it is advantageous to make assumptions about CV_i and not CU_i, $i \in \overline{1, N}$. In the case of separable local constraints (2.36), $CV_i = C_i \times H_i(V) \subseteq C_i \times U_i$, $i \in \overline{1, N}$, and they are (weakly) compact if C_i, U_i, $i \in \overline{1, N}$ and V are (weakly) compact.

The applicability conditions of PFM, formulated in Theorem 2.14, are all natural and weak (boundedness from below, continuity of mappings, compact sets, and so forth). Therefore, the method is applicable to many areas.

2.3.3. COORDINATOR STRATEGIES

Coordinator strategies are algorithms for finding a solution of the coordinator problem CP_{pty}, i.e., algorithms for finding the coordinating values of penalty coefficients ρ and for using those values to solve the parametric problem.

It was shown in Theorem 2.14 that values of dual variables ρ are coordinating if they are sufficiently large. Therefore, it is possible at the very beginning to take very large penalty coefficients ρ and to solve the parametric problem with them only once. But we have found that to solve the parametric problem once with large values of the penalty coefficients is reasonable only if the initial point in the numerical procedure lies very close to the required minimum. In other cases, it is more efficient to solve the parametric problem repeatedly with increasing values of the penalty coefficients and a rule for choosing the solution of the parametric problem, for the small value of ρ, as a starting point for the next two-level optimization. This strategy has three levels: two-level parametric problem and a higher level for adjusting ρ. Observe also that we are interested in the solution of the parametric problem only for one coordinating value of ρ. Therefore, solving that problem precisely for every succeeding ρ could lead to an unnecessary increase in computation; it is advisable to start with low accuracy and to increase it as ρ increases. This method coincides perfectly with the theoretical results given in Theorem 2.14: $\alpha^n \to 0$ means that the accuracy of the optimization increases.

A heuristic approach toward better efficiency, proposed by Franz *et al.* (1975) consists in modification of the parametric problem. It is solved repeatedly with increasing ρ, but local decision problems LP_{ptyi} are solved only one during one parametric problem minimization—only for the first (initial) value of v—and local solutions are then kept constant until ρ increases.

It should be noted that the aim of all the above strategies is first to get a satisfactory approximation of the optimal point of problem (2.32) with small values of the penalty coefficients. With larger values of ρ, the parametric problem becomes poorly conditioned and hence numerical procedures are less effective, which is well known in any application of penalty functions.

To make the coordinator strategy effective, it is also very important to evaluate the coordinator problem performance index $\hat{Q}_{pty}(\cdot, \rho)$ only rarely since each evaluation requires minimization of all N local decision problems LP_{ptyi}. That is why the properties of the performance index should be carefully taken into account. Theorems and comments concerning properties of function $\hat{Q}_{pty}(\cdot, \rho)$ can be found in Appendix A at the end of the book.

114

2.3.4. FURTHER APPLICATIONS OF PFM

It was noted in section 2.3.1. that PFM can be also applied to a more general form of the system optimization then problem (2.32). The overall system constraints of the type (2.12):

$$CU_0 = \{(c, u) \in \mathscr{C} \times \mathscr{U} : r(c, u) = w(r_1(c_1, u_1), \ldots, r_N(c_N, u_N)) \in R\} \quad (2.44)$$

where for every $i \in \overline{1, N}$

$$r_i : \mathscr{C}_i \times \mathscr{U}_i \to \mathscr{R}_i, \quad w : \overset{N}{\underset{i=1}{\times}} \mathscr{R}_i \to \mathscr{R}, \quad R \subset \mathscr{R}$$

can be easily handled by PFM. In this case, the local decision problems are formulated as follows and denoted LP'_{ptyi}

For given coordination variable $x_i = (v, r_{di}) \in \mathscr{Y} \times \mathscr{R}_i$ and penalty coefficient $\rho_i \in \mathbb{R}_+$ find control

$$\hat{c}_i(x_i, \rho_i) = \arg \min_{C_i(v)} Q_{ptyi}(\cdot, x_i, \rho_i) \quad (2.45)$$

where

$$\forall c_i \in \mathscr{C}_i \qquad Q_{ptyi}(c_i, x_i, \rho_i) \triangleq Q_i(c_i, H_i v) +$$

$$\rho_i(\|v_i - F_i(c_i, H_i v)\|^2 + \|r_{di} - r_i(c_i, H_i v)\|^2).$$

Coordinator problem CP_{pty} should be formulated as follows

Find the value $\hat{\rho}$ of $\rho = (\rho_1, \ldots, \rho_N) \in \mathbb{R}_+^N$ and the coordination variable

$$\hat{x}(\hat{\rho}) = \arg \min_x \hat{Q}_{pty}(\cdot, \hat{\rho}), \quad (2.46)$$

where

$$X = \{x \triangleq (v, r_d) = (v, r_{d1}, \ldots, r_{dN}) \in \mathscr{Y} \times \mathscr{R} : v \in V \wedge w(r_{d1}, \ldots, r_{dN}) \in R\},$$

such that $(\hat{c}(\hat{x}(\hat{\rho}), \hat{\rho}), v(\hat{\rho}))$ serves as a satisfactory approximation of the initial problem solution.

All the results obtained until now for problem (2.32) are true, even with additional constraint (2.44), provided that mappings r_i satisfy assumptions analogous to those concerning F_i, $i \in \overline{1, N}$. Suitable modifications of all formulations are rather obvious.

The penalty function method can be further generalized to completely eliminate the dependence of local decision problem constraint sets on coordination variables; troubles with the set V_0' are thereby avoided and the coordinator problem functional is made easier to differentiate. However, this can be done only at the cost of higher dimensionality of local decision problems, which should be formulated as follows and denoted $\mathrm{LP}''_{\mathrm{ptyi}}$

For given coordination variable $v \in \mathcal{Y}$ and penalty coefficient $\rho_i \in \mathbb{R}_+$, find control and interaction

$$(\hat{c}_i(v, \rho_i), \hat{u}_i(v, \rho_i)) = \arg \min_{CU_i} Q_{\mathrm{ptyi}}(\cdot, \cdot, v, \rho_i), \tag{2.47}$$

where

$$\forall (c_i, u_i) \in \mathcal{C}_i \times \mathcal{U}_i \quad Q_{\mathrm{ptyi}}(c_i, u_i, v, \rho_i) \triangleq Q_i(c_i, u_i)$$
$$+ \rho_i(\|v_i - F_i(c_i, u_i)\|^2 + \|u_i - H_i v\|^2).$$

Of course, two different penalty coefficients for output and structure equations can also be used instead of one ρ_i. Having defined local problems $\mathrm{LP}_{\mathrm{ptyi}}$, we formulate the coordinator problem as follows and denote it by $\mathrm{CP}''_{\mathrm{pty}}$

Find the coordinating value $\hat{\rho}$ of $\rho = (\rho_1, \ldots, \rho_N) \in \mathbb{R}^N_+$ and the coordination variable

$$\hat{v}(\hat{\rho}) = \arg \min_V \hat{Q}_{\mathrm{pty}}(\cdot, \rho), \tag{2.48}$$

where $V = Y$ and $\hat{Q}_{\mathrm{pty}}(\cdot, \rho)$ is defined analogously to its definiton in $\mathrm{CP}_{\mathrm{pty}}$.

All results obtained until now for the best case of separable local constraint sets (2.36) remain valid for the above formulation. We must only use, as it follows from $\mathrm{LP}''_{\mathrm{ptyi}}$, $x_i = (c_i, u_i)$ as local decision variables instead of c_i, and the sets $X_i = CU_i$ instead of C_i, $i \in \overline{1, N}$. It is also important not to increase local problem dimensions higher than necessary. For example, consider a subproblem with

$$u_i \triangleq (u_{i1}, u_{i2}, u_{i3}) \in \mathbb{R}^3, \qquad c_i \triangleq (c_{i1}, c_{i2}, c_{i3}) \in \mathbb{R}^3,$$
$$CU_i \triangleq \{(c_i, u_i): 0 \le c_{ij} \le a, \, c_{i1} + u_{i1} \le b, \, c_{i2} - \sqrt{c_{i1} + u_{i1}} \le c\},$$

where a, b, and c are scalars, and where the i-th structure equation is

$$u_i = \begin{bmatrix} u_{i1} \\ u_{i2} \\ u_{i3} \end{bmatrix} = \begin{bmatrix} H_{i1} \\ H_{i2} \\ H_{i3} \end{bmatrix} v = H_i v,$$

It is reasonable to assume that $x_i \triangleq (c_i, u_{i1}) \in \mathbb{R}^4$ instead of the general formulation $x_i \triangleq (c_i, u_i) \in \mathbb{R}^6$, because taking only u_{i1} as the additional local decision variable fully eliminates the dependence of local constraint set CU_i on coordination variables v; u_{i2} and u_{i3} do not enter the description of CU_i. Therefore, u_{i2} and u_{i3} can be kept constant at the values given by the structure equations, and the following local problem arises

Find $\qquad (\hat{c}_i(v, \rho), \hat{u}_{i1}(v, \rho)) = \arg \min Q_{\text{pty}i}(\cdot, \cdot, v, \rho),$ where

$Q_{\text{pty}i}(c_i, u_{i1}, v, \rho)$

$$= Q_i(c_i, u_{i1}, H_{i2}v, H_{i3}v) + \rho(\|v_i - F_i(c_i, u_{i1}, H_{i2}v, H_{i3}v)\|^2 + \|u_{i1} - H_{i1}v\|^2).$$

The general formulation (2.47) is defined for "the worst case," i.e., when all u_{ij}, $j \in \overline{1, \dim u_i}$ enter the description of the set CU_i; the above example explains how to adopt this formulation when the actual case is better than "the worst." Observe that the constraint set V of CP''_{pty} is extremely simple, consisting only of the admissible output set Y.

2.4. PRICE COORDINATION—THE INTERACTION BALANCE METHOD

In this section we will discuss the basic properties of the interaction balance method (IBM) (see Chapter 1). The problem statement and formalization of the method will be followed by some general comments. Next the possibilities for application of IBM will be presented together with the appropriate coordination algorithms. Finally, some different methods using the price mechanism will be presented.

2.4.1. STATEMENT OF THE PROBLEM

The interaction balance method has been introduced in Chapter 1. Here we will use the following formulation of the system optimization problem (OP):

Find control $\hat{c}$ and interaction $\hat{u}$ such that

$$Q(\hat{c}, \hat{u}) = \min_{c,u} Q(c, u) \tag{2.49}$$

subject to

1. Subsystem equations (2.4), i.e., $y_i = F_i(c_i, u_i, z_i)$, $i \in \overline{1, N}$, or, in compact notation, $y = F(c, u, z)$
2. Coupling equations

$$P(u, y) = \sum_{i=1}^{N} P_i(u_i, y_i) = 0; \tag{2.50}$$

Eqs. (2.6) represent a particular form of these couplings
3. Local constraints $(c_i, u_i) \in CU_i$, $i \in \overline{1, N}$; $(c, u) \in CU = CU_1 \times \cdots \times CU_N$—see Eq. (2.9)

We assume, that $Q(c, u) = \sum_{i=1}^{N} Q_i(c_i u_i)$, $P_i : \ell_i \times \mathcal{Y}_i \to \mathcal{P}$ and that $\mathcal{C}_i$, $\mathcal{U}_i$, $\mathcal{Y}_i$, $\mathcal{P}$ are real Hilbert spaces. Some considerations and results presented in this section can be generalized to Banach spaces (or even linear-topological spaces). However, it seems to be reasonable to restrict the considerations to real Hilbert spaces. We denote $c \triangleq (c_1, \ldots, c_N)$, $F(c, u, z) \triangleq (F_1(c_1, u_1, z_1), \ldots, F_N(c_N, u_N, z_n))$ etc. It should be noted here that Q_i, F_i can be defined only on some appropriate subsets of $\mathcal{C}_i \times \mathcal{U}_i$ and $\mathcal{C}_i \times \ell_i \times \mathcal{Z}_i$.

Let us define the Lagrangian:

$$L(c, u, \lambda, z) = Q(c, u) + \langle \lambda, P(u, F(c, u, z)) \rangle, \qquad (2.51)$$

where $\lambda \in \mathcal{P}$. In most of the following discussion, the distinction between c and u will be not necessary; thus to simplify the notation we will introduce the following definitions:

$$w_i \triangleq (c_i, u_i), \qquad i \in \overline{1, N} \quad (w \triangleq (c, u)),$$

$$Q(c, u) = Q(w) = \sum_{i=1}^{N} Q_i(w_i) \qquad (2.52)$$

$$V(w, z) \triangleq P(u, F(c, u, z)) \qquad (2.53)$$

$$L(c, u, \lambda, z) = L(w, \lambda, z) = \sum_{i=1}^{N} L_i(w_i, \lambda, z_i) \qquad (2.54)$$

Now we can formulate the infimal (lower-level) problem (IP) of IBM:

For given $\lambda \in \mathcal{P}$ find $\hat{w}(\lambda) = (\hat{c}(\lambda), \hat{u}(\lambda))$ such that

$$L(\hat{w}(\lambda), \lambda, z) = \min_{w \in CU} L(w, \lambda, z) \qquad (2.55)$$

Remark A. The notation $\hat{w}(\cdot)$ may suggest that $\hat{w}(\lambda)$ is a point-to-point function of λ. However, this notation is used here only to denote the dependence of solution $\hat{w}(\lambda)$ on λ. An assumption about the uniqueness of $\hat{w}(\lambda)$ will be introduced later ($\hat{w}_i$, and other notations in this section, should be interpreted in the same way).

The infimal problem IP can be solved as N independent local problems (LP$_i$):

For given $\lambda \in \mathcal{P}$ find $\hat{w}_i(\lambda) = (\hat{c}_i(\lambda), \hat{u}_i(\lambda))$ such that

$$L_i(\hat{w}_i(\lambda), \lambda, z_i) = \min_{w_i \in CU_i} L_i(w_i, \lambda, z_i) \qquad (2.56)$$

The set of solutions of IP, or the set of solutions of LP$_i$, for given λ are denoted by $\overline{CU}(\lambda)$ or $\overline{CU}_i(\lambda)$. Of course, $\overline{CU}(\lambda) = \overline{CU}_1(\lambda) \times \cdots \times \overline{CU}_N(\lambda)$.

118

The coordinator problem (CP) may be now defined as follows:

Find $\hat{\lambda} \in \mathscr{P}$ such that the set $\overline{CU}(\hat{\lambda})$ is nonempty and for every

$$w \in \overline{CU}(\hat{\lambda}) \quad \text{we have} \quad V(w, z) = 0. \tag{2.57}$$

The formulation of IBM originated in mathematical programming (Lasdon 1968, 1970). It was introduced in a slightly different form for system optimization (Mesarović *et al.* 1970) where the coordinator problem was stated in the following general but not very practical form:

Find $\hat{\lambda} \in \mathscr{P}$ such that there exists

$$w \in \overline{CU}(\hat{\lambda}) \quad \text{for which} \quad V(w, z) = 0. \tag{2.58}$$

It is evident that since we are not able to distinguish the different solutions of IP at the lower level, the solution $\hat{\lambda}$ of problem (2.58) may not provide us with a feasible solution of OP.

Problem IP can be formulated differently; instead of solving the minimization problem one may solve the the set of necessary optimality conditions (Titli 1972) under appropriate differentiability and regularity assumptions. Such an approach has been used for dynamic optimization; see Pearson (1971), Singh *et al.* (1975), and Tamura (1975). It may be shown that if we are able to choose an appropriate solution of the necessary optimality conditions, then the price method may provide us with an optimal solution of OP even when IBM—as presented above—fails (Ostrovskii and Volin 1975). However, it is not yet possible to specify the rule for choosing the "proper" solutions of the necessary optimality conditions. Moreover, if a solution is unique, then—at least for problems of practical interest—it reduces to the solution of IP. On the other hand, the above formulation of IBM is more natural for decision-making hierarchies and does not require the differentiability and regularity assumptions for OP.

2.4.2.   APPLICABILITY OF THE INTERACTION BALANCE METHOD

In this section, z will be eliminated from all mappings for convenience, since it is not involved explicitly in any assumption; thus, the assumptions are assumed to hold for every z of interest. The following proposition is essential for applications of IBM.

PROPOSITION.   *If there exists $\hat{\lambda}$, a solution of CP, then every $\hat{w}(\hat{\lambda}) \in \overline{CU}(\hat{\lambda})$ is a solution of OP.* $\square$

The proof is elementary and results from definitions of IP and CP, and from Eq. (2.51).

DEFINITION. *We say that* IBM *is applicable to* OP *if there exists at least one solution* $\hat{\lambda}$ *of* CP.

THEOREM 2.15. *if* IBM *is applicable to* OP *then there exists a saddle point* (w^0, λ^0) *of Lagrangian* $L(\cdot, \cdot)$ *on* $CU \times \mathscr{P}$. *If* (a) *there exists a saddle point of* $L(\cdot, \cdot)$ *on* $CU \times \mathscr{P}$ *and* (b) *every point* (w^*, λ^*) *such that* $L(w^*, \lambda^*) = max_{\lambda \in \mathscr{P}} min_{w \in CU} L(w, \lambda)$ *is a saddle point of* $L(\cdot, \cdot)$ *on* $CU \times \mathscr{P}$, *then* IBM *is applicable.*

Proof. If IBM is applicable, then any point $(\hat{\lambda}, \hat{w}(\hat{\lambda}))$, where $\hat{w}(\hat{\lambda}) \in \overline{CU}(\hat{\lambda})$, is a saddle point of $L(\cdot, \cdot)$ on $CU \times \mathscr{P}$ since $\forall w \in CU$, $\lambda \in \mathscr{P}$:

$$L(w, \hat{\lambda}) \geq L(\hat{w}(\hat{\lambda}), \hat{\lambda}) = L(\hat{w}(\hat{\lambda}), \lambda).$$

On the other hand, it is well known that if (w^0, λ^0) is a saddle point of $L(\cdot, \cdot)$ on $CU \times \mathscr{P}$, then w^0 is a solution of OP. So let us assume that (w^0, λ^0) is a saddle point of $L(\cdot, \cdot)$ on CU (according to assumption (a) at least one such point exists). Then every (w^*, λ^0) such that

$$L(w^*, \lambda^0) = \min_{w \in CU} L(w, \lambda^0),$$

i.e., every solution w^* of IP for $\lambda = \lambda^0$ has to satisfy

$$L(w^0, \lambda^0) \leq L(w^*, \lambda^0) \leq L(w^0, \lambda^0).$$

Hence

$$L(w^*, \lambda^0) = L(w^0, \lambda^0) = \max_{\lambda \in \mathscr{P}} \min_{w \in CU} L(w, \lambda)$$

and according to assumption (b) (w^*, λ^0) is a saddle point of $L(\cdot, \cdot)$ on $CU \times \mathscr{P}$. So w^* is a solution of OP. Therefore, $\hat{\lambda} = \lambda^0$ is a solution of CP; $\overline{CU}(\hat{\lambda})$ is nonempty since $w^0 \in \overline{CU}(\hat{\lambda})$. $\square$

We will now present some theorems and lemmas giving more insight into the applicability of IBM and more constructive sufficient conditions for this applicability.

Let us define first the dual function $\varphi_1 : \mathscr{P} \to \mathbb{R}$ as follows

$$\varphi_1(\lambda) = \inf_{w \in CU} L(w, \lambda) \tag{2.59}$$

One of the most general results is given by the following theorem:

THEOREM 2.16. (*applicability conditions of* IBM). *If we assume that*

1. *Set* CU *is (weakly) compact, P is (weakly) continuous on* $\mathscr{C} \times \mathscr{U}$ *(usually P is an affine mapping—see Eqs.* (2.6)), *F is (weakly) continuous on* CU, *and Q is (weakly) lower semicontinuous on* CU,

2. *There exists $\hat{\lambda} \in \mathscr{P}$, such that*

$$\varphi_1(\hat{\lambda}) = \max_{\lambda \in \mathscr{P}} \varphi_1(\lambda),$$

3. *Set $V(\overline{CU(\hat{\lambda})})$ consists of a single point,*

then $\hat{\lambda}$ is a solution of CP *and $\hat{w}(\hat{\lambda})$ is a solution of* OP.

Proof. It is easy to show that $\varphi_1(\lambda)$ is concave on $\mathscr{P}$. Indeed, $\forall \lambda^1, \lambda^2 \in \mathscr{P}$, $\forall w \in CU$, $\forall \rho \in [0, 1]$ we have

$$L(w, \rho\lambda^1 + (1-\rho)\lambda^2) = \rho L(w, \lambda^1) + (1-\rho)L(w, \lambda^2).$$

Therefore

$$\varphi_1(\rho\lambda^1 + (1-\rho)\lambda^2) = \inf_{w \in CU} L(w, \rho\lambda^1 + (1-\rho)\lambda^2)$$

$$\geq \rho \inf_{w \in CU} L(w, \lambda^1) + (1-\rho) \inf_{w \in CU} L(w, \lambda^2) = \rho\varphi_1(\lambda^1) + (1-\rho)\varphi_1(\lambda^2).$$

From assumption 1 it follows that $\forall \lambda \in \mathscr{P}$ set $\overline{CU}(\lambda)$ is nonempty and $\varphi_1(\lambda) > -\infty$. Thus, function $\varphi(\lambda) = -\varphi_1(\lambda)$ is convex and continuous on $\mathscr{P}$ and $\varphi(\lambda)$ has a nonempty, weakly compact and convex subdifferential $\partial\varphi(\lambda)$, $\forall \lambda \in \mathscr{P}$ (Ioffe and Tikhomirov 1974). The representation of $\partial(\lambda)$ can be obtained from the following:

THEOREM (*Ioffe and Tikhomirov 1974*). *Suppose that $\mathscr{X}$ is a linear locally convex topological space, $\mathscr{S}$ is a compact topological space and $f: \mathscr{S} \times \mathscr{X} \to \mathbb{R}$. Assume that $\forall s \in \mathscr{S}$ $f(s, \cdot)$ is convex on $\mathscr{X}$ and $\forall x \in \mathscr{X}$ $f(\cdot, x)$ is upper semicontinuous on $\mathscr{S}$. Let us denote*

$$f_0(x) = \sup_{s \in \mathscr{S}} f(s, x), \quad S_0(x) = \{s \in \mathscr{S} : f(s, x) = f_0(x)\}.$$

Then $\forall x \in \mathscr{X}$ we have

$$\overline{\mathrm{conv}} \left(\bigcup_{s \in S_0(x)} \partial f_s(x) \right) \subset \partial f_0(x_0).$$

If $\forall s \in \mathscr{S}$ $f(s, \cdot)$ is continuous in $x = x_0 \in \mathscr{X}$ then

$$\overline{\mathrm{conv}} \left(\bigcup_{s \in S_0(x_0)} \partial f_s(x_0) \right) = \partial f_0(x_0)$$

where $f_s(\cdot) = f(s, \cdot)$ and $\overline{\mathrm{conv}} A$ is a weak closure of the convex hull of $A \in \mathscr{X}^*$ in $\mathscr{X}^*$.*

From the above theorem and assumption 1 it follows that $\forall \lambda \in \mathscr{P}$

$$\partial\varphi(\lambda) = \mathrm{conv}\,(-V(\overline{CU(\lambda)})). \tag{2.60}$$

$\varphi(\hat{\lambda}) = \min_{\lambda \in \mathscr{P}} \varphi(\lambda)$ (assumption 2) and $\varphi(\lambda)$ is convex, so $\theta \in \partial \varphi(\hat{\lambda})$. But according to (2.60) and assumption 3 we have

$$\partial \varphi(\hat{\lambda}) = -V(\overline{CU(\hat{\lambda})}).$$

Therefore $\forall w \in \overline{CU(\hat{\lambda})}$, $V(w) = 0$ and $\overline{CU(\hat{\lambda})}$ is nonempty. So $\hat{\lambda}$ is a solution of CP and of course any $\hat{w}(\hat{\lambda})$ is a solution of OP. $\square$

The following lemma follows directly form the proof of Theorem 2.16.

LEMMA 2.9. *If assumption 1 of Theorem 2.16 holds and* $\forall \lambda \in \mathscr{P}_0 \subset \mathscr{P}$ *the set* $V(\overline{CU(\lambda)})$ *consists of one point (which is true, of course, if* $\overline{CU(\lambda)}$ *consists of a single point), then* $\varphi_1(\cdot)$ *is weakly differentiable* $\forall \lambda \in \mathscr{P}_0$. *Moreover, if* $\mathscr{P}_0$ *is open and* $V(\overline{CU(\cdot)})$ *is continuous on* $\mathscr{P}_0$ *then* $\varphi_1(\cdot)$ *has a Fréchet derivative*

$$\varphi_1'(\lambda) = V(\overline{CU(\lambda)}) \quad \forall \lambda \in \mathscr{P}_0.$$

Remark B. Assumption 1 of Theorem 2.16 is restrictive in the norm topology version with respect to the compactness requirement of $\overline{CU}$. This requirement is difficult to relax. After modifying the proof of Theorem 2.16, one can see that it is enough to assume that $\forall \lambda \in \mathscr{P}_0$, where $\mathscr{P}_0$ is such that $\hat{\lambda}$ defined in assumption 2 belongs to it, and $\mathscr{P}_0$ is open and convex, the solution set $\overline{CU(\lambda)}$ is nonempty and belongs to some compact set $A \subset \mathscr{C} \times \mathscr{U}$. In the weak topology version we would like to relax the assumption about the weak continuity of F on CU (the assumption that CU is weakly compact is not a very restrictive one). It is enough to demand that the functional $L(\cdot, \lambda)$ be weakly lower semicontinuous $\forall \lambda \in \mathscr{P}_0$ ($\mathscr{P}_0$ such that $\hat{\lambda} \in \mathscr{P}_0$ is convex and open).

The key assumptions of Theorem 2.16 are 2 and 3. The following lemmas specify some sufficient conditions under which 2 and 3 hold.

LEMMA 2.10. *If we assume that (a) assumption 1 of Theorem 2.16 holds and* Q *is bounded on* U, *and (b) for all* $h \in \mathscr{P}$, $\|h\| = k_1 > 0$ *there exist* $h_1, \|h_1\| \leq k_2 < k_1$ *and* $w \in CU$ *such that* $V(w) = h + h_1$, *then assumption 2 of Theorem 2.16 holds and* $\|\hat{\lambda}\| \leq r$, *where*

$$r = \frac{k_0' - k_0''}{k_1 - k_2}, \qquad k_0' = \sup_{CU} Q(w), \qquad k_0'' = \min_{CU} Q(w).$$

Proof. It was remarked before that $\varphi_1(\cdot)$ is concave and continuous on $\mathscr{P}$. Therefore, $\varphi_1(\cdot)$ is also weakly upper semicontinuous on $\mathscr{P}$. Now, for every $\lambda \in \mathscr{P}$ we can find $h, \|h\| = k_1$, such that $\langle \lambda, h \rangle = -k_1 \|\lambda\|$. Then, from the assumptions we have

$$\varphi_1(\lambda) \leq L(w, \lambda) = Q(w) + \langle \lambda, V(w) \rangle,$$

122

where $w \in CU$ and $V(w) = h + h_1$, $\|h_1\| \le k_2$. Therefore

$$\varphi_1(\lambda) \le k_0' - k_1\|\lambda\| + k_2\|\lambda\|.$$

Thus for $\|\lambda\| > r$ $\varphi_1(\lambda) < \varphi_1(0) = k_0''$. This means that

$$\sup_{\lambda \in \varphi} \varphi_1(\lambda) = \sup_{\lambda \in \bar{\mathscr{B}}(\theta;r)} \varphi_1(\lambda)$$

$$\mathscr{B}(0\,;r) = \{\lambda \in \mathscr{P} : \|\lambda\| < r\}, \; \bar{\mathscr{B}}(0\,;r) \text{ is a closure of } \mathscr{B}(0\,;r).$$

But ball $\bar{\mathscr{B}}(0\,;r)$ is weakly compact and since $\varphi_1(\cdot)$ is weakly upper semicontinuous then there exists $\hat{\lambda}$ such that

$$\varphi_1(\hat{\lambda}) = \max_{\lambda \in \mathscr{P}} \varphi_1(\lambda)$$

and $\|\hat{\lambda}\| \le r$. $\square$

LEMMA 2.11. *If we assume that*

1. *Set CU is convex and weakly compact,*
2. *There exists $\hat{\lambda}$ such that $\varphi_1(\hat{\lambda}) = max_{\lambda \in \mathscr{P}} \varphi_1(\lambda)$,*
3. *$\hat{\lambda} \in \mathscr{P}_0 \subset \mathscr{P}$, where $\mathscr{P}_0$ is an open and convex set, and $\forall \lambda \in \mathscr{P}_0$ the functional $L(\cdot, \lambda)$ is strictly quasi-convex on CU,*

then assumption 3 of Theorem 2.16 is satisfied and φ_1 is weakly differentiable on $\mathscr{P}_0$.

The simple proof of the above lemma is omitted.

Remark C. If Q is strictly convex on CU, and F and P are affine mappings, then $L(\cdot, \lambda)$ is strictly convex (and thus strictly quasi-convex) on CU for all $\lambda \in \mathscr{P}$.

When $L(\cdot, \lambda)$ is convex on CU for $\lambda \in \mathscr{P}_0$ but is not strictly convex, we can make the local performance criteria in LP$_i$ strictly convex by adding to them "small" terms $\varepsilon\|w_i\|^2$ ($\varepsilon > 0$). It can be shown then, that by using this modification we obtain an auxiliary optimization problem to which IBM is applicable. Moreover, for sufficiently small ε the solution of this auxiliary problem is very close to a solution of OP. It is widely known (Hestenes 1969, Polyak and Tretyakov 1973, Bertsekas 1976a) that in some cases the desirable convexity properties of a modified performance may be obtained by adding the following term to Lagrangian 2.51:

$$\rho\langle V(w), V(w)\rangle, \qquad \rho > 0 \tag{2.61}$$

to make it an augmented Lagrangian. However, such modification cannot be directly used in IBM because it is not decomposable; if we have to use an augmented Langrangian then methods different from IBM must be applied. These methods will be considered in detail in section 2.5.

The sufficient conditions of applicability of IBM (mainly the conditions that guarantee the uniqueness of the IP solutions) are given in general, more or less implicit forms. Lemma 2.11 together with Remark C gives of course the concrete but restrictive conditions (F is assumed to be affine). However, it is well known from practice that with IBM we can solve nonlinear problems. For a certain class of these problems we can formulate more concrete applicability conditions:

Let us assume, that performance index Q and system operator V (see Eqs. (2.52) and (2.53)) have the following form (with appropriate separability properties):

$$Q(w) = \langle w, Q^I w \rangle + \langle q, w \rangle \qquad (2.62)$$

$$V(w) = B(w, w) + Aw + a = 0, \qquad (2.63)$$

where $Q^I : \mathscr{C} \times \mathscr{U} \to \mathscr{C} \times \mathscr{U}$, $q \in \mathscr{C} \times \mathscr{U}$, and $B : (\mathscr{C} \times \mathscr{U}) \times (\mathscr{C} \times \mathscr{U}) \to \mathscr{P}$; B is a continuous bilinear mapping, and $A : \mathscr{C} \times \mathscr{U} \to \mathscr{P}$ is a continuous linear operator, $a \in \mathscr{P}$.

THEOREM 2.17. *If we assume that*

1. $d_1 \|w\|^2 \le \langle w, Q^I w \rangle \le d_2 \|w\|^2$, $d_2 \ge d_1 > 0$,
2. *Set CU is closed and convex,*
3. *There exist constants $k_1, k_2, k_3 \in \mathbb{R}$, $k_1 > k_2 \ge 0$, $k_3 > 0$, such that for all $h \in \mathscr{P}$, $\|h\| = k_1$ there exist $h_1 \in \mathscr{P}$, $\|h_1\| \le k_2$ and $w \in CU \cap \bar{\mathscr{B}}(0; k_3)$ such that*

$$B(w, w) + Aw + a = h + h_1;$$

in other words, set $V(CU \cap \bar{\mathscr{B}}(0; k_3))$ contains a k_2—net of sphere $\mathscr{S}(0; k_1) = \{h : \|h\| = k_1\}$,

4. $\|B(w, w)\| \le d_3 \|w\|^2$,

5. $k_0 = \dfrac{d_2(k_3)^2 + k_3\|q\| + \dfrac{1}{4d_1}\|q\|^2}{k_1 - k_2} < \dfrac{d_1}{d_3}$,

then IBM is applicable to the system optimization problem (where Q and V are given by Eqs. (2.62) and (2.63)). Moreover, $\|\hat{\lambda}\| \le k_0$, where $\hat{\lambda}$ is a solution of CP. $\square$

The proof is given in Malinowski (1977) together with the simple examples of applications of the theorem.

Assumption 5 of the above theorem plays an essential role and restricts the nonlinearity of operator B. In some cases we can avoid (or relax) this assumption. For example, the following result may be obtained.

LEMMA 2.12. *If we assume that*

1. $\mathscr{C} \times \mathscr{U} = \mathbb{R}^n$, $CU = \mathscr{C} \times \mathscr{U}$,
2. *Assumptions 1, 3 and 4 of Theorem 2.17 are in effect,*
3. *For every $\lambda \in \mathscr{P}$, $\|\lambda\| \le k_0$ (where k_0 is given by formula from assumption 5 of Theorem 2.17) such that the quadratic term of $L(\cdot, \lambda)$ is positively semidefinite, the following condition holds:*

$$\text{if } \ker\left(\langle \cdot, Q^I(\cdot)\rangle + \langle \lambda, B(\cdot, \cdot)\rangle\right) \ne \{0\}$$
$$\text{then } \ker\left(\langle \cdot, Q^I(\cdot)\rangle + \langle \lambda, B(\cdot, \cdot)\rangle\right) \backslash \ker(q + A^*\lambda) \ne \{0\},$$

then IBM *is applicable to the system optimization problem* OP *with Q and V given by Eqs. (2.62) and (2.63), and $\|\hat{\lambda}\| \le k_0$ (where $\hat{\lambda}$ is a solution of* CP*).*

The proof is given in Malinowski (1977) with applications of the lemma.

To summarize, in this section the conditions were given under which the price mechanism in the form of IBM can be used for system optimization problems. The most important practical condition is the uniqueness of IP solutions for λ belonging to some appropriate subset $\mathscr{P}_0$ of $\mathscr{P}$.

2.4.3. COORDINATION ALGORITHMS

If IBM is applicable to OP, then to find $\hat{\lambda}$, a solution of CP, we must have appropriate algorithms or coordination strategies; at the same time, the local decision units have to use appropriate, efficient, classical numerical algorithms for solving local problems LP_i. In this section we will only consider the coordination algorithms for solving CP. Following the results of the previous section, we make the practical assumption that $\forall \lambda \in \mathscr{P}_0 \subset \mathscr{P}$, the solution of IP is unique. As before, we will omit z in all mappings. We assume that all assumptions hold uniformly for every z of interest.

From Theorems 2.15 and 2.16 it follows that if IBM is applicable (and this is assumed in this section), then $\hat{\lambda} = \arg \max_{\lambda \in \mathscr{P}} \varphi_1(\lambda)$. From the assumption that $\hat{w}(\lambda)$ is unique for $\lambda \in \mathscr{P}_0$, it follows that $\nabla \varphi_1(\lambda) = V(\hat{w}(\lambda))$ $\forall \lambda \in \mathscr{P}_0$ (see Lemma 2.9). Therefore, we can use the gradient numerical optimization procedures for solving CP; e.g., the steepest ascent or conjugate gradient algorithms may be applied. Such coordination strategies are used very often and—especially when $\mathscr{P}$ is a finite-dimensional space—this approach gives good results. If $\hat{w}(\hat{\lambda})$ is unique (or if only $V(\hat{w}(\hat{\lambda}))$ is unique—see Theorem 2.16) but $\varphi_1(\cdot)$ is not differentiable at some points in an interesting neighborhood $\mathscr{P}_0$ of $\hat{\lambda}$, then we may apply subgradient techniques (Auslender 1970, Shor 1972, and Wolfe 1975) to solve CP. Usually, especially when $\mathscr{P}$ is an infinite-dimensional space, it is difficult to prove the convergence of the subgradient algorithms, unless a special choice of step length is made in the subsequent iterations (the small-step step method). With this choice of a step length, the algorithms are known to converge slowly. We will not discuss the subgradient procedures here but will retain our assumption, justified in practice, that $\hat{w}(\lambda)$ is unique for all λ in $\mathscr{P}_0$, where the set $\mathscr{P}_0$ is such that $\hat{\lambda} \in \operatorname{int} \mathscr{P}_0$. It should be noted that if the optimization procedures are used by the coordinator, then the values of $\varphi_1(\lambda)$ become the main information for the coordinator. Therefore, these values have to be sent to the coordinator from the local decision units along with the values of $V_i(\hat{w}_i(\lambda))$. This is not an essential drawback, but there are two important difficulties connected with the application of optimization procedures as coordination algorithms.

First, convergence properties of the coordination strategies depend on whether $\mathscr{P}$ is a finite-dimensional space or an infinite-dimensional space. If $\mathscr{P}$ is a finite-dimensional space, then the uniqueness of $\hat{w}(\lambda)$ and concavity of $\varphi_1(\cdot)$ imply that $\varphi_1(\cdot)$ is Fréchet differentiable. This follows from Lemma 2.9 and the well-known fact (Rockafellar 1970) that a concave or convex function that is Gâteaux differentiable on an open set in $\mathbb{R}^n$ is also Fréchet differentiable on that set. In most cases (e.g., under assumption 1 of Theorem 2.16), the uniqueness of $\hat{w}(\lambda)\forall \lambda \in \mathscr{P}_0 \subset \mathbb{R}^n$ implies the continuity of $V(\hat{w}(\cdot))$ on $\mathscr{P}_0$. From this we can prove the convergence of the steepest ascent algorithm (applied as a coordination strategy) if the level sets of $\varphi_1(\cdot)$ are compact (which is true, for example, under the assumptions of Lemma 2.10). In many interesting cases the dual function $\varphi_1(\cdot)$ is not differentiable twice, and it is therefore difficult to prove the convergence of a conjugate gradient algorithm, for example. In most cases, though, this algorithm gives good practical results. If $\mathscr{P}$ is an infinite-dimensional space, then it is difficult to obtain and prove the strong convergence of maximizing sequences produced by the numerical procedures for dual function maximization. The assumptions that are necessary to demonstrate such convergence allow also for proving the convergence of more general coordination algorithms which will be examined later in this section.

Second, in some situations we want to use the coordination algorithms not only for solving CP, but also as an *on-line* coordination strategy in various control structures (see Chapters 3 and 4). In those cases we often do not want the coordination algorithm to solve the equation

$$W(\lambda) = 0 \tag{2.64}$$

where

$$W(\lambda) \triangleq - V(\hat{w}(\lambda)), \tag{2.65}$$

but to solve the modified equation

$$T(\lambda) = 0 \tag{2.66}$$

where

$$T(\lambda) = W(\lambda) + s(\lambda), \qquad s : \mathscr{P} \to \mathscr{P}. \tag{2.67}$$

The meaning of function $s(\lambda)$ will be explained in Chapter 3; if CP is being solved, then $s(\lambda) \equiv 0$ and algorithm (2.68) below reduces to a gradient algorithm that performs, of course, the maximization of the dual function. Also, in many cases we want the value of an appropriate norm of $W(\lambda^{(n)})$, or $T(\lambda^{(n)})$, to be decreased for each subsequent value of $\lambda^{(n)}$ generated by the coordination algorithm.

Because of the above reasons, but mainly the second one, we will present and examine one possible coordination algorithm for solving Eq. 2.66 in set $\mathcal{P}_0$. Let us consider the following algorithm, which generates a sequence $\{\lambda^{(n)}\}$ of values of the coordination variable λ:

$$\lambda^{(n+1)} = \lambda^{(n)} - \varepsilon_n A T(\lambda^{(n)}), \tag{2.68}$$

where $\varepsilon_n > 0$, and $A : \mathcal{P} \to \mathcal{P}$ is a linear continuous operator. When we use algorithm (2.68) to solve Eq. (2.64), we should set $s(\lambda) \equiv 0$. The following lemma is of basic importance.

LEMMA 2.13.　*Assume that*

1.　$\forall \lambda, \lambda + h \in \mathcal{P}_1 \subset \mathcal{P}_0$ *the following conditions are satisfied:*
    (i)　$\|W(\lambda + h) - W(\lambda)\| \le \sigma_I \|h\|$,

    (ii)　$\langle W(\lambda + h) - W(\lambda), h \rangle \ge \sigma \|h\|^2$ *(of course $\sigma_I \ge \sigma$),*

    (iii)　$\|s(\lambda + h) - s(\lambda)\| \le \sigma_{II} \|h\|$.

2.　*Operator A in Eq. 2.68 is self-conjugated and $\forall \lambda \in \mathcal{P}$*

$$\mu_1 \|\lambda\|^2 \le \langle \lambda, A\lambda \rangle \le \mu_2 \|\lambda\|^2, \qquad \mu_1 > 0$$

In other words, A is strongly positive definite.

3.　$\sigma\mu_1 > \sigma_{II}\mu_2$; *this assumption is always satisfied when $s(\lambda) \equiv \theta$.*

Let us define a new norm $\|\cdot\|_A$ in $\mathcal{P}$:

$$\|\lambda\|_A \triangleq \sqrt{\langle \lambda, A\lambda \rangle}$$

There exists, then, a nonempty set $[q_0, 1) \subset \mathbb{R}_+$ such that for any $q \in [q_0, 1)$ there exists set $[\varepsilon'_q, \varepsilon''_q] \subset \mathbb{R}_+$ such that if $\varepsilon_n \in [\varepsilon'_q, \varepsilon''_q]$ and $\lambda^{(n)}, \lambda^{(n+1)}$ given by Eq. (2.68) belong to $\mathcal{P}_1$, then

$$\|T(\lambda^{(n+1)})\|_A \le q \|T(\lambda^{(n)})\|_A. \tag{2.69}$$

The proof is given in Appendix A, section A.2. Lemma 2.13 can be used for proving the convergence of algorithm (2.68).

THEOREM 2.18.　*Suppose that the assumptions of Lemma 2.13 are fulfilled and that the starting point $\lambda^{(0)}$ for algorithm (2.68) has been chosen in such a* *way that for some* $q \in [q_0, 1)$ *we have*

$$\bar{\mathscr{B}}_1 = \bar{\mathscr{B}}(\lambda^{(0)}; \ \varepsilon''_q \mu_2 \frac{1}{1-q} \|T(\lambda^{(0)})\|) \subset \mathscr{P}_1.$$

The sequence $\{\lambda^{(n)}\}_{n=0}^{\infty}$ *generated by algorithm* (2.68) *belongs then to* $\bar{\mathscr{B}}_1$ *and converges (in norm topology) to* $\tilde{\lambda}$, *a unique solution of Eq.* (2.66) *in* $\mathscr{P}_1$.

THEOREM 2.19.   *Suppose that the assumptions of Lemma 2.13 are satisfied, there exists coordinating price* $\tilde{\lambda}$ *for which* $T(\tilde{\lambda}) = \theta$, *and the starting point* $\lambda^{(0)}$ *for algorithm* (2.68) *has been chosen in such a way that for some* $q \in [q_0, 1)$ *we have*

$$\bar{\mathscr{B}}_2 \triangleq \bar{\mathscr{B}}\left(\tilde{\lambda}; \ \left[\max \left(\frac{q}{\delta} \sqrt{\frac{\mu_2}{\mu_1}}, 1 \right) + \varepsilon''_q \mu_2 \right] \|T(\lambda^{(0)})\| \right) \subset \mathscr{P}_1$$

(where $\delta = \sigma - \sigma_{II}$*).*
  Then the sequence $\{\lambda^{(n)}\}_{n=0}^{\infty}$ *generated by algorithm* (2.68) *belongs to* $\bar{\mathscr{B}}_2$ *and converges (in norm topology) to* $\tilde{\lambda}$ *(and* $\tilde{\lambda}$ *is a unique solution of* (2.66) *in* $\mathscr{P}_1$*).* $\square$

The proof of Theorem 2.18 is given in Appendix A, section A.3. The proof of Theorem 2.19 is left to the reader.

Remark D.   In the assumptions of Theorem 2.18 the existence of $\tilde{\lambda}$ is not explicitly required. However, the existence of $\tilde{\lambda}$ is implicitly required to be able to make the assumption that $\bar{\mathscr{B}}_1 \subset \mathscr{P}_1$. In the most important case, when $s(\lambda) \equiv 0$, we have $\tilde{\lambda} = \hat{\lambda}$ and the results of the previous section, which concentrate on the conditions that guarantee the existence of $\hat{\lambda}$, may be used. These results have a global character. The above convergence results, are, however, only local.
  The above results reveal an appealing feature of the proposed algorithm: in each iteration, the value of $\|T(\lambda)\|_A$, or $\|W(\lambda)\|_A$, is reduced (see Eq. (2.69)). Therefore, we can estimate before we begin, if we know the appropriate constants, the number of steps (iterations) required to reduce $\|T(\lambda)\|$ beyond a desired threshold.

Now we will establish some conditions that guarantee the fulfillment of assumptions 1(i) and 1(ii) of Lemma 2.13.

LEMMA 2.14. *If we assume that*

  1.   *There exists a convex set* $\mathscr{P}_1 \subset \mathscr{P}_0$, *such that* $\forall \lambda \in \mathscr{P}_1$ *the solution of* IP $\hat{w}(\lambda)$ *belongs to a convex set* $D \subset CU$.

2. *Mapping $V(\cdot)$ is Lipschitz continuous on D, i.e.,*

$$\forall w^1, w^2 \in D : \|V(w^1) - V(w^2)\| \le \sigma_2 \|w^1 - w^2\|,$$

3. $\forall \lambda^1, \lambda^2 \in \mathscr{P}_1 \quad \|\hat{w}(\lambda^1) - \hat{w}(\lambda^2)\| \ge \sigma_1 \|\lambda^1 - \lambda^2\|, \quad \sigma_1 > 0.$

4. *Lagrangian $L(\cdot, \cdot)$ satisfies the following condition:*

$$\forall \lambda \in \mathscr{P}_1, \quad \forall w^1, w^2 \in D, \quad \text{and} \quad \forall \rho \in [0, 1]$$

$$\rho L(w^1, \lambda) + (1 - \rho) L(w^2, \lambda) \ge L(\rho w^1 + (1 - \rho) w^2, \lambda)$$
$$+ \sigma_3 \rho (1 - \rho) \|w^1 - w^2\|^2, \quad \sigma_3 > 0$$

(This means that $L(\cdot, \lambda)$ is strongly convex on D uniformly in $\lambda \in \mathscr{P}_1$),

then assumptions 1(i) and 1(ii) of Lemma 2.13 are satisfied on $\mathscr{P}_1$ with constants σ_I, σ $(\sigma > 0)$.

The proof is given in Appendix A, section A.4.

Fulfillment of condition 1(iii) of Lemma 2.13 has to be examined whenever a nonzero term $s(\lambda)$ is introduced.

In Lemma 2.14, the essential assumption is the third one. In some cases, we may formulate the conditions under which this assumption is satisfied.

LEMMA 2.15.   *If we assume that*

1.   *Set $S = S_1 \times \cdots \times S_N$, where $CU = \{(c, u) \in \mathscr{C} \times \mathscr{U} : G(c, u) \in S\}$ (see Eq. (2.9)) is a closed, convex cone in $\mathscr{S}$, G is concave with respect to $\mathscr{S}$ on $\mathscr{C} \times \mathscr{U}$, and int $S \ne \varnothing$,*

2.   *Performance function Q and mappings V and G are twice continuously differentiable (with bounded second derivative) on a convex open set $D_0 \subset \mathscr{C} \times \mathscr{U}$,*

3. $\forall \lambda \in \mathscr{P}_1$, *the solution $\hat{w}(\lambda) \in D$, where $D \subset D_0$ and D is a convex set,*

4. $\forall w \in D, \forall \lambda \in \mathscr{P}_1, \forall x \in \mathscr{C} \times \mathscr{U}$

$$\mu_3 \|x\|^2 \le \langle x, L''_{ww}(w, \lambda)x \rangle \le \mu_4 \|x\|^2, \qquad \mu_3 > 0,$$

5. $\forall w \in D, \forall (v_1, v_2) \in \mathscr{U} \times (-S^*)$ *the following condition is satisfied:*

$$\|[V'_w(w)]^* v_1 + [G'_w(w)]^* v_2\| \ge \nu \|v_1\|, \qquad \nu > 0$$
$$\text{and} \quad \|[G'_w(w)]^* v_2\| \ge \nu_1 \|v_1\|, \qquad \nu_1 > 0. \quad (2.70)$$

then the assumptions of Lemma 2.13 are satisfied on $\mathscr{P}_1$ with constants σ_I, σ $(\sigma > 0)$ (if σ_{II} is sufficiently small).

The proof is given in Appendix A, section A.5.

Remark E.   If CU is given in the form (for $\mathscr{S} = \mathbb{R}^{n_s}$):

$$CU = \{(c, u) \in \mathscr{C} \times \mathscr{U} : g_j(c, u) \le 0, j = 1, \cdots, n_s\}, \qquad (2.71)$$

where $g_j : \mathscr{C} \times \mathscr{U} \to \mathbb{R}$, then in condition (2.70) it is enough to consider only those constraints g_j that are active in IP for $\lambda \in \mathscr{P}_1$.

Let us now consider a case of system optimization. Suppose that we have a problem with a quadratic (strongly convex) performance index, linear system equations, and convex set CU. In such a case, the restrictions on set $\mathscr{P}_1$ are imposed only because of assumption 3 of Lemma 2.14. It is possible, of course, to satisfy this assumption on the whole space $\mathscr{P}$, but only if set CU is unbounded. If CU is bounded, then assumption 3 can be satisfied only on some bounded subset of $\mathscr{P}$. In a general case, this assumption may be given the following rather rough but practical interpretation: from Eq. (2.70) it follows, that the number of active constraints in IP cannot be too large. If $\mathscr{C} \times \mathscr{U} = \mathbb{R}^{n_c} \times \mathbb{R}^{n_u}$ and $\mathscr{S} = \mathbb{R}^{n_s}$, then this number should not be greater than n_c.

In this section we have considered algorithms which may be used as coordination strategies (for solving the CP of IBM). The detailed analysis was done for the very important algorithm (2.68). The conditions which guarantee the convergence of this algorithm are not very easy to satisfy. Nevertheless, it should be noted that the differentiability of $\hat{w}(\lambda)$ *is not required*; nor is it necessary to fulfill the strict complementarity condition in $\hat{w}(\lambda)$ with respect to active constraints in IP in a finite-dimensional case. Operator A in algorithm (2.68) may be chosen as $A = I$; a more sophisticated and sometimes more effective choice of A is presented in Appendix A, section A.6. In section 2.6.4 we will present some computational results to illustrate the behavior of the coordination strategies described above.

2.4.4. GLOBAL SYSTEM CONSTRAINTS AND IMPLICIT SUBSYSTEM EQUATIONS

Let us assume now that in addition to the local constraints (3) in OP we have a global system constraint (see section 2.1):

$$r(c, u) = \sum_{i=1}^{N} r_i(c_i, u_i) \leq r_0, \tag{2.72}$$

where

$$r(c, u), r_0 \in \mathbb{R}^{n_r}.$$

It is possible in such a case to apply IBM to the system optimization problem if instead of Lagrangian (2.51) we use the following extended Lagrangian:

$$L_r(c, u, \lambda, \lambda_r) = L(c, u, \lambda) + \langle \lambda_r, r(c, u) - r_0 \rangle \tag{2.73}$$

where L is given by Eq. (2.51), $\lambda_r \in \mathbb{R}^{n_r}_+$, and z is omitted for convenience. The infimal problem will remain the same as the IP of IBM, with $L(w, \lambda)$ replaced by $L_r(w, \lambda, \lambda_r)$, where $w = (c, u)$, but the coordinator problem will have to be redefined in the following way, where by $\overline{CU}(\lambda, \lambda_r)$ we denote the solution set of the new infimal problem:

CP$_r$: Find $\hat{\lambda} \in \mathscr{P}$, $\hat{\lambda}_r \in \mathbb{R}_+^{n_r}$ such that the set $\overline{CU}(\hat{\lambda}, \hat{\lambda}_r)$ is nonempty and for every $w = (c, u) \in \overline{CU}(\hat{\lambda}, \hat{\lambda}_r)$ we have:

$$V(w) = 0$$
$$r(w) - r_0 \leq 0 \tag{2.74}$$
$$\langle \hat{\lambda}_r, r(w) - r_0 \rangle = 0$$

The applicability conditions of the method may be established (Malinowski 1974) in the same way that the applicability conditions for the basic version of IBM were established in section 2.4.2.

It should be remarked that constraint (2.72) creates less severe problems than equality constraint $V(w) = \theta$. For example, the simplest sufficient conditions that allow $L_r(\cdot, \lambda, \lambda_r)$ to be strictly convex on CU (compare with Remark C) are the strict convexity of Q, linearity (or affinity) of V, and convexity of r on CU.

On the other hand, the coordination strategies will become more complicated when we want to solve CP$_r$: the numerical procedures for maximization of the dual function will have to deal with the constrained problem $(\lambda_r \in \mathbb{R}_+^{n_r})$ and there is a need to suitably modify algorithm (2.68) in order to handle this case.

Very often the subsystem equations are given in implicit form as in Eq. (2.5)

$$y_i = F_i^0(c_i, u_i, y_i), \qquad (y = F^0(c, u, y)) \tag{2.75}$$

instead of the explicit form of Eq. (2.4) that we have been using in this section. We assume of course that $\forall (c_i, u_i) \in CU_i$ $(\forall (c, u) \in CU)$ Eq. (2.75) has a unique solution:

$$y_i = F_i(c_i, u_i), (y = F(c, u)).$$

If we use IBM to solve the system optimization problem, then the necessity to solve Eq. (2.75) for every $(c, u) \in CU$ (these equations are not "cut") may create additional computational difficulties. However, if we use methods presented in section 2.5, where the subsystem equations are "cut" (despite their form), then we can deal with Eq. (2.75) as easily as with Eq. (2.4).

2.5. MIXED METHODS

2.5.1. THE CONCEPT OF THE AUGMENTED LAGRANGIAN

The extensive theoretical research on penalty function methods, which was stimulated by the many possible applications, has led to the equivalent concepts of shifted penalty function (Powell 1969) and augmented Lagrangians (Hestenes 1969). Theoretically, the augmented Lagrangian has proved more convenient and is therefore more often used. The augmented Lagrangian for the optimization problem

minimize $f(x)$

subject to $g(x) = 0$, where $f : \mathcal{X} \to \mathbb{R}$, $g : \mathcal{X} \to \mathcal{G}$,

and $\mathcal{X}$ and $\mathcal{G}$ are Hilbert spaces,

is called the function $L_a : \mathcal{X} \times \mathcal{G}^* \times \mathbb{R}_+ \to \mathbb{R}$

$$L_a(x, \lambda, \rho) \triangleq f(x) + \langle \lambda, g(x) \rangle + \tfrac{1}{2}\rho\|g(x)\|^2,$$

where $\lambda \in \mathcal{G}^*$, and $\rho \in \mathbb{R}_+$ is the penalty coefficient. Functional L_a could appear as an "augmentation" of the functional used in the penalty function method by addition of the linear term $\langle \lambda, g(x) \rangle$. On the other hand, however, functional L_a can be regarded as an "augmentation" of the normal Lagrange functional because of the addition of the quadratic penalty term (to convexify the problem).

The applicability conditions of the methods based on augmented Lagrangians are only slightly less general than those for penalty function methods, and the optimum is achieved by small values of the penalty coefficient; the need to use large coefficients, the most important defect of the penalty function method, is thereby eliminated. But these applicability conditions are much weaker than those of the pure price methods using only Lagrangian multipliers because the quadratic penalty term convexifies the problem. For a large class of nonconvex problems, normal Lagrangians do not yield suitable saddle points, which correspond to the optimal solution of a problem and characterize the optimal conditions for its solution. But because of the convexification, augmented Lagrangians yield saddle points for these nonconvex problems if ρ is chosen appropriately. These saddle points are of local or global nature depending on the properties of the problem. Thus, optimization techniques based on the existence or search for saddle points can be applied for this class of nonconvex problems; note that the applicability of the interaction balance method described in the previous section requires the existence of the saddle point of the normal Lagrangian (Theorem 2.15).

The most important drawback of the augmented Lagrangian approach used for coordination purposes is the loss of full separability of the additively decomposed optimization problems because of the quadratic penalty term. This problem will be presented more thoroughly in the following sections, together with some efficient strategies to handle it.

2.5.2. THE LINEARIZED AUGMENTED LAGRANGIAN METHOD

Throughout this section, the following additively decomposed overall optimization problem will be considered

Find control $\hat{c}$ giving interaction input $\hat{u}$, such that

$$(\hat{c}, \hat{u}) = \arg \min \sum_{i=1}^{N} Q_i(c_i, u_i) \tag{2.76}$$

subject to

$$y_i = F_i(c_i, u_i, z_i), \qquad u_i = H_i y, \qquad (c_i, u_i) \in CU_i, \qquad i \in \overline{1, N}.$$

where $c_i \in \mathscr{C}_i$, $u_i \in \mathscr{U}_i$, $y_i \in \mathscr{Y}_i$; the spaces $\mathscr{C}_i, \mathscr{U}_i, \mathscr{Y}_i$ will be described later. The disturbances z_i, $i \in \overline{1, N}$ are assumed to be constant throughout the optimization and will be omitted for convenience. We will denote $\mathscr{C} = \times_{i=1}^{N} \mathscr{C}_i$, $\mathscr{U} = \times_{i=1}^{N} \mathscr{U}_i$, $\mathscr{Y} = \times_{i=1}^{N} \mathscr{Y}_i$, $F(c, u) = (F_1(c_1, u_1), \ldots, F_N(c_N, u_N))$ and $CU = \times_{i=1}^{N} CU_i$. We will use an approach similar to that used for the interaction balance method (IBM) for problem (2.76), but we will use the following augmented Lagrangian:

$$L_a(c, u_1 \lambda, \rho) \triangleq Q(c, u) + \langle \lambda, u - HF(c, u) \rangle + \tfrac{1}{2}\rho \| u - HF(c, u) \|^2. \tag{2.77}$$

The function (2.77) is not decomposable like a normal Lagrangian because of the nondecomposable quadratic penalty term:

$$\| u - HF(c, u) \|^2 = \| u \|^2 + \| HF(c, u) \|^2 - 2\langle u, HF(c, u) \rangle$$

$$= \sum_{i=1}^{N} (\| u_i \|^2 + \| F_i(c_i, u_i) \|^2) - 2\langle u, HF(c, u) \rangle,$$

and the cross-term $\langle u, HF(c, u) \rangle$ cannot be directly transformed into a summation of "local" terms, each of which is dependent only on (c_i, u_i) for some $i \in \overline{1, N}$. To overcome this difficulty, Stephanopoulos and Westerberg (1975) proposed the following expansion of the cross-term $\langle u, F(c, u) \rangle$ into a series around some point $(\underline{c}, \underline{u}) \in CU$:

$$\langle u, HF(c, u) \rangle \approx \langle \underline{u}, HF(\underline{c}, \underline{u}) \rangle + \langle u, - \underline{u}, HF(\underline{c}, \underline{u}) \rangle + \langle u, HF(\underline{c}, \underline{u}) - HF(\underline{c}, \underline{u}) \rangle$$

$$= -\langle \underline{u}, HF(\underline{c}, \underline{u}) \rangle + \langle u, HF(\underline{c}, \underline{u}) \rangle + \langle \underline{u}, HF(c, u) \rangle. \tag{2.78}$$

Observe that the linearization is done with respect to subsystem functions, not variables. That is justified by the fact that nonlinearities belonging to one subsystem should be taken into account locally while one is solving the local problem corresponding to this subsystem. Denoting by Λ the function

(2.77) with the expansion (2.78), we get

$$\Lambda(c, u, \underline{c}, \underline{u}, \lambda, \rho) \triangleq Q(c, u) + \langle \lambda, u - HF(c, u)\rangle$$
$$+ \tfrac{1}{2}\rho\|u\|^2 + \tfrac{1}{2}\rho\|F(c, u)\|^2$$
$$+ \rho(\langle \underline{u}, HF(\underline{c}, \underline{u})\rangle - \langle u, HF(\underline{c}, \underline{u})\rangle - \langle \underline{u}, HF(c, u)\rangle)$$

$$= \sum_{i=1}^{N} \left[Q_i(c_i, u_i) + \langle \lambda_i, u_i\rangle - \sum_{j=1}^{N} \langle \lambda_j, H_{ji}F_i(c_i, u_i)\rangle \right.$$
$$+ \tfrac{1}{2}\rho\|u_i\|^2 + \tfrac{1}{2}\rho\|F_i(c_i, u_i)\|^2 + \rho(\langle \underline{u}_i, H_iF(\underline{c}, \underline{u})\rangle$$
$$\left. - \langle u_i, H_iF(\underline{c}, \underline{u})\rangle - \sum_{j=1}^{N} \langle \underline{u}_j, H_{ji}F_i(c_i, u_i)\rangle\right]$$

$$= \sum_{i=1}^{N} \Lambda_i(c_i, u_i, \underline{c}, \underline{u}, \lambda, \rho). \tag{2.79}$$

Thus we can approximate augmented Lagrangian (2.77) by the function (2.79) in some neighborhood of a point $(\underline{c}, \underline{u})$, and function (2.79) is decomposable like a normal Lagrangian in the interaction balance method.

To present properly the two-level optimization method based on Eq. (2.79), together with its informational structure, the functions Λ_i should be reformulated as follows

$$\Lambda_i(c_i, u_i, \underline{c}, \underline{u}, \lambda, \rho) = \sum_{i=1}^{N} \left[Q_i(c_i, u_i) + \langle \lambda_i, u_i\rangle - \sum_{j=1}^{N} \langle \lambda_j, H_{ji}F_i(c_i, u_i)\rangle \right.$$
$$+ \tfrac{1}{2}\rho\|u_i\|^2 + \tfrac{1}{2}\rho\|F_i(c_i, u_i)\|^2 + \rho(\langle \underline{u}_i, H_i\underline{y}\rangle$$
$$\left. - \langle u_i, H_i\underline{y}\rangle - \sum_{j=1}^{N} \langle \underline{u}_j, H_{ji}F_i(c_i, u_i)\rangle\right]$$
$$= \Lambda_{ai}(c_i, u_i, \underline{u}_i, \underline{y}, \lambda, \rho), \quad \text{where} \quad \underline{y} \triangleq F(\underline{c}, \underline{u}).$$

Each local function i depends only on the i-th subsystem model (Q_i, F_i), and information concerning other subsystems consists of the outputs of the whole system $\underline{y} = F(\underline{c}, \underline{u})$ evaluated at the expansion point $(\underline{c}, \underline{u})$; or, more precisely, only those components $H_i\underline{y}$ of $\underline{y}$ that are connected to the inputs of the subsystem.

The algorithm is a two-level one; the lower level or infimal problem consists of the following *local decision problems* (LP_{ai}^k):

For given values $\underline{u}_i^k, \underline{y}^k, \lambda^k$, and ρ find both control and interaction

$$(c^{k+1}, u^{k+1}) = \arg\min_{CU_i} \Lambda_{ai}(\cdot, \cdot, \underline{u}_i^k, \underline{y}^k, \lambda^k, \rho). \tag{2.80}$$

The main job of the *upper level* or *coordinator* is to adjust the multipliers according to the formula

$$\lambda_i^{k+1} = \lambda_i^k + \tfrac{1}{2}\rho(u_i^k - H_i y^k), \quad y^k = F(c^k, u^k), \quad i \in \overline{1, N}. \tag{2.81}$$

Formula (2.81), similar to the multiplier rule of Hestenes and Powell, was proposed and justified by Stoilov (1977). Without a delay, formula (2.81) would be $\lambda^{k+1} = \lambda^k + \tfrac{1}{2}\rho(u^{k+1} + Hy^{k+1})$.

Formula (2.81) is efficient only for sufficiently large values of penalty coefficient ρ, which will be shown by the theorem given below. To prove this theorem, the following form of the constraint sets CU_i will be assumed

$$CU_i = \{(c_i, u_i) \in \mathscr{C}_i \times \mathscr{U}_i : G_i(c_i, u_i) \leqslant 0\}, \quad G_i : \mathscr{C}_i \times \mathscr{U}_i \to \mathbb{R}^{n_i}, \quad i \in \overline{1, N}. \tag{2.82}$$

The constraints (2.82) are in the form of Eq. (2.9), with finite dimensional spaces $\mathscr{G}_i = \mathbb{R}^{n_i}$ and closed negative orthants as cones S_i.

Recall that for the optimal point $(\hat{c}, \hat{u})$, $\hat{u} = H\hat{y}$, of problem (2.76) *second order sufficient optimality conditions with strict complementarity* are satisfied if (a) first order necessary conditions are satisfied at $(\hat{c}, \hat{u})$, i.e., there are multipliers $\hat{\lambda} = (\hat{\lambda}_1, \ldots, \hat{\lambda}_N) \in \mathscr{Y}^*$, $\hat{\eta} = (\hat{\eta}_1, \ldots, \hat{\eta}_N) \in \mathbb{R}_+^n = \times_{i=1}^N \mathbb{R}_+^{n_i}$ such that

$$L'_{(c,u)}(\hat{c}, \hat{u}, \hat{\lambda}, \hat{\eta}) = 0,$$

$$\langle \hat{\eta}_i, G_i(\hat{c}_i, \hat{u}_i) \rangle = 0, \quad i \in \overline{1, N},$$

where $L(c, u, \lambda, \eta) = Q(c, u) + \sum_{i=1}^N (\langle \lambda_i, u_i - H_i F(c_i, u_i) \rangle + \langle \eta_i, G_i(c_i, u_i) \rangle)$ is the normal Lagrange function; (b) Q_i, F_i, G_i, and hence the functional $L(\cdot, \cdot, \lambda, \eta)$ are continuously differentiable twice in some neighborhood of the optimum, and there exists $\delta > 0$ such that

$$\langle L''_{(c,u)(c,u)}(\hat{c}, \hat{u}, \hat{\lambda}, \hat{\eta})(\bar{c}, \bar{u}), (\bar{c}, \bar{u}) \rangle \geqslant \delta \|(\bar{c}, \bar{u})\|^2$$

for every point $(\bar{c}, \bar{u}) \in \mathscr{C} \times \mathscr{U}$ satisfying

$$\bar{u}_i - H_i(F_i)_{(c_i, u_i)}(\hat{c}_i, \hat{u}_i)(\bar{c}, \bar{u}) = 0, \quad i \in \overline{1, N},$$

$$(G_{ij})'_{(c_i, u_i)}(\hat{c}_i, \hat{u}_i)(\bar{c}_i, \bar{u}_i) = 0, \qquad j \in A_i = \{j \in \overline{1, n_i} : G_{ij}(\hat{c}, \hat{u}) = 0\}, \qquad i \in \overline{1, N};$$

and (c) for every $j \in A_i$ and $i \in \overline{1, N}$, $\hat{\eta}_{ij} > 0$, i.e., all multipliers corresponding to active constraints are positive (the strict complementarity condition).

THEOREM 2.20. (*convergence theorem*). *If we assume that*

1. *Spaces $\mathscr{C} = \times_{i=1}^N \mathscr{C}_i$, $\mathscr{U} = \times_{i=1}^N \mathscr{U}_i$ are finite dimensional,*
2. *Functions Q_i, F_i, G_i are continuously differentiable twice, $i \in \overline{1, N}$,*

3. *Gradients of the active through all constraints (i.e., $u_i - HF_i(c, u) = 0$ and $G_i(c_i, u_i) \leq 0$, $i \in \overline{1, N}$) are linearly independent at the optimal point $(\hat{c}, \hat{u})$ of the initial problem* (2.76),

4. *Second order sufficient optimality conditions with strict complementarity are satisfied at $(\hat{c}, \hat{u})$,*

5. *For every sufficiently great ρ and any point (c^k, u^k, λ^k) from some neighborhood of $(\hat{c}, \hat{u}, \hat{\lambda})$, there exist solutions of problems (LP_{ai}^k), $i \in \overline{1, N}$,*

then there exists $\hat{\rho} \in \mathbb{R}_+$ such that for every $\rho \geq \hat{\rho}$ the algorithm (2.80), (2.81) *is linearly convergent to the optimal point.*

The proof is omitted here, but it can be found in Stoilov (1977).

Theorem 2.20 was proved for the finite-dimensional case, and is probably true for the more general case as well.

Based on the results of Theorem 2.20, the coordinator should adjust multipliers according to rule (2.81) and ensure that the penalty coefficients are sufficiently great. This can be done by the following algorithm (2.83):

1. Set initial values $k = 1$, c^1, u^1, y^1, λ^1, ρ^1, δ, ε_1, ε_2, ε_3.
2. Evaluate (c_i^{k+1}, u_i^{k+1}), $i \in \overline{1, N}$, as a solution of local problems (LP_{ai}^k); denote $y_i^{k+1} = F_i(c_i^{k+1}, u_i^{k+1})$, $i \in \overline{1, N}$.
3. If

$$(\|c^{k+1} - c^k\| \leq \varepsilon_1) \wedge (\|u^{k+1} - u^k\| \leq \varepsilon_2) \wedge (\|u^{k+1} - Hy^{k+1}\| \leq \varepsilon_3)$$

then stop.
4. Calculate

$$\lambda_i^{k+1} = \lambda_i^k + \tfrac{1}{2}\rho(u_i^k - H_i y^k), \qquad i \in \overline{1, N},$$

and

$$p^{k+1} = (\|c^{k+1} - c^k\|^2 + \|u^{k+1} - u^k\|^2 + \|\lambda^{k+1} - \lambda^k\|^2)^{\frac{1}{2}}.$$

5. If $k = 1$ or $p^{k+1} < p^k$, then go to step 6.
5a. If $\rho^k = \rho^{k+1}$ then set $p^{k+1} = \delta p^k$, $k = k + 1$, and go to step 2.
6. Set $\rho^{k+1} = \rho^k$, $k = k + 1$, and go to step 2.

The initial parameters of the algorithm must be arbitrarily chosen from experience. The value of ρ^1 should not be chosen too large because we would like the local decision problems to be numerically well conditioned. If it is too small to achieve convergence, the algorithm will increase it automatically; multiplying by $\delta > 1$, and setting the value of $\delta = 5, \ldots, 10$ is recommended from experience.

2.5.3. THE INPUT PREDICTION AND BALANCE METHOD

The direct coordination mechanism used in the so-called interaction prediction method (see Mesarovič *et al.* 1970) can be combined with the price mechanism used in the interaction balance method. The combination (see, e.g., Smith and Sage 1973, Singh *et al.* 1975, and Singh 1977) will be referred to as the *input prediction and balance method* (IPBM) and can be based on the normal Lagrangian only, as in the literature cited above. However, we will use the more general augmented Lagrangian in order to make IPBM more universally applicable.

IPBM will be presented for the optimization problem (2.76), with disturbances z omitted as before. It will be assumed throughout this section that $\mathscr{C}_i, \mathscr{U}_i, \mathscr{Y}_i$ are real Hilbert spaces. To formulate IPBM, we define the following *infimal (lower-level) problem* (IP_a):

For given input prediction u, multiplier $\lambda \in \mathscr{U}^*$ and penalty coefficient $\rho \in \mathbb{R}_+$ find control

$$\hat{c}(u, \lambda, \rho) = \arg\min_{C(u)} L_a(\cdot, u, \lambda, \rho), \tag{2.84}$$

where

$$C(u) \triangleq \mathop{\times}_{i=1}^{N} (C_i(u) \triangleq \{c_i \in \mathscr{C}_i : (c_i, u_i) \in CU_i\})$$

and

$$L_a(c, u, \lambda, \rho) \triangleq Q(c, u) + \langle \lambda, u - HF(c, u) \rangle + \tfrac{1}{2}\rho \|u - HF(c, u)\|^2$$

is the augmented Lagrangian.

Since

$$L_a(c, u, \lambda, \rho) = \sum_{i=1}^{N} [L_{ai}(c_i, u, \lambda, \rho) = Q_i(c_i, u_i) + \langle \lambda_i, u_i \rangle$$
$$- \sum_{j=1}^{N} \langle \lambda_j, H_{ji}F_i(c_i, u_i) \rangle + \tfrac{1}{2}\rho(\|u_i\|^2 + \|F_i(c_i, u_i)\|^2$$
$$- \sum_{j=1}^{N} \langle u_j, H_{ji}F_i(c_i, u_i) \rangle)],$$

the infimal problem IP_a can be solved as N independent *local problems* (LP_{ai}):

For given input prediction u, multiplier λ, and penalty coefficient ρ, find control

$$\hat{c}_i(u, \lambda, \rho) = \arg\min_{C_i(u)} L_{ai}(\cdot, u, \lambda, \rho). \tag{2.85}$$

Let us denote the set of solutions of IP_a by $\hat{C}(u, \lambda, \rho) \triangleq \times_{i=1}^{N} \hat{C}_i(u, \lambda, \rho)$ and the set of admissible input predictions by

$$U_0 \triangleq \{u \in \mathcal{U}: \ C(u) \neq \varnothing\}. \tag{2.86}$$

Being rigorously precise, we should restrict our attention to the subset $U_r = \pi_u D$ of the set U_0, where

$$D = \{(u, \lambda, \rho) \in \mathcal{U} \times \mathcal{U}^* \times \mathbb{R}_+: \ \hat{C}(u, \lambda, \rho) \neq \varnothing\}.$$

We will, however, equate U_0 with U_r for convenience, since this equality can be ensured by assumptions that are easily satisfied; e.g., when $L(\cdot, u, \lambda, \rho)$ is (weakly) lower semicontinuous and $C(u)$ (weakly) compact. We can now formulate the following *coordinator problem* (CP_a):

Find a saddle point $(\hat{u}; \hat{\lambda})$ of the primal–dual function $\varphi(\cdot, \cdot, \rho)$ on $U_0 \times \mathcal{U}^*$, where

$$\varphi(u, \lambda, \rho) \triangleq \min_{C(u)} L_a(\cdot, u, \lambda, \rho), \tag{2.87}$$

such that the balance condition $\forall \hat{c} \in \hat{C}(\hat{u}, \hat{\lambda}, \rho) \ \hat{u}\text{-}HF(\hat{c}, \hat{u}) = 0$ holds.

The saddle point in CP_a is of course understood as a minimum with respect to primal variable u, and a maximum with respect to dual variable λ, i.e.,

$$\forall (u, \lambda) \in U_0 \times \mathcal{U}^* \qquad \varphi(\hat{u}, \lambda, \rho) \leq \varphi(\hat{u}, \hat{\lambda}, \rho) \leq \varphi(u, \hat{\lambda}, \rho). \tag{2.88}$$

It should be noted that we must require in CP_a the condition $\hat{u}\text{-}HF(\hat{c}, \hat{u}) = 0$ to be true for all $\hat{c} \in \hat{C}(\hat{u}, \hat{\lambda}, \rho)$ because we are not able to distinguish between various $\hat{c} \in \hat{C}(\hat{u}, \hat{\lambda}, \rho)$.

LEMMA 2.16.

    1. *If $(\hat{u}; \hat{\lambda})$ is a solution to CP_a, then for any $\hat{c} \in \hat{C}(\hat{u}, \hat{\lambda}, \rho)$, $(\hat{c}, \hat{u}; \hat{\lambda})$ is a saddle point of $L_a(\cdot, \cdot, \cdot, \rho)$ on $CU \times \mathcal{U}^*$.*

    2. *If $(\hat{c}, \hat{u}; \hat{\lambda})$ is a saddle point of $L_a(\cdot, \cdot, \cdot, \rho)$ on $CU \times \mathcal{U}^*$ then $(\hat{u}; \hat{\lambda})$ is a saddle point of $\varphi(\cdot, \cdot, \rho)$ on $U_0 \times \mathcal{U}^*$ and is also a solution to CP_a if $\forall c \in \hat{C}(\hat{u}, \hat{\lambda}, \rho) \ \hat{u}\text{-}HF(c, \hat{u}) = 0$.*

Proof. (1) Since $\hat{u}\text{-}HF(\hat{c}, \hat{u}) = 0$,

$$\forall \lambda \in \mathcal{U}^* \qquad L_a(\hat{c}, \hat{u}, \lambda, \rho) \leq L_a(\hat{c}, \hat{u}, \hat{\lambda}, \rho).$$

On the other hand, we have

$$(\forall u \in U_0)(\forall \hat{c}(u, \hat{\lambda}, \rho) \in \hat{C}(u, \hat{\lambda}, \rho) \ L_a(\hat{c}, \hat{u}, \hat{\lambda}, \rho) \leq L_a(\hat{c}(u, \hat{\lambda}, \rho), u, \hat{\lambda}, \rho),$$

hence

$$\forall (c, u) \in CU \qquad L_a(\hat{c}, \hat{u}, \hat{\lambda}, \rho) \leq L_a(c, u, \hat{\lambda}, \rho).$$

(2) If $(\hat{c}, \hat{u}; \hat{\lambda})$ is a saddle point of $L_a(\cdot, \cdot, \cdot, \rho)$ on $CU \times \mathcal{U}^*$, then $\hat{u} - HF(\hat{c}, \hat{u}) = 0$, since in another case it would be possible to choose such $\tilde{\lambda} \in \mathcal{U}^*$ that $L_a(\hat{c}, \hat{u}, \tilde{\lambda}, \rho) > L_a(\hat{c}, \hat{u}, \hat{\lambda}, \rho)$. Then the proof follows from the relations

$$\forall \lambda \in \mathcal{U}^* \qquad \varphi(\hat{u}, \lambda, \rho) = L_a(\hat{c}(\hat{u}, \lambda, \rho), \hat{u}, \lambda, \rho) \leq L_a(\hat{c}, \hat{u}, \lambda, \rho) \leq L_a(\hat{c}, \hat{u}, \hat{\lambda}, \rho),$$

$$\forall u \in U_0 \qquad \varphi(\hat{u}, \hat{\lambda}, \rho) = L_a(\hat{c}, \hat{u}, \hat{\lambda}, \rho) \leq L_a(\hat{c}(u, \hat{\lambda}, \rho), u, \hat{\lambda}, \rho) = \varphi(u, \hat{\lambda}, \rho). \quad \square$$

COROLLARY. *If $(\hat{u}; \hat{\lambda})$ is a solution to* CP_a, *then for any point $\hat{c} \in \hat{C}(\hat{u}, \hat{\lambda}, \rho)$, the point $(\hat{c}, \hat{u})$ is a solution to the initial problem* (2.76).

Thus, we will say that IPBM *is applicable* if at least one solution to CP_a exists. The formulation of the coordinator problem in the form CP_a is not a common one for IPBM. Usually, only specific coordination strategies are defined, i.e., a specific algorithm for finding $\hat{u}$ and $\hat{\lambda}$ (see Smith and Sage 1973 and Singh *et al.* 1975). Therefore, it may be that formulation CP_a is too strong—if one can find $\hat{u}$, the optimal input in problem (2.76), some other way, then the existence of a saddle point of only $L_a(\cdot, \hat{u}, \cdot, \rho)$ on $C(\hat{u}) \times \mathcal{U}^*$ satisfying the balance condition would yield the desired solution (and not a saddle point of $\varphi(\cdot, \cdot, \rho)$ on $U_0 \times \mathcal{U}^*$ or, equivalently, from Lemma 2.16, a saddle point of $L_a(\cdot, \cdot, \cdot, \rho)$ on $CU \times \mathcal{U}^*$). However the question of reasonable *coordination strategies* arises immediately. Some of them will be discussed later; but there is no strategy known to the authors that would be able to choose $\hat{u}$ independently of the existence of a saddle point of $L_a(\cdot, \cdot, \cdot, \rho)$. Singh *et al.* (1975) are of the same opinion.

Summarizing the above discussion, we would like to point out that in using the normal Lagrangian (i.e., $\rho = 0$) in IPBM, we require the existence of the saddle point $(\hat{c}, \hat{u}; \hat{\lambda})$ of this normal Lagrangian for its applicability. This implies the applicability of the interaction balance method (IBM-section 2.4) if only the balance condition of this method

$$\forall (\hat{c}, \hat{u}) \in \hat{C}U(\hat{\lambda}) \qquad \hat{u} - HF(\hat{c}, \hat{u}) = 0$$

is satisfied; this condition is only slightly stronger than the one used in CP_a. Observe, however, that IPBM, but not IBM, can be applied to system optimization problems slightly more general than (2.76), namely, with $Q(c, u) = \sum_{i=1}^{N} Q_i(c_i, u)$, and it has lower-dimensional local problems than IBM.

The situation looks quite different when we use the augmented Lagrangian ($\rho > 0$) in IPBM. As was mentioned in section 2.5.1, there is a large class of nonconvex problems for which suitable saddle points $((\hat{c}, \hat{u}; \hat{\lambda})$, where $(\hat{c}, \hat{u})$ is a solution to (2.76) and $\hat{\lambda}$ is a normal Lagrange multiplier) do not exist for normal Lagrangians, but do exist for augmented Lagrangians if ρ is appropriately chosen. Thus there is a large class of problems to which IPBM with $\rho > 0$ is applicable. We will formulate below the conditions under which saddle points exist as solutions to CP_a; these are the applicability conditions of IPBM. We first define the following *quadratic growth condition* (Rockafellar 1974)

$$(\exists \bar{\rho} \in \mathbb{R}_+)(\exists \alpha \in \mathbb{R})(\forall (c, u) \in CU) \qquad L_a(c, u, 0, \bar{\rho}) \geq \alpha,$$

and recall that a point $(\hat{c}, \hat{u})$ is the *unique solution in the strong sense* to the problem (2.76) if for every neighborhood N of $(\hat{c}, \hat{u})$, there exists $\varepsilon > 0$ such that

$$(\forall (c, u) \in CU)\, [(Q(c, u) - Q(\hat{c}, \hat{u}) \leq \varepsilon \wedge \|u - HF(c, u)\|^2 \leq \varepsilon) \;\Rightarrow\; (c, u) \in N].$$

If we assume that the constraint sets CU_i are defined by (2.82), the following theorem can be formulated.

THEOREM 2.21 (*applicability conditions of* IPBM). *If we assume that*

1. At the optimal point $(\hat{c}, \hat{u})$ of the problem (2.76) (with $CU_i \triangleq \{(c_i, u_i) : G_i(c_i, u_i) \leq 0\}$, and $Q(c, u) = \sum_{i=1}^{N} Q_i(c_i, u)$ also admissible) second order sufficient optimality conditions with strict complementarity (for definitions, see section 2.5.2) are satisfied and Fréchet derivative $h'_{(c, u)}(\hat{c}, \hat{u})$ of the operator $h(c, u) \triangleq u - HF(c, u)$ has its image closed in $\mathcal{U}$,
2. Problem (2.76) satisfies the quadratic growth condition and $(\hat{c}, \hat{u})$ is its unique solution in the strong sense,

then there exists $\hat{\rho} \in \mathbb{R}_+$ such that for every $\rho \geq \hat{\rho}$

$$\forall (u, \lambda) \in U_0 \times \mathcal{U}^* \qquad \varphi(\hat{u}, \lambda, \rho) \leq \rho(\hat{u}, \hat{\lambda}, \rho) \leq \varphi(u, \hat{\lambda}, \rho)$$

and the set $\hat{C}(\hat{u}, \hat{\lambda}, \rho)$ consists of a single point $\hat{c}$ ($\hat{\lambda}$ is the normal Lagrange multiplier for the constraint $h(c, u) = 0$).

The proof follows the argument given in Rockafellar (1974) for the finite dimensional case (assumptions about strict complementarity and closed range of $h'(\hat{c}, \hat{u})$ are then superfluous); for Hilbert spaces, the theorem can be proved analogously to the more complex Theorem 2.22 from the next section.

Since $\hat{u} - HF(\hat{c}, \hat{u}) = 0$, Theorem 2.21 states conditions under which a solution of the coordinator problem CP_a exists. One of the most important results of the theorem is the easy manner of "adjusting" the penalty coefficients—they should only be sufficiently large; these sufficiently large values of ρ have been found to be moderate or even small in practice and by no means comparable with the large penalty coefficients needed in the penalty function method described in section 2.3. An aribtrary choice of one fixed value of ρ for the whole optimization process is a common technique that is almost always successful. That is why we treated ρ as a given, fixed parameter in formulating the coordinator problem CP_a. If ρ happens to be too small, which can be easily observed from the rate of decrease of $\|u\text{-}HF(c, u)\|$, it should be increased, e.g., by multiplying by some constant greater than one. Such simple protection can be easily placed in every coordinator algorithm, as it was done in algorithm (2.83) in the previous section.

Algorithms for solving the coordinator problem CP_a are called *coordinator strategies*. They can be roughly divided into two groups: two-level strategies and one-level strategies.

Two-level (double-loop) coordination strategies

For a given λ, the minimum of $\varphi(\cdot, \lambda, \rho)$ is sought and λ is adjusted so as to approach the balance condition, and so forth. Lambda can be adjusted by using, for example, the following step formula

$$\lambda^{k+1} = \lambda^k + \rho[\hat{u}^k - HF(\hat{c}(\hat{u}^k, \lambda^k, \rho), \hat{u}^k)], \tag{2.89}$$

where $\hat{u}^k = \hat{u}(\lambda^k, \rho)$ is a point minimizing $\varphi(\cdot, \lambda^k, \rho)$. To justify (2.89), we rewrite the formula for the gradient of the functional $\varphi(u, \lambda, \rho)$ with respect to u, assuming that $CU = C \times U$, the gradient exists, and using the results of Appendix A, section A.1

$$\varphi'_u(u, \lambda, \rho) = (L_a)'_u(\hat{c}(u, \lambda, \rho), u, \lambda, \rho) = Q'_u(\hat{c}(u, \lambda, \rho), u)$$

$$+ (I - HF'_u(\hat{c}(u, \lambda, \rho), u))^*\lambda + \rho(I - HF'_u(\hat{c}(u, \lambda, \rho), u))^*(u - HF(\hat{c}(u, \lambda, \rho), u))$$

$$= Q'_u(\hat{c}(u, \lambda, \rho), u) + (I\text{-}HF'_u(\hat{c}(u, \lambda, \rho), u))^*(\lambda + \rho(u\text{-}HF(\hat{c}(u, \lambda, \rho), u))).$$

If we could find a fixed point $\bar{\lambda}$ of the adjusting rule (2.89), we would have

$$\bar{u}\text{-}HF(\hat{c}(\bar{u}, \bar{\lambda}, \rho), \bar{u}) = 0,$$

where $\bar{u} = \hat{u}(\bar{\lambda}, \rho)$ is the point minimizing $\varphi(\cdot, \bar{\lambda}, \rho)$. Assuming that $\bar{u} \in \text{int } U$, we would therefore get

$$\varphi'_u(\bar{u}, \bar{\lambda}, \rho) = Q'_u(\hat{c}(\bar{u}, \bar{\lambda}, \rho), \bar{u}) + (I\text{-}HF'_u(\hat{c}(\bar{u}, \bar{\lambda}, \rho), \bar{u}))^*\bar{\lambda} = 0,$$

and $\bar{\lambda}$ would be the normal Lagrange multiplier $\hat{\lambda}$ and $(\hat{c}(\bar{u}, \bar{\lambda}, \rho), \bar{u})$ would be the solution $(\hat{c}, \hat{u})$ of the intial problem since it minimizes $L_a(\cdot, \cdot, \hat{\lambda}, \rho)$ on CU. The argument mentioned above was first presented by Hestenes (1969) for the optimization problem discussed in section 2.5.1 and formula (2.89) is usually called the Hestenes multiplier rule. The restriction $u \in \text{int } U$ was only used to simplify the above discussion and can be easily removed without changing formula (2.89). Moreover, the result can be shown to be true for the general case of the set CU, i.e., the condition $CU = C \times U$ is not necessary. An application of strategy (2.89) can be found in section 2.6.

One-level (single-loop) coordination strategies

The saddle point of $\varphi(\cdot, \cdot, \rho)$ is found by making simultaneous iterations of u and λ: if, in the k-th step, a point $c^k = \hat{c}(u^k, \lambda^k, \rho)$ minimizing the infimal problem IP_a is found, both u^k and λ^k are adjusted using some step formulae, and so on.

The single-loop adjusting formulae for u and λ can be derived from the necessary conditions for a saddle point. If $\varphi(\cdot, \cdot, \rho)$ is differentiable and $CU = C \times \mathcal{U}$, and $\mathcal{U}$ is the whole space, then these conditions are

$$\varphi'_{\lambda_i}(u, \lambda, \rho) = (L_a)'_{\lambda_i}(\hat{c}(u, \lambda, \rho), u, \lambda, \rho) = 0$$

$$\varphi'_{u_i}(u, \lambda, \rho) = (L_a)'_{u_i}(\hat{c}(u, \lambda, \rho), u, \lambda, \rho) = 0, \quad i \in \overline{1, N}. \tag{2.90}$$

Formulae of derivatives of upper-level functions are discussed in Appendix A.

The ordinary gradient strategy would have the form

$$\lambda_i^{k+1} = \lambda_i^k + \varepsilon_{i1}^k \cdot \varphi'_{\lambda_i}(u^k, \lambda^k, \rho), \qquad \varepsilon_{i1}^k > 0, \tag{2.91λ}$$

$$u_i^{k+1} = u_i^k - \varepsilon_{i2}^k \cdot \varphi'_{u_i}(u^k, \lambda^k, \rho), \qquad \varepsilon_{i2}^k > 0, \qquad i \in \overline{1, N}, \tag{2.91u}$$

and the main application difficulty is choosing the step coefficients ε_{i1}^k and ε_{i2}^k; see, e.g., Smith and Sage (1973). Using the multiplier rule of Hestenes, we will derive an efficient version of the above strategy. Recall that

$$L_a(c, u, \lambda, \rho) \triangleq Q(c, u) + \langle \lambda, u\text{-}HF(c, u) \rangle + \tfrac{1}{2}\rho\|u\text{-}HF(c, u)\|^2$$

$$= \sum_{j=1}^{N} \left(Q_j(c_j, u_j) + \langle \lambda_j, u_j - H_j F(c, u) \rangle + \tfrac{1}{2}\rho\|u_j - H_j F(c, u)\|^2 \right)$$

$$= \sum_{j=1}^{N} \left(Q_j(c_j, u_j) + \langle \lambda_j, u_j - \sum_{k=1}^{N} H_{jk} F_k(c_k, u_k) \rangle + \tfrac{1}{2}\rho\left\| u_j - \sum_{k=1}^{N} H_{jk} F_k(c_k u_k) \right\|^2 \right).$$

Therefore, the formula for the derivative with respect to u_i is

$$(L_a)'_{u_i}(c, u, \lambda, \rho) = (Q_i)'_{u_i}(c_i, u_i) + \lambda_i + \rho(u_i - H_i F(c, u))$$

$$- \sum_{j=1}^{N} (H_{ji}(F_i)'_{u_i}(c_i, u_i))^*[\lambda_j + \rho(u_j - H_j F(c, u))], \qquad i \subset 1, N.$$

Demanding $(L_a)'_{u_i}(c, u, \lambda, \rho) = 0$ and using the Hestenes multiplier rule for adjusting λ, we get the following adjusting formulae

$$\lambda_i^{k+1} = \lambda_i^k + \rho(u_i^k - H_i F(c^k, u^k)), \tag{2.92λ}$$

$$u_i^{k+1} = H_i F(c^k, u^k) - \frac{1}{\rho}(Q_i)'_{u_i}(c_i^k, u_i^k) + \lambda_i^k$$
$$- \sum_{j=1}^{N} (H_{ji}(F_i)'_{u_i}(c_i^k, u_i^k))^* \lambda_j^{k+1}), \quad i \in \overline{1, N}. \tag{2.92u}$$

The coordinator strategy (2.92) will be called the *multiplier strategy*. Since

$$(L_a)'_\lambda(c, u, \lambda, \rho) = u - HF(c, u)$$

it can be easily observed that this strategy is a special case of the ordinary gradient with $\varepsilon_{i1}^k = \rho$ and $\varepsilon_{i2}^k = 1/\rho$. The above choice of step coefficients coincides with a known property of augmented Lagrangians: the larger the value of ρ, the better conditioned the maximization, with respect to dual variable λ, and the the worse conditioned the minimization, with respect to primal variable u. In (2.92), the larger the value of ρ is, the greater the stepsize for λ and the smaller the stepsize for u.

Another, more sophisticated algorithm was derived from the necessary conditions for a saddle point (2.90) by Wierzbicki (1976). He built two variable matrices based on appropriate gradients for variables u and λ. Generally, the algorithm requires directional minimization with respect to u for each adjustment of λ, which places it between single- and double-loop strategies.

It should be realized that of the coordination strategies discussed, only the two-level ones are global in nature, i.e., a global saddle point of $\varphi(\cdot, \cdot, \rho)$ can be found by using them. But to ensure the existence of a global saddle point, a global minimization of infimal problem IP_a for every u and λ sent from coordinator, and a global minimization of $\varphi(\cdot, \lambda, \rho)$ in the coordinator must be performed. In another case, the strategies could optimize locally like all single-loop strategies. These strategies seek a saddle point or, more precisely, the point satisfying the necessary conditions for a saddle point, in the neighborhood in which the initial values of the primal and dual variables were chosen. Of course, it does not matter if, besides the global saddle point that exists as a result of the assumptions of Theorem 2.21, there are no other local saddle points of the primal–dual functional $\varphi(\cdot, \cdot, \rho)$. Such a situation is typical if the initial problem does not have local minima. By a local minimum of (2.76), we understand a point $(\tilde{c}, \tilde{u}) \in CU \cap CU^F$ such that for every $(c, u) \in CU \cap CU^F \cap \tilde{N}$ $Q(\tilde{c}, \tilde{u}) \leq Q(c, u)$ and $Q(\tilde{c}, \tilde{u}) > Q(\hat{c}, \hat{u})$, where $(\hat{c}, \hat{u})$ is the global minimum of (2.76), $CU^F \triangleq \{(c, u): u - HF(c, u) = 0\}$ and $\tilde{N}$ denotes some neighborhood of $(\tilde{c}, \tilde{u})$. If problem (2.76) has a local minimum $(\tilde{c}, \tilde{u})$, then this point can be found by every local coordinator strategy. This fact results immediately from Theorem 2.21, which can be formulated for point $(\tilde{c}, \tilde{u})$ instead of $(\hat{c}, \hat{u})$ if the set $CU \cap \tilde{N}$ instead of CU is considered.

The proof of Theorem 2.21 consists of two parts. The first, using only assumptions (1), shows, that there is a neighborhood $\hat{N}$ of $(\hat{c}, \hat{u})$ such that $(\hat{u}; \hat{\lambda})$ is the saddle point of $\varphi(\cdot, \cdot, \varphi)$ if the set $CU \cap \hat{N}$ instead of CU is considered; then, using additional assumptions (2), the saddle point is extended to the whole feasible set. Recall that all the valuable ordinary optimization procedures, such as conjugate gradients or variable metric methods, are, as a rule, local procedures.

The necessary conditions for a saddle point, Eq. (2.90), and the resulting strategies (2.91) and (2.92) were formulated for the case when $CU = C \times \mathcal{U}$. If $CU = C \times U$ then formulae (2.90) remain valid for derivatives of $\varphi(\cdot, \cdot, \rho)$. Only the necessary condition $\varphi'_u(u, \lambda, \rho) = 0$ and the adjusting rule (2.92u) resulting from it must be appropriately modified to take the constraints given by the set U into account; e.g., when these constraints are linear we can use the projected gradient. The situation is more difficult if $CU \neq C \times U$, i.e., if the lower-level constraint set $C(u)$ really depends on u. The existence conditions of the gradient $\varphi'_u(\cdot, \cdot, \rho)$ are then stronger and the formula describing it is more complicated (see Appendix A for formulae for derivatives). Moreover, set U_0, Eq. (2.86), can be extremely difficult to identify (compare with the identical set V'_0 in the penalty function method, section 2.3.1). Formulae for adjusting u in single-loop strategies are then difficult to derive, since U_0 cannot be violated (the set $C(u)$ cannot be empty; in the previous case when $CU = C \times U$ we always have $C(u) = C \neq \varnothing$).

When the global constraint in the form of Eq. (2.16)

$$r(c, u) = \sum_{i=1}^{N} r_i(c_i, u_i) \leq r_o, \qquad r_0 \in \mathbb{R}^{n_r}, \tag{2.93}$$

is added to the initial problem (2.76), it may be handled by using IPBM. Constraint (2.93) should then be reformulated as constraint (2.44) was reformulated in PFM (section 2.3):

$$r_i(c_i, u_i) = r_{d_i}, \qquad i \in \overline{1, N}, \qquad \sum_{i=1}^{N} r_{d_i} \leq r_0,$$

where $r_d \triangleq (r_{d_1}, \ldots, r_{d_N}) \in \mathbb{R}^{N \cdot n_r}$ are additional coordination variables. The equality constraints can now be used in the augmented Lagrangian L_a in the same way that the interconnection equations $u - HF(c, u) = 0$ were, and the inequality constraint should be taken into account in the coordinator. A similar approach to constraint (2.93), also based on the use of agumented Lagrangians, can be found in Telle (1975).

2.5.4. THE OUTPUT PREDICTION AND BALANCE METHOD

We return now to the most general form of the system optimization problem:

Find control $\hat{c}$ giving output $\hat{y}$ (and interaction input $\hat{u}$ since $\hat{u} = H\hat{y}$), such that

$$(\hat{c}, \hat{y}) = \arg \min \, \psi(Q_1(c_1, u_1), \ldots, Q_N(c_N, u_N)) \tag{2.94}$$

subject to

$$y_i = F_i(c_i, u_i); \qquad u_i = H_i y, \qquad (c_i, u_i) \in CU_i, \qquad i \in \overline{1, N}, \qquad y \in Y,$$

where $\psi : \mathbb{R}^N \to \mathbb{R}$ is a function strictly preserving order on some set $\Omega \subseteq \mathbb{R}^N$. The disturbances z_i are omitted since they are assumed to be constant during the optimization process.

The direct method of coordination, together with various versions of it, was discussed for the optimization problem in the form of (2.94) in section 2.2. Let us briefly summarize the advantages and drawbacks of this method. The pure direct approach preserves all the constraints during the coordination process; its application is however limited to the cases with no fewer controls than outputs in each subsystem, and to cases in which it is possible to determine and preserve set V_0 (V_0 is defined in section 2.2.). The modified direct approach in which one knows only the subset V_F of V_0 has a complicated structure and restrictive applicability conditions. On the other hand, the penalty function method (section 2.3) has very weak applicability assumptions, but may need large values of the penalty coefficients, which can make it ineffective. The method presented in this section can be regarded as a generalization of the penalty function approach, where linear–quadratic modifications are used instead of pure quadratic modifications in order to create an appropriate augmented Lagrangian.

The augmented Lagrangian cannot be introduced for problem (2.94) in a classical manner, i.e., as

$$\psi(Q_1(c_1, u_1), \ldots, Q_N(c_N, u_N)) + \sum_{i=1}^{N} (\langle \lambda_i, u_i - H_i F(c, u) \rangle$$
$$+ \tfrac{1}{2}\rho \|u_i - H_i F(c, u)\|^2), \tag{2.95}$$

the way the augmented Lagrangian L_a was introduced in IPBM in the previous section. Nor can it be introduced as

$$\psi(Q_1(c_1, H_1 v), \ldots, Q_N(c_N, H_N v)) + \sum_{i=1}^{N} (\langle \lambda_i, v_i - F_i(c_i, H_i v) \rangle$$
$$+ \tfrac{1}{2}\rho \|v_i - F_i(c_i, H_i v)\|^2), \tag{2.96}$$

where $v \in \mathcal{Y}$ are in the nature of predicted outputs identical to those in the direct approach described in section 2.2. If the augmented Lagrangian is introduced in these ways, the resulting functions (2.95) or (2.96) are nondecomposable if ψ is nonadditive. Therefore, we will introduce augmented Lagrangians locally and construct the following *augmented ψ-Lagrangian* (Tatjewski 1977)

$$L_\psi(c, v, \lambda, \rho) \triangleq \psi(Q_{a1}(c_1, v, \lambda_1, \rho), \ldots, Q_{aN}(c_N, v, \lambda_N, \rho)), \qquad (2.97)$$

where $\forall i \in \overline{1, N}$

$$Q_{ai}(c_i, v, \lambda_i, \rho) \triangleq Q_i(c_i, H_i v) + \langle \lambda_i, v_i - F_i(c_i, H_i v) \rangle$$
$$+ \tfrac{1}{2}\rho \| v_i - F_i(c_i, H_i v) \|^2. \qquad (2.98)$$

The coordination method based on the augmented ψ-Lagrangian (2.97) will be called the *output prediction and balance method* (OPBM). It will be assumed throughout this section that all spaces $\mathscr{C}_i$, $i \in \overline{1, N}$, $\mathcal{U} = \times_{i=1}^N \mathcal{U}_i$ and $\mathcal{Y} = \times_{i=1}^N \mathcal{Y}_i$ are real Hilbert spaces, as they were for problem (2.76) in section 2.5.3.

Formulating OPBM, we define the *infimal (lower-level) problem* as the minimization of the following N *local problems* ($\mathrm{LP}_{\psi i}$)

For given output prediction $v \in \mathcal{Y}$, multiplier $\lambda_i \in \mathcal{Y}_i^*$ and penalty coefficient $\rho \in \mathbb{R}_+$, find control

$$\hat{c}_i(v, \lambda_i, \rho) = \arg \min_{C_i(v)} Q_{ai}(\cdot, v, \lambda_i, \rho), \qquad (2.99)$$

where

$$C_i(v) \triangleq \{ c_i \in \mathscr{C}_i : (c_i, H_i v) \in CU_i \}.$$

Let us denote the set of solutions of each local problem $\mathrm{LP}_{\psi i}$ by $\hat{C}_i(v, \lambda_i, \rho)$ and hence the set of solutions of the infimal problem by $\hat{C}(v, \lambda, \rho) \triangleq \times_{i=1}^N \hat{C}_i(v, \lambda_i, \rho)$, $\lambda = (\lambda_1, \ldots, \lambda_N)$. By $V = Y \cap V_0'$, we will denote the set of admissible output predictions, where

$$V_0' \triangleq \left\{ v \in \mathcal{Y} : C(v) = \overset{N}{\underset{i=1}{\times}} C_i(v) \neq \varnothing \right\}.$$

Let us denote also

$$CV \triangleq \bigcup_{v \in V} (C(v) \times \{v\}) = \{ (c, v) \in \mathscr{C} \times \mathcal{Y} : (c, Hv) \in CU \wedge y \in Y \}, \qquad (2.100)$$

$$Q_a(c, v, \lambda, \rho) \triangleq (Q_{a1}(c_1, v, \lambda_1, \rho), \ldots, Q_{aN}(c_N, v, \lambda_N, \rho)) \qquad (2.101)$$

and define

$$\Lambda(\rho) \triangleq \{ \lambda \in \mathcal{Y}^* : \forall (c, v) \in CV \quad Q_a(c, v, \lambda, \rho) \in \Omega \}, \qquad (2.102)$$

146

where $\Omega \subseteq \mathbb{R}^N$ is the set on which the function ψ strictly preserves order. We will assume for every $v \in V$, $\rho \in \mathbb{R}_+$, and $\lambda \in \Lambda(\rho)$ that $\hat{C}(v, \lambda, \rho) \neq \varnothing$.

The following *coordinator problem* (CP_ψ) can now be formulated:

Find a saddle point $(\hat{v}; \hat{\lambda})$ of the function $\hat{\psi}(\cdot, \cdot, \rho)$ on $V \times \Lambda(\rho)$, where

$$\hat{\psi}(v, \lambda, \rho) \triangleq \psi\left(\min_{C_1(v)} Q_{a1}(\cdot, v, \lambda_1, \rho), \ldots, \min_{C_N(v)} Q_{aN}(\cdot, v, \lambda_N, \rho) \right), \quad (2.103)$$

such that the balance condition $\forall \hat{c} \in \hat{C}(\hat{v}, \hat{\lambda}, \rho)$ $\hat{v} - F(\hat{c}, H\hat{v}) = 0$ holds.

The saddle point in CP_ψ is of course understood as the minimum with respect to v and a maximum with respect to λ; i.e.,

$$\forall (v, \lambda) \in V \times \Lambda(\rho) \qquad \hat{\psi}(\hat{v}, \lambda, \rho) \leq \hat{\psi}(\hat{v}, \hat{\lambda}, \rho) \leq \hat{\psi}(v, \hat{\lambda}, \rho).$$

The condition $\hat{v} - F(\hat{c}, H\hat{v}) = 0$ is required in CP_ψ for all $\hat{c} \in \hat{C}(\hat{v}, \hat{\lambda}, \rho)$ since we are not able to distinguish between various $\hat{c} \in \hat{C}(\hat{v}, \hat{\lambda}, \rho)$ and hence all of them should be equally good.

The connections between the solutions of CP_ψ and the initial problem (2.94) result from the following lemma.

LEMMA 2.17.

1. If $(\hat{v}; \hat{\lambda})$ is a solution to CP_ψ, then for any $\hat{c} \in \hat{C}(\hat{v}, \hat{\lambda}, \rho)$ $(\hat{c}, \hat{v}; \hat{\lambda})$ is a saddle point of $L_\psi(\cdot, \cdot, \lambda, \rho)$ on $CV \times \Lambda(\rho)$.

2. If $(\hat{c}, \hat{v}; \hat{\lambda})$ is a saddle point of $L_\psi(\cdot, \cdot, \cdot, \rho)$ on $CV \times \Lambda(\rho)$ and $\hat{v} - F(\hat{c}, H\hat{v}) = 0$, then $(\hat{v}; \hat{\lambda})$ is a saddle point of $\hat{\psi}(\cdot, \cdot, \rho)$ on $V \times \Lambda(\rho)$ and is also a solution to CP_ψ if $\forall c \in \hat{C}(\hat{v}, \hat{\lambda}, \rho)$ $\hat{v} - F(c, H\hat{v}) = 0$.

The proof is identical to the one for Lemma 2.16 in section 2.5.3.

COROLLARY. If $(\hat{v}; \hat{\lambda})$ is a solution to CP_ψ then for any point $\hat{c} \in \hat{C}(\hat{v}, \hat{\lambda}, \rho)$, the point $(\hat{c}, \hat{u})$, $\hat{u} = H\hat{v}$, is a solution to the initial problem (2.94).

Thus, we will say that OPBM *is applicable* if at least one solution $(\hat{v}; \hat{\lambda})$ to CP_ψ exists.

The way that the function $\hat{\psi}$ is defined is interesting and important for further discussion. Observe that if $v \in V$ and $\lambda \in \Lambda(\rho)$ then

$$\hat{\psi}(v, \lambda, \rho) = \varphi_\psi(v, \lambda, \rho), \qquad (2.104)$$

where

$$\varphi_\psi(v, \lambda, \rho) \triangleq \min_{C(v)} L_\psi(\cdot, v, \lambda, \rho)$$

is the primal–dual function. The equality (2.104) holds for $v \in V$ and $\lambda \in \Lambda(\rho)$ because for all $c \in C(v)$, $Q_a(c, v, \lambda, \rho) \in \Omega$ and ψ strictly preserves order on Ω; hence $\min_{C(v)} \psi(Q_a(\cdot, v, \lambda, \rho)) = \psi(\min_{C_1(v)} Q_{a1}(\cdot, v, \lambda_1, \rho), \ldots, \min_{C_N(v)} Q_{aN}(\cdot, v, \lambda_N, \rho))$. Thus, instead of evaluating $\varphi_\psi(v, \lambda, \rho)$, which cannot be done directly using decomposition, we evaluate $\hat{\psi}(v, \lambda, \rho)$, which is equivalent to evaluating $\varphi_\psi(v, \lambda, \rho)$ if $(v, \lambda) \in V \times \Lambda(\rho)$ and can be done using decomposition. But is the set $\Lambda(\rho)$ "sufficiently large," i.e., does it introduce any essential restrictions on the existence of the solutions of CP_ψ? The following theorem is of great value here.

THEOREM 2.22. *If the set $\Omega \subseteq \mathbb{R}^N$ on which the utility function ψ strictly preserves order is such that*

$$(\forall a \in \Omega)(\forall a' \in \mathbb{R}^N) \qquad a' \geq a \;\Rightarrow\; a' \in \Omega, \tag{2.105}$$

and problem (2.94) satisfies the following boundedness condition

$$(\exists \bar{\rho} \in \mathbb{R}_+)(\exists \bar{a} \in \mathrm{int}\,\Omega)(\forall (c, v) \in CV) \qquad Q_a(c, v, 0, \bar{\rho}) \geq \bar{a}, \tag{2.106}$$

then for every bounded subset Λ of $\mathcal{Y}^*$ there exists $\rho^* \in \mathbb{R}_+$ such that

$$\forall \rho \geq \rho^* \qquad \Lambda \subseteq \Lambda(\rho).$$

Proof. Condition (2.106) states that

$$(\forall (c, v) \in CV)(\forall i \in \overline{1, N}) \quad Q_i(c_i, H_i v) + \tfrac{1}{2}\bar{\rho}\|v_i - F_i(c_i, H_i v)\|^2 \geq \bar{a}_i. \tag{1}$$

Since $\bar{a} = (\bar{a}_1, \ldots, \bar{a}_N) \in \mathrm{int}\,\Omega$, there exists $\varepsilon > 0$ such that

$$\bar{a}_\varepsilon \triangleq (\bar{a}_1 - \varepsilon, \ldots, \bar{a}_N - \varepsilon) \in \Omega.$$

Denoting

$$\|\bar{\lambda}\| \triangleq \max_{i \in \overline{1, N}} \left(\sup_{\lambda \in \Lambda} \|\lambda_i\| \right)$$

we get

$$(\forall \lambda \in \Lambda)(\forall (c, v) \in CV) \quad \langle \lambda_i, v_i - F_i(c_i, H_i v)\rangle + \tfrac{1}{2}\rho'\|v_i - F_i(c_i, H_i v)\|^2 \rangle$$

$$\geq -\|\bar{\lambda}\|\|v_i - F_i(c_i, H_i v)\| + \tfrac{1}{2}\rho'\|v_i - F_i(c_i, H_i v)\|^2 \geq -\varepsilon, \quad \text{if} \quad \rho' \geq \frac{\|\bar{\lambda}\|^2}{2\varepsilon}, i \in \overline{1, N}.$$

Hence, using (1) and taking $\rho^* = \bar{\rho} + \rho'$ we obtain

$$(\forall \lambda \in \Lambda)(\forall (c, v) \in CV) \qquad Q_a(c, v, \lambda, \rho^*) \geq a_\varepsilon \in \Omega,$$

which implies, because of (2.105), that

$$(\forall \lambda \in \Lambda)(\forall (c, v) \in CV) \qquad Q_a(c, v, \lambda, \rho^*) \in \Omega.$$

Finally, we note that $\forall (c, v, \lambda) \in CV \times \mathcal{Y}^*$ functions $Q_{ai}(c_i, v, \lambda_i, \cdot)$ are nondecreasing, $i \in \overline{1, N}$. $\square$

Theorem 2.22, proved under weak assumptions, states that every $\lambda \in \mathcal{Y}^*$ together with any member of its neighborhood can be made feasible for CP_ψ by choosing ρ sufficiently large. Hence the set $\Lambda(\rho)$ need not be evaluated at all; if at some coordination step the value of $Q_a(\hat{c}(v, \lambda, \rho), v, \lambda, \rho)$ does not belong to Ω, ρ should be increased and the solution of the local problems repeated as long as the optimization results do not belong to Ω. Such protection can be easily placed into any coordination algorithm.

To formulate applicability conditions of OPBM, formula (2.82) of the sets CU_i

$$CU_i = \{(c_i, u_i) \in \mathcal{C}_i \times \mathcal{U}_i : G_i(c_i, u_i) \leq 0\}, \qquad G_i : \mathcal{C}_i \times \mathcal{U}_i \to \mathbb{R}^{n_i}, \qquad i \in \overline{1, N},$$

and the following form of set Y

$$Y = \{y \in \mathcal{Y} : G_0(y) \leq 0\}, \qquad G_0 : \mathcal{Y} \to \mathbb{R}^{n_0},$$

will be used. We denote $\mathbb{R}^n = \times_{i=0}^{N} \mathbb{R}^{n_i}$ and rewrite the *second-order sufficient optimality conditions with strict complementarity* for the optimal point $(\hat{c}, \hat{v})$, $\hat{v} = \hat{y} = H^{-1}\hat{u}$, of problem (2.94):

- First-order necessary conditions are satisfied at $(\hat{c}, \hat{v})$, i.e., there are multipliers $\hat{\lambda}_L = (\hat{\lambda}_{L1}, \ldots, \hat{\lambda}_{LN}) \in \mathcal{Y}^*$, $\hat{\eta} = (\hat{\eta}_0, \hat{\eta}_1, \ldots, \hat{\eta}_N) \in \mathbb{R}^n_+$ such that

$$L'_{(c,u)}(\hat{c}, \hat{v}, \hat{\lambda}_L, \hat{\eta}) = 0,$$

$$\langle \hat{\eta}_0, G_0(\hat{v}) \rangle = 0, \qquad \langle \hat{\eta}_i, G_i(\hat{c}_i, H_i\hat{v}) \rangle = 0, \qquad i \in \overline{1, N},$$

where

$$L(c, v, \lambda_L, \eta) \triangleq Q(c, Hv) + \sum_{i=1}^{N} (\langle \lambda_{Li}, v_i - F_i(c_i, H_iv) \rangle$$

$$+ \langle \eta_i, G_i(c_i, H_iv) \rangle) + \langle \eta_0, G_0(v) \rangle \quad (2.107)$$

is the normal Lagrange function;

- Q_i, F_i, G_i, ψ, and hence the function $L(\cdot, \cdot, \lambda_L, \eta)$, are twice continuously differentiable (in the neighborhood of the optimum), and there exists $\delta > 0$ such that

$$\langle L''_{(c,u)(c,u)}(\hat{c}, \hat{v}, \hat{\lambda}_L, \hat{\eta})(\bar{c}, \bar{v}), (\bar{c}, \bar{v}) \rangle \geq \delta \|(\bar{c}, \bar{v})\|^2 \qquad (2.108)$$

for every point $(\bar{c}, \bar{v}) \in \mathcal{C} \times \mathcal{Y}$ satisfying

$$\bar{v}_i - (F_i)'_{(c_i, H_iv)}(\hat{c}_i, H_i\hat{v})(\bar{c}, H_i\bar{v}) = 0, \qquad i \in \overline{1, N},$$

$$(G_{ij})'_{(c_i, H_iv)}(\hat{c}_i, H_i\hat{v})(\bar{c}, H_i\bar{v}) = 0, \qquad j \in A_i \triangleq \{j \in \overline{1, n_i} : G_{ij}(\hat{c}_i, H_i\hat{v}) = 0\}, \qquad i \in \overline{1, N},$$

$$(G_{0j})'_v(\hat{v})\bar{v} = 0, \qquad j \in A_0 \triangleq \{j \in \overline{1, n_0} : G_{0j}(\hat{v}) = 0\};$$

• For every $j \in A_i$ and $i \in \overline{0, N}$, $\hat{\eta}_{ij} > 0$, i.e., all multipliers corresponding to active constraints are positive—the strict complementarity condition.

THEOREM 2.23. (*applicability conditions of* OPBM). *If we assume that*

1. *For optimal point* $(\hat{c}, \hat{v})$ *of problem* (2.94), *with* $CU_i = \{(c_i, u_i): G_i(c_i, u_i) \leq 0\}$, $Y = \{y: G_0(y) < 0\}$, (*and* $Q(c, u) = \psi(Q_1(c_1, u), \ldots, Q_N(c_N, u))$ *also admissible*), *second-order sufficient optimality conditions with strict complementarity are satisfied, partial derivatives of* ψ *taken at* $\bar{Q}(\hat{c}, H\hat{v}) \triangleq (Q_1(c_1, H_1\hat{v}), \ldots, Q_N(\hat{c}_N, H_N\hat{v}))$ *and denoted by* $\psi_i'(\bar{Q}(\hat{c}, H\hat{v}))$ *are positive,* $i \in \overline{1, N}$, *and Fréchet derivative* $h_{(c, v)}'(\hat{c}, \hat{v})$ *of the operator* $h(c, v) \triangleq v - F(c, Hv)$ *has its image closed in* $\mathcal{Y}$;

2. *The assumptions of Theorem 2.22 are satisfied along with conditions* (2.109) *and* (2.110):

$$(\forall \{a^n\} \subset \Omega)(\forall a \in \Omega) \ \overline{\lim_{n \to \infty}} \|a^n\| = \infty \Rightarrow \overline{\lim_{n \to \infty}} \psi(a^n) > \psi(a), \quad (2.109)$$

where $\{a^n\}$ *is any sequence from* Ω *bounded from below,*

$$(\forall N \subset \mathcal{C} \times \mathcal{Y})(\exists \varepsilon > 0)(\forall \varepsilon \geq \alpha > 0)(\forall (c, v) \in \underline{X}_\alpha)$$

$$\psi(Q_1(c_1, H_1 v) - \alpha, \ldots, Q_N(c_N, H_N v) - \alpha) \leq \psi(\bar{Q}(\hat{c}, H\hat{v})) \Rightarrow (c, v) \in N,$$
$$(2.110)$$

where $\underline{X}_\alpha \triangleq \{(c, v) \in CV: \|v - F(c, Hv)\| \leq \alpha\}$, *and* N *denotes some neighborhood of* $(\hat{c}, \hat{v})$;

then there exists $\hat{\rho} \in \mathbb{R}_+$ *such that for every* $\rho \geq \hat{\rho}$

$$\forall (v, \lambda) \in V \times \Lambda(\rho) \qquad \hat{\psi}(\hat{v}, \lambda, \rho) \leq \hat{\psi}(\hat{v}, \hat{\lambda}, \rho) \leq \hat{\psi}(v, \hat{\lambda}, \rho)$$

and $\hat{\lambda} = (\hat{\lambda}_1, \ldots, \hat{\lambda}_N) \in \Lambda(\rho)$, *where*

$$\hat{\lambda}_i = \hat{\lambda}_{Li} / \psi_i'(\bar{Q}(\hat{c}, H\hat{v})), \qquad i \in \overline{1, N}, \quad (2.111)$$

and $\hat{\lambda}_L = (\hat{\lambda}_{L1}, \ldots, \hat{\lambda}_{LN})$ *is the normal Lagrangian multiplier for the constraint* $h(c, v) = 0$. *Moreover, the set* $\hat{C}(\hat{v}, \hat{\lambda}, \rho)$ *consists of a single point* $\hat{c}$.

Proof. In the major part of the proof, the distinction between c and v will be superfluous, so we denote $(c, v) \triangleq x \in X = CV$, $X \subset \mathcal{X} = \mathcal{C} \times \mathcal{Y}$. By $Q_i(x)$, $h_i(x)$, $G_i(x), \ldots$, we will understand $Q_i(c_i, H_i v)$, $h_i(c_i, v)$, $G_i(c_i, H_i v)$, etc. We will also simplify the notation of derivatives in the proof; in cases of one argument only we will use $G'(x)$, $Q_i'(x)$, etc., instead of $G_x'(x)$, $(Q_i)_x'(x)$, etc.

Let us expand $L(x, \hat{\lambda}_L, \hat{\eta})$ into a Taylor series around the point $\hat{x} = (\hat{c}, \hat{v})$:

$$L(x, \hat{\lambda}_L, \hat{\eta}) = Q(\hat{x}) + \langle L_x'(\hat{x}, \hat{\lambda}, \hat{\eta}), \bar{x} \rangle + \langle L_{xx}''(\hat{x}, \hat{\lambda}_L, \hat{\eta})\bar{x}, \bar{x} \rangle + o(\|\bar{x}\|^2),$$

where $\bar{x} = x - \hat{x}$,

$$\langle L_x'(\hat{x}, \hat{\lambda}_L, \hat{\eta}), \bar{x} \rangle = \sum_{i=1}^{N} [\psi_i'(\bar{Q}(\hat{x}))\langle Q_i'(\hat{x}), \bar{x} \rangle + \langle \hat{\lambda}_{Li}, h_i'(\hat{x})\bar{x} \rangle] + \langle \hat{\eta}, G'(\hat{x})\bar{x} \rangle$$

150

where

$$G \triangleq (G_0, G_1, \ldots, G_N), \quad h \triangleq (h_1, \ldots, h_N), \quad h_i(x) \triangleq v_i - F_i(c_i, H_i v),$$

$$\langle L''_{xx}(\hat{x}, \hat{\lambda}_L, \hat{\eta})\bar{x}, \bar{x}\rangle = \sum_{i=1}^{N} [\psi'_i(\bar{Q}(\hat{x}))\langle Q''_i(\hat{x})\bar{x}, \bar{x}\rangle + \langle \hat{\lambda}_{Li}, h''_i(\hat{x})(\bar{x}, \bar{x})\rangle$$

$$+ \sum_{j=1}^{N} \psi''_{ij}(\bar{Q}(\hat{x}))\langle Q'_j(\hat{x}), \bar{x}\rangle\langle Q'_i(\hat{x}), \bar{x}\rangle] + \langle \hat{\eta}, G''(\hat{x})(\bar{x}, \bar{x})\rangle.$$

We have assumed for every $i \in \overline{1, N}$ that the partial derivatives $\psi'_i(\bar{Q}(\hat{x}))$ are positive; hence, the multipliers (2.111) $\hat{\lambda} = (\hat{\lambda}_1, \ldots, \hat{\lambda}_N)$ are well defined. Let us formulate the normal Lagrangian for the augmented ψ-Lagrangian (2.97) $L_\psi(\cdot, \hat{\lambda}, \rho)$ with the constraints $G(x) \leqslant 0$:

$$\mathscr{L}(x, \hat{\lambda}, \rho, \eta) \triangleq L_\psi(x, \hat{\lambda}, \rho) + \langle \eta, G(x)\rangle.$$

Let us take $\eta = \hat{\eta}$ and expand $\mathscr{L}(\cdot, \hat{\lambda}, \rho, \hat{\eta})$ into a Taylor series around the point $\hat{x}$

$$\mathscr{L}(x, \hat{\lambda}, \rho, \hat{\eta}) = Q(\hat{x}) + \langle \mathscr{L}'_x(\hat{x}, \hat{\lambda}, \rho, \hat{\eta}), \bar{x}\rangle + \langle \mathscr{L}''_{xx}(\hat{x}, \hat{\lambda}, \rho, \hat{\eta})\bar{x}, \bar{x}\rangle \quad o(\|\bar{x}\|^2),$$

where

$$\langle \mathscr{L}'_x(\hat{x}, \hat{\lambda}, \rho, \hat{\eta}), \bar{x}\rangle = \sum_{i=1}^{N} \psi'_i(\bar{Q}(\hat{x}))[\langle Q'_i(\hat{x}), \bar{x}\rangle + \langle \hat{\lambda}_i, h'_i(\hat{x})\bar{x}\rangle] + \langle \hat{\eta}, G'(\hat{x})\bar{x}\rangle,$$

$$\langle \mathscr{L}''_{xx}(\hat{x}, \hat{\lambda}, \rho, \hat{\eta})\bar{x}, \bar{x}\rangle = \sum_{i=1}^{N} \psi'_i(\bar{Q}(\hat{x}))[\langle Q''_i(\hat{x})\bar{x}, \bar{x}\rangle + \langle \hat{\lambda}_i, h''_i(\hat{x})(\hat{x}, \hat{x})\rangle$$

$$+ \sum_{j=1}^{N} \psi''_{ij}(\bar{Q}(\hat{x}))(\langle Q'_j(\hat{x}), \bar{x}\rangle + \langle \hat{\lambda}_j, h'_j(\hat{x})\bar{x}\rangle)(\langle Q'_i(\hat{x})\bar{x}\rangle$$

$$+ \langle \hat{\lambda}_i, h'_i(\hat{x})\bar{x}\rangle) + \rho\langle h'_i(\hat{x})\bar{x}, h'_i(\hat{x})\bar{x}\rangle] + \langle \hat{\eta}, G''(\hat{x})(\bar{x}, \bar{x})\rangle.$$

Looking at the second-order derivatives of L and $\mathscr{L}$, we can see that

$$\langle \mathscr{L}''_{xx}(\hat{x}, \hat{\lambda}, \rho, \hat{\eta})\bar{x}, \bar{x}\rangle = \langle L''_{xx}(\hat{x}, \hat{\lambda}_L, \hat{\eta})\bar{x}, \bar{x}\rangle + k(\bar{x}) + B(\bar{x}, \bar{x}),$$

where

$$k(\bar{x}) = \rho \sum_{i=1}^{N} \psi'_i(\bar{Q}(\hat{x})) \|h'_i(\hat{x})\bar{x}\|^2,$$

$$B(\bar{x}, \bar{x}) = \sum_{i=1}^{N} \sum_{j=1}^{N} \psi''_{ij}(\bar{Q}(\hat{x}))(\langle Q'_j(\hat{x}), x\rangle\langle \hat{\lambda}_i, h'_i(\hat{x})\bar{x}\rangle$$

$$+ \langle \hat{\lambda}_j, h'_j(\hat{x})\bar{x}\rangle(\langle Q'_i(\hat{x})\bar{x}\rangle + \langle \hat{\lambda}_i, h'_i(\hat{x})\bar{x}\rangle).$$

Let us denote by G_A the active part at point $\hat{x}$ of the constraints G, $G_A = (G_{0A}, G_{1A}, \ldots, G_{NA})$. The kernel of the operator $G'_A(\hat{x})$, being a closed subspace of the Hilbert space $\mathscr{X}$, is also a Hilbert space. Denote it by

$\mathscr{X}_A$, $\mathscr{X}_A = \ker G'_A(\hat{x})$. The operator $h'(\hat{x})$ has, by assumption, its image $\operatorname{im} h'(\hat{x}) = h'(\hat{x})(\mathscr{X})$ closed in $\mathscr{Y}$. Hence, the image $h'(\hat{x})(\mathscr{X}_A)$ is also closed in $\mathscr{Y}$. All further considerations will be in space $\mathscr{X}_A$; by $h'(\hat{x})$, the restriction $h'(\hat{x}) \mid \mathscr{X}_A$ should be understood. The kernel $\ker h'(\hat{x})$ is a closed subspace of the Hilbert space $\mathscr{X}_A$; hence, owing to the orthogonal projection theorem (Dunford and Schwartz 1958)

$$(\forall \bar{x} \in \mathscr{X}_A)(\exists ! \ \bar{x}' \in (\ker h'(\hat{x}))^{\perp})(\exists ! \bar{x}'' \in \ker h'(\hat{x})) \quad \bar{x} = \bar{x}' + \bar{x}''.$$

The fact that $\operatorname{im} h'(\hat{x})$ is closed implies that (Dunford and Schwartz 1958)

$$(\exists \tilde{\varepsilon} > 0)(\forall \bar{x} \in \mathscr{X}_A) \qquad \|h'(\hat{x})\bar{x}\| \geqslant \tilde{\varepsilon} \|\bar{x}'\|.$$

Let us denote

$$\psi_m = \min \{\psi'_1(\bar{Q}(\hat{x})), \dots, \psi'_N(\bar{Q}(\hat{x}))\},$$

from the assumption that $\psi_m > 0$, since all $\psi'_i(\bar{Q}(\hat{x}))$ are positive. Combining the above relations, we get for $\varepsilon = \psi_m \cdot \tilde{\varepsilon}$

$$(\exists \varepsilon > 0)(\forall \bar{x} \in \mathscr{X}_A) \qquad k(\bar{x}) = \rho \sum_{i=1}^{N} \psi'_i(\bar{Q}(\hat{x})) \|h'_i(\hat{x})\bar{x}\|^2$$

$$\geqslant \rho \psi_m \sum_{i=1}^{N} \|h'_i(\hat{x})\bar{x}\|^2 = \psi_m \|h'(\hat{x})\bar{x}\|^2 \geqslant \rho \varepsilon \|\bar{x}'\|^2.$$

Observe that

$$\forall \bar{x}'' \in \ker h'(\hat{x}) \qquad B(\bar{x}'', \bar{x}'') = 0,$$

$$\forall \bar{x} \in \mathscr{X}_A \qquad k(\bar{x}) = k(\bar{x}' + \bar{x}'') = k(\bar{x}').$$

Denoting $\gamma = \|\bar{x}''\|^2 / \|\bar{x}'\|^2$ and using inequality (2.108), we get

$$\langle \mathscr{L}''_{xx}(\hat{x}, \hat{\lambda}, \rho, \hat{\eta})\bar{x}, \bar{x} \rangle = \langle L''_{xx}(\hat{x}, \hat{\lambda}_L, \hat{\eta})(\bar{x}' + \bar{x}''), (\bar{x}' + \bar{x}'') \rangle$$

$$+ B(\bar{x}' + \bar{x}'', \bar{x}' + \bar{x}'') + k(\bar{x}' + \bar{x}'')$$

$$\geqslant - \|L''_{xx}(\hat{x}, \hat{\lambda}_L, \hat{\eta})\| \|\bar{x}'\|^2 - 2\|L''_{xx}(\hat{x}, \hat{\lambda}_L, \hat{\eta})\| \|\bar{x}'\| \|\bar{x}''\| + \delta \|\bar{x}''\|^2$$

$$- \|B\| \|\bar{x}'\|^2 - 2\|B\| \|\bar{x}'\| \|\bar{x}''\| + \rho \varepsilon \|\bar{x}'\|^2$$

$$= \|\bar{x}'\|^2 (\delta \gamma^2 - 2(\|L''_{xx}(\hat{x}, \hat{\lambda}_L, \hat{\eta})\| + \|B\|)\gamma - \|L''_{xx}(\hat{x}, \hat{\lambda}_L, \hat{\eta})\| - \|B\| + \rho \varepsilon)$$

$$= \|\bar{x}'\|^2 (\delta \gamma^2 - \beta \gamma - \delta + \rho \varepsilon) \geqslant \|\bar{x}'\|^2 \varepsilon'(1 + \gamma^2) = \varepsilon' \|\bar{x}\|^2$$

only if

$$0 < \varepsilon' < \delta \quad \text{and} \quad \rho \geqslant \frac{\beta^2 + 4(\delta - \varepsilon')(\delta + \varepsilon')}{4\varepsilon(\delta - \varepsilon')}.$$

We have proved the following result

$$(\exists \varepsilon' > 0)(\exists \tilde{\rho} \in \mathbb{R}_+)(\forall \rho \geqslant \tilde{\rho})(\forall x \in \mathscr{X}_A) \quad < \mathscr{L}''_{xx}(\hat{x}, \hat{\lambda}, \rho, \hat{\eta})\bar{x}, \bar{x} \rangle \geqslant \varepsilon' \|\bar{x}\|^2.$$

From the necessary optimality conditions

$$L'_x(\hat{x}, \hat{\lambda}_L, \hat{\eta}) = 0,$$

hence, from equality (2.111)

$$\mathscr{L}'_x(\hat{x}, \hat{\lambda}, \rho, \hat{\eta}) = 0.$$

Because the strict complementarity condition is satisfied at point $\hat{x}$, the relations that have just been proved mean that at point $\hat{x}$ the second-order sufficient optimality conditions for the minimum of function $L_\psi(\cdot, \hat{\lambda}, \rho)$ subject to constraint $G(x) \leq 0$ are satisfied. Hence, there exists neighborhood $\hat{N}$ of point $\hat{x}$ such that

$$(\forall \rho \geq \tilde{\rho})(\forall x \in X \cap \hat{N}) \qquad L_\psi(x, \hat{\lambda}, \rho) \geq L_\psi(\hat{x}, \hat{\lambda}, \rho) = \psi(\bar{Q}(\hat{x}))$$

and the equality can be obtained only for the unique point $x = \hat{x}$ (remember also that $h(\hat{x}) = 0$).

Let us observe that we have used only assumption (1). Using assumption (2) as well, we will show that $\hat{x}$ is the unique global minimum point of $L_\psi(\cdot, \hat{\lambda}, \rho)$ on set X. By virtue of (2.106) there exists $\beta > 0$ such that

$$\bar{a}_\beta = (\bar{a}_1 - \beta, \ldots, \bar{a}_N - \beta) \in \Omega.$$

From assumption (2.110) there exists α, $0 < \alpha \leq \beta/2$, such that

$$(x \in \underline{X}_\alpha \wedge \psi(Q_1(x) - \alpha, \ldots, Q_N(x) - \alpha) \leq \psi(\bar{Q}(\hat{x}))) \Rightarrow x \in (\hat{N} \cap X).$$

Therefore, if $x \notin \hat{N} \cap X$ then $x \in \overline{X}_\alpha$ or $x \in \tilde{X}_\alpha$, where

$$\overline{X}_\alpha = \{x \in X : \|h(x)\| \geq \alpha\},$$
$$\tilde{X}_\alpha = \{x \in \underline{X}_\alpha : \psi(Q_1(x) - \alpha, \ldots, Q_N(x) - \alpha) > \psi(\bar{Q}(\hat{x}))\}.$$

We will now show that if ρ is sufficiently large, then for every x from $\overline{X}_\alpha$ or $\tilde{X}_\alpha$, $L_\psi(x, \hat{\lambda}, \rho) > L_\psi(\hat{x}, \hat{\lambda}, \rho)$. Let us suppose first that $x \in \overline{X}_\alpha$. Arguing as in the proof of Theorem 2.22 (for $\Lambda = \{\hat{\lambda}\}$), we obtain from (2.105) and (2.106) that

$$(\exists \rho^* \geq \bar{\rho})(\forall \rho \geq \rho^*)(\forall x \in X) \qquad Q_a(x, \hat{\lambda}, \rho) \geq \bar{a}_\beta \in \Omega,$$

which implies that

$$(1) \qquad (\forall \gamma > 0)(\exists \rho_\gamma \in \mathbb{R}_+)(\forall \rho \geq \rho_\gamma)(\forall x \in \overline{X}_\alpha) \quad \|Q_a(x, \hat{\lambda}, \rho)\| \geq \gamma,$$

since $Q_a(\cdot, \hat{\lambda}, \rho^*)$ is bounded below by $\bar{a}_\beta$ and $\|h(x)\| \geq \alpha$ on $\overline{X}_\alpha$. Thus, by enlarging $\rho = \rho^* + \Delta\rho$ we can make at least one penalty term

$$\rho\|h_i(x)\| = \rho^*\|h_i(x)\|^2 + \Delta\rho\|h_i(x)\|^2$$

as large as we want on the set $\overline{X}_\alpha$. Let us denote by $\bar{x}$ a point from the set

$\underline{X}_0 = \{x \in X : \|h(x)\| = 0\}$ and suppose that

$$(2) \qquad \liminf_{\substack{\rho \to \infty \\ \overline{X}_\alpha}} L_\psi(\cdot, \hat{\lambda}, \rho) \leqslant L_\psi(\hat{x}, \hat{\lambda}, \rho) = \psi(\bar{Q}(\hat{x})),$$

which means that

$$(\forall \varepsilon > 0)(\exists \{(x^n, \rho^n)\} \subset \overline{X}_\alpha \times \mathbb{R}_+, \rho^n \to \infty)$$
$$(\exists n_0 \geqslant 1)(\forall n \geqslant n_0) \ L_\psi(x^n, \hat{\lambda}, \rho^n) \leqslant L_\psi(\hat{x}, \hat{\lambda}, \rho) + \varepsilon.$$

On the other hand, (1) implies that

$$\|Q_a(x^n, \hat{\lambda}, \rho^n)\| \to \infty.$$

Since the sequence $\{Q_a(x^n, \hat{\lambda}, \rho^n)\}$ is for $\rho \geqslant \rho^*$ bounded below by $\bar{a}_\beta \in \Omega$ and contained in Ω, and $\bar{Q}(\hat{x}) \in \Omega$, by virtue of (2.109):

$$\varlimsup_{n \to \infty} L_\psi(x^n, \hat{\lambda}, \rho^n) > \psi(\bar{Q}(\hat{x})) = L_\psi(\hat{x}, \hat{\lambda}, \rho),$$

which contradicts supposition (2). The reversed inequality (2) means that

$$(\exists \rho' \in \mathbb{R}_+)(\forall \rho \geqslant \rho')(\forall x \in \overline{X}_\alpha) \ L_\psi(x, \hat{\lambda}, \rho) > L_\psi(\hat{x}, \hat{\lambda}, \rho).$$

Let us suppose now that $x \in \tilde{X}_\alpha$. If we take $\rho \geqslant \rho'' = \|\hat{\lambda}\|/2\alpha$ then

$$(\forall x \in X)(\forall i \in \overline{1, N}) \ \langle \hat{\lambda}_i, h_i(x) \rangle + \tfrac{1}{2}\rho\|h_i(x)\|^2 \geqslant -\alpha.$$

Since $\alpha \leqslant \beta/2$, $\bar{Q}(X_\alpha) - \tilde{\alpha} \in \Omega$, where $\tilde{\alpha} \triangleq (\alpha, \ldots, \alpha) \in \mathbb{R}_+^N$. Hence, because the function ψ strictly preserves order on Ω, we get

$$(\forall \rho \geqslant \rho'')(\forall x \in \tilde{X}_\alpha) \ L_\psi(x, \hat{\lambda}, \rho) > L(\hat{x}, \hat{\lambda}, \rho) = \psi(\bar{Q}(\hat{x})).$$

Taking $\hat{\rho} = \max \{\tilde{\rho}, \rho', \rho'', \bar{\rho}\}$, we get finally

$$(\forall \rho \geqslant \hat{\rho})(\forall x \in X_\alpha) \ L_\psi(x, \hat{\lambda}, \rho) > L_\psi(\hat{x}, \hat{\lambda}, \rho).$$

and the equality can be obtained only for the unique point $\hat{x} = (\hat{c}, \hat{v})$. Since $h(\hat{x}) = 0$, $(\hat{x}; \hat{\lambda})$ is a saddle point

$$(\forall x \in X)(\forall \lambda \in \mathcal{Y}^*) \qquad L_\psi(\hat{x}, \lambda, \rho) \leqslant L_\psi(\hat{x}, \hat{\lambda}, \rho) \leqslant L_\psi(x, \hat{\lambda}, \rho).$$

We have also

$$\forall \rho \geqslant \hat{\rho} \qquad \hat{\lambda} \in \Lambda(\rho)$$

since $h(\hat{x}) = 0$ implies that $\forall \rho \in \mathbb{R}_+ \ Q_a(\hat{x}, \hat{\lambda}, \rho) = Q_a(\hat{x}, 0, \bar{\rho})$ and $Q_a(\hat{x}, 0, \bar{\rho}) \in \Omega$ by virtue of (2.106). Returning now to the distinction between c and v, we see that the uniqueness of $\hat{x} = (\hat{c}, \hat{v})$ implies that the set $\hat{C}(\hat{v}, \hat{\lambda}, \rho)$ consists of a single pont $\hat{c}$. To complete the proof it is sufficient now to apply Lemma 2.17, part 2, since $\forall \rho \geqslant \hat{\jmath} \ \Lambda(\rho) \subseteq \mathcal{Y}^*$. $\quad\square$

Some of the assumptions of Theorem 2.23 need comments. Let us notice first that the requirement that the partial derivatives $\psi'_i(\bar{Q}(\hat{c}, H\hat{v}))$ be positive, and conditions (2.109) and (2.105) are needed only for general, nonadditive functions ψ (they are satisfied if $\psi = \sum_{i=1}^{N}$). It can be easily shown that it is a common feature of the functions that strictly preserve order on some set Ω that their partial derivatives are positive on Ω (and $\bar{Q}(\hat{c}, H\hat{v}) \in \Omega$ is guaranteed by the boundedness condition (2.106)). On the other hand, assumption (2.109) means that any local function Q_i cannot override the others, which seems to be true in practice. Also, assumption (2.105) is commonly satisfied by a wide class of important functions ψ, for example, multiplicative and mixed multiplicative–additive functions.

Assumption (2.110) also looks complicated and needs further discussion. This assumption is in fact a generalization to the nonadditive case of the requirement that $\hat{x}$ be the unique solution in the strong sense to the initial optimization problem (see the previous section just before Theorem 2.21). Assumption (2.110) can be ensured by some more immediate conditions.

Remark. If $(\hat{c}, \hat{v})$ is the unique point minimizing problem (2.94) globally, Q_i are lower semicontinuous and F_i continuous, set CV is compact and function ψ is continuous, then (2.110) is satisfied.

Proof. If (2.110) is not satisfied, then there exists neighborhood N^0 of $(\hat{c}, \hat{v})$ such that

$$(\forall \alpha^n > 0)(\exists (c^n, v^n) \in CV) \quad (\psi(Q_1(c_1^n, H_1 v^n) - \alpha^n, \ldots, Q_N(c_N^n, H_N v^n) - \alpha^n)$$

$$\leq \psi(\bar{Q}(\hat{c}, H\hat{v})) \wedge (c^n, v^n) \notin N^0).$$

Let us take $\alpha^n = 1/n$. Owing to the compactness of $CV \backslash N^0$, it can be assumed without loss of generality that $(c^n, v^n) \to (c^0, v^0) \in CV \backslash N^0$. Because of the continuity of F_i, we get $v^0 - F(c^0, Hv^0) = 0$. Taking into account the semicontinuity of Q_i and the continuity of ψ, we have $\psi(\bar{Q}(c^0, Hv^0)) \leq \psi(\bar{Q}(\hat{c}, H\hat{v}))$, which contradicts the assumption that $(\hat{c}, \hat{v})$ is the unique optimal point. $\square$

Theorem 2.21, which was formulated in the previous section, is like Theorem 2.23 (it is in fact a special case of Theorem 2.23 with $\psi = \Sigma$ and operator $u - HF(c, u)$ instead of $v - F(c, Hv)$, which does not introduce any trouble). Thus, all the comments following Theorem 2.21 and concerning IPBM are also true for OPBM (the discussion of the adjustment of ρ performed just after Theorem 2.21 is an example). The theorems are similar because the coordinator problems CP_a and CP_ψ are similar. Thus, the discussion of coordinator strategies of IPBM also applies to OPBM, with the obvious modifications resulting from the different forms of the performance functions.

The *coordinator strategies* of OPBM can be roughly divided into two groups:

Two-level (double-loop) coordination strategies

For given λ the minimum of $\hat{\psi}(\cdot, \lambda, \rho)$ is found and λ is adjusted in order to approach the balance condition, and so on. The adjustment of λ can be done using Hestenes multiplier rule

$$\lambda^{k+1} = \lambda^k + \rho[\hat{v}^k - F(\hat{c}(\hat{v}^k, \lambda^k, \rho), \hat{v}^k)], \tag{2.112}$$

where $\hat{v}^k$ is the point minimizing $\hat{\psi}(\cdot, \lambda^k, \rho)$. The multiplier rule holds in its classical form (2.112) also for the augmented ψ-Lagrangian since λ are distinct from the normal Lagrangian multipliers λ_L and connected with them (in the optimum) through relation (2.111). Assuming that $CU = C \times U$ (hence $CV = C \times V$) and $\lambda \in \Lambda(\rho)$ (hence 2.104 holds) we have, using the derivative theorem from Appendix A,

$$\hat{\psi}'_{v_i}(v, \lambda, \rho) = (L_\psi)'_{v_i}(\hat{c}(v, \lambda, \rho), v, \lambda, \rho)$$

$$= \sum_{j=1}^{N} \psi'_j(Q_a(\hat{c}(v, \lambda, \rho), v, \lambda, \rho)) \cdot [Q_j)'_{v_i}(\hat{c}_j(v, \lambda, \rho), H_j v)$$

$$+ (\delta_{ji}I - (F_j)'_{v_i}(\hat{c}_j(v, \lambda, \rho), H_j v))^*(\lambda_j + \rho(v_j - F_j(\hat{c}_j(v, \lambda, \rho), H_j v)))],$$

where δ_{ji} is the Kronecker delta and $\psi'_i(\cdot) \triangleq \partial \psi / \partial Q_i(\cdot)$, $i \in \overline{1, N}$. Observe now that if we could find a fixed point $\bar{\lambda}$ of the adjusting rule (2.112) then we would have

$$\bar{v}_j - F_j(\hat{c}_j(\bar{v}, \bar{\lambda}, \rho), H_j \bar{v}) = 0, \qquad j \in \overline{1, N},$$

where $\bar{v} = (\bar{v}_1, \ldots, \bar{v}_N) = \hat{v}(\bar{\lambda}, \rho)$ is the point minimizing $\hat{\psi}(\cdot, \bar{\lambda}, \rho)$. Hence

$$\psi'_j(Q_a(\hat{c}(\bar{v}, \bar{\lambda}, \rho), v, \lambda, \rho)) = \psi'_j(\bar{Q}(\hat{c}(\bar{v}, \bar{\lambda}, \rho), H\bar{v})), \qquad j \in \overline{1, N},$$

and, assuming $v \in \text{int } V$, we would get

$$\hat{\psi}_{v_i}(\bar{v}, \bar{\lambda}, \rho) = \sum_{j=1}^{N} \psi'_j(\bar{Q}(\hat{c}(\bar{v}, \bar{\lambda}, \rho), H\bar{v})) \cdot (Q_j)'_{v_i}(\hat{c}_j(\bar{v}, \bar{\lambda}, \rho), H_j \bar{v})$$

$$+ (\delta_{ji}I - (F_j)'_{v_i}(\hat{c}_j(\bar{v}, \bar{\lambda}, \rho), H_j \bar{v}))^* \bar{\lambda}_j] = 0, \qquad j \in \overline{1, N}.$$

Therefore,

$$\bar{\lambda}_L = (\bar{\lambda}_{L1}, \ldots, \bar{\lambda}_{LN}), \quad \bar{\lambda}_{Lj} = \psi'_j(\bar{Q}(\hat{c}(\bar{v}, \bar{\lambda}, \rho), H\bar{v})) \cdot \bar{\lambda}_j$$

for every $j \in \overline{1, N}$, would be the normal Lagrangian multiplier $\hat{\lambda}_L$ and $(\hat{c}(\bar{v}, \bar{\lambda}, \rho), \bar{v})$ would be the solution $(\hat{c}, \hat{v})$ of the initial problem since it minimizes $L_\psi(\cdot, \cdot, \hat{\lambda}, \rho)$ on CV. The above argument can also be made when $v \in \text{int } V$ and $CV = C \times V$ are not necessarily true, but it becomes more

156

complicated. The result, formula (2.112) for the adjustment of λ, remains unchanged, however. The strict proof of the convergence of the algorithm created by this formula can also be performed, but is rather arduous (Tatjewski 1976).

One-level (single-loop) coordination strategies

The saddle point of $\hat{\psi}(\cdot, \cdot, \rho)$ is found by making simultaneous iterations of v and λ: if, in step k, $c^k = \hat{c}(v^k, \lambda^k, \rho)$ is found, then both v^k and λ^k are adjusted using some step formulae, and so on. The single-loop adjusting formulae for v and λ can be derived from the necessary conditions for a saddle point as formulae (2.91) and (2.92) were derived from (2.90) in the previous section. However, now

$$(L_\psi)'_{v_i}(c, v, \lambda, \rho) = \psi'_i(Q_a(c, v, \lambda, \rho))[(Q_i)'_{v_i}(c_i, H_i v) + \lambda_i + \rho(v_i - F_i(c_i, H_i v))]$$

$$+ \sum_{\substack{j=1 \\ j \neq i}}^{N} \psi'_j(Q_a(c, v, \lambda, \rho))(Q_j)'_{v_i}(c_j, H_j v)$$

$$- (F_j)'^*_{v_i}(c_j, H_j v)(\lambda_j + \rho(v_j - F_j(c_j, Hv)))], \qquad i \in \overline{1, N},$$

and the *multiplier strategy* takes the form

$$\lambda_i^{k+1} = \lambda_i^k + \rho(v_i^k - F_i(c_i^k, H_i v^k)), \tag{2.113λ}$$

$$v_i^{k+1} = F_i(c_i^k, H_i v^k) - \frac{1}{\rho} \left(\sum_{j=1}^{N} \frac{\psi'_j(k)}{\psi'_i(k)} (Q_j)'_{v_i}(c_j^k, H_j v^k) + \lambda_i^k \right.$$

$$\left. - \sum_{j=1}^{N} \frac{\psi'_j(k)}{\psi'_i(k)} (F_j)'^*_{v_i}(c_j^k, H_j v^k)\lambda_j^{k+1} \right), \qquad i \in \overline{1, N}, \quad (2.113\text{v})$$

where

$$\psi'_j(k) \triangleq \psi'_j(Q_a(c^k, v^k, \lambda^k, \rho)) \triangleq \frac{\partial \psi}{\partial Q_j}(Q_a(c^k, v^k, \lambda^k, \rho)), \qquad i \in 1, N.$$

Formulae (2.113λ) and (2.113v) are a special case of the ordinary gradient with variable step coefficients $\varepsilon_{i1}^k = \rho/\psi'_i(k)$ and $\varepsilon_{i2}^k = 1/\rho \cdot \psi'_i(k)$, respectively; i.e., they are equivalent to

$$\lambda_i^{k+1} = \lambda_i^k + \frac{\rho}{\psi'_i(k)} \hat{\psi}'_{\lambda_i}(v^k, \lambda^k, \rho),$$

$$v_i^{k+1} = v_i^k - \frac{1}{\rho \cdot \psi'_i(k)} \hat{\psi}'_{v_i}(v^k, \lambda^k, \rho), \qquad i \in \overline{1, N}.$$

The two-level strategy based on (2.112) and the multiplier strategy have been applied, and the results are presented in section 2.6.

The discussion concerning the global and local nature of coordinator strategies made for IPBM in the previous section remains valid here for OPBM, especially the discussion of the constrained choice of v for various cases of the set CU (notice that the sets V'_0 and U_0 are equivalent). However, we can always formulate the local problems of OPBM to obtain the case $V'_0 = \mathcal{Y}$ (the whole space) as we did for the penalty function method in section 2.3.4; the cost, however, is an increase in the dimensionality of the local decision problems where the controls and all or only some of the interaction inputs are treated as local decision variables. We obtain then the case $CV = C \times V$, and V can be treated in any standard way: e.g., by using projected gradients when possible, or generally forming an additional augmented Lagrangian term for the constraints describing V. The reader is referred to section 2.6, where some aspects of these problems are illustrated.

2.6. SIMULATION RESULTS

2.6.1. SAMPLE PROBLEM

Let us consider the steady-state system shown in Figure 2.4. The subsystem models are as follows:

Subsystem 1

$$y_1 = F_1(c_1, u_1) = c_{11} - c_{12} + 2u_1, \quad \text{where} \quad c_1 = (c_{11}, c_{12}),$$
$$Q_1(c_1, u_1) = (u_1 - 1)^4 + 5(c_{11} + c_{12} - 2)^2,$$
$$CU_1 = \{(c_1, u_1) \in \mathbb{R}^3 : (c_{11})^2 + (c_{12})^2 \leq 1 \wedge 0 \leq u_1 \leq 0.5\}.$$

Subsystem 2

$$c_2 = (c_{21}, c_{22}, c_{23}), \qquad u_2 = (u_{21}, u_{22}), \qquad y_2 = (y_{21}, y_{22}),$$
$$y_{21} = F_{21}(c_2, u_2) = c_{21} - c_{22} + u_{21} - 3u_{22},$$
$$y_{22} = F_{22}(c_2, u_2) = 2c_{22} - c_{23} - u_{21} + u_{22},$$
$$Q_2(c_2, u_2) = 2(c_{21} - 2)^2 + (c_{22})^2 + 3(c_{23})^2 + 4(u_{21})^2 + (u_{22})^2,$$
$$CU_2 = \{(c_2, u_2) \in \mathbb{R}^5 : 0.5c_{21} + c_{22} + 2c_{23} \leq 1 \wedge 4(c_{21})^2$$
$$+ 2c_{21}u_{21} + 0.4u_{21} + c_{21}c_{23} + 0.5(c_{23})^2 + (u_{21})^2 \leq 4\}.$$

Subsystem 3

$$y_3 = F_3(c_3, u_3) = c_{31} + 2.5c_{32} - 4u_3, \quad \text{where} \quad c_3 = (c_{31}, c_{32}),$$
$$Q_3(c_3, u_3) = (c_{31} + 1)^2 + (u_3 - 1)^2 + 2.5(c_{32})^2,$$
$$CU_3 = \{(c_3, u_3) \in \mathbb{R}^3 : c_{31} + u_3 + 0.5 \geq 0 \wedge 0 \leq c_{32} \leq 1\}.$$

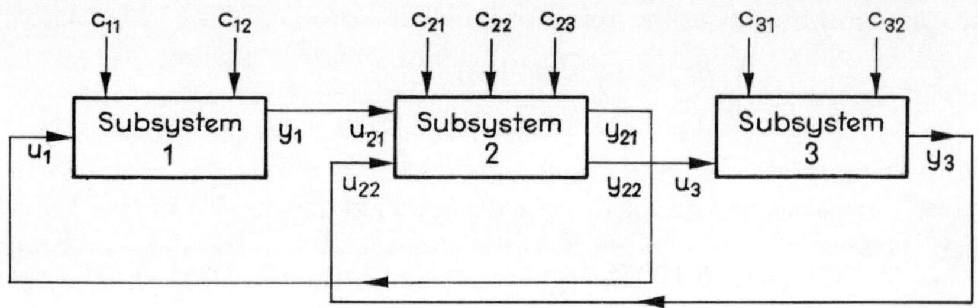

FIGURE 2.4   System structure used in the example.

All of the subsystem output equations are linear and all of the subsystem performance indices are convex, though not strictly in the first subsystem. The subsystems constraints are nonlinear in the first and second subsystems.

The goal of the optimization is to minimize the overall system performance index $Q(c, u)$ subject to subsystem outputs and constraints, where

$$Q(c, u) = Q_1(c_1, u_1) + Q_2(c_2, u_2) + Q_3(c_3, u_3), \tag{2.114}$$

$$Q(c, u) = Q_1(c_1, u_1) \cdot Q_2(c_2, u_2) + Q_3(c_3, u_3), \tag{2.115}$$

i.e., some methods will be tested using the overall performance index (2.114), and some using (2.115). In both cases Q is composed of Q_i, $i = 1, 2, 3$, through the following function ψ that strictly preserves order on some set Ω:

$$\psi(Q_1, Q_2, Q_3) = \sum_{i=1}^{3} Q_i, \quad \text{hence} \quad \Omega = \mathbb{R}^3, \quad \text{for} \quad (2.114);$$

$$\psi(Q_1, Q_2, Q_3) = Q_1 \cdot Q_2 + Q_3, \quad \text{hence} \quad \Omega = \mathbb{R}^2_{++} \times \mathbb{R}, \quad \text{for} \quad (2.115),$$

where

$$\mathbb{R}^2_{++} = \{(a_1, a_2) \in \mathbb{R}^2 : a_1 > 0, a_2 > 0\}.$$

Of course, the structure of the system must also be taken into account in the optimization. It can be easily seen from Figure 2.4 that the structure matrix H, where $u = Hy$ ($u \triangleq (u_1, u_2, u_3) \in \mathbb{R}^4$ and $y \triangleq (y_1, y_2, y_3) \in \mathbb{R}^4$), has the form

$$H = \left[\begin{array}{cccc} 0 & 1 & 0 & 0 \\ \hline 1 & 0 & 0 & 0 \\ 0 & 0 & 0 & 1 \\ \hline 0 & 0 & 1 & 0 \end{array} \right] = \left[\begin{array}{c} H_1 \\ \hline H_2 \\ \hline H_3 \end{array} \right]$$

The optimal solutions of the above optimization problem for (2.114) are:

$$\hat{c}_1 = (0.53054, 0.84766)$$
$$\hat{c}_2 = (0.99094, -0.14728, 0.00307)$$
$$\hat{c}_3 = (-0.50504, 0.34261)$$
$$\hat{u} = (0.17288, 0.02864, 0.33133, 0.00504)$$
$$Q(\hat{c}, \hat{u}) = 6.10075,$$

and for (2.115):

$$\hat{c}_1 = (0.53160, 0.84699)$$
$$\hat{c}_2 = (0.99710, -0.15238, -0.01001)$$
$$\hat{c}_3 = (-0.52438, 0.38192)$$
$$\hat{u} = (0.16456, 0.01374, 0.33288, 0.02438)$$
$$Q(\hat{c}, \hat{u}) = 6.73310,$$

where $Q_1(\hat{c}_1, \hat{u}_1) = 2.41784$, $Q_2(\hat{c}_2, \hat{u}_2) = 2.14671$; hence, $(Q_1(\hat{c}_1, \hat{u}_1),$ $Q_2(\hat{c}_2, \hat{u}_2),\ Q_3(\hat{c}_3, \hat{u}_3)) \in \text{int } \Omega$.

The problem was solved on an ODRA-1325 computer using various multilevel methods described in the previous sections. All local-level problems and coordinator problems were solved numerically with specified but not high accuracy. Therefore, the approximations of the accurate optimum, especially the values of the optimal controls c_i, differ slightly among the various methods.

2.6.2. THE DIRECT METHOD

In the direct method, the output variables y are taken as coordination variables v, i.e.,

$$v = (v_1, v_{21}, v_{22}, v_3) \triangleq y \in \mathbb{R}^4.$$

The local decision problems of the pure direct method are as follows:

First local problem

$$\text{minimize } [Q_1(c_1, H_1 v) = (v_{21} - 1)^4 + 5(c_{11} + c_{12} - 2)^2]$$
$$\text{subject to } c_1 \in C_1(v) = \{c_1 \in \mathbb{R}^2 : v_1 - F_1(c_1, H_1 v)$$
$$= v_1 - c_{11} - c_{12} + 2v_{21} = 0 \wedge (c_{11})^2 + (c_{22})^2 \leqslant 1\}.$$

Second local problem

$$\text{minimize } [Q_2(c_2, H_2 v) = 2(c_{21} - 2)^2 + (c_{22})^2 + 3(c_{23})^2 + 4(v_1)^2 + (v_3)^2]$$

160

subject to $c_2 \in C_2(v) = \{c_2 \in \mathbb{R}^3 : v_{21} - F_{21}(c_2, H_2 v)$

$$= v_{21} - c_{21} + c_{22} - v_1 + 3v_3 = 0 \wedge$$

$$v_{22} - F_{22}(c_2, H_2 v) = v_{22} - 2c_{22} + c_{23} + v_1 - v_3 = 0 \wedge$$

$$0.5c_{21} + c_{22} + 2c_{23} \leqslant 1 \wedge$$

$$4(c_{21})^2 + 2c_{21}v_1 + 0.4v_1 + c_{21}c_{23} + 0.5(c_{23})^2 + v_1^2 \leqslant 4\}.$$

Third local problem

minimize $[Q_3(c_3, H_3 v) = (c_{31} + 1)^2 + (v_{22} - 1)^2 + 2.5(c_{32})^2]$
subject to

$$c_3 \in C_3(v) = \{c_3 \in \mathbb{R}^2 : v_3 - F_3(c_3, H_3 v)$$

$$= v_3 - c_{31} - 2.5c_{32} + 4v_{22} = 0 \wedge$$

$$c_{31} + v_{22} + 0.5 \geqslant J \wedge 0 \leqslant c_{32} \leqslant 1\}.$$

It is possible to solve the above local decision problems analytically for given values of coordination variables v; the solution is difficult, though, especially in the second problem. The local decision sets $C_i(v)$ depend strongly on coordination variables v, hence, the set V_0,

$$V_0 = \{v \in \mathbb{R}^4 : C_i(v) \neq \varnothing, i = 1, 2, 3\},$$

must be taken into account in the coordinator problem. Because local problems have explicit analytical solutions, set V_0 may be determined, but it is not a simple job:

$$V_0 = \{v \in \mathbb{R}^4 : 0 \leqslant v_{21} \leqslant 0.5 \wedge 2v_{21} - v_1 - \sqrt{2} \leqslant 0 \wedge$$

$$-2v_{21} + v_1 - \sqrt{2} \leqslant 0 \wedge v_1 \leqslant \tfrac{7}{20}(-0.8 + \sqrt{0.64 + \tfrac{320}{7}}) \wedge$$

$$\Delta \geqslant 0 \wedge -D - \sqrt{\Delta} \leqslant 0 \wedge -4v_{22} - 2v_3 - 0.5 \leqslant 0\},$$

where

$$\Delta = D^2 - 32E, \qquad D = 10x_{1p} + 2v_1 + 3x_{3p},$$

$$E = 4x_{1p}^2 + 2x_{1p}v_1 + 0.4v_1 + x_{1p}x_{3p} + 0.5x_{3p}^2,$$

$$x_{1p} = \frac{1}{5.5}(-5A - 2B + 1).$$

$$x_{2p} = \frac{1}{5.5}(0.5A - 2B + 1),$$

$$x_{3p} = \frac{1}{5.5}(A + 1.5B + 2),$$

$$A = v_1 - 3v_3 - v_{21}, \qquad B = -v_1 + v_3 - v_{22}.$$

The coordinator problem is therefore as follows

$$\text{minimize } [\hat{Q}(v) = \psi(\hat{Q}_1(v), \hat{Q}_2(v), \hat{Q}_3(v))]$$

$$\text{subject to } v \in V_0,$$

where $\hat{Q}_i(v)$, $i = 1, 2, 3$, denote the solutions of local decision problems for a given value of v, $\psi = \sum_{i=1}^{3}$ for (2.114), and $\psi(\hat{Q}_1, \hat{Q}_2, \hat{Q}_3) = \hat{Q}_1 \cdot \hat{Q}_2 + \hat{Q}_3$ for (2.115).

Since the set V_0 cannot be violated during the optimization run, the coordinator problem was programmed using the interior penalty function technique (barrière function method), with the Powell procedure for unconstrained minimization. The gradient procedure would not be used here since the function $Q(\cdot)$ is not differentiable. The point $v = (0, 0, 0, 0)$ was taken as a starting point.

The unconstrained minimization of the barrière function was performed only once with the penalty coefficient $\rho = 10^{-5}$, which was satisfactory since the optimal point happened to lie in the interior of the set V_0. The final results for (2.114) are as follows:

$$\hat{c}_1 = (0.51322, 0.85825)$$

$$\hat{c}_2 = (0.98530, -0.13116, 0.00521)$$

$$\hat{c}_3 = (-0.50863, 0.34620)$$

$$\hat{u} = (0.19559, 0.04615, 0.32234, 0.00863)$$

$$Q(\hat{c}, \hat{u}) = 6.1076$$

and for (2.115)

$$\hat{c}_1 = (0.53604, 0.84418)$$

$$\hat{c}_2 = (0.99841, -0.15653, -0.01044)$$

$$\hat{c}_3 = (-0.05230, 0.38009)$$

$$\hat{u} = (0.15886, 0.00958, 0.33522, 0.02300)$$

$$Q(\hat{c}, \hat{u}) = 6.7347$$

The optimization runs up to the tenth Powell iteration (the computations were continued but without improvement of the results) are presented in Table 2.1. The columns of the table are the Powell procedure iteration number k, the value of the performance function after iteration k (Q_a^k for (2.114) or Q_b^k for (2.115)), and the number of iterations of the coordinator algorithm including iteration k (n_a^k for (2.114) or n_b^k for (2.115)). The number of iterations of the coordinator algorithm (i.e., local-level calls) is one of the best measures of the computational effort when multilevel methods are used.

TABLE 2.1 Simulation Results for the Direct Method

k	Q_a^k	n_a^k	Q_b^k	n_b^k
1	6.4242	27	7.0319	28
2	6.2840	44	6.8415	46
3	6.2499	65	6.8270	65
4	6.2225	85	6.8266	86
5	6.1245	108	6.7370	111
6	6.1227	124	6.7366	126
7	6.1221	143	6.7364	145
8	6.1220	161	6.7363	161
9	6.1079	188	6.7348	186
10	6.1076	201	6.7347	200

2.6.3. THE PENALTY FUNCTION METHOD

In the penalty function method, as in the direct method, output variables y are taken as coordination variables v, i.e.,

$$v = (v_1, v_{21}, v_{22}, v_3) \in \mathbb{R}^4.$$

Three local problems should be formulated, and we are free to choose the variables that will be the local decision variables. We can formulate the local decision problems in the form of $\mathrm{LP}_{\mathrm{pty}i}$ from section 2.3.1. It is the only reasonable way to formulate the first local problem, since CU_1 is separable

$$CU_1 = C_1 \times U_1 = \{c_1 \in \mathbb{R}^2 : (c_{11})^2 + (c_{12})^2 \leqslant 1\} \times \{u_1 \in \mathbb{R} : 0 \leqslant u_1 \leqslant 0.5\}.$$

Thus we obtain

First local problem

$$\text{minimize } [Q_{\mathrm{pty}1}(c_1, v, \rho_1) = Q_1(c_1, H_1 v) + \rho_1 \|v_1 - F_1(c_1, H_1 v)\|^2$$
$$= (v_{21} - 1)^4 + 5(c_{11} + c_{12} - 2)^2 + \rho_1(v_1 - c_{11} + c_{12} - 2v_{21})^2]$$

subject to $c_1 \in C_1$.

The formulation is not so evident with the second subsystem, since the set CU_2 is not separable. If we formulate the second local problem according to $\mathrm{LP}_{\mathrm{pty}2}$ (i.e., with c_2 as local decision variables only), we would get a local decision set dependent on v, and the nice properties of the coordinator problem are thus lost (see Appendix A). Therefore, we will formulate the second local problem according to $\mathrm{LP}''_{\mathrm{pty}i}$ (see section 2.3.4), and take only one interaction input (namely u_{21}) as the additional local decision variable to get the desired separability.

Second local problem

$$\text{minimize } [Q_{pty2}(c_2, u_{21}, v, \rho_2) = Q_2(c_2, u_{21}, v) + \rho_2(\|v_2 - F_2(c_2, u_{21}, v)\|^2$$
$$+ \|u_{21} - v_1\|^2) = 2(c_{21} - 2)^2 + (c_{22})^2 + 3(c_{23})^2$$
$$+ 4(u_{21})^2 + (v_3)^2 + \rho_2((v_{21} - c_{21}$$
$$+ c_{22} - u_{21} + 3v_3)^2 + (v_{22} - 2c_{22}$$
$$+ c_{23} + u_{21} - v_3)^2 + (u_{21} - v_1)^2)]$$

subject to $(c_2, u_{21}) \in CU_3$.

The third local problem will also be formulated according to LP''_{ptyi}:

Third local problem

$$\text{minimize } [Q_{pty3}(c_3, u_3, v, \rho_3) = Q_3(c_3, u_3) + \rho_3(\|v_3 - F_3(c_3, u_3)\|^2$$
$$+ \|u_3 - v_{22}\|^2) = (c_{31} + 1)^2 + (u_3 - 1)^2 + 2.5(c_{32})^2$$
$$+ \rho_3((v_3 - c_{31} - 2.5c_{32} + 4u_3)^2 + (u_3 - v_{22})^2)]$$

subject to $(c_3, u_3) \in CU_3$.

Observe that taking u_3 as the decision variable in the above local problem only endows the coordinator problem function with simple differentiability properties (see Appendix A). Defining the third local problem in the form of LP_{ptyi} (with c_3 as the local decision variable only) does not create any additional constraints on the coordination variables because

$$(\forall v \in \mathbb{R}^4) \qquad C_3(v) = \{c_3 \in \mathbb{R}^2 : c_{31} + v_{22} + 0.5 \geqslant 0 \land 0 \leqslant c_{32} \leqslant 1\} \neq \varnothing.$$

Having defined the local decision problems, we get the following *supremal problem*

$$\text{minimize } [\hat{Q}_{pty}(v, \rho) = \psi(\hat{Q}_{pty1}(v, \rho_1), \hat{Q}_{pty2}(v, \rho_2), \hat{Q}_{pty3}(v, \rho_3))]$$
$$\text{subject to } v \in V_0' = V_L = \{v \in \mathbb{R}^4 : 0 \leqslant v_{21} \leqslant 0.5\},$$

where $\hat{Q}_{ptyi}(v, \rho_i)$ denote the results of local problem minimizations for given values of v and ρ_i. The constraint set V_0' can be violated during the optimization runs and is therefore denoted by V_L (see section 2.3.1); hence, it can be treated by any standard method.

The sample problem was solved numerically using the penalty function method with local and supremal problems formulated as above. The values of the penalty coefficients were the same in each local decision problem, i.e., $\rho_1 = \rho_2 = \rho_3 = \rho$ and only Eq. (2.115) was considered, i.e., $\psi(Q_1, Q_2, Q_3) = Q_1 \cdot Q_2 + Q_3$. The optimization of the supremal problem (see section 2.3.1) was performed for increasing values of ρ, which gives a good insight into the nature of the method. Local problems were solved numerically using the shifting penalty function method (see, e.g., Wierzbicki and Kurcyusz 1977), and in the supremal problem the derivative evaluated according to formula (A.13) (Appendix A) was used to build conjugate gradients. The main results are shown in Table 2.2. In the rows of the table are optimization results corresponding to increasing values of the penalty coefficient $\rho = \rho^k$, $k = 1, 2, 3, 4$. Only the first row ($k = 0$) is different—it shows the starting point. The other columns of the table are:

$\hat{Q}^k$—The optimal value of the performance index of the supremal problem with $\rho = \rho^k$

$\hat{Q}_n^k$—The optimal value of the nonpenalized performance index (the penalty terms are set to zero)

n^k—The number of solutions of the local problems (number of lower-level calls) during the minimization of the supremal problem with $\rho = \rho^k$

$\hat{v}^k = (\hat{v}_1^k, \hat{v}_{21}^k, \hat{v}_{22}^k, \hat{v}_3^k)$—The optimal values of the coordination variables obtained by $\rho = \rho^k$

$\|e^k\|$—The norm of the coordination error at the optimal point for $\rho = \rho^k$

The optimal value $\hat{v}^k$ obtained from optimization with $\rho = \rho^k$ ($k = 0, 1, 2, 3$) was taken as the starting point for optimization with the next larger value of ρ. The coordination error is as follows:

$$e_1(c_1, v) = v_1 - F_1(c_1, H_1 v) = v_1 - c_{11} + c_{12} - 2v_{21},$$

$$e_2(c_2, u_{21}, v) = \begin{bmatrix} v_{21} - F_{21}(c_2, u_{21}, v) = v_{21} - c_{21} + c_{22} - u_{21} + 3v \\ v_{22} - F_{22}(c_2, u_{21}, v) = v_{22} - 2c_{22} + c_{23} + u_{21} - v_3 \\ u_{21} - v_1 \end{bmatrix},$$

$$e_3(c_3, u_3, v) = \begin{bmatrix} v_3 - F_3(c_3, u_3) = v_3 - c_{31} - 2.5c_{32} + 4u_3 \\ u_3 - v_{22} \end{bmatrix},$$

$$e(c, u, v) = \begin{bmatrix} e_1(c_1, v) \\ e_2(c_2, u_{21}, v) \\ e_3(c_3, u_3, v) \end{bmatrix} \tag{2.116}$$

TABLE 2.2 Simulation Results for the Penalty Function Method

k	ρ^k	$\hat{Q}^k$	$\hat{Q}_n^k$	n^k	$\hat{v}_1^k$	$\hat{v}_2^k$	$\hat{v}_3^k$	$\hat{v}_4^k$	$\|e^k\|$
0 (initial point)	10	16.698		11	0.0	0.0	0.0	0.0	0.3289
1	10	6.6382	6.44443	33	0.06260	0.20077	−0.03799	0.32029	0.0993
2	10^2	6.7609	6.7325	24	0.04337	0.17502	−0.03149	0.34165	0.0124
3	10^3	6.7899	6.7869	28	0.03042	0.17140	−0.04335	0.34220	0.00125
4	10^4	6.7933	6.7933	3^a	0.03040	0.17140	−0.04448	0.34195	0.00012

[a] The line search was performed succesfully only in one direction, the first.

Looking at Table 2.2, we can see the main features of the penalty function method: the greater the penalty coefficient the lower the coordination error and the greater the optimal value of the performance index for given ρ. The convergence of the performance index values $\hat{Q}^k$ and $\hat{Q}_n^k$ to themselves, which means that the penalty terms converge to zero, can also be observed. The results are a good illustration of Theorem 2.14 in section 2.3.

To be complete, let us give the optimal values of controls and interactions obtained in the last iteration ($k = 4$):

$$\hat{c}_1 = (0.53348, 0.84815)$$
$$\hat{c}_2 = (0.99043, -0.17651, 0.00304)$$
$$\hat{c}_3 = (-0.45136, 0.24860)$$
$$\hat{u} = (0.17140, 0.03033, 0.34195, -0.04300).$$

2.6.4. THE PRICE METHOD

The application of the price method (IBM) requires a reformulation of the problem (see section 2.4). It is necessary to introduce the following local problems (the price method can be applied when performance $Q(c, u)$ is in additive form (2.114)):

First local problem

minimize $[L_1(c_1, u_1, \lambda) = (u_1 - 1)^4 + 5(c_{11} + c_{12} - 2)^2 + \lambda_1 u_1 - \lambda_{21}(c_{11} - c_{12} - 1)$

subject to $(c_1, u_1) \in CU_1$.

Second local problem

$$\begin{aligned}
\text{minimize } L_2(c_2, u_2, \lambda) = {}&2(c_{21} - 2)^2 + (c_{22})^2 + 3(c_{23})^2 + 4(u_{21})^2 + (u_{22})^2 \\
&+ \lambda_{21} u_{21} + \lambda_{22} u_{22} - \lambda_1(c_{21} - c_{22} + u_{21} - 3u_{22}) \\
&- \lambda_3(2c_{22} - c_{23} - u_{21} + u_{22})
\end{aligned}$$

subject to $(c_2, u_2) \in CU_2$

Third local problem

$$\begin{aligned}
\text{minimize } L_3(c_3, u_3, \lambda) = {}&(c_{31} + 1)^2 + (u_3 - 1)^2 \\
&+ 2.5(c_{32})^2 + \lambda_3 u_3 - \lambda_{22}(c_{32} + 2.5c_{32} - 4u_3)
\end{aligned}$$

subject to $(c_3, u_3) \in CU_3$.

The dimensions of the local problems are increased when compared with the direct method. Yet, the local constraints do not depend on the coordinating inputs λ.

The coordinator task consists (see section 2.4) in finding $\hat{\lambda}$ for which the local solutions satisfy the interconnection equations. To accomplish this task for our problem, two coordination algorithms have been tried. In the first case the dual function

$$\varphi(\lambda) = \max_{(c,u)\in CU} \sum_{i=1}^{3} L_i(c_i, u_i, \lambda)$$

obtained through local problem solutions was maximized with the standard conjugate gradient procedure for unconstrained optimization. The algorithm was stopped after the condition $\|\nabla\varphi(\lambda)\| \leq 10^{-3}$ was satisfied (where $\nabla\varphi(\lambda) = -W(\lambda)$).

In the second case, algorithm (2.68) from section 2.4 was applied and we set $s(\lambda) = 0$. Therefore, the coordination strategy was as follows:

$$\lambda^{k+1} = \lambda^k - \varepsilon A W(\lambda^k).$$

Two possiblities for matrix A were tried: $A = I$ (with $\varepsilon = 0.8$), and $A = A_0^{-1}$ (with $\varepsilon = 0.8$), where A_0 was chosen according to the method described in Appendix A. The stop criterion $\|w(\lambda)\| \leq \delta$ was used, with δ ranging from 10^{-2} to 10^{-4}.

In both cases the local problems were solved with the shifting penalty function method using the conjugate gradient algorithm for unconstrained optimization. In both cases the coordination algorithms were started from $\lambda^0 = 0$.

The major numerical results are displayed in Table 2.3. $\hat{Q}_a$ denotes the optimal value of the performance index and n denotes the number of iterations of the coordination algorithm. The maximization of the dual function gave slightly better results (smaller n) than the application of algorithm (2.68). This algorithm, however, is to be used mainly for on-line coordination purposes (see Chapter 3).

TABLE 2.3   Simulation Results for the Price Method

	$\hat{Q}_a$	Stop Criterion	n	$\lambda = (\lambda_1, \lambda_{21}, \lambda_{22}, \lambda_3)$
Case I ($\mathscr{C}\mathscr{G}$)	6.101	$\|W(\lambda)\| \leq 10^{-3}$	66	$(-0.604, -1.473, 0.659, -0.411)$
Case II ($\mathscr{A}$)	6.100	$\|W(\lambda)\| \leq 10^{-2}$	58	$(-0.602, -1.643, 0.679, -0.440)$
$A = I$				
Case II ($\mathscr{A}$)	6.100	$\|W(\lambda)\| \leq 10^{-2}$	48	
$A = A_0^{-1}$	6.101	$\|W(\lambda)\| \leq 10^{-3}$	73	
	6.101	$\|W(\lambda)\| \leq 10^{-4}$	97	$(-0.619, -1.654, 0.679, -0.440)$

2.6.5. THE LINEARIZED AUGMENTED LAGRANGIAN METHOD

The linearized augmented Lagrangian method, like the price method, is applicable only in the additive case (2.114) and uses all controls and interaction inputs as local decision variables.

The linearized augmented Lagrangian (2.79) for the sample problem has the following form

$$\Lambda(c, u, \underline{c}, \underline{u}, \lambda, \rho) \;=\; \sum_{i=1}^{3} Q_i(c_i, u_i) + \lambda_1(u_1 - F_{21}(c_2, u_2))$$

$$+ \lambda_{21}(u_{21} - F_1(c_1, u_1)) + \lambda_{22}(u_{22} - F_3(c_3, u_3)) + \lambda_3(u_3 - F_{22}(c_2, u_2))$$
$$+ \tfrac{1}{2}\rho((u_1)^2 + (u_{21})^2 + (u_{22})^2 + (u_3)^2 + (F_1(c_1, u_1))^2 + (F_{21}(c_2, u_2))^2 + (F_{22}(c_2, u_2))^2$$
$$+ (F_3(c_3, u_3))^2) + \rho(\underline{u}_1 y_{21} + \underline{u}_{21} \underline{y}_1 + u_{22} y_3 + \underline{u}_3 y_{22} - u_1 \underline{y}_{21} - u_{21} \underline{y}_1 - u_{22} y_3$$
$$- u_3 y_{22} - \underline{u}_1 F_{21}(c_2, u_2) - \underline{u}_{21} F_1(c_1, u_1) - \underline{u}_{22} F_3(c_3, u_3) - \underline{u}_3 F_{22}(c_2, u_3)),$$

where $\underline{y} \triangleq F(\underline{c}, \underline{u})$, $\lambda = (\lambda_1, \lambda_{21}, \lambda_{22}, \lambda_3) \in \mathbb{R}^4$ is the vector of multipliers, and $\rho \in \mathbb{R}_+$ is the penalty coefficient. Hence, local decision problems LP_{ai}^{k} are as follows

First local problem

minimize $Q_1(c_1, u_1) + \lambda_1^k u_1 - \lambda_{21}^k F_1(c_1, u_1) + \tfrac{1}{2}\rho((u_1)^2 + (F_1(c_1, u_1))^2)$
$$- \rho(u_1 y_{21}^k + u_{21}^k F_1(c_1, u_1))$$

subject to $(c_1, u_1) \in CU_1$,

Second local problem

minimize $Q_2(c_2, u_2) + \lambda_{21}^k u_{21} + \lambda_{22}^k u_{22} - \lambda_1^k F_{21}(c_2, u_2)$
$$- \lambda_3^k F_{22}(c_2, u_2) + \tfrac{1}{2}\rho((u_{21})^2 + (u_{22})^2 + (F_{21}(c_2, u_2))^2$$
$$+ (F_{22}(c_2, u_2)^2) - \rho(u_{21} y_1^k + u_{22} y_3^k + u_1^k F_{21}(c_2, u_2)$$
$$+ u_3^k F_{22}(c_2, u_2))$$

subject to $(c_2, u_2) \in CU_2$,

Third local problem

minimize $Q_3(c_3, u_3) + \lambda_3^k u_3 - \lambda_{22}^k F_3(c_3, u_3)$
$$+ \tfrac{1}{2}\rho((u_3)^2 + (F_3(c_3, u_3))^2 - \rho(u_3 y_{22}^k + u_{22}^k F_3(c_3, u_3))$$

subject to $(c_3, u_3) \in CU_3$,

Denoting the solutions of local problems as

$$c_i^{k+1} = \hat{c}_i(u^k, y^k, \lambda^k, \rho), \qquad u_i^{k+1} = \hat{u}_i(u^k, y^k, \lambda^k, \rho), \qquad i \in \overline{1, N},$$

168

and taking $y_i^{k+1} \triangleq F_i(c_i^{k+1}, u_i^{k+1})$, $i \in \overline{1, N}$, we get the following coordinator problem:

Adjust the Lagrangian multipliers according to the formula

$$\lambda_i^{k+1} = \lambda_i^k + \tfrac{1}{2}\rho(u_i^k - H_i y^k), \qquad i \in \overline{1, N}.$$

In fact, the more complicated algorithm (2.83) was applied instead of the above simple coordinator problem, but it was not needed because it was not necessary to increase the initial value of the penalty coefficient $\rho = 1$. The initial values of all variables (multipliers, inputs, controls) were set to zero. The optimization run is presented in Table 2.4. The columns of the table are:

k—The iteration number,

Q^k—The value of the linearized augmented Lagrangian after iteration k,

Q_n^k—The value of the original, unmodified performance index after iteration k,

u^k—The optimal values of the interaction inputs,

λ^k—The values of the multipliers.

For completeness, the optimal values of the controls after 14 iterations are given below:

$$\hat{c}_1 = (0.53110, 0.84730)$$

$$\hat{c}_2 = (0.99100, -0.14855, 0.00322)$$

$$\hat{c}_3 = (-0.50524, 0.34238).$$

2.6.6. THE INPUT PREDICTION AND BALANCE METHOD

The classical augmented Lagrangian for the sample problem has the form (compare section 2.5.3)

$$\begin{aligned}
L_a(c, u, \lambda, \rho) = \sum_{i=1}^3 Q_i(c_i, u_i) &+ \lambda_1(u_1 - F_{21}(c_2, u_2)) + \lambda_{21}(u_{21} - F_1(c_1, u_1)) \\
&+ \lambda_{22}(u_{22} - F_3(c_3, u_3)) + \lambda_3(u_3 - F_{22}(c_2, u_2)) \\
&+ \tfrac{1}{2}\rho((u_1)^2 + (u_{21})^2 + (u_{22})^2 + (u_3)^2 + (F_1(c_1, u_1))^2 \\
&+ (F_{21}(c_2, u_2))^2 + (F_{22}(c_2, u_2))^2 + (F_3(c_2, u_2))^2) - \rho(u_1 F_{21}(c_2, u_2) \\
&+ u_{21} F_1(c_1, u_1) + u_{22} F_3(c_3, u_3) + u_3 F_{22}(c_2, u_2)),
\end{aligned}$$

which is separable with respect to c_i, $i \in \overline{1, N}$. We cannot treat variables u_{21} and u_3 as local decision variables, as in the penalty function method, owing to mixed terms in the last parentheses. Hence, the local decision problems

TABLE 2.4 Simulation Results for the Linearized Augmented Lagrangian Method

k	Q^k	Q_n^k	u_1^k	u_{21}^k	u_{22}^k	u_3^k	λ_1^k	λ_{21}^k	λ_{22}^k	λ_3^k
1	5.39999	5.05776	0.32192	−0.07719	0.10465	0.00978	0.0	0.0	0.0	0.0
2	5.51102	5.22349	0.30938	−0.04790	0.08797	0.07324	0.00173	−0.30960	0.35643	−0.05644
3	5.69315	5.41012	0.28922	−0.02503	0.13197	0.04851	−0.14893	−0.59711	0.56682	−0.13626
4	5.83604	5.59615	0.26509	−0.00739	0.19460	0.02910	−0.34019	−0.83509	0.67523	−0.20544
5	5.91249	5.76415	0.24052	0.00556	0.25116	0.01558	−0.49630	−1.02210	0.72434	−0.26534
6	5.94094	5.89878	0.21888	0.01459	0.29199	0.00734	−0.59180	−1.16178	0.74087	−0.32029
7	5.94757	5.99354	0.20208	0.02057	0.31675	0.00322	−0.63364	−1.26029	0.73961	−0.36934
8	5.94821	6.05153	0.19037	0.02430	0.32921	0.00188	−0.64090	−1.32575	0.72932	−0.40873
9	5.94895	6.08162	0.18296	0.02646	0.33391	0.00212	−0.63192	1.36687	0.71583	0.43580
10	5.95091	6.09402	0.17865	0.02760	0.33459	0.00301	−0.61908	−1.39149	0.70306	−0.45079
11	5.95359	6.09723	0.17631	0.02812	0.33371	0.00394	−0.60848	−1.40581	0.69320	−0.45633
12	5.95620	6.09691	0.17508	0.02832	0.33255	0.00465	−0.60182	−1.41414	0.68690	−0.45603
13	5.95826	6.09608	0.17441	0.02837	0.33166	0.00507	−0.59853	−1.41919	0.68376	−0.45315
14	5.95959	6.09586	0.17400	0.02839	0.33114	0.00525	−0.59737	−1.42248	0.68282	−0.44993

must be formulated with LP_{ai} that was introduced in section 2.5.3:

First local problem

minimize $Q_1(c_1, u_1) + \lambda_1 u_1 - \lambda_{21} F_1(c_1, u_1)$

$$+ \tfrac{1}{2}\rho((u_1)^2 + (F_1(c_1, u_1))^2) - \rho u_{21} F_1(c_1, u_1)$$

subject to $c_1 \in C_1 = \{c_1 \in \mathbb{R}^2: (c_{11})^2 + (c_{12})^2 \leq 1\}$,

Second local problem

minimize $Q_2(c_2, u_2) + \lambda_{21} u_{21} + \lambda_{22} u_{22} - \lambda_1 F_{21}(c_2, u_2) - \lambda_3 F_{22}(c_2, u_2)$

$$+ \rho((u_{21})^2 + (u_{22})^2 + (F_{21}(c_2, u_2))^2 + (F_{22}(c_2, u_2))^2)$$

$$- \rho(u_1 F_{21}(c_2, u_2) + u_3 F_{22}(c_2, u_2))$$

subject to $c_2 \in C_2(u) = \{c_2 \in \mathbb{R}^3: 0.5c_{21} + c_{22} + 2c_{23} \leq 1 \wedge$

$$4(c_{21})^2 + 2c_{21} u_{21} + 0.4u_{21} + c_{21} c_{23} + 0.5(c_{23})^2 + (u_{21})^2 \leq 4\},$$

Third local problem

minimize $Q_3(c_3, u_3) + \lambda_3 u_3 - \lambda_{22} F_3(c_3, u_3)$

$$+ \tfrac{1}{2}\rho((u_3)^2 + (F_3(c_3, u_3))^2) - \rho u_{22} F_3(c_3, u_3)$$

subject to $c_3 \in C_3(u) = \{c_3 \in \mathbb{R}^2: c_{31} + u_3 + 0.5 \geq 0 \wedge 0 \leq c_{32} \leq 1\}$.

The method uses interaction inputs $u = (u_1, u_{21}, u_{22}, u_3) \in \mathbb{R}^4$ as coordination variables. Only in the first local problem is the constraint set not dependent on the coordination variables because of the separability of the set $CU_1 = C_1 \times U_1$. The constraint sets in the second and third local problems depend continuously on the values of u. Luckily, the constraint set in the third problem $C_3(u)$ is not empty for all values of u because there are no absolute value constraints on c_{31}; this is not a typical formulation. But the set $C_2(u)$ may be empty for some values of u, namely, when u does not belong to the set

$$U_{20} = \{u \in \mathbb{R}^4: -2.662939361 \leq u_{21} \leq 2.102939361\}.$$

Thus, the admissible input prediction set U_0 consists of two parts:

$$U_0 = U_{20} \cap U_1 = U_{20} \cap \{u \in \mathbb{R}^4: 0 \leq u_1 \leq 0.5\},$$

and the set U_{20} cannot be violated during the optimization run.

Because of the dependence of the local constraint sets on u, the gradient formula (A.13) derived in Appendix A cannot be applied to the coordinator problem functional

$$\varphi(\cdot, \lambda, \rho) = \sum_{i=1}^{3} \hat{L}_{ai}(\cdot, \lambda, \rho),$$

where $\hat{L}_{ai}(u, \lambda, \rho)$ denotes the optimization result of the local decision problem i, $i = 1, 2, 3$. There are two possible approaches from this point:

- To apply a two-level coordinator strategy to minimize the function $\varphi(\cdot, \lambda, \rho)$ by means of a nongradient procedure;
- To evaluate the gradient of $\varphi(\cdot, \lambda, \rho)$ using Theorem A.3 from Appendix A, and then to apply two-level coordinator strategy using a gradient procedure for minimizing $\varphi(\cdot, \lambda, \rho)$, or to apply a one-level strategy to the obtained gradient. The gradient of $\varphi(\cdot, \lambda, \rho)$ exists only in the interior of set U_{20}.

For the simulation problem the first possibility was chosen. The multipliers were adjusted according to Hestenes's formula with $\rho = 100$. The function $\varphi(\cdot, \lambda, \rho)$ was minimized by the Powell procedure. The coordinator algorithm was started from $\lambda^0 = 0$, $u^0 = 0$. The local problems were solved numerically using the shifting penalty function method for constrained minimization. The results are given in Table 2.5. The columns of the table are:

k—The iteration number corresponding to changes in λ,
n^k—The number of local-level calls from iteration 1 to k, inclusive,
$\hat{Q}^k$—The value of the performance index after iteration k,
$\|e^k\|$—The norm of the coordination error after iteration k, where

$$e^k \triangleq u^k - HF(c^k, u^k).$$

The final values of the price interactions and controls were:

$$\hat{\lambda} = (-0.5655, -1.509, 0.6802, -0.4199),$$

$$\hat{u} = (0.17503, 0.03099, 0.32888, 0.01225),$$

$$\hat{c}_1 = (0.5286, 0.8489),$$

$$\hat{c}_2 = (0.9904, -0.1413, 0.0033),$$

$$\hat{c}_3 = (-0.5122, 0.3559).$$

TABLE 2.5 Simulation Results for the Input Prediction and Balance Method

k	n^k	$\hat{Q}^k$	$\|e^k\|$
1	44	6.1158	0.00032
2	65	6.101	7×10^{-6}

2.6.7. THE OUTPUT PREDICTION AND BALANCE METHOD

The output prediction and balance method (OPBM), described in section 2.5.4, is a generalization of the penalty function method (PFM). The local problems of OPBM are simple modifications of the local problems of PFM that are obtained by using linear and quadratic modifications of local performance indices instead of only quadratic modifications. Applying OPBM to the sample problem, we get:

First local problem

$$\text{minimize}\,[Q_{a1}(c_1, v, \lambda_1, \rho) = Q_1(c_1, H_1 v) + \rho\|v_1 - F_1(c_1, H_1 v)\|^2$$
$$+ \langle \lambda_1, v_1 - F_1(c_1, H_1 v)\rangle = (v_{21} - 1)^4 + 5(c_{11} + c_{12} - 2)^2$$
$$+ \rho(v_1 - c_{11} + c_{12} - 2v_{21})^2 + \lambda_1(v_1 - c_{11} + c_{12} - 2v_{21})$$

subject to $c_1 \in C_1$.

The second and third local problems for OPBM are also similar to those of PFM(see section 2.6.3) and therefore will not be given here.

The simulation problem was solved using two different coordination strategies of OPBM:

- The two-level strategy with multipliers adjusted according to formula (2.112),
- The one-level strategy described by formulae (2.113). Only case (2.115) was considered, i.e., $\psi(Q_1, Q_2, Q_3) = Q_1 \cdot Q_2 + Q_3$. Output variables $v \in \mathbb{R}^4$ and multipliers $\lambda \in \mathbb{R}^6$ were set to zero initially. The value of the penalty coefficient was fixed and equal to 100 and 200 for the one-level and two-level strategies, respectively. In both cases, local problems were solved numerically using the shifting penalty function method for constrained minimization.

Using the two-level strategy, the minimization of the function

$$\varphi(\cdot, \lambda^k, \rho) \triangleq \hat{Q}_{a1}(\cdot, \lambda_1^k, \rho) \cdot \hat{Q}_{a2}(\cdot, \lambda_2^k, \rho) + \hat{Q}_{a3}(\cdot, \lambda_3^k, \rho),$$

was performed by means of the conjugate gradient method, where $\hat{Q}_{ai}(v, \lambda_i^k, \rho)$ denotes the ith local problem solution for given values of v, λ^k, and ρ, $i = 1, 2, 3$. The constraint $v \in V_0' = \{v \in \mathbb{R}^4 : 0 \le v_{21} \le 0.5\}$ was not essential since it was inactive (it could be included in a penalty term or an augmented Lagrangian term). The computation results are given in Table 2.6.

The columns of the table are the same as in Table 2.5 except that the coordination error is now defined as in Eq. (2.116). Satisfactory results were obtained after the fifth iteration. The following iterations do not improve the value of the performance index because we are near the optimum. The final

TABLE 2.6 Simulation Results for the Two-Level Strategy of the Output Prediction and Balance Method

k	n^k	$\hat{Q}^k$	$\|e^k\|$
1	25	6.7073	0.0114
2	30	6.7266	0.0059
3	35	6.7315	0.0029
4	41	6.7328	0.0051
5	42	6.7331	0.0008
6	43	6.7332	0.00042
7	44	6.7332	0.00022
8	45	6.7332	0.00012
9	46	6.7332	0.00006
10	47	6.7332	0.00003

values of the controls and interactions (recall $u = Hy = Hv$) were

$$\hat{c}_1 = (0.53177, 0.84688)$$

$$\hat{c}_2 = (0.99637, -0.14966, 0.01002)$$

$$\hat{c}_3 = (-0.52674, 0.3863)$$

$$\hat{u} = (0.16562, 0.01613, 0.33229, 0.02674).$$

For the one-level strategy (2.113), the computations were performed with the penalty coefficient $\rho = 200$, since with $\rho = 100$ there was no convergence (the results oscillated). The results for $\rho = 200$ are given in Table 2.7. The columns of the table are:

k—The iteration number corresponding to changes in λ and v,
Q^k—The actual value of the performance index after iteration k,
$\|e^k\|$—The norm of the coordination error, as defined by Eq. (2.116).

The final values of the controls and interactions were:

$$\hat{c}_1 = (0.53172, 0.84691)$$

$$\hat{c}_2 = (0.99835, -0.15431, 0.00964)$$

$$\hat{c}_3 = (-0.52480, 0.38296)$$

$$\hat{u} = (0.16233, 0.00944, 0.3320, 0.02482).$$

TABLE 2.7 Simulation Results for the One-Level Strategy of the Output Prediction and Balance Method

k	Q^k	$\|e^k\|$
1	20.222	0.0214
2	8.9090	0.0043
3	6.7932	0.0128
4	6.7826	0.00338
5	6.7736	0.00299
6	6.7666	0.00088
7	6.7610	0.00071
8	6.7565	0.00027
9	6.7529	0.00021
10	6.7501	0.00013
15	6.7419	0.000071
20	6.7382	0.000046
30	6.7349	0.000027
40	6.7338	0.000017
52	6.7333	0.000010

2.7. DECOMPOSED OPTIMIZATION IN ON-LINE CONTROL

The computational methods and algorithms for coordinated solution of decomposed optimization problems that have been described in this chapter can be used for model-based control decisions for complex systems. The control structure would be an open-loop structure because the algorithms cannot accommodate any feedback from the real system before a calculation is finished. In Chapter 3, algorithms are elaborated that accept feedback information from the real system during the solution iterations for steady-state control, and Chapter 4 presents methods using feedback to improve dynamic control.

Open-loop optimization algorithms are used in the optimizing layer of a multilayer control structure (see Figure 1.4). Thus, the optimization algorithms are not used to directly determine the values of manipulated variables (control actions) of a process, but rather to determine the tasks of first-layer controllers, for example, the so-called set points or desired values c_d of some chosen variables c.

This two-layer structure, described in some detail in Chapter 1, makes extensive use of feedback in the direct control layer to perform the follow-up task (forcing c to equal c_d) as accurately as possible. It is the determination of the optimal values of c_d which would be done in an open-loop structure using the algorithms of Chapter 2 (closed-loop algorithms are used in Chapters 3 and 4).

The two-layer structure to control a complex system will look like the one shown in Figure 2.5, where the direct control layer is sometimes not entirely decentralized, that is, some links between the controllers are possible and may be necessary. The essential feature of the structure is that the first-layer controllers do not optimize the process but perform a regulatory action, that is, they enforce some given trajectories of c_d. Whether the enforcement can be done in some optimal way, for example, by minimizing a quadratic performance functional is a different question. The design of that part of the system, that is the choice of the structure and parameters of direct controllers, as well as the choice of a few intercontroller linkages, belongs to the area of multivariable control systems. This problem is mentioned in section 4.6 of Chapter 4, but the area is very broad and beyond the scope of this book.

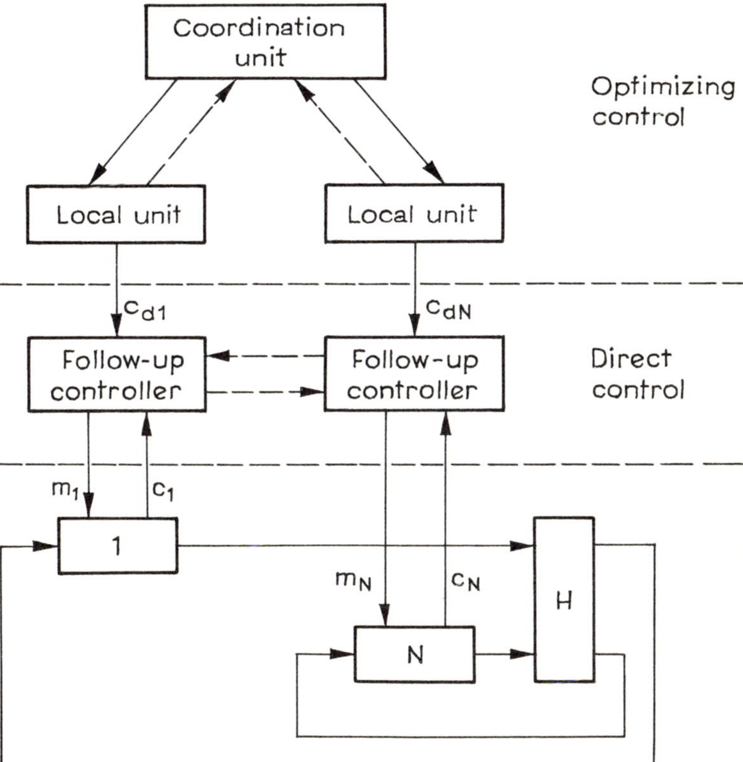

FIGURE 2.5   A two-layer structure with on-line open-loop optimizing control.

Coming back to the open-loop optimizing layer, we may ask why a model-based solution of the optimization problem should be done in a decomposed fashion; a straightforward, one-level solution algorithm with the same system model would be—in principle—identical. We have hinted at that question in Chapter 1; see Figure 1.6. What are the advantages and disadvantages of a two-level or multilevel solution algorithm in on-line, open-loop optimizing control?

The advantages can be seen if we consider that the decomposed solution can be programmed into a hierarchy of computers rather than only one computer. The immediate results could be:

- Reduced requirements for transmission of information since more optimizing commands will be generated locally than the number of coordinating commands to be transmitted from the control unit;
- Increased reliability of the system: if one of the local computers fails, the rest of the system can continue to operate;
- Reduced overall computational requirements due to parallel computations in lower-level computers and simple computations in higher-level computers.

Note that the above features, in particular the information transmission aspect, remain valid if we consider human decision makers instead of computers as the control units.

The informational aspect gains in importance if we consider that parts of the controlled system itself may be located at a large distance from one another. The reliability aspect gains in importance when parts of the system are relatively weakly coupled in the sense that a lack of optimizing control in one part does not affect the performance of the system as a whole very much.

There may be cases where the choice of one computer or several computers has already been made. The subsystems may simply have their own computers for performing the task of direct control or for other purposes. In that case a divided, or decomposed, optimizing control algorithm is a natural solution.

If there is one central computer in which the on-line, open-loop optimization problem is to be solved, then the only reason for using decomposition would be for more efficient computation. In our experience, however, mostly with nonlinear programming applied to complex chemical processes, decomposition did not increase computational efficiency. Thus, in open-loop optimizing control, we recommend decomposed methods of solution when the decision units (the control computers) are divided, even though the computations may be less efficient. We also recommend that for appropriate conditions and system layout one should use several computers for optimization rather than a single computer.

The open-loop optimizing algorithms can be used to solve either static or dynamic optimization problems. There is no difference in the principles of the methods of coordination and for that reason Chapter 2 has been written in a general notation to cover both cases. However, in the application to on-line optimizing control, a distinction between static and dynamic problems has to be made. In the first case, we deal with on-line, steady-state optimization. More precisely, the optimal trajectory of c_d is determined as a sequence of values that are constant over some intervals T_i, and moreover, the conditions of the problem permit us to determine each value by static optimization, that is, in the steady state (see Chapter 1, Figure 1.15).

There should be some reasonable incentive, as well as a data base, for recalculating the c_d in the on-line control system, even if it is an open-loop optimizing structure; otherwise, one solution would do forever. There are two groups of such reasons:

1. Information about changes in external factors, i.e., disturbances such as changes in prices, loads, desired production rate, raw material properties, ambient temperature;

2. Information about a change in the system model, for example, a decrease in catalyst activity or malfunction of a system element; the source of such information may be on-line model identification or a knowledge of its time-varying or disturbance-related properties.

Let us note that case 1 means a change of some exogenous variables in the model and case 2 a change of model parameters. Both cases call for a recomputation of the optimal values of the endogenous variables in the model, in particular of the control decision $\hat{c}_d$, to the same degree. It is not relevant, in principle, whether the recomputations are done routinely at constant time intervals, or on demand, that is, whenever a significant change in exogenous variables or parameters is observed. Some discussion of this problem can be found in Donoghue and Lefkowitz (1972) and Tsuji and Lefkowitz (1975).

The picture will be slightly changed if the on-line, open-loop optimizing control is concerned with a dynamic optimization problem. In that case the trajectory $\hat{c}_d$ may also be represented as a sequence of values that are constant over some intervals T_i. However, this sequence of values has to be determined at $t = 0$ for the whole optimization horizon (compare Chapter 1, Figure 1.10) and the computation must take into account the dynamic properties of the controlled system as well as the future disturbances. The sequence of values of $\hat{c}_d$ will then be a solution to a dynamic optimization problem.

For the steady-state case, two conditions were given for a repetition of the optimization computation: a change of exogenous variables or a change in model parameters. The same changes may require a repetition of the open-loop dynamic solution computation. However, the meaning of intervals T_i over which $\hat{c}_d(\cdot)$ is kept constant is now different: these intervals result from a discretization of the continuous dynamic problem and have little to do with the need to recalculate c_d for new data. In other words, this recalculation can be done for example, every $5T_i$ or every $10T_i$, or on demand. In the steady-state case, the intervals T_i were exclusively intervals of recalculation.

The computation of the solution for a dynamic optimization problem must include the initial state $x(0)$ of the controlled system; it must be assessed and introduced into the open-loop control program. What should be done each time the dynamic optimal control is recalculated? If we use a guessed value of the state, for example, the state that was prescribed for that time by the computation done at $t = 0$, then our control is indeed on open-loop one, and recalculation is caused and justified by external factors only. Should, to the contrary, a measured value of the system state be used as the initial condition for each recalculation, we have a closed-loop dynamic structure referred to as repetitive control. This kind of structure is discussed extensively in Chapter 4.

The last point we wish to make in this brief survey of problems lying at the interface of optimization and on-line control relates to the question of whether optimization models, algorithms, and procedures used in on-line control should be identical or similar to those that would be used for a single, unrepeated solution of an optimization problem, for example, the kind of optimization problem that arises in system design, where one wants to determine the optimal operating conditions of a process. Generally, they should differ, first, because accuracy may be sacrificed for ease of computation. It will seldom be the case that both the models and the disturbance data are very accurate, so for that reason alone the solution iterations may be stopped earlier than for many other scientific or engineering calculations. Second, the on-line solution will have to be repeated many times for changed values of a few, and always the same few, variables or parameters. And last, the recalculations in on-line control start in the vicinity of the solution that was found for the previous interval of time. This may also have an influence on our choice of iterative procedure.

REFERENCES

Arrow, K. J., L. Hurwicz, and H. Uzawa (1958). *Studies in Linear and Non-Linear Programming*. Stanford University Press, Stanford, California.

Auslender, A. (1970). "Recherche des points de selle d'une fonction." *Lecture Notes in Mathematics*, vol. 132. Springer-Verlag, Berlin, pp. 37–52.

Berge, C. (1963). *Topological Spaces*. Oliver & Boyd, Edinburgh.

Bertsekas, D. P. (1976a). "Multiplier methods: A survey." *Automatica* 12: 133–145.

Bertsekas, D. P. (1976b). "On penalty and multiplier methods for constrained minimization." *SIAM J. Control and Optimization* 14: 216–235.

Brosilow, C. B., L. S. Lasdon, and J. D. Pearson (1965). "Feasible optimization method for interconnected systems." *Proceedings, Joint Automatic Control Conference.*

Céa, J. (1971). *Optimisation: theorie et algorithmes.* Dunod, Paris.

Donoghue, J. F., and I. Lefkowitz (1972). "Economic tradeoffs associated with a multilayer control strategy for a class of static systems." *IEEE Trans. Autom. Control* AC-17 (1): 7–15.

Dunford, N., and R. Schwartz (1958). *Linear Operators, Part I.* Wiley-Interscience, New York.

Ekeland, I., and J. T. Temam (1976). *Convex Analysis and Variational Problems.* North-Holland, Amsterdam.

Findeisen, W. (1968). "Parametric optimization by primal method in multilevel systems." *IEEE Trans. Syst. Sci. Cybern.,* SSC-4 (2).

Findeisen, W. (1974). *Wielopoziomowe uklady sterowania [Multilevel Control Systems].* PWN Warsaw. German translation: *Hierarchische Steuerungssysteme.* Verlag Technik, Berlin, 1977.

Findeisen, W. (1976). "Lectures on hierarchical control systems." *Report,* Center for Control Sciences, University of Minnesota, Minneapolis, Minnesota.

Findeisen, W., J. Szymanowski, and A. Wierzbicki (1977). *Computational Methods of Optimization.* PWN, Warsaw. (in Polish)

Franz, S., W. Liniger, and B. Wutke (1975). "Optimization and direct coordination of undisturbed and disturbed processes." *Proceedings,* Workshop Discussion on Multilevel Control, Technical University of Warsaw, pp. 95–102.

Geoffrion, M. (1970). "Elements of large-scale mathematical programming." *Manage Sci.* 16 (11).

Gol'shteĭn, E. G. (1971). *Duality Theory in Mathematical Programming.* Nauka, Moscow. (in Russian)

Grabowski, W. (1969). "Decomposition of linear program." *Report.* Central School of Planning and Statistics, Warsaw. (in Polish)

Hakkala, L., and J. Hirvonen (1977). "Gradient-based dynamical coordination strategies for interaction prediction method." *Report B38,* Helsinki University of Technology, Helsinki.

Heescher, A., K. Reinisch, and R. Schmitt (1975). "On multilevel optimization of nonconvex static problems—application to water distribution of a river system." *Proceedings, 6th IFAC Congress,* Pt. IIIC, Boston, Massachusetts.

Hestenes, M. R. (1969). "Multiplier and gradient methods." *Journal of Optimization Theory and Applications* 4: 303–320.

Himmelblau, D. M., ed. (1973). *Decomposition of Large-Scale Problems.* North-Holland, Amsterdam.

Ioffe, A. D., and W. M. Tikhomirov (1974). *Theory of Extremal Problems.* Nauka, Moscow. (in Russian)

Kuratowski, K. (1966). *Topology,* vol. 1. Academic Press, New York.

Kuratowski, K. (1968). *Topology,* vol. 2. Academic Press, New York.

Lasdon, L. S. (1968). "Duality and decomposition in mathematical programming." *IEEE Trans. Syst. Sci. Cybern.* SSC-4 (2): 86–111.

Lasdon, L. S. (1970). *Optimization Theory for Large Systems.* MacMillan, London.

Malinowski, K. (1974). "Properties of multilevel methods of mathematical programming and optimization of complex systems." *Doctoral dissertation,* Technical University of Warsaw. (in Polish)

Malinowski, K. (1975). "Properties of two balance methods of coordination." *Bull. Pol. Acad. Sci. Ser. Tech. Sci.* 23, (9).

Malinowski, K. (1976). "Lectures on hierarchical optimization and control." *Report,* Center for Control Sciences, University of Minnesota, Minneapolis, Minnesota.

Malinowski, K. (1977). "Applicability of the Lagrange multiplier method to optimization of quadratic systems." *IEEE Trans. Autom. Control* AC-22 (4): 648–651.

Malinowski, K., and J. Szymanowski (1976). "Theoretical and practical aspects of coordination by primal method," in J. Céa, ed., *Optimization Technique*, part 2, Springer-Verlag, Berlin.

Mesarović, M. D., D. Macko, and Y. Takahara (1970). *Theory of Hierarchical, Multilevel Systems*. Academic Press, New York.

Opoytsev, V. I. (1976). "Topological methods in theory of complex systems." *Autom. Remote Control* 3.

Ostrovskii, G. M., and Yu. M. Volin (1975). *Modelling of Complex Chemical Technological Systems*. Khimiya, Moscow. (in Russian)

Pearson, J. D. (1971). "Dynamic decomposition techniques," in D. A. Wismer, ed., *Optimization Methods for Large-Scale Problems*. McGraw-Hill, New York.

Pervozvanski, A. A. (1975). *Mathematical Models in Production Planning and Control*. Nauka, Moscow. (in Russian)

Polyak, B. T., and M. V. Tretyakov (1973). "Method of penalty terms for constrained optimization problems." *Journ. of Comp. Math. and Math. Physics.* 13 (1): 34–46.

Polak, E. (1971). *Computational Methods in Optimization*. Academic Press, New York.

Powell, M. I. D. (1969). "A method for nonlinear constraints in minimization problems," in R. Fletcher, ed., *Optimization*. Academic Press, New York.

Rockafellar, R. T. (1970). *Convex Analysis*. Princeton University Press, Princeton, New Jersey.

Rockafellar, R. T. (1974). "Augmented Lagrange multiplier functions and duality in nonconvex programming." *SIAM J. Control* 12: 268–285.

Schoeffler, J. D. (1971). "Static multilevel systems," in D. A. Wismer, ed., *Optimization Methods for Large-Scale Systems*. McGraw-Hill, New York.

Shor, N. Z. (1972). "About a method for minimization of almost differentiable functions." *Kibernetika* 4: 65–70. (in Russian)

Singh, M. G. (1977). *Dynamical Hierarchical Control*. North-Holland, Amsterdam.

Singh, M. G., S. A. W. Drew, and J. F. Coales (1975). "Comparisons of practical hierarchical control methods for interconnected dynamical systems." *Automatica* 11: 331–350.

Singh, M. G., M. F. Hassan, and A. Titli (1976). "Multilevel feedback control for interconnected dynamical systems using the prediction principle." *IEEE Trans. Syst. Man. Cybern.* SMC-6: 233–239.

Smith, N. J., and A. P. Sage (1973). "An introduction to hierarchical systems theory." *Computers and Electrical Engineering* 1: 55–71.

Stephanopoulos, G., and A. W. Westerberg (1975). "The use of Hestenes' method of multipliers to resolve dual gaps in engineering system optimization." *Journal of Optimization Theory and Applications* 15: 285–309.

Stoilov, E. (1977). "Augmented Lagrangian method for two-level static optimization." *Arch. Autom. Telemech* 22: 210–237. (in Polish)

Tamura, H. (1975). "Decentralised optimization for distributed-lag models of discrete systems." *Automatica* 11: 593–602.

Tatjewski, P. (1976). "Properties of multilevel dual optimization methods." *Doctoral dissertation*, Technical University of Warsaw. (in Polish)

Tatjewski, P. (1977). "Dual methods of multilevel optimization." *Bull. Pol. Acad. Sci., Ser. Tech. Sci.* 25: 247–254.

Tatjewski, P. (1978). "A penalty function approach to coordination in multilevel optimization problems." *R.A.I.R.O. Automatique/Systems Analysis and Control* 12: 221–244.

Telle, W. (1975). "The use of a modified Lagrange function in block programming." *Ekonomika: Matematicheskie Metody* 11 (5). (in Russian)

Titli, A. (1972). "Contribution a l'étude des structures de commande hierarchisées en vue de l'optimization de processus complexes." *Doctoral dissertation*, Université Paul Sabatier, Toulouse. Dunod, Paris.

Tsuji, K., and I. Lefkowitz (1975). "On the determination of an on-demand policy for a multilayer control system." *IEEE Trans. Autom. Control* AC-20: 464–472.

Wierzbicki, A. P. (1976). "A primal-dual large-scale optimization method based on augmented Lagrange functions and interaction shift prediction." *Ricerche di Automatica* 7.

Wierzbicki, A. P., and S. Kurcyusz (1977). "Projection on a cone, penalty functionals and duality theory for problems with inequality constraints in Hilbert space." *SIAM J. Control* 15 (1).

Wismer, D. A., ed. (1971). *Optimization Methods for Large-Scale Problems.* McGraw-Hill, New York.

Wolfe, P. (1975). "A method of conjugate subgradients for minimizing nondifferentiable functions," *Mathematical Programming, Study* 3. North-Holland, Amsterdam.

Woźniak, A. (1973). "Applicability conditions of two-level parametric optimization." *Podstawy Sterowania* 3: 65–76. (in Polish).

Woźniak, A. (1975). "Control of complex systems and parametric coordination." *Doctoral dissertation.* Technical University of Warsaw. (in Polish)

Woźniak, A. (1976). "Parametric method of coordination using feedback from the real process." *Proceedings, IFAC Symposium on Large-Scale Systems Theory and Applications.* Udine, Italy.

3 Iterative Coordination for Steady-State Control

3.1. PROBLEM DESCRIPTION

In this chapter we shall be concerned with hierarchical control structures. They will involve local decision-making units and a supremal coordinating unit. The control problem will be to optimize the steady state in the controlled system. There is a suitable direct control layer, so we determine, for example, the optimal set-points for the stabilizing controllers (see section 1.2). The control structures will be closed-loop, that is, they will use feedback from the real system. Those in Chapter 2 were open-loop.

As this chapter will show, it is possible to develop a body of theoretical principles and iterative procedures of control and coordination applicable to steady-state or static systems, with due attention paid to model-reality differences. These iterative procedures cannot be practically applied to dynamic control, except for cyclic or batch processes (we present some computational examples in section 3.6). On the other hand, steady-state optimization is the current method of control in many industrial processes and its applications certainly deserve much attention.

The conditions under which steady-state optimization acting on a direct control layer is a reasonable approximation to the true dynamic problem were given in section 1.2: the approach is valid for slow disturbances acting on a fast system.

The hierarchical structure that we are going to discuss was presented in general terms in Figure 1.7 in Chapter 1. There are two ways of thinking about the local decision units presented there: they may be computer algorithms or human decision makers. We shall use formal models for these units, that is, we shall present their control decisions as the solutions of some

182

appropriately defined optimization problems. This approach can be used directly for computer-based control decisions. However, we may also consider these algorithms to be models of rational decision making performed by human operators who want to achieve well-defined goals optimally. The assumption that human operators would optimize in a rational way, and moreover that they would optimize goals which were imposed on them by system design or by the coordinator, is very simplistic. It is only a rough approximation of the real processes of decision making.

There are two principal modes by which the coordinator could intervene in the local problems. The use of direct instruments (section 3.2) means prescribing the values of some of the subsystem variables, e.g., the outputs, while the use of price instruments (section 3.3 and 3.4) leaves the assignment of values to the local decision units. With the presence of constraints and model-reality differences, the two modes differ significantly in both applicability and performance.

In most of our discussion, the local algorithms will be based on models of the controlled subsystems. The models would use available estimates of real disturbances acting on the subsystems. In most of this chapter we assume that disturbances, as well as the disturbance estimates, are slow and consequently can be treated as constant parameters during coordination. In section 3.5, however, we present a discussion of the performance of iterative algorithms applied to a time-varying system. Such an application corresponds to the assumption that the controlled system does not yet have to be treated as dynamic (for example, the optimal state could be constant irrespective of disturbances, or only slowly varying), and therefore we can still use iterative procedures; at the same time we cannot assume that the disturbance is constant during a full sequence of the iterations.

THE USE OF FEEDBACK INFORMATION

Most of the discussion in this chapter concerns control structures where feedback information is used only in the coordinating algorithm and the local decisions are computed on the basis of subsystem models. Thus, the feedback information reaches the local units indirectly through the intervention of the coordinator. This structure is primarily applicable to computer-based control hierarchies. The advantage of the structure with feedback to the coordinator is the ease with which stability can be obtained: there is only one loop of iterations in the structure. This structure is the first step of improvement over open-loop control. The iterations that involve feedback should always be initiated from a starting point supplied by the solution of the model-based optimization, that is, from the open-loop optimal control.

It is also possible to use feedback information directly in the local steady-state optimization algorithms. We devote section 3.4 to a discussion of structures allowing the direct use of feedback that use price coordination (see also section 1.3). The use of both kinds of feedback is conceivable, though difficult. Section 3.3 describes one of the possibilities, where the coordinator uses feedback in the form of measured output values for his iterations of the prices, while the same measurements are used in the local control algorithms to compensate for the inadequacy of the models.

FEASIBLE CONTROLS

If we recognize that a difference exists between the real system and its model, we have to pay attention to the feasibility of model-based control with respect to the real system constraints. We should consider this problem for every coordination method. As a rule, we require that the control obtained from an iterative procedure should be feasible. If this is not enough, that is, if we want feasibility for each iteration, we have to resort to some special way of generating feasible controls. A discussion of this subject and some algorithms are given in section 3.6.

FORMULATION OF THE CONTROL PROBLEM

We follow the formulation given in section 2.1, emphasizing that the case is finite dimensional, and that there are model-reality differences.

The controlled system including its direct controls (compare Figure 2.5), will be described as follows:

$$y_1 = F_{*1}(c_1, u_1, z_1), \qquad u_1 = H_1 y$$
$$\cdots \tag{3.1}$$
$$y_N = F_{*N}(c_N, u_N, z_N), \qquad u_N = H_N y,$$

where y_i are subsystem outputs, u_i are interconnection inputs, c_i are controls, z_i are disturbances. $F_{*i} : \mathscr{C}_i \times \mathscr{U}_i \times \mathscr{Z}_i \to \mathscr{Y}_i$ is subsystem input-output mapping i, $y \triangleq (y_i, \ldots, y_N) \in \mathscr{Y}_1 \times \ldots \times \mathscr{Y}_N \triangleq \mathscr{Y}$, and H_i is interconnection matrix i composed of zeros and ones. The subscript $*$ will be added later to all mappings related to real (not model) objects.

The couplings are separable:

$$u_i = \sum_{j=1}^{N} H_{ij} y_j.$$

When we denote, for the whole system, $c \triangleq (c_1, \ldots, c_N) \in \mathscr{C}_1 \times \ldots \times \mathscr{C}_N \triangleq \mathscr{C}$, $u \triangleq (u_1, \ldots, u_N) \in \mathscr{U}_1 \times \ldots \times \mathscr{U}_N \triangleq \mathscr{U}$ and

$$z \triangleq (z_1, \ldots, z_N) \in \mathscr{Z}_1 \times \ldots \times \ldots \times \mathscr{Z}_N \triangleq \mathscr{Z},$$

then (3.1) can be written in compact form as

$$y = F_*(c, u, z), \qquad u = Hy, \tag{3.2}$$

where

$$F_*(c, u, z) \triangleq \begin{bmatrix} F_{*1}(c_1, u_1, z_1) \\ .. \\ F_{*N}(c_N, u_N, z_N) \end{bmatrix} \quad \text{and} \quad H \triangleq \begin{bmatrix} H_1 \\ ... \\ H_N \end{bmatrix}.$$

In many places in the chapter, the disturbance inputs are not important, since they are assumed to be constant, though unknown. We shall therefore often use (3.1) and (3.2) with an abbreviated notation, as

$$y_1 = F_{*1}(c_1, u_1), \qquad u_1 = H_1 y,$$
$$... \tag{3.1'}$$
$$Y_N = F_{*N}(c_N, u_N), \qquad u_N = H_N y,$$

and

$$y = F_*(c, u), \qquad u = Hy. \tag{3.2'}$$

The interconnected system reacts to the control input c as defined by (3.1) or (3.2). We shall always assume that for each $(c, z) \in \mathscr{C} \times \mathscr{Z}$ there exists exactly one output $y = (y_1, \ldots, y_N)$ in the system. The system may then be generally described by mapping

$$K_* : \mathscr{C} \times \mathscr{Z} \to \mathscr{Y},$$

which means the following input–output equation of the interconnected system

$$y = K_*(c, z). \tag{3.3}$$

The argument z may be dropped in appropriate circumstances, as it was in (3.1') and (3.2'), giving

$$y = K_*(c) \tag{3.3'}$$

Mapping K_* and Eq. (3.1) or (3.2) describe the same object and are closely related. This relation is the following

$$\forall (c, z) \in \mathscr{C} \times \mathscr{Z} \;\; [y = K_*(c, z) \Leftrightarrow \exists u \in \mathscr{U} \;\; (y = F_*(c, u, z) \wedge u = Hy)]$$

We mentioned at the beginning of the chapter that the system relations are not known exactly; we only know models that are some approximation of reality

$$y_1 = F_1(c_1, u_1, z_1), \qquad u_1 = H_1 y,$$
$$... \tag{3.4}$$
$$y_N = F_N(c_1, u_1, z_1), \qquad u_N = H_N y,$$

186

where $F_i : \mathscr{C}_i \times \mathscr{U}_i \times \mathscr{Z}_1 \to \mathscr{Y}_i$ is a model of subsystem input-output mapping i, and H_i, $i \in \overline{1, N}$, are interconnection matrices that are assumed to be the same in the model and reality.

Mapping $F : \mathscr{C} \times \mathscr{U} \times \mathscr{Z} \to \mathscr{Y}$, composed of F_i, and mapping $K : \mathscr{C} \times \mathscr{Z} \to \mathscr{Y}$, representing the model system including interconnections, can be defined as they were for the real system relations.

The abbreviated form of (3.4) can be used where appropriate:

$$y = F(c, u), \qquad u = Hy. \tag{3.4'}$$

In the general case, Eqs. (3.1) and (3.4) differ, that is, F_{*i} and F_i are not the same. Only in some of the considerations in this chapter, for example, in section 3.6, will we assume that F_{*i} and F_i are the same mappings, and consequently that the model differs from reality only in the value of parameter z.

We now describe the constraints. The *local constraints* are assumed to be given explicitly as

$$(c_i, u_i) \in CU_i \triangleq \{(c_i, u_i) \in \mathscr{C}_i \times \mathscr{U}_i : G_i(c_i, u_i) \leq 0\}, \qquad i \in \overline{1, N}, \tag{3.5}$$

where $G_i : \mathscr{C}_i \times \mathscr{U}_i \to \mathbb{R}^{m_i}$, or in the form involving the output:

$$(c_i, u_i, y_i) \in CUY_i \triangleq \{(c_i, u_i, y_i) \in \mathscr{C}_i \times \mathscr{U}_i \times \mathscr{Y}_i : G_i^0(c_i, u_i, y_i) \leq 0\}, \tag{3.6}$$

where G_i^0 is another mapping with values in the real space.

The constraints (3.5) or (3.6) can be conveniently described jointly as (3.7) or (3.8), respectively:

$$(c, u) \in CU \triangleq \{(c, u) \in \mathscr{C} \times \mathscr{U} : G(c, u) \leq 0\} \tag{3.7}$$

$$(c, u, y) \in CUY \triangleq \{(c, u, y) \in \mathscr{C} \times \mathscr{U} \times \mathscr{Y} : G^0(c, u, y) \leq 0\} \tag{3.8}$$

Except in some places in sections 3.3 and 3.6, we assume that (3.5) or (3.6) are the same for the real system and the model. In other words, we assume that we know the constraint functions exactly. However, even under this assumption, if the feasible set is defined primarily in the form of (3.6), we can reduce it to (3.5) by substituting the subsystem equations for y_i. We then obtain different feasible sets for the real system:

$$(c_i, u_i) \in CU_{*i} = \{(c_i, u_i) \in \mathscr{C}_i^| \times \mathscr{U}_i : G_i^0(c_i, u_i, F_{*i}(c_i, u_i)) \leq 0\}, \tag{3.9}$$

and for the model

$$(c_i, u_i) \in CU_i \triangleq \{(c_i, u_i) \in \mathscr{C}_i \times \mathscr{U}_i : G_i^0(c_i, u_i, F_i(c_i, u_i)) \leq 0\}. \tag{3.10}$$

An optimal control for the model that is feasible according to (3.10), may be infeasible for the real system, that is, it may violate (3.9).

The range of *global constraints* in addition to (3.7) or (3.8) that will be discussed in this chapter is restricted. In some of the sections we consider an overall constraint on the output

$$y \in Y \triangleq \{y \in \mathcal{Y} : G(y) \leq 0\}, \tag{3.11}$$

where $G : \mathcal{Y} \to R^{m_0}$.

In other sections we consider the following resource constraint:

$$(c, u) \in CU_0 \triangleq \{(c, u) \in \mathcal{C} \times \mathcal{U} : \sum_{i-1}^{N} r_i(c_i, u_i) \leq r_0\}, \tag{3.12}$$

where $r_i : \mathcal{C}_i \times \mathcal{U}_i \to \mathbb{R}^m$, $r_0 \in \mathbb{R}^m$.

A known local performance function (that is, one that is the same in reality and in the model) is associated with each subsystem. It may be explicit in (c_i, u_i), that is,

$$Q_i : \mathcal{C}_i \times \mathcal{U}_i \to \mathbb{R}, \tag{3.13}$$

or may also involve the output

$$Q_i : \mathcal{C}_i \times \mathcal{U}_i \times \mathcal{Y}_i \to \mathbb{R}. \tag{3.14}$$

As was the case with constraint relations, the form of (3.14) can be reduced to the form of (3.13), but the reduction leads to differences between the local performance functions in the model and those in the real system.

The global performance functions

$$Q : \mathcal{C} \times \mathcal{U} \to \mathbb{R}, \quad \text{or} \quad Q : \mathcal{C} \times \mathcal{U} \times \mathcal{Y} \to \mathbb{R},$$

are composed of the local performance indices (3.13) or (3.14) respectively, in the following way:

$$Q = \psi \circ (Q_1, \ldots, Q_N), \tag{3.15}$$

where function $\psi : \mathbb{R}^N \to \mathbb{R}$ is assumed to be strictly order preserving.

For the price coordination methods we are going to assume that (3.15) is of additive form

$$Q = \sum_{i=1}^{N} Q_i \tag{3.16}$$

3.2. COORDINATION BASED ON THE DIRECT APPROACH

3.2.1. THE BASIC CONCEPT

In this section we describe the use of the desired values of subsystem outputs as coordination instruments. Thus, we will use the direct (parametric, feasible, or image) mode of coordination (cf. section 2.2.1.).

Let us first consider the application of pure direct coordination to real system control (Findeisen 1974a,b). The structure of the two-level control

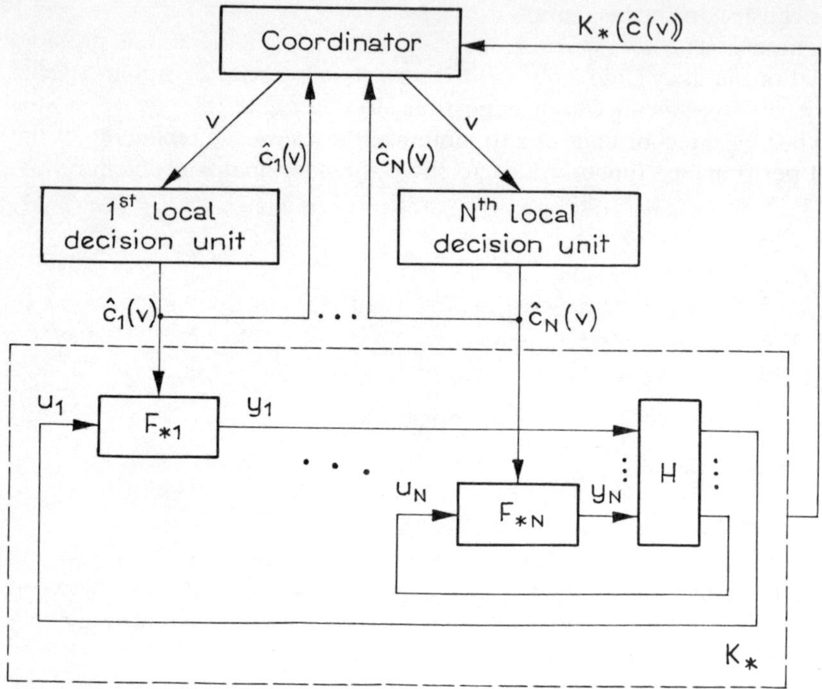

FIGURE 3.1   The structure of direct coordination with feedback.

system with feedback in which decision problems are specified with the direct method is shown in Figure 3.1.

Each local decision problem (LP_i) is formulated as follows:

For a given value of coordination variable $v \in \mathcal{Y}$ find the control

$$\hat{c}_i(v) = \arg \min_{C_i(v)} Q_i(\,\cdot\,, H_i v) \tag{3.17}$$

where

$$C_i(v) \triangleq \{c_i \in \mathcal{C}_i : (\exists u_i, y_i)\ [(c_i, u_i) \in CU_i \wedge y_i$$
$$= F_i(c_i, u_i) \wedge u_i = H_i v \wedge y_i = v_i]\}$$
$$= \{c_i : (c_i, H_i v) \in CU_i \wedge v_i = F_i(c_i, H_i v)\}.$$

It is a model-based problem that is the same as Eq. (2.22), in which the coordination variable v prescribes the subsystem outputs and interactions. Notice that in the above LP_i, $y_i = v_i$ and $u_i = H_i v$ are strict equalities, which is possible only when

$$v \in V_0 \triangleq \{v \in \mathcal{Y} : \forall i \in \overline{1, N}\ \ C_i(v) \neq \varnothing\},$$

i.e., when the coordination instruments belong to the set of outputs feasible in the model. The set V_0, not specified explicitly in the initial problem formulation, is associated only with direct coordination; except in special cases, it is very difficult, if not impossible, to determine this set.

The task of the coordinator is to minimize the measured real value of the overall performance function, i.e., to solve the coordinator problem (CP_0):

Find a coordination variable

$$\hat{v} = \arg \min_{V_* \cap V_0} Q(\hat{c}(\,\cdot\,), HK_*(\hat{c}(\,\cdot\,)))$$

where

$$V_* \cap V_0 \triangleq \{v \in \mathcal{Y} : K_*(\hat{c}(v)) \in Y \wedge \forall i \in \overline{1, N} \ (\hat{c}_i(v), H_i K_*(\hat{c}(v))) \in CU_i\} \cap V_0,$$

As before, we assume that $\hat{c}(v) \triangleq (\hat{c}_1(v), \ldots, \hat{c}_N(v))$ exists for every v in $V_* \cap V_0$, i.e., $V_0 = V_r$ (cf. section 2.2.4.) and $K_*(\hat{c}(v)) \in \mathcal{Y}$ denotes the measured real system outputs corresponding to the implemented controls $\hat{c}(v)$ (cf. 3.3′).

The coordinator problem introduced above differs from the one described in section 2.2.1. since it uses the values of outputs measured in the real system in the definitions of the performance index and the feasible set. The definitions formalize in the terms of a decision problem the idea of using feedback information that proceeds from measurements in the real system made by a supremal decision unit.

The structure now being described performs a kind of peak-holding control, where the coordinator uses subsystem outputs as decision variables. We should note that we do not allow the coordinator to set the controls c_i directly, that is, to override the local decisions, or to perform the hill-climbing on all decisions. It is assumed that the number of output variables that the coordinator deals with is much less than the total number of local decision variables (controls). Otherwise the structure may not have practical sense.

The local decision units solve typical off-line optimization problems based on subsystem models, and the coordinator solves the on-line problem based on measurements performed in the real system. The structure can thus ensure that at least at the end of the supremal unit's hill-climbing, the inequality constraints in the real system will not be violated. When the models used in local problems are accurate (i.e., $F_{*i} = F_i$ for every $i \in \overline{1, N}$), the result obtained with this structure is strictly optimal.

As noted earlier, in the general case we are not able to determine the set V_0. This difficulty is a major drawback of the pure direct approach. Another weakness is the need for the coordinator to obtain complete information about the local feasible sets CU_i. The above shortcomings are not quite so

190

severe when controls and interactions are constrained separately, that is, for each $i \in \overline{1, N}$

$$CU_i = C_i \times U_i,$$

where $C_i \subseteq \mathscr{C}_i$ and $U_i \subseteq \mathscr{U}_i$. In this case the local feasible set is only parameterized by equations of the subsystem model. Consequently, the separated local problem (SLP_i) has the form:

For a given v in $\mathscr{Y}$ find a control

$$\hat{c}_i(v) = \arg \min_{C_i'(v)} Q_i(\,\cdot\,, H_i v) \tag{3.18}$$

where

$$C_i'(v) \triangleq \{c_i \in \mathscr{C}_i : c_i \in C_i \wedge v_i = F_i(c_i, H_i\, v)\}.$$

In the supremal problem the feasible set has a simpler form as well. The set V_* does not depend on constraints imposed on controls

$$V_* = \{v \in \mathscr{Y} : K_*(\hat{c}(v)) \in Y \cap (H_1^{-1}(U_1) \times \cdots \times H_N^{-1}(U_N))\}.$$

In the following sections we shall describe some methods that allow the application of direct coordination with feedback when V_0 is only partially known or entirely ignored. The methods are based on appropriate modification of the local problem. The coordination variables and the control structure (cf. Figure 3.1) will be unchanged.

3.2.2. COORDINATION WITH PARTIAL KNOWLEDGE OF THE FEASIBLE SET V_0

Even though we may not know set V_0, we are in some cases, able to determine the set V_F resulting from all local constraints but without interactions:

$$V_F \triangleq \mathop{\mathbf{X}}_{i=1}^{N} F_i(CU_i)$$

where $F_i(CU_i)$ is the image of the set CU_i. Knowledge of the set V_F means that for each model of a subsystem we know the range of the subsystem output variation corresponding to the feasible variation (belonging to the set CU_i) of subsystem inputs (controls and interactions). The way to use the additional information given by knowledge of the V_F was presented in sections 2.2.4. and 2.2.8. Arguing as in section 2.2.8. for the selection of the coordinator strategy, we obtain the following modification of pure direct coordination.

Local decision problem i ($m\text{LP}_i$) has the form:

For a given coordination variable $v \in \mathcal{Y}$, first find an interaction $u_i^m(v) \in U_i(v_i)$, such that

$$u_i^m(v) = \arg \min_{u_i \in U_i(v_i)} \|u_i - H_i v\|, \tag{3.19}$$

and then find a control $\hat{c}_i(v)$, such that

$$\hat{c}_i(v) = \arg \min_{C_i^m(v)} Q_i(\cdot, H_i v) \tag{3.20}$$

where

$$U_i(v_i) \triangleq \{u_i \in \mathcal{U}_i : \exists c \ [(c, u_i) \in CU_i \wedge v_i = F_i(c, u_i)]\} \tag{3.21}$$

$$C_i^m(v) \triangleq \{c_i \in \mathcal{C}_i : (c_i, u_i^m(v)) CU_i \wedge v_i = F_i(c_i, u_i^m(v))\}.$$

The task of the coordinator, as before, is minimization of performance function

$$\hat{Q}(\cdot) \triangleq Q(\hat{c}(\cdot), HK_*(\hat{c}(\cdot)))$$

based on implemented (real) values of controls $\hat{c}(v)$ and corresponding outputs $K_*(\hat{c}(v))$ measured in the real system and subject to the real constraints. We get the following coordinator problem (CP):

Find a coordination variable

$$\hat{v} = \arg \min_{V_* \cap V_F} Q(\hat{c}(\cdot), HK_*(\hat{c}(\cdot))) \tag{3.22}$$

where

$$V_* \cap V_F = \{v \in \mathcal{Y} : K_*(\hat{c}(v)) \in Y \wedge \forall i \in \overline{1, N} \ (\hat{c}_i(v), H_i K_*(\hat{c}(v))) \in CU_i\} \cap V_F.$$

In the above formulation, the problems (3.19) are introduced in order to relax the previous stiff requirement that the local optimal control be found under the condition that the interaction input is given as $u_i = H_i v$. The above formulation of the local problem leaves both subsystem inputs to be chosen by the local decision unit. Nevertheless, after the iterations, the interactions have to match the coordinator's desire. Therefore, in (3.19) we ask that $u_i^m(v)$ be as close as possible to $H_i v$, i.e., the value of the interaction set by the coordinator.

In mathematical terms (cf. section 2.2.8., problem $m\text{LP}_i$) the above modification of the local problem extends mapping $\hat{c}(\cdot)$, whose original domain was the set V_0, onto the set $V_F \supseteq V_0$. Lemma 2.7 shows that when set CU_i is compact and subsystem model F_i is continuous, problem ($m\text{LP}_i$)

192

has a solution $(u_i^m(v), \hat{c}_i(v))$ for all coordination instruments v in V_F. Unlike the extended local problem, the supremal problem CP remains in principle unchanged—only set V_0 is replaced by set V_F.

Before investigating the applicability conditions, we should notice that the two-level problem $(\{m\text{LP}_i\}_{i=1}^N, \text{CP})$ just described is equivalent to the following optimization problem (OP):

Find $\tilde{c}$, such that

$$Q(\tilde{c}, HK_*(\tilde{c})) = \min_{C_f} Q(c, HK_*(\,\cdot\,)), \qquad (3.23)$$

where

$$C_f \triangleq \{c \in \mathscr{C} : K_*(c) \in Y \wedge \forall i \in \overline{1, N} \ \ (c_i, H_i K_*(c)) \in CU_i\} \cap \hat{c}(V_F).$$

The equivalence means that

$$\hat{Q}(\hat{v}) = Q(\tilde{c}, HK_*(\tilde{c})).$$

This problem is an approximation of the optimizing controller problem for the real system (IOP):

Find c_*, such that

$$Q(c_*, HK_*(c_*)) = \min_{C_*} Q(\,\cdot\,, HK_*(\cdot)) \qquad (3.24)$$

where

$$C_* \triangleq \{c \in \mathscr{C} : K_*(c) \in Y \wedge \forall i \in \overline{1, N} \ \ (c_i, H_i K_*(c)) \in CU_i\}.$$

The only difference is in the definition of the feasible sets C_f and C_*. If we know the mapping K_* exactly (i.e., the model F_i, $i \in \overline{1, N}$, is accurate) the constraint $c \in \hat{c}(V_F)$ is not essential; it would be satisfied for each control c that is a solution of problem IOP. Therefore, problems IOP and OP are equivalent. We assumed that function ψ is strictly order preserving, so problems $(\{m\text{LP}_i\}_{i=1}^N, \text{CP})$ and IOP are also equivalent (Theorem 2.1 treats a similar problem). Because we do not have exact knowledge of mapping K_* representing the real system as a whole, the value of the performance functional corresponding to two-level direct coordination with feedback is generally worse than the value associated with the solution of the ideal optimizing controller problem. We should remember, however, that the number of iterated variables is much less. In principle, the more inaccurate the models, the greater the loss of performance. In all simulations, however, the loss was much smaller than for the model-based, open-loop control.

In formulating the coordinator problem, we implicitly assumed that mappings $u_i^m(\,\cdot\,)$ and $\hat{c}_i(\,\cdot\,)$ are well defined and that feasible set V_* is not empty.

Let us briefly consider the nonemptiness conditions first. Comparing CP and IOP, one sees that

$$v \in V_* \iff \hat{c}(v) \in C_*,$$

hence,

$$C_* \neq \varnothing \quad \text{and} \quad \exists \bar{v} \in V_F \ \hat{c}(\bar{v}) \in C_* \tag{3.25}$$

are the conditions that we are seeking. The set C_* is not empty if the constraints are not contradictory for the real system. Similarly $\bar{v}$ exists when the difference between the real system equations and the models of the equations is not very large.

When these requirements are not satisfied, the system model $F_i, H_i,$ $i \in \overline{1, N}$, is too inaccurate or the local feasible sets CU_i are too tight and both must be improved. We restrict our attention to cases in which requirements (3.25) are fulfilled.

We now discuss the applicability conditions of the method considered. Applicability means that the solutions of local problems and coordinator problems exist. Assume that $\mathscr{C}_i, \mathscr{U}_i, i \in \overline{1, N}$ and $\mathscr{Y}$ are real Hilbert spaces.

THEOREM 3.1 *If the controlled system and its model are such that:*

1. *Mapping K_* representing the real system including interactions is continuous,*
2. *The overall constraint set Y is closed,*
For each $i \in \overline{1, N}$:
3. *Local constraint set CU_i is compact,*
4. *Mapping F_i is continuous and open on CU_i and such that mapping*

$$(c_i, u_i) \mapsto (u_i, F_i(c_i, u_i))$$

is open on CU_i,
5. *For all v in V_F, local problem mLP_i has no more than one solution $(u_i^m(v), \hat{c}_i(v))$,*
6. *Functions Q_i and ψ are continuous,*

then for all $i \in \overline{1, N}$, a solution of local problem mLP_i exists and there exists a solution of coordinator problem CP, i.e., the direct method with feedback is applicable.

Proof. The proof will be based on the well-known Weierstrass theorem which says that a continuous function defined on a compact set achieves its minimum and maximum.

Because functions F_i and Q_i are assumed to be continuous, one can infer from assumptions 3 and 5 that the mappings

$$V_F \ni v \mapsto u_i^m(v) = \arg \min_{u_i \in U_i(v_i)} \|u_i - H_i v\| \in \mathscr{U}_i,$$

194

and

$$V_F \ni v \mapsto \hat{c}_i(v) = \arg\min_{C_i^m(v)} Q_i(\,\cdot\,, H_i v) \in \mathscr{C}_i$$

are well defined, which means, among other things, that for all v in V_F, a solution of problem $m\mathrm{LP}_i$ exists.

Similarly, cp has a solution whenever the function

$$\hat{Q}(\,\cdot\,) = Q(\hat{c}(\,\cdot\,), HK_*(\hat{c}(\,\cdot\,))) = \psi \circ (Q_1(\hat{c}_1(\,\cdot\,), H_1 K_*(\hat{c}(\,\cdot\,))), \dots,$$
$$Q_N(\hat{c}_N(\,\cdot\,), H_N K_*(\hat{c}(\,\cdot\,))))$$

is continuous and the feasible set of the coordinator $V_* \cap V_F$ is compact.

The definition of the set $V_* \cap V_F$ can be rewritten as

$$V_* \cap V_F = \hat{c}^{-1}(C_*) \cap \underset{i=1}{\overset{N}{\times}} F_i(CU_i)$$

where

$$C_* = K_*^{-1}(Y) \cap \bigcap_{i=1}^{N} D_{*i}^{-1}(CU_i),$$

and the mappings D_{*i} are defined as

$$\mathscr{C} \ni c \mapsto D_{*i}(c) = (c_i, H_i K_*(c)) \in \mathscr{C}_i \times \mathscr{U}_i.$$

The mappings D_{*i} are continuous because the mappings H_i and K_* are assumed to be continuous. The sets CU_i and Y are closed, so the set C_* is closed. From the compactness of CU_i and the continuity of F_i, we obtain the compactness of $\times_{i=1}^{N} F_i(CU_i)$. Therefore, the set $V_* \cap V_F$ is compact whenever the mapping $\hat{c}(\cdot)$ is continuous. The definition of $\hat{Q}$ and assumptions 1 and 6 imply that $\hat{Q}$ is continuous whenever mapping $(\hat{c}_1(\cdot), \dots, \hat{c}_N(\cdot)) = \hat{c}(\cdot)$ is continuous. Therefore, the continuity of $\hat{c}_i(\cdot)$ for all i in $\overline{1, N}$ is sufficient to prove the theorem. In order to prove this continuity we introduce the following mappings:

$$V_F \ni v \mapsto \dot{Q}_i^m(v) = \min_{C_i^m(v)} Q_i(\,\cdot\,, H_i v) = Q_i(\hat{c}_i(v), H_i v) \in \mathbb{R},$$

and

$$V_F \ni v \mapsto B_i(v) = \{c_i \in C_i^0 : \hat{Q}^m(v) - Q_i(c_i, H_i v) \geq 0\} \in C\mathscr{P}(C_i^0)$$

where

$$C_i^0 \triangleq \pi_{\mathscr{C}_i}(CU_i).$$

In the proof of Theorem 2.11 in section 2.2.8. (cf. also the proof of Theorem 2.9) it was shown that assumptions 3, 4, 5, and 6 imply that mapping $\hat{Q}_i^m(\cdot)$ is continuous. So, the set

$$\{(v, c_i) \in V_F \times C_i^0 : c_i \in B_i(v)\} = \{(v, c_i) : \hat{Q}^m(v) - Q(c_i, H_i v) \geq 0\}$$

is closed. The set $V_F \times C_i^0$ is compact by assumption 3, hence, from Kuratowski (1968, Theorem 4 of §43, I), mapping $B_i(\cdot)$ is upper semicontinuous with exponential topology introduced in $C\mathscr{P}(C_i^0)$.

We can now prove the continuity of $\hat{c}_i(\cdot)$ as we proved the continuity of mapping $u_i^m(\cdot)$ in the proof of Theorem 2.11. Mapping $B_i(\cdot)$ will play the same role as mapping $W_i(\cdot)$ and the uniqueness assumption of (3.20) will play the same role as assumption 3 of Theorem 2.11. We therefore omit the details. Since the continuity of $\hat{c}_i(\cdot)$ for each i in $\overline{1, N}$ is sufficient to prove the existence of the coordinator problem solution, the theorem is proved. $\square$

We shall now discuss the applicability conditions. The assumptions that mappings K_*, F_i, Q_i, and ψ are continuous, the sets CU_i compact, and the set Y is closed are natural and satisfied in most applications. The first crucial assumption is the requirement that the mappings F_i and $(c_i, u_i) \mapsto f_i(c_i, u_i) = (u_i, F_i(c_i, u_i))$ restricted to the set CU_i be open. Unfortunately, these assumptions are vital. Consider the following example.

Let $\mathscr{C}_1 = \mathscr{U}_1 = \mathscr{Y}_1 = \mathscr{Y}_2 = \mathbb{R}$. The first subsystem model is given by the equation

$$\mathbb{R}^2 \ni (c_1, u_1) \mapsto F_1(c_1, u_1) = -c_1 + u_1 \in \mathbb{R}$$

and coupled according to the equation

$$(y_1, y_2) \mapsto H_1(y_1, y_2) = y_2.$$

The local constraint set is given in $\mathbb{R}^2$ as

$$CU_1 = ([1, 4] \times [2, 5]) \cup \{(c_1, u_1) : 0 \leq c_1 \leq 1 \wedge u_1 = c_1 + 1\}.$$

Let $v_2 = 1$; from (3.19) we obtain (see Figure 3.2)

$$u_1^m(v_1, 1) = \begin{cases} 2 & \text{for} \quad v_1 \in [-2, 1) \\ 1 & \text{for} \quad v_1 = 1 \\ v_1 + 1 & \text{for} \quad v_1 \in (1, 4]. \end{cases}$$

So mapping $u_1^m(\cdot)$ is discontinuous because mapping $F_1|CU_1$ is not open. Consequently, mappings $C_1^m(\cdot)$ and $\hat{c}_1(\cdot)$ are discontinuous too.

In Woźniak (1976) there is an example that shows that when mapping f_i is not open, mapping $\hat{Q}(\cdot)$ is not even lower semicontinuous.

Stating sufficient conditions for openness is very difficult because, as we saw in the example, apart from the behavior of mapping F_i, we must take

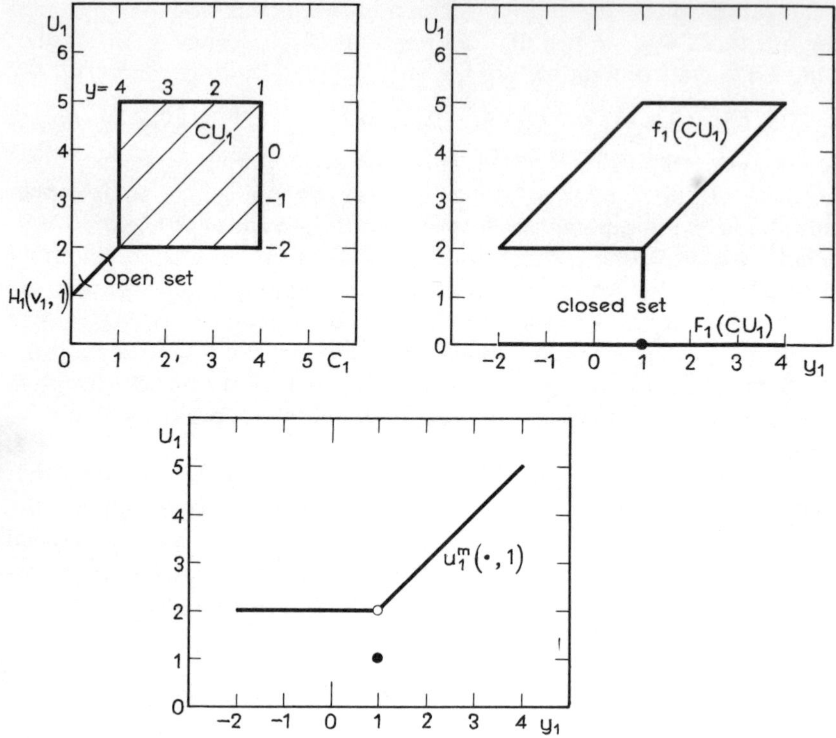

FIGURE 3.2   The sets CU_1, $f_1(CU_1)$, $F_1(CU_1)$, and mapping $u_1^m(\cdot, 1)$ in the example.

the properties ("shape") of set CU_i into consideration. The following lemma (Schwartz 1967, Theorem 30, Chapter III) can help to check the openness assumption.

LEMMA 3.1.   *A continuously differentiable mapping* $F_i : \mathscr{C}_i \times \mathscr{U}_i \to \mathscr{Y}_i$ *is open on the open set* $X \subseteq \mathscr{C}_i \times \mathscr{U}_i$ *if for all* $(c_i, u_i) \in X$, *Fréchet derivative* $F_i'(c_i, u_i) \in \mathscr{L}(\mathscr{C}_1 \times \mathscr{U}_i, \mathscr{Y}_i)$ *is a surjection.*

In the finite-dimensional case, the above lemma takes the following simpler form.

COROLLARY 3.2.   *If* $\mathscr{C}_i$, $\mathscr{U}_i$, *and* $\mathscr{Y}_i$ *are finite-dimensional Hilbert spaces, and mapping* $F_i : \mathscr{C}_i \times \mathscr{U}_i \to \mathscr{Y}_i$ *is continuously differentiable, then*

1.   *It is open on open set* $X \subseteq \mathscr{C}_i \times \mathscr{U}_i$ *whenever*

$$\forall (c, u) \in X \quad \operatorname{rank}[F_i'(c, u)] = \dim \mathscr{Y}_i,$$

2. *Mapping $(c_i, u_i) \mapsto (u_i, F_i(c_i, u_i))$ is open on open set $X \subseteq \mathscr{C}_i \times \mathscr{U}_i$ whenever*

$$\forall (c, u) \in X \quad \text{rank}[F'_{ic}(c, u)] = \dim \mathscr{Y}_i,$$

where $[F'_i(\cdot)]$ ($[F'_{ic}(\cdot)]$) denotes the derivative—Jacobi matrix—(the partial derivative with respect to $c_i \in \mathscr{C}_i$) of the mapping F_i.

The rank conditions of the corollary have a clear meaning in engineering: each subsystem must have an adequate number of controls $c_i \in \mathscr{C}_i$—namely, $\dim \mathscr{C}_i \geq \dim \mathscr{Y}_i$—and they must be appropriately located in the subsystem.

The second restrictive assumption is that for each v in V_F the local problem $m\text{LP}_i$ must have a unique solution. Without this assumption, $u_i^m(\cdot)$ and $\hat{c}_i(\cdot)$ as defined above could be point-to-set mappings and the applicability conditions as stated in Theorem 3.1 would be insufficient.

If for all $i \in \overline{1, N}$ and for all $v_i \in F_i(CU_i)$ the set $U_i(v_i)$ defined by (3.21) is convex, then the well-known theorem about the existence of a unique element with a minimal norm in a convex set in a Hilbert space implies the uniqueness of the solution to (3.19). In the finite-dimensional case, the convexity of the set $U_i(v_i)$ is not only sufficient but also necessary to ensure the uniqueness required above (cf. Rice 1969). For the nonconvex case, we consider the following lemma.

LEMMA 3.2.   *If the set $U \subseteq \mathscr{U}_i$ is compact then the set*

$$NU \triangleq \mathscr{U}_i \setminus \{u_i : \exists! u^m \in \mathscr{U}_i \quad u^m = \arg \min_{u \in U} \|u - u_i\|\}$$

has an empty interior.

Proof.   We prove first that if $\bar{u} \in \mathscr{U}_i$ and $0 \neq u^0 \in \bar{\mathscr{B}}(\bar{u}, r)$, then

$$(1) \qquad \text{fr}\,\bar{\mathscr{B}}(\bar{u}, r) \cap \text{fr}\,\bar{\mathscr{B}}(\bar{u} + u^0, |r - \|u^0\||) = \left\{\bar{u} + \frac{r}{\|u^0\|} u^0\right\}.$$

For convenience we assume that $\bar{u} = \theta$. Let $u' \in R_{u^0} \cap \text{fr}\,\bar{\mathscr{B}}(0, r)$ where $R_{u^0} \triangleq \{u_i : \exists \eta \geq 0 \quad u_i = \eta u^0\}$. Since $u' \in R_{u^0}$, then there exists $\delta_1 \geq 0$ such that $u' = \delta_1 u^0$. The inequality $\|u^0\| \leq r = \|u'\| = \delta_1 \|u^0\|$ implies that $\delta_1 \geq 1$. The norm of the difference $u^0 - u'$ equals

$$\|u^0 - u'\| = |1 - \delta_1| \cdot \|u^0\| = (\delta_1 - 1)\|u^0\| = \|u'\| - \|u^0\| = r - \|u^0\|,$$

hence $u' \in \text{fr}\,\bar{\mathscr{B}}(0, r) \cap \text{fr}\,\bar{\mathscr{B}}(u^0, |r - \|u^0\||)$.

Let $u'' \in \text{fr}\,\bar{\mathscr{B}}(0, r) \cap \text{fr}\,\bar{\mathscr{B}}(u^0, |r - \|u^0\||)$, therefore, $\|u''\| = \|u'\|$ and $\|u^0 - u''\| = \|u^0 - u'\|$. So,

$$\|u'' - u^0\| + \|u^0\| = \|u' - u^0\| + \|u^0\| = \|u'\| = \|u''\|.$$

198

We assumed that $\mathcal{U}_i$ is a Hilbert space, hence, the equality

$$\|u'' - u^0\| + \|u^0\| = \|u'' - u^0 + u^0\|$$

implies that there exists $\delta_2 \geq 0$ such that $u'' - u^0 = \delta_2 u^0$. Because

$$\delta_2\|u^0\| = \|u'' - u^0\| = \|u' - u^0\| = (\delta_1 - 1)\|u^0\|$$

we obtain

$$\delta_2 = \delta_1 - 1,$$

which means that $u'' = (\delta_2 + 1)u^0 = \delta_1 u^0 = u'$, and proves assertion (1).

We now prove the lemma. Let $\bar{u} \in NU$ and $\min_{u \in U}\|u - \bar{u}\| = r$. We assume that $u^* \in \{u_i : \exists 0 < \alpha < 1 \; u_i = (1 - \alpha)\bar{u} + \alpha u^m\} \triangleq S$, where $u^m \in \arg\min_{u \in U}\|u = \bar{u}\|$. From (1), after algebraic transformations, we get

$$\mathrm{fr}\,\bar{\mathcal{B}}(\bar{u}, r) \cap \mathrm{fr}\,\bar{\mathcal{B}}(u^*, |r - \|u^* - \bar{u}\||) = \{u^m\},$$

which means that $u^* \notin NU$. Now, from the definition of segment S it follows that in every nonempty open set there exists a point that does not lie in NU. Hence, the set NU has an empty interior. $\square$

The lemma means that the set of points in which the uniqueness assumption is not satisfied has an empty interior. Therefore, in a numerical search CP, we should be able to find the solution $\hat{v}$ of CP even though Eq. (3.19) may not have a unique solution.

The uniqueness property of problem (3.20) is a necessary requirement in most multilevel methods, and except for the problems with a strictly quasi-convex performance index and convex feasible set, we must check for this property case by case.

3.2.3. COORDINATION BASED ON PENALTY FUNCTIONS

The main difficulty of the methods described in the previous sections is the necessity to know and, moreover, not to violate the sets V_0 (in the pure direct approach) or V_F during the optimization process. This requirement is all the more troublesome because the only reason for preserving V_0 or V_F in the coordinator problem is the need to have nonempty constraint sets in the local decision problems, which have been constructed from inadequate models. If we could formulate local problems so that their decision sets would always be nonempty, we could eliminate the requirement. Of the direct optimization methods described in Chapter 2, the penalty function method allowed such a formulation. We will now apply the local problems of this method to our coordination structure with feedback.

We do not want to have any set of the type V_0 in the coordinator problem, so we will use the generalized local problems $\mathrm{LP}''_{\mathrm{ptyi}}$ defined in section 2.3.4. and not the local problems $\mathrm{LP}_{\mathrm{ptyi}}$ defined in section 2.3.1. This approach gives us a simpler coordinator problem, which makes the iterations on the real system easier. Thus, the local decision problems (LP_i^ρ) are as follows:

For given coordination variable v find both control and interaction

$$(\hat{c}_i(v, \rho_i), \hat{u}_i(v, \rho_i)) = \arg \min_{CU_i} Q_{\mathrm{ptyi}}(\,\cdot\,,\,\cdot\,, v, \rho_i), \qquad (3.26)$$

where

$$Q_{\mathrm{ptyi}}(c_i, u_i, v, \rho_i) \triangleq Q_i(c_i, u_i) + \rho_i(\|v_i - F_i(c_i, u_i)\|^2 + \|u_i - H_i v\|_0^2).$$

$\rho_i \in \mathbb{R}_+$ is an appropriate penalty coefficient and the norms $\|\cdot\|$ and $\|\cdot\|_0$ must not be the same, e.g., $\|\cdot\| = \kappa_i \|\cdot\|_0$, where $\kappa_i \in \mathbb{R}_+$ reintroduces the same weighting between outputs and structure equations.

The interaction variables u_i are treated as additional local decision variables only in order to have the constraint sets of the local problems independent of coordination variables v. Thus, only those elements of the interaction input vector u_i which are constrained together with the local controls c_i (see example in section 2.3.4.) should be treated as local decision variables. Hence, if the sets CU_i are separable, i.e.,

$$CU_i = C_i \times U_i,$$

then the local decision problems would only have controls as decision variables. Consequently, the separated local problem (SLP_i^ρ) is defined as follows:

For given coordination variable v find control

$$\hat{c}_i(v, \rho_i) = \arg \min_{C_i} Q_{\mathrm{ptyi}}(\,\cdot\,, v, \rho_i), \qquad (3.27)$$

where

$$Q_{\mathrm{ptyi}}(c_i, v, \rho_i) \triangleq Q_i(c_i, H_i v) + \rho_i \|v_i - F_i(c_i, H_i v)\|^2.$$

Separable sets are most favorable for the methods described in this section, since the coordinator problem constraints are simple; this was mentioned in section 3.2.1.

Using local decision problems LP_i^ρ, we get the following coordinator problem, CP^ρ:

Find coordination variable

$$\hat{v} = \arg \min_{V_*^\rho} Q(\hat{c}(\,\cdot\,, \rho), HK_*(\hat{c}(\,\cdot\,, \rho))), \qquad (3.28)$$

$$V_*^\rho \triangleq \{v \in \mathcal{Y} : K_*(\hat{c}(v, \rho)) \in Y \wedge \forall i \in 1, N \ \ (\hat{c}_i(v, \rho), H_i K_*(\hat{c}(v, \rho))) \in CU_i\}$$

where $\rho \triangleq (\rho_1, \ldots, \rho_N) \in \mathbb{R}_+^N$ and $\hat{c}(v, \rho) \triangleq (\hat{c}_1(v, \rho), \ldots, \hat{c}_N(v, \rho))$ (cf. problem (3.22)).

Coordinator problem CP^ρ is a reasonable optimization problem only when mapping $\hat{c}(\cdot, \rho)$ is a point-to-point mapping (both local and coordinator problems are solved with a constant value of ρ)—which implies that local problem LP_i^ρ should have a unique minimum. It is also desirable for $\hat{c}(\cdot, \rho)$ to be a continuous mapping. This requirement is satisfied by the assumptions also necessary to ensure reasonable features of the local decision problems.

LEMMA 3.3   *If function Q_i and mapping F_i are continuous, set CU_i is compact and local problem LP_i^ρ has a unique minimum for every $v \in \mathcal{Y}$, then $\hat{c}_i(\cdot, \rho_i)$ is continuous on $\mathcal{Y}$.*

The proof is standard and therefore omitted (cf. the proof of Theorem 3.1). Observe that local problem LP_i^ρ has a unique minimum when CU_i is convex, Q_i strictly quasi-convex, and F_i affine.

To make possible and justify the application of the penalty approach, the adjustment of penalty coefficient ρ and the result of control should be discussed. The theorem formulated below contributes to this discussion.

THEOREM 3.3.   *If the assumptions of Lemma 3.3 are satisfied and $\{v^n\}_{n=1}^\infty$ is any sequence from $\mathcal{Y}$ convergent to some v^0, then for every sequence $\{\rho_i^n\}_{n=1}^\infty$ tending to infinity*

$$\lim_{n' \to \infty} (\hat{c}_i^{n'}, \hat{u}_i^{n'}) \in M_i(v^0),$$

where

$$M_i(v^0) \triangleq \underset{(c_i, u_i) \in CU_i}{\text{Arg} \ min} \ (\|v_i^0 - F_i(c_i, u_i)\|^2 + \|u_i - H_i v^0\|_0^2)$$

is the set of points minimizing the function $\|v_i^0 - F_i(c_i, u_i)\|^2 + \|u_i - H_i v^0\|_0^2$ on CU_i and $\{(\hat{c}_i^{n'}, \hat{u}_i^{n'})\}$ is any convergent subsequence of the sequence $\{(\hat{c}_i^n, \hat{u}_i^n)\} \triangleq \{(\hat{c}_i(v^n, \rho_i^n), \hat{u}_i(v^n, \rho_i^n))\}$. If additionally the mapping $M_i(\cdot)$ is continuous at v^0 then

$$\lim_{n' \to \infty} (\hat{c}_i^{n'}, \hat{u}_i^{n'}) \in \hat{M}_i(v^0),$$

where

$$\hat{M}_i(v^0) = \underset{M_i(v^0)}{\text{Arg} \ min} \ Q_i(\cdot, \cdot)$$

is the set of points minimizing Q_i on $M_i(v^0)$.

Proof. Denote by $(\hat{c}_i, \hat{u}_i) \in CU_i$ the limit of the sequence $\{(\hat{c}_i^{n'}, \hat{u}_i^{n'})\}$, and suppose that $(\hat{c}_i, \hat{u}_i) \notin M_i(v^0)$. The function Q_i is bounded on CU_i since it is continuous and CU_i is compact. The set $M_i(v^0)$ is nonempty since the function $k_i(\,\cdot\,,\,\cdot\,, v^0)$, is continuous on compact set CU_i where

$$(c_i, u_i, v) \mapsto k_i(c_i, u_i, v) = \|v_i - F_i(c_i, u_i)\|^2 + \|u_i - H_i v\|_0^2.$$

Denote by $(\bar{c}_i, \bar{u}_i)$ any point of $M_i(v^0)$. It follows from the boundedness of Q_i on CU_i and the continuity of k_i on $CU_i \times \mathcal{Y}$ that for sufficiently large n'

$$Q_{\mathrm{ptyi}}(\hat{c}_i^{n'}, \hat{u}_i^{n'}, v^{n'}, \rho_i^{n'}) > Q_{\mathrm{ptyi}}(\bar{c}_i, \bar{u}_i, v^{n'}, \rho^{n'}),$$

which contradicts the fact that $(\hat{c}_i^{n'}, \hat{u}_i^{n'})$ is a point minimizing $Q_{\mathrm{ptyi}}(\,\cdot\,,\,\cdot\,, v^{n'}, \rho_i^{n'})$ on CU_i for every n'. Hence $(\hat{c}_i, \hat{u}_i) \in M_i(v^0)$.

For every $v \in \mathcal{Y}$ the set $M_i(v)$ is compact since $k_i(\,\cdot\,,\,\cdot\,, v)$ is continuous and CU_i is compact; hence, $\hat{M}_i(v)$ is nonempty and compact because Q_i is continuous. Let us denote by $(\hat{c}_i(v^0), \hat{u}_i(v^0))$ a point from $\hat{M}_i(v^0)$, and by $\{(c_i^{n'}, u_i^{n'})\}$ a sequence convergent to $(\hat{c}_i(v^0), \hat{u}_i(v^0))$ such that for every n': $(c_i^{n'}, u_i^{n'}) \in M_i(v^{n'})$. By virtue of the assumed continuity of $M_i(\,\cdot\,)$, such a sequence exists, and we have for every n'

$$Q_i(\hat{c}_i^{n'}, \hat{u}_i^{n'}) + \rho_i^{n'} k_i(\hat{c}_i^{n'}, \hat{u}_i^{n'}, v^{n'}) \leq Q_i(c_i^{n'}, u_i^{n'}) + \rho_i^{n'} k_i(c_i^{n'}, u_i^{n'}, v^{n'})$$

because the points $(\hat{c}_i^{n'}, \hat{u}_i^{n'})$ minimize Q_i. Of course

$$k_i(\hat{c}_i^{n'}, \hat{u}_i^{n'}, v^{n'}) \geq k_i(c_i^{n'}, u_i^{n'}, v^{n'}),$$

which implies that for every n'

$$Q_i(\hat{c}_i^{n'}, \hat{u}_i^{n'}) \leq Q_i(c_i^{n'}, u_i^{n'}).$$

Taking into account the continuity of Q_i, we get in the limit

$$Q_i(\hat{c}_i, \hat{u}_i) \leq Q_i(\hat{c}_i(v^0), \hat{u}_i(v^0)),$$

from which it follows that $(\hat{c}_i, \hat{u}_i) \in \hat{M}_i(v^0)$. $\square$

The above theorem provides that for sufficiently large values of ρ the solutions of local decision problems LP_i^ρ approximate the solutions of the following so-called limit local problems LP_i^L:

For given coordination variable v find control and interaction

$$(\hat{c}_i^L(v), \hat{u}_i^L(v)) = \arg \min_{M_i(v)} Q_i(\,\cdot\,,\,\cdot\,). \tag{3.29}$$

202

The coordinator problem (CP^L) corresponding to the above local problems is as follows:

Find coordination variable

$$\hat{v} = \arg \min_{V_*^L} Q(\hat{c}^L(\,\cdot\,), HK_*(\hat{c}^L(\,\cdot\,))), \tag{3.30}$$

$$V_*^L \triangleq \{v \in \mathcal{Y} : K_*(\hat{c}^L(v)) \in Y \wedge \forall i \in \overline{1,N}\ (\hat{c}_i^L(v), H_i K_*(\hat{c}^L(v))) \in CU_i\},$$

where $\hat{c}^L(v) \triangleq (\hat{c}_1^L(v), \dots, \hat{c}_N^L(v))$. It will be called the limit coordinator problem.

As in the two-level problem ($\{m\text{LP}_i\}_{i=1}^N, \text{CP}$) in the previous section, the two-level problem ($\{\text{LP}_i^L\}_{i=1}^N, \text{CP}^L$) is equivalent to the following optimization problem (OP^L)

Find control

$$\tilde{c} = \arg \min_{C^L} Q(\,\cdot\,, HK_*(\,\cdot\,)), \tag{3.31}$$

$$C^L \triangleq \{c \in \mathcal{C} : K_*(c) \in Y \wedge \forall i \in \overline{1,N}\ (c_i, H_i K_*(c)) \in CU_i\} \cap \hat{c}^L(\mathcal{Y}).$$

Problems OP (Eq. 3.23) and OP^L are nearly the same; the only difference is that the set $\hat{c}^L(\mathcal{Y})$ in OP^L replaces $\hat{c}(V_F)$ in OP. As for $\hat{c}(V_F)$, the constraint $c \in \hat{c}^L(\mathcal{Y})$ is not essential when the models are accurate (i.e., $F_i = F_{*i}$ for every $i \in \overline{1,N}$), and, moreover, the whole optimality discussion made earlier for OP and IOP can also be made for OP^L and IOP.

We will consider the applicability conditions of the penalty method with feedback. If we take into account approximation Theorem 3.3, it is sufficient to discuss only requirements that ensure that a solution of CP^L exists. First of all, its feasible set V_*^L should not be empty. The required nonemptiness condition is similar to (3.25):

$$C_* \neq \varnothing \quad \text{and} \quad \exists \bar{v} \in \mathcal{Y}\ c^L(\bar{v}) \in C_*,$$

and the whole discussion made earlier for (3.25) can also be made here to ensure that V_*^L is not empty. The required existence conditions can now be stated as follows.

THEOREM 3.4. *If the controlled system and its model are such that:*

1. *The mapping K_* is continuous, the constraint set Y is closed, the function ψ is continuous, and $\mathcal{Y}$ is some compact subset of the system output space,*

2. *For all $i \in \overline{1,N}$: the local sets CU_i are compact, functions Q_i and mappings F_i are continuous and such that the point-to-set mappings $M_i(\,\cdot\,)$ are*

continuous on V_*^L, and limit local problems LP_i^L have a unique solution for all $v \in V_*^L$,

then a solution of a limit coordinator problem CP^L exists.

Proof. Let us show first that the mapping $\hat{c}^L(\cdot)$, i.e., each of the mappings $\hat{c}_i^L(\cdot)$, $i \in \overline{1, N}$, is continuous on $\mathcal{Y}$. Let us take any point $v^0 \in \mathcal{Y}$ and any sequence $\{v^n\}_{n=1}^\infty \subset \mathcal{Y}$ converging to v^0. Since CU_i is compact we can assume, without loss of generality, that $\{(\hat{c}_i^n, \hat{u}_i^n)\} \triangleq \{(\hat{c}_i^L(v^n), \hat{u}_i^L(v^n))\} \subset CU_i$ converges to some point $(c_i^L, u_i^L) \in CU_i$. Suppose that $(c_i^L, u_i^L) \neq (\hat{c}_i^L(v^0), \hat{u}_i^L(v^0))$. It follows from the continuity of $M_i(\cdot)$ that $(c_i^L, u_i^L) \in M_i(v^0)$, hence, owing to the assumed uniqueness of the solutions of LP_i^L,

$$Q_i(c_i^L, u_i^L) > Q_i(\hat{c}_i^L(v^0), \hat{u}_i^L(v^0)).$$

The continuity of $M_i(\cdot)$ implies that there exists a sequence $\{(c_i^n, u_i^n)\} \subset CU_i$ such that for every n: $(c_i^n, u_i^n) \in M_i(v^n)$ and $(c_i^n, u_i^n) \rightarrow (\hat{c}_i^L(v^0), \hat{u}_i^L(v^0))$. For every $n = 1, 2, \ldots$, we have by definition

$$Q_i(\hat{c}_i^n, \hat{u}_i^n) \leq Q_i(c_i^n, u_i^n),$$

which contradicts in the limit the previous strict inequality. Thus for each $i \in \overline{1, N}$: $(c_i^L, u_i^L) = (\hat{c}_i^L(v^0), \hat{c}_i^L(v^0))$, which implies the continuity of $\hat{c}^L(\cdot)$.

The above result ensures the continuity of $Q(\hat{c}^L(\cdot), HK_*(\hat{c}^L(\cdot)))$, since ψ, K_*, and Q_i, $i \in \overline{1, N}$, are assumed to be continuous. The assumptions that the sets CU_i, $i \in \overline{1, N}$, are compact, the set Y closed, and $\mathcal{Y}$ compact are now sufficient to guarantee that V_*^L is compact owing to the continuity of $K_*(\hat{c}^L(\cdot))$. The existence of a solution of CP^L follows directly from the Weierstrass theorem. $\square$

Let us discuss briefly the assumptions of the above theorem. Demanding that mappings should be continuous and sets closed or compact is not unusual. The most restrictive assumptions seem to be that mappings $M_i(\cdot)$ be continuous and that problems LP_i^L have a unique solution; they are needed to guarantee that $\hat{c}^L(\cdot)$ is a point-to-point, continuous mapping. We must check these assumptions on a case-by-case basis. Note, however, that they concern only the model of the system, i.e., the known, and as a rule simplified, mathematical description of reality.

The whole optimality and applicability discussion has been mainly for the limit problem $(\{\text{LP}_i^L\}_{i=1}^N, \text{CP}^L)$, which is justified by the approximation Theorem 3.3. The assumptions of both Lemma 3.3 and Theorem 3.4 are sufficient for the existence of solutions of the coordinator problem CP^ρ as long as set V_*^ρ is not empty—which should be the case for appropriately large ρ. Owing to Theorem 3.3, these solutions should tend to the solution of CP^L as ρ tends to infinity, the optimality of that solution was discussed earlier (see Eq. (3.31)).

It is no trouble to apply the penalty methods when local performance indices Q_i and/or local constraint sets depend explicitly on outputs y, or when there are global resource constraints. The generalization of the methods to such cases is rather obvious and is left to the reader.

3.2.4. COORDINATION STRATEGIES

We are now in a position to choose the method of solving coordinator problem CP or CP$^\rho$, the so-called coordinator strategy. These problems are formulated as optimization problems in which we do not know the analytical definition of the minimized functional and the feasible set because they are defined in terms of mappings $\hat{c}(\cdot)$ or $\hat{c}(\cdot, \rho)$ and $K_*(\cdot)$. We can only compute the values of the former functions and measure the values of the latter in the real system. Under these circumstances, the penalty function method provides a powerful way to find the coordinator problem solution.

The coordinator feasible sets $V_* \cap V_F$ or V_*^ρ were defined such that for all feasible v the values $\hat{c}(v)$ or $\hat{c}(v, \rho)$, respectively, do not violate the real constraints. Therefore, we should choose a barrière (interior penalty) function of the set $V_* \cap V_F$ or V_*^ρ as the penalty function. Unfortunately, it seems that a statement of the sufficient conditions ensuring the nonemptiness of the interiors of the above sets, which is absolutely necessary in order to use the interior penalty function approach, is in general impossible. Consequently, a mixed or even an exterior penalty function could be used.

We propose the following experimental method of verifying the nonemptiness of the interior of $V_* \cap V_F$ or V_*^ρ. We shall consider the set $V_* \cap V_F$ only; the set V_*^ρ can be considered in the same manner. The set $V_* \cap V_F$ equals

$$V_F \cap V_* = \underset{i=1}{\overset{N}{\times}} F_i(CU_i) \cap c^{-1}\left(\bigcap_{i=1}^{N} D_{*i}^{-1}(CU_i)\right) \cap \hat{c}^{-1}(K_*^{-1}(Y))$$

where $c \mapsto D_{*i}(c) = (c_i, H_i K_*(c))$. The interiors of the sets $K_*^{-1}(Y)$, $D_{*i}^{-1}(CU_i)$, $F_i(CU_i)$, $i \in \overline{1, N}$, are not empty whenever the interiors of sets Y and CU_i are not empty, mappings K_*, D_{*i}, F_i, $i \in \overline{1, N}$, are continuous, and mappings F_i are open. For continuous mapping $f: E_1 \to E_2$ and $E \subseteq E_2$, $\forall e \in E_1$ $[f(e) \in \operatorname{int} E \Rightarrow e \in \operatorname{int} f^{-1}(E)]$. Hence the following chain of implications is true whenever mapping $\hat{c}(\cdot)$ is continuous:

$$\forall i \in \overline{1, N} \ [v_i \in F_i(\operatorname{int} CU_i) \wedge (\hat{c}_i(v), H_i K_*(\hat{c}(v))) \in \operatorname{int} CU_i] \wedge K_*(\hat{c}(v)) \in \operatorname{int} Y$$

$$\Updownarrow$$

$$\forall \in \operatorname{int} \underset{i=1}{\overset{N}{\times}} F_i(CU_i) \wedge \forall i \in \overline{1, N} \ D_{*i}(\hat{c}(v)) \in \operatorname{int} CU_i \wedge K_*(\hat{c}(v)) \in \operatorname{int} Y$$

$$\Updownarrow$$

$$v \in \operatorname{int} V_F \wedge \forall i \in \overline{1, N} \ \hat{c}(v) \in \operatorname{int} D_{*i}^{-1}(CU_i) \wedge \hat{c}(v) \in \operatorname{int} K_*^{-1}(Y)$$

$$\Updownarrow$$

$$v \in \text{int } V_F \wedge \hat{c}(v) \in \bigcap_{i=1}^{N} \text{int } D_{*i}^{-1}(CU_i) \cap \text{int } K_{*}^{-1}(Y)$$

$$\Downarrow$$

$$v \in \text{int } V_F \cap \text{int } \hat{c}^{-1}\left(\bigcap_{i=1}^{N} D_{*i}^{-1}(CU_i) \cap K_{*}^{-1}(Y)\right) =$$

$$\text{int } (V_F \cap V_*).$$

Thus, when the assumptions in Theorem 3.1 hold, implying, in particular, that mapping $\hat{c}(\cdot)$ is continuous, and

$$\text{int } Y \neq \varnothing \text{ and } \forall i \in \overline{1, N} \ \ \text{int } CU_i \neq \varnothing,$$

yjr nonemptiness of int $(V_* \cap V_F)$ can be ascertained by searching for a v in int V_F such that

$$\forall i \in \overline{1, N} \ \ (\hat{c}_i(v), H_i K_{*}(\hat{c}(v))) \in \text{int } CU_i$$

and

$$K_{*}(\hat{c}(v)) \in \text{int } Y$$

where $\hat{c}(v)$ is a compound solution of local problems $\{m\text{LP}_i\}_{i=1}^{N}$ and $K_{*}(\hat{c}(v))$ is a measured real system output corresponding to $\hat{c}(v)$. To find v we can use the method of Fiacco and McCormick (1968) for finding an interior point.

When the point v^0 in int $(V_* \cap V_F)$ or int V_*^o is found, we can proceed to numerical solution of CP or CP$^\rho$ with v^0 as a starting point. The techniques of applying penalty functions to these problems are standard and can be found, e.g., in Fiacco and McCormick (1968) and Polak (1971). However, when a mixed or exterior penalty function is used, the inequality constraints in the real system may be violated in the search for the optimal value of the coordination variable.

In penalty coordination with feedback the proper value of the penalty coefficient ρ must be chosen. A suitable first approximation is the value obtained from the solution of the model-based, off-line problem. We now briefly, consider the choice of numerical procedures for solving local and coordinator problems. For a fixed value of the coordination variable v and the penalty coefficient ρ, local problems $m\text{LP}_i$ or LP$_i^\rho$ are standard problems of nonlinear programming. Hence, the choice of a procedure depends on properties like differentiability, convexity, and so on. As far as the coordinator problems are concerned, even for simple cases the function to be minimized may not be differentiable (cf. section 2.2). Since the statement of differentiability conditions is a rather formidable problem, we have to apply nongradient search methods to solve CP or CP$^\rho$.

3.2.5. SIMULATION RESULTS

Pure direct coordination with feedback was tested using a modification of the system described in section 2.6.1, which is composed of three subsystems for which the performance functions, constraints, model, and real input–output mappings are as follows:

Subsystem 1

$$Q_1(c_1, u_1) = (u_1 - 1)^4 + 5(c_{.1} + c_{12} - 2)^2$$
$$CU_1 = \{(c_1, u_1) \in \mathbb{R}^3 : (c_{11})^2 + (c_{12})^2 \le 1 \wedge 0 \le u_1 \le 0.5\}$$
$$y_1 = F_{*1}(c_1, u_1) = 1.3c_{11} - c_{12} + 2u_1 + 0.15u_1c_{11}$$
$$y_1 = F_1(c_1, u_1) = c_{11} - c_{12} + 2u_1$$

Subsystem 2

$$Q_2(c_2, u_2) = 2(c_{21} - 2)^2 + (c_{22})^2 + 3(c_{23})^2 + 4(u_{21})^2 + (u_{22})^2$$
$$CU_2 = \{(c_2, u_2) \in \mathbb{R}^5 : 0.5c_{21} + c_{22} + 2c_{23} \le 1 \wedge 4(c_{21})^2 + 2c_{21}u_{21} + 0.4u_{21} + c_{21}c_{23}$$
$$+ 0.5(c_{23})^2 + (u_{21})^2 \le 4\}$$

$$y_{21} = F_{*21}(c_2, u_2) = c_{21} - c_{22} + 1.2u_{21} - 3u_{22} + 0.1(c_{22})^2$$
$$y_{22} = F_{*22}(c_2, u_2) = 2c_{22} - 1.25c_{23} - u_{21} + u_{22} + 0.25c_{22}c_{23} + 0.1$$
$$y_{21} = F_{21}(c_2, u_2) = c_{21} - c_{22} + u_{21} - 3u_{22}$$
$$y_{22} = F_{22}(c_2, u_2) = 2c_{22} - c_{23} - u_{21} + u_{22}$$

Subsystem 3

$$Q_3(c_3, u_3) = (c_{31} + 1)^2 + (u_3 - 1)^2 + 2.5(c_{23})^2$$
$$CU_3 = \{(c_3, u_3) \in \mathbb{R}^3 : c_{31} + u_3 + 0.5 \ge 0 \wedge 0 \le c_{32} \le 1\}$$
$$y_3 = F_{*3}(c_3, u_3) = 0.8c_{31} + 2.5c_{32} - 4.2u_3$$
$$y_3 = F_3(c_3, u_3) = c_{31} + 2.5c_{32} - 4u_3.$$

where Q_i is the performance function, CU_i is the constraint set, F_{*i} is the real input–output mapping, and F_i is the model-based input–output mapping. The structure of the system described in Figure 2.4 has the following matrix form:

$$u_1 = H_1 y = [0 \quad 1 \quad 0 \quad 0][y_1 y_{21} y_{22} y_3]^T$$

$$u_2 = [u_{21} u_{22}]^T = H_2 y = \begin{bmatrix} 1 & 0 & 0 & 0 \\ 0 & 0 & 0 & 1 \end{bmatrix} [y_1 y_{21} y_{22} y_3]^T$$

$$u_3 = H_3 y = [0 \quad 0 \quad 1 \quad 0][y_1 y_{21} y_{22} y_3]^T.$$

The system considered in section 2.6.1 is here assumed to be a model of a controlled real process.

As in the model example, the task of the coordinator is to influence the local decision units so as to make the global performance function

$$Q(c, u) = Q_1(c_1, u_1) + Q_2(c_2, u_2) + Q_3(c_3, u_3) \tag{2.114}$$

or

$$Q(c, u) = Q_1(c_1, u_1) \cdot Q_2(c_2, u_2) + Q_3(c_3, u_3) \tag{2.115}$$

as small as possible.

The local decision problems are the model-based problems, hence, they are the same as in section 2.6.1. The coordinator solves the problem

$$\text{minimize } \hat{Q}(v) = \psi(\hat{Q}_1(v), \hat{Q}_2(v), \hat{Q}_3(v))$$

$$\text{subject to } v \in V_* \cap V_0$$

where $\hat{Q}_i(v) = Q_i(\hat{c}_i(v), H_i K_*(\hat{c}(v)))$ denotes the real local optimal performance, and $\psi(Q_1, Q_2, Q_3)$ equals $Q_1 + Q_2 + Q_3$ for Eq. (2.114) and $Q_1 \cdot Q_2 + Q_3$ for Eq. (2.115).

Since set $V_* \cap V_0$ cannot be ignored during the coordination process, the problem was programmed using the interior penalty function technique. Because the function $\hat{Q}(\cdot)$ is not differentiable, the Powell procedure for unconstrained minimization was used.

The starting point was $v = (0, 0, 0, 0)$ and the unconstrained minimization was performed once with the penalty coefficient set equals to 10^4. The results are shown in Table 3.1.

TABLE 3.1 Results of the Simulation for Pure Direct Coordination with Feedback

	Eq. (2.114)	Eq. (2.115)
$\hat{v}_1$	0.0531	-0.0450
$\hat{v}_{21}$	0.2574	0.2102
$\hat{v}_{22}$	0.0411	0.2722
$\hat{v}_3$	0.2691	0.2265
$\hat{c}_{11}$	0.4375	0.4350
$\hat{c}_{12}$	0.8992	0.9004
u_1	0.1857	0.1403
$\hat{c}_{21}$	0.9716	0.9968
$\hat{c}_{22}$	0.0400	0.0623
$\hat{c}_{23}$	0.0949	0.1280
u_{21}	0.0531	0.0450
u_{22}	0.2966	0.2469
$\hat{c}_{31}$	-0.5411	0.3385
$\hat{c}_{32}$	0.3898	0.6615
u_3	0.0455	0.2624
$\hat{Q}(\hat{v})$	6.3841	7.9465
$\hat{Q}_*$	6.3142	7.2245
$\hat{Q}_m$	6.8998	14.3618

208

$\hat{Q}_*$ is the real optimal value of the performance, and $\hat{Q}_m$ is the model-based (open-loop) value of the performance.

The generated optimal controls in combination with the corresponding real outputs in both cases do not violate the constraints. This is contrary to the case when model-based controls are applied and the constraints in the real system may be violated. As expected, the losses in performance are much smaller than for open-loop control.

The numerical example using the direct method described in section 3.2.2 can be found in Woźniak (1976) and the example using the direct penalty method in section 3.2.3 in Findeisen *et al.* (1978).

3.3. COORDINATION BASED ON PRICE INSTRUMENTS

3.3.1. THE BASIC CONTROL STRUCTURE: THE INTERACTION BALANCE METHOD WITH FEEDBACK (IBMF)

We now use price instruments for coordination and allow the local decision units to solve the local problems of the interaction balance method (see section 2.4). The local units use only mathematical models. The collection of all N local problems, referred to as the infimal problem, IP, has the following form, where the price vector λ is fixed:

Find control $\hat{c}(\lambda)$ and interaction input $\hat{u}(\lambda)$ for the disconnected system, such that

$$Q_{\mathrm{mod}}(\hat{c}(\lambda), \hat{u}(\lambda), \lambda) = \min_{(c,u)\in CU} Q_{\mathrm{mod}}(c, u, \lambda) \qquad (3.32)$$

$$CU \triangleq \sum_{i=1}^{N} CU_i, \quad CU_i = \{(c_i, u_i) : G_i(c_i, u_i) \in S_i\}$$

where

$$Q_{\mathrm{mod}}(c, u, \lambda) = \sum_{i=1}^{N} Q_i(c_i, u_i) + \langle \lambda, u - HF(c, u) \rangle.$$

It is assumed in this section that $\mathscr{C}$ and $\mathscr{U}$ are real Hilbert spaces, although IBMF is applicable mainly to static systems; however, it can also be used in the control of batch processes which are periodic (see section 1.3).

The infimal problem IP is, of course, separable and we can solve it by solving N local problems IP_i independently.

Find control $\hat{c}_i(\lambda)$ and interaction input $\hat{u}_i(\lambda)$ for subsystem i such that

$$Q_{\mathrm{mod}i} \, \hat{c}_i(\lambda), \hat{u}_i(\lambda), \lambda) = \min_{(c_i,u_i)\in CU_i} Q_{\mathrm{mod}i}(c_i, u_i, \lambda) \qquad (3.33)$$

where

$$Q_{\mathrm{modi}}\,(c_i,\,u_i,\,\lambda) = Q_i(c_i,\,u_i) + \langle \lambda_i,\,u_i \rangle - \sum_{j=1}^{N} \langle \lambda_j,\,H_{ji}F_i(c_i,\,u_i) \rangle.$$

We assume here that the solution $(\hat{c}(\lambda),\,\hat{u}(\lambda))$ of the infimal problem is unique for every $\lambda \in \Lambda \subset \mathscr{U}$.

The solutions $(\hat{c}_i(\lambda),\,\hat{u}_i(\lambda))$ of the local problems IP_i, $i \in \overline{1,\,N}$, for given λ are used in the following way (Figure 3.3): the controls $\hat{c}_i(\lambda)$ are applied to the subsystems, and the interaction inputs $u_*(\lambda) \triangleq HK_*(\hat{c}(\lambda))$ are measured and transmitted to the coordinator. His task (CP) is defined as follows:

Find $\tilde{\lambda} = (\tilde{\lambda}_1,\,\ldots,\,\tilde{\lambda}_N)$ such that

$$\hat{u}(\tilde{\lambda}) = u_*(\tilde{\lambda}). \tag{3.34}$$

Note that the coordination condition in CP is arbitrary: it is evident that the control $\hat{c}(\tilde{\lambda})$, if $\tilde{\lambda}$ exists, obtained by IBMF is in general not optimal in the model or in the real system.

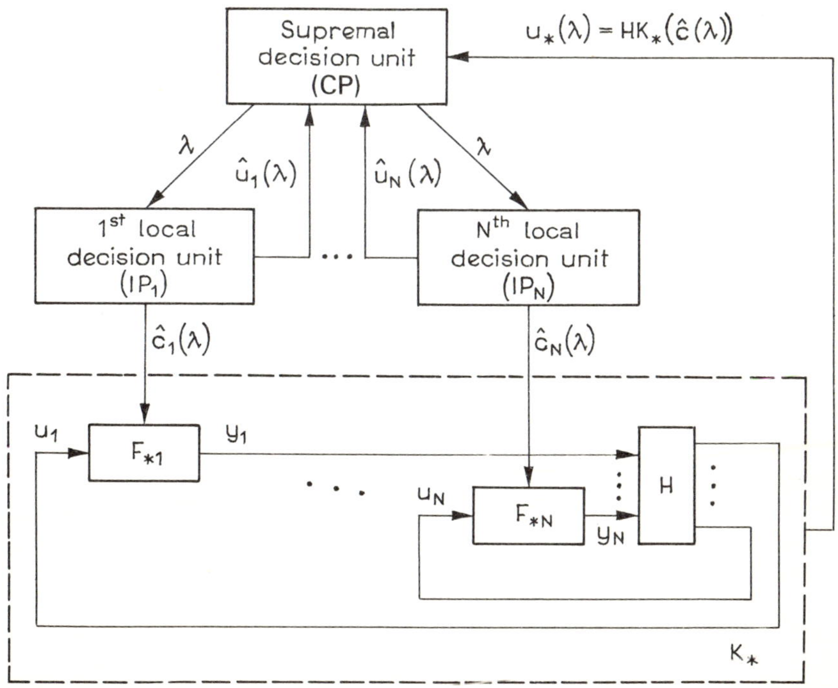

FIGURE 3.3 The structure of price coordination.

210

IBMF was described by Findeisen (1974) and its main properties have been investigated by Malinowski (1976) and Malinowski and Ruszczyński (1975), where it was shown that if $F_* = F + \delta$ ($\delta = $ const) then $\hat{c}(\tilde{\lambda})$ is the optimal solution of the system optimization problem. The important questions concerning such properties as: the existence of $\tilde{\lambda}$, the feasibility of $\hat{c}(\tilde{\lambda})$, the suboptimality of $\hat{c}(\tilde{\lambda})$, and coordination strategies will be discussed in the following sections. Possibilities for modifying IBMF to enlarge the class of problems for which it gives feasible control solutions will be discussed in sections 3.3.4. and 3.6.

3.3.2. THE EXISTENCE AND FEASIBILITY OF THE SOLUTION $(\tilde{\lambda})$

THEOREM 3.5 (*existence of $\tilde{\lambda}$*).   *If we assume that*

1. *CU is a weakly compact and convex subset of $\mathscr{C} \times \mathscr{U}$,*
2. *F_i and F_{*i} are weakly continuous on $\mathscr{C} \times \mathscr{U}$ and Q_i is bounded on CU_i and weakly lower semicontinuous on $\mathscr{C} \times \mathscr{U}$ for each $i \in 1, N$,*
3. *For every $s \in S$ the following system optimization problem:*
find $c^m(s), u^m(s)$ such that

$$Q(c^m(s), u^m(s)) = \min_{(c,u)} Q(c, u) \qquad (3.35)$$

subject to $u = HF(c, u) + s$ and $(c, u) \in CU$
can be solved (i.e., can be coordinated) by the interaction balance method IBM with coordination price $\hat{\lambda}_s \in \Lambda$, where

$$S \triangleq \{s \in \mathscr{U} : \exists (c, u) \in CU \ \ s = HF_*(c, u) - HF(c, u))\},$$

4. *For every $s \in S$ the solution of problem (3.35) is unique and the mappings $c^m(\cdot)$, $u^m(\cdot)$ are weakly continuous on S;*

then there exists at least one solution $\tilde{\lambda}$ of the coordinator problem of IBMF.

The proof is given in Appendix B.1.

The above theorem is founded on assumptions 3 and 4. Note that S is a set depending on the model-reality difference. Coordinability by IBM (assumption 3) has been extensively discussed in section 2.4. The following theorem can be proved with the information in that section.

THEOREM 3.6 (*Coordinability conditions for problem (3.35)*).   *Suppose that assumptions 1 and 2 of Theorem 3.5 hold and*
1. *There exists $k_1 > 0$, such that for every $u^0 \in \mathscr{U}, \|u^0\| \leq k_1$, there exists $(c, u) \in CU$ such that $u = HF(c, u) + u^0$,*

2. *For every $(c, u) \in CU$ following inequality is satisfied*

$$\|HF_*(c, u) - HF(c, u)\| \leq k_2,$$

where

$$0 \leq k_2 < k_1,$$

3. $(\hat{c}(\lambda), \hat{u}(\lambda))$ *is unique for*

$$\lambda \in \Lambda \triangleq \left\{ \lambda \in \mathcal{U} : \|\lambda\| \leq \frac{k_0' - k_0''}{k_1 - k_2} \right\}$$

where

$$k_0' = \sup_{(c,u) \in CU} Q(c, u), \qquad k_0'' = \min_{(c,u) \in CU} Q(c, u),$$

then assumption 3 of Theorem 3.5 will be satisfied and moreover $c^m(s)$ and $u^m(s)$ are unique.

The proof is given in Appendix B.2. Note that assumption 2 means that the model-reality difference is bounded.

In general, it is difficult to give explicit conditions under which assumption 4 of Theorem 3.5 is satisfied, and, in particular, conditions under which the uniqueness property is guaranteed. Some special cases (convex problems) have been considered in detail in Chapter 2. The continuity of solutions to the optimization problem with respect to the parameters of the problem is discussed in Chapter 4.

The feasibility of $\hat{c}(\tilde{\lambda})$ follows directly from (3.34) (CP) if we assume that the constraints have the form $(c, u) \in CU$ and that they are the same for the model and the real system, i.e., $CU = CU_*$. IP ensures that $(\hat{c}(\lambda), \hat{u}(\lambda)) \in CU$. We always have $c_* = \hat{c}(\lambda)$ and for $\lambda = \tilde{\lambda}$ we have $u_* = \hat{u}(\lambda)$.

3.3.3. OPTIMALITY OF THE SOLUTION $(\tilde{\lambda})$

We now discuss the suboptimality properties of the basic version of IBMF. If the assumptions of Theorem 3.6 are satisfied, $CU = CU_*$ and solution $\tilde{\lambda}$ of coordinator problem CP exists, then the following inequality holds:

$$0 \leq Q(\hat{c}(\tilde{\lambda}), u_*(\tilde{\lambda})) - Q(\hat{c}_*, HK_*(\hat{c}_*)) \leq 2k_2 \frac{k_0' - k_0''}{k_1 - k_2} \tag{3.36}$$

where $\hat{c}_*$ is the real optimal control vector and the parameters k_1, k_2, k_0', k_0'' are specified in the assumptions of Theorem 3.6. Indeed, the left side of inequality (3.36) is obvious since $\hat{c}(\tilde{\lambda})$ and $u_*(\tilde{\lambda})$ are feasible control and

212

interaction inputs. From (3.32) it follows that we can write

$$Q(\hat{c}(\tilde{\lambda}), u_*(\tilde{\lambda})) + \langle \tilde{\lambda}, u_*(\tilde{\lambda}) - HF(\hat{c}(\tilde{\lambda}), u_*(\tilde{\lambda})) \rangle$$
$$\leq Q(\hat{c}_*, HK_*(\hat{c}_*)) + \langle \tilde{\lambda}, HK_*(\hat{c}_*) - HF(\hat{c}_*, HK_*(\hat{c}_*)) \rangle$$

Using the above inequality and the assumptions of Theorem 3.6, we obtain

$$Q(\hat{c}(\tilde{\lambda}), u_*(\tilde{\lambda})) - Q(\hat{c}_*, HK_*(\hat{c}_*))$$
$$\leq \langle \tilde{\lambda}, [HF_*(\hat{c}_*, HK_*(\hat{c}_*)) - HF(\hat{c}_*, HK_*(\hat{c}_*))]$$
$$- [HF_*(\hat{c}(\tilde{\lambda}), u_*(\tilde{\lambda})) - HF(\hat{c}(\tilde{\lambda}), u_*(\tilde{\lambda})] \quad (3.37)$$

which yields the required bound. In (3.36), the coefficient k_2 depends on differences between the mathematical model and the description of the real system. Therefore, if $\|F_*(c, u) - F(c, u)\| \to 0$ uniformly on CU (i.e., $k_2 \to 0$), then the loss of performance decreases to zero.

The upper bound on performance loss given by (3.36) is large. Indeed, it follows from (3.37) that if $\forall (c, u) \in CU$ $HF_*(c, u) - HF(c, u) = \beta$, that is, when the model-reality difference reduces to a constant vector β, then

$$Q(\hat{c}(\tilde{\lambda}), u_*(\tilde{\lambda})) = Q(\hat{c}_*, HK_*(\hat{c}_*)),$$

that is, there is no loss of performance and $\hat{c}(\tilde{\lambda})$ is the real optimal control. This conclusion does not follow directly from (3.36). In such a situation, the optimal, open-loop control from the model can be very far from the optimal value.

The simple geometric example in Figure 3.4 presents generally the operation and advantages of IBMF. The continuous line is the set of real system operating points (c, u) satisfying an equation $u = F_*(c, u)$. The dashed lines are sets of system model operating points, $u = F(c, u)$, drawn for two models. The performance index is assumed to be $Q(c, u) = -c$, and the optimal performance is at point B. The control value c_B is then the real optimal solution. The model-based solutions are c_A and c_D. When the optimal control c_A is applied to the real system, infeasible operating point A^* is obtained. When c_D is applied, point D^* is obtained. It is feasible, but far from the optimal point B. Both models give inadequate open-loop control, but IBMF would give the real optimal solution exactly. To show this, we formulate the infimal problem IP:

$$\underset{(c,u) \in CU}{\text{minimize}} [Q(c, u) + \lambda(u - F(c, u))]$$

and ask the supremal level to change λ until $\lambda = \tilde{\lambda}$. For arbitrary λ, IP generates result $\hat{c}(\lambda)$, $\hat{u}(\lambda)$, which will not satisfy $u = F(c, u)$, but can be thought of as satisfying the equation

$$\hat{u}(\lambda) = F(\hat{c}(\lambda), \hat{u}(\lambda)) - v(\lambda).$$

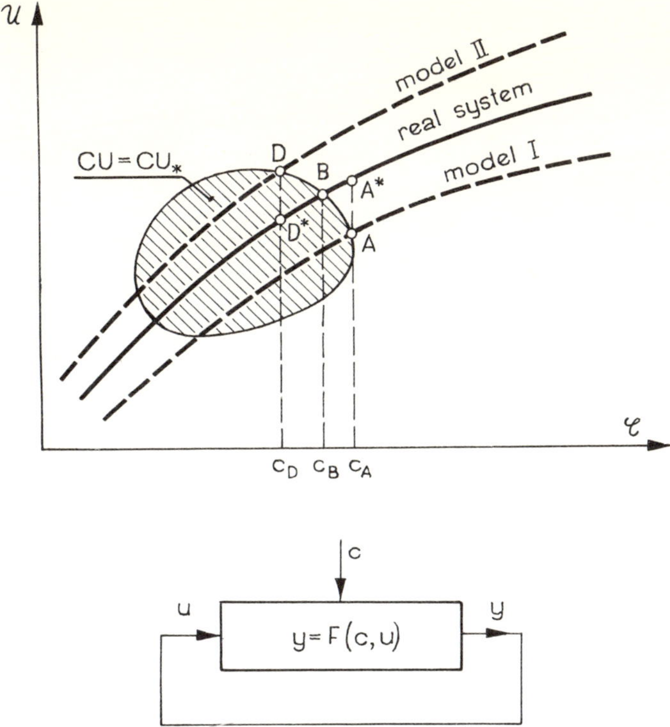

FIGURE 3.4   Example illustrating IBMF.

This is equivalent to saying that IP has solved a model-based optimal problem for a shifted model.

If the model is like I or II in Figure 3.4, and with $Q(c, u)$ as in the figure, the solutions $\hat{c}(\lambda)$, $\hat{u}(\lambda)$ satisfying the shifted model will still lie on the boundary of the set CU. While varying λ, we change the shift $v(\lambda)$ and allow the coordinator to move point A or D along the CU boundary. Each $\hat{c}(\lambda)$ will be applied to the system, that is, it will generate a $u_*(\lambda)$ according to the real system line in Figure 3.4. Achieving $\lambda = \tilde{\lambda}$ where $\hat{u}(\lambda) = u_*(\lambda)$ is in this case equivalent to arriving at point B.

3.3.4.   THE MODIFIED INTERACTION BALANCE METHOD WITH FEEDBACK (MIBMF)

It has been assumed that inequality constraints (see the definition of CU in IP at the beginning of section 3.3) were given on the values of controls and interaction inputs only. This, together with the coordinating condition of

214

IBMF, Eq. (3.34), guaranteed the feasibility of controls $\hat{c}(\tilde{\lambda})$ generated by IBMF. We should also consider a more general case where the original inequality constraints have the following form:

$$G_i^0(c_1, u_i, y_i) \in S_i, \qquad i \in \overline{1, N} \qquad (G^0(c, u, y,) \in S). \qquad (3.38)$$

The sets CU_i used in the formulation of IP will now be defined as follows:

$$CU_i = \{(c_i, u_i): G_i^0(c_i, u_i, F_i(c_i, u_i)) \in S_i\}, \qquad i \in \overline{1, N}.$$

Due to model-reality differences, the real feasible sets

$$CU_{*i} = \{(c_i, u_i): G_i^0(c_i, u_i, F_{*i}(c_i, u_i)) \in S_i\}, \qquad i \in \overline{1, N} \qquad (3.39)$$

are different from CU_i, and the controls $\hat{c}(\tilde{\lambda})$, along with $u_*(\tilde{\lambda})$ as generated by IBMF, may violate the constraints.

We can satisfy the constraints (3.38) in the real system by introducing a modified version of IBMF. The infimal set of local problems, MIP, of the modified interaction balance method with feedback (MIBMF) is defined as follows for given and fixed (λ, v):

Find control $\hat{c}(\lambda, v)$ and interaction input $\hat{u}(\lambda, v)$ for the disconnected system, such that

$$Q_{\mathrm{mod}}(\hat{c}(\lambda, v)\hat{u}(\lambda, v), \lambda) = \min_{(c,u)\in CU(v)} Q_{\mathrm{mod}}(c, u, \lambda)$$

$$CU(v) = \sum_{i=1}^{N} CU_i(v), \qquad (3.40)$$

$$CU_i(v) \triangleq \{(c_i, u_i): G_i^0(c_i, u_i, F_i(c_i, u_i) + v_i) \in S_i\},$$

$$v = (v_1, \ldots, v_N).$$

We assume for every $\lambda \in \Lambda \subset \mathcal{U}$ and $v \in V \subset \mathcal{Y}$ that the infimal problem, which may be solved as N independent local problems has a unique solution $\hat{c}(\lambda, v)$, $\hat{u}(\lambda, v)$, and we define $\hat{y}(\lambda, v) \triangleq F(\hat{c}(\lambda, v), \hat{u}(\lambda, v)) + v$. Each time that the infimal problem is solved, the control vector $\hat{c}(\lambda, v)$ is applied to the real system and the information about interaction inputs and outputs, $u_*(\lambda, v) \triangleq HK_*(\hat{c}(\lambda, v))$, $y_*(\lambda, v) \triangleq K_*(\hat{c}(\lambda, v))$ is transmitted to the supremal decision unit. Its task, MSP, will be defined as follows:
Find $\tilde{\lambda} \in \Lambda$, $\tilde{v} \in V$, such that:

$$\hat{u}(\tilde{\lambda}, \tilde{v}) = u_*(\tilde{\lambda}, \tilde{v}) \qquad (3.41)$$

$$\hat{y}(\tilde{\lambda}, \tilde{v}) = y_*(\tilde{\lambda}, \tilde{v}).$$

If the solution of the coordinator problem exists, then the control vector $\hat{c}(\tilde{\lambda}, \tilde{v})$ is feasible for the real system when real constraints are given by (3.38). The feasibility is obtained at the cost of introducing an additional coordination variable v. If G^0 does not depend explicitly on y, then the results of an application of MIBMF to the real system are virtually the same as when the basic version of IBMF is applied.

The conditions under which solution $(\tilde{\lambda}, \tilde{v})$ exists are given in the following theorem.

THEOREM 3.7 (*existence of* $(\tilde{\lambda}, \tilde{v})$). *If we assume that*

1. *Set* $CUY \triangleq \{(c, u, y): G^0(c, u, y) \in S\}$ *is a weakly compact convex subset of* $\mathscr{C} \times \mathscr{U} \times \mathscr{Y}$,

2. F_i, F_{*i} *are weakly continuous and* Q_i *is weakly lower semicontinuous on* $\mathscr{C}_i \times \mathscr{U}_i \times \mathscr{Y}_i$ *for each* $i \in \overline{1, N}$,

3. *For every* $v \in V$ *there is the following optimization problem* (A): *Find* $c^m(v), u^m(v)$ *such that*

$$Q(c^m(v), u^m(v)) = \min_{(c, u)} Q(c, u),$$

subject to $u = HF(c, u) + Hv$ *and* $(c, u) \in CU(v)$, *can be solved by* IBM *with the resulting optimal price* $\hat{\lambda}_v \in \Lambda$, *where*

$$V \triangleq \{v \in \mathscr{Y} : \exists (c, u, y) \in CUY : v = F_*(c, u) - F(c, u)\},$$

4. *For every* $v \in V$ *the solution of problem A is unique and the mappings* $c^m(\cdot), u^m(\cdot)$ *are weakly continuous on* V,

then there exists at least one solution $(\tilde{\lambda}, \tilde{v})$ *of the coordinator problem of* MIBMF.

The proof is similar to the proof of Theorem 3.5.

More detailed conditions of the existence of $(\tilde{\lambda}, \tilde{v})$ can be formulated, for example, for convex problems. The output shifts v can also be manipulated at the infimal level if appropriate local feedback is applied. This creates new structural possibilities and is the subject of current research.

3.3.5. COORDINATION STRATEGIES

In this section, we present some ways of solving the coordinator problem of IBMF. With some modifications, these algorithms may also be used for solving the coordinator problem of MIBMF.

In IBMF the task of the coordinator is to find the price satisfying the condition (see CP):

$$\hat{u}(\lambda) - u_*(\lambda) \triangleq R_*(\lambda) = 0. \tag{3.42}$$

When we know very little about the properties of R_* besides perhaps its continuity, it is difficult to propose a method for solving (3.42) other than direct minimization of the following function:

$$J(\lambda) \triangleq \|R_*(\lambda)\|^2. \tag{3.43}$$

It should be noted that even when the functions Q_i, F_1, F_{*i}, G_i have reasonable differentiability properties, the mapping R_* may be nondifferentiable. Thus, J is also nondifferentiable in most cases. Even if J were differentiable, we would not know ∇J since it depends on the unknown F_*. Local minima of J can also make the minimization of J difficult.

The difficulties associated with the minimization of J make it worthwhile to consider for solving Eq. (3.42) a modification of the Newton method for solving nonlinear equations. Unfortunately, the direct application of Newton's algorithm is not possible since we do not know the Jacobian $R'_*(\lambda)$. In most cases it does not even exist. However, the structure of Newton's algorithm suggests to us how we can try to generate new values of λ. This coordination strategy (CSI) can be formulated as follows (Zinchenko 1973):

$$\lambda^{k+1} = \lambda^k - [R'(\lambda^k)]^{-1}R_*(\lambda^k), \qquad k = 0, 1, 2, \ldots \tag{3.44}$$

where R is a certain differentiable approximation of R_* and R' is the Jacobian of R.

The general conditions under which algorithm (3.44) can generate sequence $\{\lambda^k\}$ convergent to the solution of (3.42) can be given by using the theorems proved by Zinchenko (1973) and Lika (1975), without assuming the differentiability of R_*.

THEOREM 3.8. *If*

1. *Fréchet derivative $R'(\cdot)$ of $R(\cdot)$ and mapping $R_0(\lambda) = R_*(\lambda) - R(\lambda)$ satisfy the Lipschitz conditions on a certain set $U_0 \subset \mathcal{U}$, i.e., for every $(\lambda^1, \lambda^2) \in U_0$*

$$\|R'(\lambda^1) - R'(\lambda^2)\| \le L\|\lambda^1 - \lambda^2\|$$

$$\|R_0(\lambda^1) - R_0(\lambda^2)\| \le L_1\|\lambda^1 - \lambda^2\|$$

2. *There exists $\Gamma_0 = [R'(\lambda^0)]^{-1}$ and*

$$\|\Gamma_0\| \le \rho_0$$

$$\|\Gamma_0 R_*(\lambda^0)\| \le \eta_0$$

3. $\rho_0 L_1 < 1$
4. $h_0 = \rho_0 L \eta_0 \le \frac{1}{2}(1 - \rho_0 L)^2$
5. *Ball $\mathcal{B}(\lambda^0; r_0) \subset U_0$,*

where

$$r_0 = (1 - \rho_0 L_1 - \sqrt{(1 - \rho_0 L_1)^2 - 2h_0})\eta_0/h_0 = \alpha \cdot \eta_0$$

then the solution of Eq. (3.42) is $\tilde{\lambda} \in \mathcal{B}(\lambda^0; r_0)$ and the sequence $\{\lambda^k\}_{k=0}^{\infty}$ generated by algorithm (3.44), where λ^0 is the starting point, is well defined and converges in norm to $\tilde{\lambda}$.

The theorem specifies not only the conditions under which (3.44) converges, but also the coordinability conditions. However, the crucial assumptions 4 and 5 can easily be satisfied by appropriate selection of λ^0 (see assumption 2) only if the real system is coordinable by IBMF. Therefore, the assumption about coordinability is hidden in 4 and 5. Also, the differentiability of $R_*(\cdot)$ is not required in the assumptions of Theorem 3.8, but assumption 3 shows the importance of the proper construction of $R(\cdot)$—the value of L_1 depends on how well $R(\cdot)$ approximates $R_*(\cdot)$.

Since we can measure $u_*(\lambda)$ in the real system, the coordinator can obtain the values of $R_*(\lambda)$. To apply strategy (3.44) for solving the coordinator problem (3.34), we have to construct approximation $R(\cdot)$ of $R_*(\cdot)$; we will use, of course, only $R'(\cdot)$.

Let us make a temporary assumption that for a unique given λ, solutions $\hat{c}(\lambda)$, $\hat{u}(\lambda)$ of the infimal problem (3.32) are differentiable functions of λ and that there exists an inverse operator to $[I - HF'_u(c, HK_*(c))]$. In such a case, it is possible to construct operator $R(\cdot)$ such that its derivative has the following form:

$$R'(\lambda) = \hat{u}'(\lambda) - [I - HF'_u(\hat{c}(\lambda), u_*(\lambda))]^{-1} HF'_c(\hat{c}(\lambda)), u_*(\lambda))\hat{c}'(\lambda)$$

$$\triangleq D(\lambda) \cdot \hat{w}'(\lambda), \quad (3.45)$$

where $w = (c, u)$, $\hat{w}(\lambda) = (\hat{c}(\lambda), \hat{u}(\lambda))$ and F'_c and F'_u are Fréchet derivatives of F with respect to c and u. $R'(\lambda)$ is the most direct approximation of $R'_*(\lambda)$—if we could use F_{*c}, F_{*u} in (3.45) instead of F'_c, F'_u, then (3.45) would just give us the derivative of $R_*(\cdot)$ in λ.

Since it is difficult to compute $\hat{w}'(\lambda)$ and it may not even exist at some points, we should rather compute a suitable approximation of $\hat{w}'(\lambda)$. This can be done by using the approach in section A.6 of Appendix A—this will give us the approximation of $\hat{w}'(\lambda)$ in the form:

$$\hat{w}'(\lambda) \simeq -[(Q_{\mathrm{mod}})''_{ww}(\hat{w}(\lambda), \lambda)]^{-1}[V'_w(\hat{w}(\lambda))]^* \quad (3.46)$$

where

$$V(w) = V(c, u) = u - HF(c, u).$$

By substituting into (3.45) we obtain the approximation of $R'(\lambda)$. However, the application of this approximation would require a large amount of on-line computation. Further simplification of (3.44) can be made by using the constant value of $R'(\cdot)$ computed at $\lambda = \lambda^0$, i.e.,

$$R'(\lambda) = -\tfrac{1}{\varepsilon}D(\lambda^0)[(Q_{\mathrm{mod}})''_{ww}(\hat{w}(\lambda^0), \lambda^0)]^{-1}[V'_w(\hat{w}(\lambda^0)]^*, \quad (3.47)$$

218

where we have introduced a scaling factor $1/\varepsilon > 0$. Assumption 3 of Theorem 3.8 can be satisfied by adjusting ε. Formula (3.47) can be further simplified if instead of $[(Q_{\mathrm{mod}})''_{ww}(\hat{w}(\lambda^0), \lambda^0)]^{-1}$ we use some positively defined operator B (e.g., $B = I$). We then obtain

$$R'(\lambda) = -\tfrac{1}{\varepsilon}D(\lambda^0)B[V'_w(\hat{w}(\lambda^0))]^*. \tag{3.48}$$

To compute the above approximation R of R_*, the coordinator has to have information about the subsystem equation before the IBMF algorithm has started. Therefore, the application of (3.48) destroys to some extent the decentralized prior information pattern of price coordination. The convergence properties of algorithm (3.44) with $R'(\lambda)$ computed according to (3.48) will be investigated in section 3.6. Solving Eq. (3.44) directly may not be the best way to achieve the desired balance condition. Assume that if $s = HF_*(c, u) - HF(c, u)$ for some $(c, u) \in CU$ then $\forall c \in \Pi_{\mathscr{C}}(CU)$ the equation

$$u - HF(c, u) = s$$

has the unique solution $u \in \mathscr{U}$. In order to achieve condition (3.42), we may solve the following equation:

$$-\hat{u}(\lambda) + HF(\hat{c}(\lambda), \hat{u}(\lambda)) + HF_*(\hat{c}(\lambda), u_*(\lambda)) - HF(\hat{c}(\lambda), u_*(\lambda)) = 0. \tag{3.49}$$

We can write the left side of the above equation in the following form

$$T(\lambda) = W(\lambda) + s(\lambda), \tag{3.50}$$

where

$$W(\lambda) = -\hat{u}(\lambda) + HF(\hat{c}(\lambda), \hat{u}(\lambda)) \tag{3.51}$$

and

$$s(\lambda) = HF_*(\hat{c}(\lambda), u_*(\lambda)) - HF(\hat{c}(\lambda), u_*(\lambda)). \tag{3.52}$$

The appealing feature of Eq. (3.50) is that (3.51) is a gradient of some functional (see section 2.4), and (3.52) depends only on the difference between the model and the real system.

To generate subsequent prices λ^k we can use algorithm (2.68), which was developed and thoroughly examined in section 2.4. The coordination strategy (CSII) will have then the following form

$$\lambda^{k+1} = \lambda^k - \varepsilon A T(\lambda^k), \qquad k = 0, 1, 2, \ldots \tag{3.53}$$

where A satisfies assumption 2 of Lemma 2.13 in section 2.4. $T(\lambda^k)$ can be easily computed by parts at the local control stations if the measurements of $u_*(\lambda)$ are available; $(u_*(\lambda) = HF_*(\hat{c}(\lambda), u_*(\lambda)))$. The amount of information sent to the coordinator in each iteration will be the same as when coordination strategy CSI, Eq. (3.44), is used.

The convergence properties of algorithm (3.53) have been studied in section 2.4. Theorems 2.18 and 2.19 of section 2.4 specify the sufficient conditions for this convergence. If the real system is coordinable by IBMF, then by finding appropriate λ^0 for (3.53) it is possible to satisfy the assumptions of Theorems 2.18 and 2.19 if the assumptions of Lemma 2.13 are satisfied. First, we are interested in satisfying condition (iii) in assumption 1 of Lemma 2.13 by operator $s(\cdot)$ as defined in Eq. (3.52). $W(\cdot)$ given by (3.51) depends on the model data only and its properties have been investigated in section 2.4.

LEMMA 3.4. *If the assumptions of Lemma 2.14 from section 2.4. are satisfied and*

(1) $$\forall c^1, c^2 \in \Pi_{\mathscr{C}} CU \ \|HK_*(c^1) - HK_*(c^2)\| \leq \sigma_4 \|c^1 - c^2\|,$$

(2) $$\forall (c^1, u^1), (c^2, u^2) \in (\Pi_{\mathscr{C}} CU) \times \mathscr{U}$$

$$\|[HF_*(c^1, u^1) - HF(c^1, u^1)] - [HF_*(c^2, u^2) - HF(c^2, u^2)]\|$$

$$\leq \sigma_6 \|(c^1, u^1) - (c^2, u^2)\|.$$

then condition (iii) of Lemma 2.13 is satisfied for λ, $\lambda + h \in \mathscr{P}_1$ and $\sigma_2 = \sigma_6 \cdot \sigma_7$, where $\sigma_7 > 0$ and σ_7 does not depend on the difference between the model and the real system. Therefore, if σ_6 is sufficiently small, assumption 3 of Lemma 2.13 is satisfied.

The proof of this lemma is straightforward if we use the proof of Lemma 2.14 from section 2.4 (see Appendix A).

It is usually very difficult to check whether the theoretical convergence conditions for algorithms (3.44) or (3.53) are satisfied for a practical problem and therefore numerical simulation is necessary to evaluate the coordination strategies.

Finally, the question concerning the practical choice of λ^0—the starting point for algorithms (3.44) or (3.53)—should be answered. Intuitively, it seems reasonable to take the model coordinating price $\hat{\lambda}$ (see section 2.4) as λ^0 since for small differences between the model and the real system $\hat{\lambda}$ should be close to $\tilde{\lambda}$. Consider the following theorem:

THEOREM 3.9. *Suppose that we are given a family $\mathscr{F}$ of the subsystem models, where $F_i^{(n)}(\cdot, \cdot)$, $i \in \overline{1, N}$, $n = 1, 2, \ldots$, are members of this family, and a family of real systems $\mathscr{F}_*$, with members $F_{*i}^{(n)}(\cdot, \cdot)$. If*

1. *Assumption 1 of Lemma 2.13 is satisfied for operators $W^{(n)}(\cdot)$, $s^{(n)}(\cdot)$ uniformly with respect to n, where $W^{(n)}(\cdot)$, $s^{(n)}(\cdot)$ are defined by Eqs. (3.51) and (3.52) with $F(c, u) = F^{(n)}(c, u)$ and $F_*(c, u) = F_*^{(n)}(c, u)$,*

2. *Using each model $F^{(n)}(\cdot, \cdot)$, the real system (with $F_*^{(n)}(\cdot, \cdot)$) is coordinable by IBMF and $\tilde{\lambda}^{(n)} \in \mathscr{P}_1[\tilde{\lambda}^{(n)}$ solves Eq. (3.49) with $T^{(n)}(\lambda) = W^{(n)}(\lambda) + s^{(n)}\lambda)]$,*

3. *The optimization problem for the system model, with $F^{(n)}(\cdot, \cdot)$, can be solved by* IBM *(see section 2.4) and $\hat{\lambda}^{(n)} \in \mathscr{P}_1$,*

4. $\forall (c, u) \in CU$

$$\|HF_*^{(n)}(c, u) - HF^{(n)}(c, u)\| \leq \beta_n$$

then $\|\tilde{\lambda}^{(n)} - \hat{\lambda}^{(n)}\| \leq \beta_n/\sigma$ where σ is defined in assumption 1 of Lemma 2.13 in section 2.4. Therefore, if $\beta_n \to 0^+$, then

$$\|\tilde{\lambda}^{(n)} - \hat{\lambda}^{(n)}\| \to 0^+.$$

The proof is given in Appendix B, section B.3.

We have then, that if some reasonable assumptions are satisfied and if the difference between the model and the real system decreases uniformly on CU, then the optimal prices from the model approach the IBMF balance prices $\tilde{\lambda}^{(n)}$. This justifies the choice of λ^0 as $\hat{\lambda}$ in the coordination strategies CSI and CSII, Eqs. (3.44) and (3.53).

3.3.6. APPLICATION OF AUGMENTED LAGRANGIANS

In section 2.5 we showed that the class of optimization problems to which price coordination is applicable can be considerably enlarged when we use the augmented Lagrangian instead of the normal Lagrangian in the infimal problem. Of course, the augmented Lagrangian does not allow easy separation of the infimal problem (3.32) into independent local problems. But with on-line price coordination of the real system, in some cases we can use the augmented Lagrangian

$$L_a(c, u, \lambda) = Q(c, u) + \langle \lambda, u - HF(c, u) \rangle$$
$$+ \tfrac{1}{2}\rho\langle u - HF(c, u), u - HF(c, u) \rangle, \qquad \rho > 0 \quad (3.54)$$

instead of Lagrangian $L(c, u, \lambda)$ in the infimal problem (3.32). If (3.54) is used in (3.32) then, without changing the coordinator problem (3.34), we can apply the on-line price coordination mechanism. The only trouble is that in this case problem (3.32) can no longer be solved by solving N independent local problems (3.33)—term $\rho\langle u, HF(c, u) \rangle$ cannot be additively separated into N terms depending on u_i, c_i. To overcome this difficulty we can use the approach described in section 2.5, which involves specific linearization of the nonseparable term in (3.54) and provides us with the approximation $\bar{L}_a$ of (3.54) in the form:

$$\bar{L}_a(c, u, \lambda; \bar{u}, \bar{y}) = Q(c, u) + \langle \lambda, u - HF(c, u) \rangle$$
$$+ \tfrac{1}{2}\rho\langle u, u \rangle + \tfrac{1}{2}\rho\langle F(c, u), F(c, u) \rangle \qquad (3.55)$$
$$+ \rho[\langle \bar{u}, H\bar{y} \rangle - \langle \bar{u}, HF(c, u) \rangle - \langle u, H\bar{y} \rangle],$$

where $\bar{y} = F(\bar{c}, \bar{u})$ and we have assumed that H is a matrix with one nonzero entry (equal to 1) in each row and in each column. Function $\bar{L}_a(\cdot, \cdot, \lambda; \bar{u}, \bar{y})$ is additively separable and using it we can define the interaction balance method with feedback and augmented Lagrangian (IBMFAL).

The infimal problem (IPA), separable into independent local problems, will have the form:

For given $\bar{c}, \bar{u}, \lambda$ find control $\hat{c}(\bar{u}, \bar{y}, \lambda)$ and interaction input $\hat{u}(\bar{u}, \bar{y}, \lambda)$ for the disconnected system, such that

$$\bar{L}_a(\hat{c}(\bar{u}, \bar{y}, \lambda), \hat{u}(\bar{u}, \bar{y}, \lambda), \lambda; \bar{u}, \bar{y}) = \min_{(c,u)\in CU} \bar{L}_a(c, u, \lambda, \bar{u}, \bar{y}), \qquad (3.56)$$

where $\bar{y} = F(\bar{c}, \bar{u})$.

The coordinator problem (CPA) that defines the coordinator's task is as follows:

Find $\tilde{c}, \tilde{u}, \tilde{\lambda}$ such that

$$\tilde{c} = \hat{c}(\tilde{u}, \tilde{y}, \tilde{\lambda}) \qquad (3.57)$$

$$\tilde{u} = \hat{u}(\tilde{u}, \tilde{y}, \tilde{\lambda}) = HK_*(\hat{c}(\tilde{u}, \tilde{y}, \tilde{\lambda}))$$

where $\tilde{y} = F(\tilde{c}, \tilde{u})$

The coordinator must find not only the balance prices $\tilde{\lambda}$ but also the controls and interaction inputs. Therefore, in order to preserve the decentralized character of the decision process and the (prior) information pattern, it is necessary to develop very simple rules for the coordination strategy.

Stoilov (1977) has proposed the following coordination strategy

$$\bar{u}^{k+1} = \hat{u}(\bar{u}^k, \bar{y}^k, \lambda^k)$$

$$\bar{c}^{k+1} = \hat{c}(\bar{u}^k, \bar{y}^k, \lambda^k) \qquad (3.58)$$

$$\lambda^{k+1} = \lambda^k + \frac{\rho}{2}[(\bar{u}^k - HF(\bar{c}^k, \bar{u}^k))$$

$$-(HK_*(\bar{c}^k) - HF(\bar{c}^k, HK_*(\bar{c}^k)))]$$

in which the coordinator has to receive and transmit only the information concerning the interaction inputs and outputs values.

The convergence properties of coordination strategy (3.58) have been studied by Stoilov (1977) for a finite-dimensional case and under typical assumptions (involving second-order local sufficient conditions of optimality, a strict complementarity condition concerning the local constraints, and so on). Perhaps the most important result is that under reasonable assumptions

$$\tilde{\lambda} = \tilde{\lambda} + \rho[HF_*(\tilde{c}, \tilde{u}) - HF(c, u)] \qquad (3.59)$$

where $\tilde{\lambda}$ is the balance price for IBMF. Therefore, by increasing ρ we convexify the initial problem, but at the same time we move the solution of IBMFAL away from $\tilde{\lambda}$ and, hence from the model-based optimal price $\hat{\lambda}$ which is used as the starting point for the coordination strategy (see the previous section). Furthermore, for large values of ρ the coordination strategy (3.58) may lose even the local convergence property (unlike in the optimization case considered in section 2.5). Still, a reasonable choice of ρ may allow the solving of some coordination problems with IBMFAL which are not coordinable with the basic price mechanism, IBMF.

3.3.7. SIMULATION RESULTS

To test IBMF and MIBMF, the following examples were simulated:

Example 1

Subsystem 1

$$y_1 = c_{11} - c_{12} + 2u_1 - 0.5c_{11}^2 + 0.5(c_{11} + c_{12} - 2)u_1 = F_{*1}(c_1, u_1)$$

$$y_1 = 1.4375c_{11} - 0.1875c_{12} + 1.75u_1 - 0.6872 = F_1(c_1, u_1)$$

$$Q_1(c_1, u_1) = (u_{11} - 1)^2 + c_{11}^2 + (c_{12} - 2)^2$$

$$G_1(c_1, u_1) = c_{11} + u_{11} - 1.006 \le 0.$$

Subsystem 2

$$\begin{bmatrix} y_{21} \\ y_{22} \end{bmatrix} = \begin{bmatrix} c_{21} - c_{22} + u_{21} - 3u_{22} \\ 2c_{22} - c_{23} - u_{21} + u_{22} \end{bmatrix} = F_{*2}(c_2, u_2) = F_2(c_2, u_2)$$

$$Q_2(c_2, u_2) = 2(c_{21} - 2)^2 + c_{22}^2 + 3c_{23}^2 + 4u_{21}^2 + u_{22}^2.$$

Subsystem 3

$$y_3 = c_3 - 4u_3 + 0.5c_3 u_2 = F_{*3}(c_3, u_3)$$

$$y_3 = 1.25c_3 - 3.75u_3 - 0.125 = F_3(c_3, u_3)$$

$$Q_3(c_3, u_3) = (c_3 + 1)^2 + (u_3 - 1)^2$$

$$G_3(c_3, u_3) = -c_3 - u_3 - 0.5 \le 0.$$

Structure equations

$$\begin{bmatrix} u_1 \\ u_{21} \\ u_{22} \\ u_3 \end{bmatrix} = \begin{bmatrix} 0 & 1 & 0 & 0 \\ 1 & 0 & 0 & 0 \\ 0 & 0 & 0 & 1 \\ 0 & 0 & 1 & 0 \end{bmatrix} \cdot \begin{bmatrix} y_1 \\ y_{21} \\ y_{22} \\ y_3 \end{bmatrix}.$$

The mappings F_{*1} and F_{*3} describe the real subsystems, and F_1 and F_3 are model equations. They are obtained by approximation of the real nonlinear equations around some chosen points (c, u) that satisfy the inequality constraints. It was assumed that $F_2 = F_{*2}$.

IBMF was applied to the above example and two coordinating strategies were used to adjust the four-dimensional λ:

1.   The direct Powell method for unconstrained minimization of $J(\lambda)$ (see Eq. (3.43)),
2.   Algorithm (3.44) with $R'(\lambda)$ computed from (3.48), where we set $B = I$ and $\varepsilon = 0.8$.

Algorithm 1 was run once with $\lambda^0 = 0$ and once with $\lambda^0 = \hat{\lambda}$, the optimal price vector from the model.

Starting from $\lambda^0 = 0$, algorithm 1 converged slowly to solution $\tilde{\lambda}$. After 55 interations at the supremal level the algorithm was stopped. Starting from $\lambda^0 = \hat{\lambda}$, algorithm 1 converged to $\tilde{\lambda}$ with an accuracy to 10^{-2} in 23 iterations.

Algorithm 2, starting from $\lambda^0 = 0$, converged to $\tilde{\lambda}$ with the same accuracy as in algorithm 1 in 8 iterations. Starting from $\lambda^0 = \hat{\lambda}$, the solution was found in 2 iterations.

The above results confirmed our intuitive expectations that a direct minimization of J at the supremal level may cause difficulties and that in any case it makes sense to begin iterations of the IBMF algorithm from the optimal price vector $\lambda^0 = \hat{\lambda}$ computed from the model.

When we assumed that constraint $G_3(c_3, u_3)$ of subsystem 3 was obtained by substituting model $F_3(c_3, u_3)$ into the following real constraint:

$$G_3^0(c_3, u_3, y_3) = -2.25c_3 + 2.75u_3 + y_3 - 0.375 \leq 0,$$

then this constraint was seriously violated after $\hat{c}(\tilde{\lambda})$ was applied to the real system. It was necessary to use MIBMF in order to obtain feasible control $\hat{c}(\tilde{\lambda}, \tilde{v})$. It is easy to see that only the application of output shift v_3 in subsystem 3 is significant and v_1 and v_2 do not have to be used (G_1 does not depend explicitly on y_1 and there is no G_2 constraint). Therefore, the task of the coordinator problem of MIBMF is to solve the following equations:

$$\hat{u}(\lambda, v_3) - u_*(\lambda, v_3) = 0$$
$$\hat{y}_3(\lambda, v_3) - y_{*3}(\lambda, v_3) = 0.$$

We denote both of them as

$$R_*(\lambda, v_3) = 0.$$

Since algorithm 2 proved to be effective, the approximation of R similar to (3.48), with $\varepsilon = 0.8$, was used in the following coordination strategy:

$$\begin{bmatrix} \lambda^{k+1} \\ v_3^{k+1} \end{bmatrix} = \begin{bmatrix} \lambda^k \\ v_3^k \end{bmatrix} - [R'(\lambda^k, v_3^k)]^{-1} R_*(\lambda^k, v_3^k).$$

224

This algorithm was run once with $\lambda^0 = 0$ and $v_3^0 = 0$, and once with $\lambda^0 = \tilde{\lambda}^0$ and $v_3^0 = 0$, where $\tilde{\lambda}^0$ is the price vector obtained by application of IBMF in the previous case.

Starting from $\lambda^0 = 0$, the algorithm converged to the point $(\tilde{\lambda}, \tilde{v}_3)$ and control vector $\hat{c}(\tilde{\lambda}, \tilde{v}_3)$ satisfying the constraints in 20 iterations; starting from $\lambda^0 = \tilde{\lambda}^0$ it took 10 iterations.

Example 2

The system described in section 2.6 can be used to simulate the application of IBMF. That system was assumed to be a mathematical model of the controlled static process. The mappings F_{*1}, F_{*2}, F_{*3} that describes the real subsystems had the following form:

Subsystem 1

$$y_{11} = 1.3c_{11} - c_{12} + 2u_{11} + 0.15u_{11}c_{11} = F_{*1}(c_1, u_1)$$

Subsystem 2

$$y_{21} = c_{21} - c_{22} + 1.2u_{21} - 3u_{22} + 0.1(c_{22})^2 = F_{*2}(c_2, u_2)$$

$$y_{22} = 2c_{22} - 1.25c_{23} - u_{21} + u_{22} + 0.25c_{23}c_{22} + 0.1$$

Subsystem 3

$$y_3 = 0.8c_{31} + 2.5c_{32} - 4.2u_3.$$

When IBMF was applied to the above example, the infimal problem was the same as the infimal problem of IBM in section 2.6. Two coordination strategies were used to achieve the desired balance condition. Algorithm (3.44) with $R'(\lambda)$ computed from (3.48), where we set $B = I$ and $\varepsilon = 0.8$; and algorithm (3.53) with (a) $A = I$ and $\varepsilon = 0.1$, and (b) A computed according to the procedure in Appendix A.2 and $\varepsilon = 0.8$. The application of algorithm (3.53) with $A = I$ does not require the coordinator to have prior information concerning the models of the subsystems. Also, the computation of A according to the approach presented in Appendix A is much simpler than the computation of $R'(\lambda)$ for algorithm (3.44).

Both algorithms were run from the starting point $\lambda^0 = 0$, which would not be reasonable in a practical application but creates rigorous test conditions for numerical simulation. The following stop criterion was used: $\|\hat{u}(\lambda) - u_*(\lambda)\| \leq \alpha$. The results of the simulation are summarized in Table 3.2.

For $\alpha = 0.0001$, the real system constraints have been satisfied with very high accuracy (better than 10^{-4}). It should be noted that some of these constraints were badly violated after application of the model optimal controls $\hat{c}$. As in Example 1, changing λ^0 from 0 to $\hat{\lambda}$ creates an advantage for the on-line coordination strategy since the number of iterations required to achieve the balance condition decreases considerably.

TABLE 3.2   Results of the Simulation for IBMF and MIBMF

	Number of iterations $\alpha = 0.01$	Performance value (real)	Number of iterations $\alpha = 0.0001$	Performance value (real)
Algorithm (3.44)	41	6.320	92	6.323
Algorithm (3.53) (case a)	59	6.323		
Algorithm (3.53) (case b)	52	6.323	78	6.323

3.4.   DECENTRALIZED CONTROL WITH PARTIAL OR FULL COORDINATION

3.4.1.   THE CONTROL STRUCTURE

The decentralized control structure has been explained and briefly presented in section 1.3 (see Figure 1.20). In this subsection it is described in detail. We start with an infimal problem formulation. The ith local decision problem (LP_i) has the following form:

For given coordination variable λ and interaction variable u_i find control $\hat{c}_i(u_i, \lambda)$ such that

$$Q_{\text{modi}}(\hat{c}_i(u_i, \lambda), u_i, \lambda) = \min_{c_i \in C_i(u_i)} Q_{\text{modi}}(c_i, u_i, \lambda) \tag{3.60}$$

where

$$Q_{\text{modi}}(c_i, u_i, \lambda) \triangleq Q_i(c_i, u_i) + \langle \lambda_i, u_i \rangle - \sum_{j=1}^{N} \langle \lambda_j, H_{ji} F_i(c_i, u_i) \rangle$$

and

$$C_i(u_i) \triangleq \{c_i \in \mathscr{C}_i : (c_i, u_i) \in CU_i\}.$$

Let us suppose that the collection of N local decision problems has been solved and control $\hat{c}(u, \lambda) = (\hat{c}_1(u_1, \lambda), \ldots, \hat{c}_N(u_N, \lambda))$ has been found. Next, the controls $\hat{c}_i(u_i, \lambda)$, $i \in \overline{1, N}$, are applied to the real interconnected subsystems and a new value $HK_*(\hat{c}(u, \lambda))$ of the interaction variable in the real system is realized. If this value of interaction equals the one that was used to obtain $\hat{c}(u, \lambda)$, then we say that the *equilibrium state* in the control structure is obtained. In this case we also say that the infimal problem is solved. Otherwise, the new value of the interaction is sent to the local decision units and the new controls are produced as before. Let us assume that for each

226

value of the interaction variable from set

$$U^0 \triangleq \{u \in \mathcal{U} : C(u) \neq \varnothing\}, \tag{3.61}$$

where

$$C(u) = \underset{i=1}{\overset{N}{\times}} C_i(u_i),$$

and for each value of the coordination variable λ from $\mathcal{U}$, any local decision problem (see (3.60)) has a unique solution. This means that a mapping $\hat{c}(\cdot, \lambda)$ is well defined on U^0 for any $\lambda \in \mathcal{U}$. The set U^0 is decomposable, i.e.,

$$U^0 = \underset{i=1}{\overset{N}{\times}} U_i^0, \tag{3.62}$$

where

$$U_i^0 = \{u_i \in \mathcal{U}_i : C_i(u_i) \neq \varnothing\}.$$

It follows directly from the above that the problem of proving the existence of and finding the infimal problem solution (or the equilibrium state of the considered control structure) is equivalent to the problem of proving the existence and finding a fixed point of the following mapping

$$\mathcal{U} \supset U^0 \ni u \rightarrow HK_*(\hat{c}(u, \lambda)) \in \mathscr{C}.$$

The infimal problem-solving (for a given λ) consists in finding the fixed point of this mapping using the following iterative scheme:

$$u^{k+1} = HK_*(\hat{c}(u^k, \lambda)) \tag{3.63}$$

This iterative scheme will be called a lower-level iterative scheme.

To find the fixed point of the mapping $HK_*(\hat{c}(\cdot, \lambda))$ or equivalently, to solve the operator equation

$$u = HK_*(\hat{c}(u, \lambda)), \tag{3.64}$$

we could use an iterative scheme different from Eq. (3.63); however, Eq. (3.63) has two appealing features:

- The interaction variable value which is actually realized in the real system is directly introduced into the local decision problems, so the measurements used in the local problems are, from the point of view of the local decision units, used in the most natural manner.
- It is completely decentralized.

The use of an iterative scheme similar to Newton's for solving equality (3.64) is discussed in Brdyś and Michalak (1978) for the linear–quadratic case. This kind of algorithm requires the existence of partial communication between the local decision units. It makes the information structure more expensive than in the structure required by algorithm (3.63). However, the completely decentralized lower-level iterative scheme given by (3.63) may fail in some cases and partial communication between the local decision units must be introduced to solve Eq. (3.64).

For a given λ let us denote the infimal problem solution by $u_b(\lambda)$ and the corresponding control by $c_b(\lambda)$ where

$$c_b(\lambda) = \hat{c}(u_b(\lambda), \lambda). \tag{3.65}$$

Let us also denote by Λ_b a set of all coordination variable values for which $u_b(\lambda)$ exists.

A coordinator problem (CP) for this control structure is defined in the following way:

Find $\hat{\lambda} = (\hat{\lambda}_1, \ldots, \hat{\lambda}_N)$ such that

$$Q_*(\hat{\lambda}) = \min_{\lambda \in \Lambda_b} Q_*(\lambda) \tag{3.66}$$

where

$$Q_*(\lambda) \triangleq \sum_{i=1}^{N} Q_i(c_{bi}(\lambda), u_{bi}(\lambda))$$

Owing to equality

$$u_b(\lambda) = HF_*(c_b(\lambda), u_b(\lambda)) \quad \text{for any} \quad \lambda \in \Lambda_b,$$

the coordinator task is to minimize the performance of the real system with respect to the controls that are chosen by the local decision units. If the coordinator problem is formulated as above, then we say that we deal with decentralized control with full coordination. This kind of coordination was first proposed by Findeisen (1976) and its main properties have been investigated by Brdyś and Michalak (1978) and Brdyś and Ulanicki (1978).

Another way to formulate the coordinator problem is the following:

Find $\lambda^0 = (\lambda_1^0, \ldots, \lambda_N^0)$ by using IBM for coordination of the mathematical model of the system.   (3.67)

In this case we deal with decentralized control with partial coordination. A solution obtained by using partial coordination is, of course, not better than that obtained by using full coordination, but it is easier to find. If the solution of (3.67) is not good enough, it can be used as a starting point for solving CP defined by (3.66).

3.4.2. OPTIMALITY OF THE SOLUTION

Since the control structure considered above produces a suboptimal solution of the control problem, we will now discuss the suboptimality properties of this solution. We will formulate the conditions under which the control structure with full coordination is consistent (see section 3.33). A control structure is consistent if and only if the solution found with this structure equals the real optimal solution when the mathematical model of the system fully describes reality.

THEOREM 3.10 (*consistency conditions*). *If we assume that (a) the mathematical model of the system is coordinable by* IBM, *and (b) for* $\lambda = \lambda^0$ *the infimal problem solution is unique, then the control structure with partial or full coordination is consistent.*

Proof. Let us consider full coordination, and suppose that $F = F_*$. From assumption (a)

$$(\exists \lambda^0 \in \Lambda_b)(\exists(\hat{c}(\lambda^0), \hat{u}(\lambda^0)) \in \mathscr{C} \times \mathscr{U}) \ (\hat{c}(\lambda^0), \hat{u}(\lambda^0)) = \arg \min_{(c,u) \in CU} Q(c, u)$$
$$+ \langle \lambda^0, u - HF(c, u) \rangle$$

and

$$\hat{u}(\lambda^0) = HF(\hat{c}(\lambda^0), \hat{u}(\lambda^0)).$$

Let us consider the infimal problem for $\lambda = \lambda^0$. Since $\lambda^0 \in \Lambda_b$, the solution of this problem exists and it is unique. According to the definition of this solution (see Eqs. (3.60), (3.64), and (3.65)) the following holds:

$$c_b(\lambda^0) = \arg \min_{c \in C(u_b(\lambda^0))} [Q(c, u_b(\lambda^0)) + \langle \lambda^0, u_b(\lambda^0) - HF(c, u_b(\lambda^0)) \rangle]$$

and

$$u_b(\lambda^0) = HF(c_b(\lambda^0), u_b(\lambda^0)).$$

Hence, from assumption (b)

$$c_b(\lambda^0) = \hat{c}(\lambda^0)$$

and

$$u_b(\lambda^0) = \hat{u}(\lambda^0).$$

Since

$$(\hat{c}(\lambda^0), \hat{u}(\lambda^0)) = \arg \min_{(c,u)} Q(c, u)$$

subject to

$$(c, u) \in CU,$$

$$u = HF(c, u)$$

and

$$\forall \lambda \in \Lambda_b \quad Q(\hat{c}(\lambda^0), \hat{u}(\lambda^0)) \le Q(c_b(\lambda), u_b(\lambda)),$$

the proof is completed. The proof with partial coordination is the same. $\square$

It follows from the theorem that if we want to apply the control structure to a real system coordination, we should check whether the model of the system is coordinable by IBM and whether the uniqueness assumption is satisfied. Let us note that the consistency conditions are more restrictive as compared with IBMF.

It is interesting to compare the suboptimality of the solution obtained by IBMF with the suboptimality of the solution obtained by decentralized control with full coordination:

THEOREM 3.11. *If we assume that (a) the system is coordinable by* IBMF *with* $\lambda = \tilde{\lambda}$, *and for* $\lambda = \tilde{\lambda}$ *the infimal problem solution is unique, then*

$$Q(c_b(\hat{\lambda}), u_b(\hat{\lambda})) \le Q(\hat{c}(\tilde{\lambda}), \hat{u}(\tilde{\lambda})),$$

where $\hat{\lambda}$ *denotes the solution of* CP *defined by* (3.66), *and* $\hat{c}(\tilde{\lambda})$ *and* $\hat{u}(\tilde{\lambda})$ *denote the infimal problem solution in* IBMF *with* $\lambda = \tilde{\lambda}$.

Proof. Taking into account the same arguments as in the proof of Theorem 3.10, we can show that

$$\hat{c}(\tilde{\lambda}) = c_b(\tilde{\lambda})$$

and

$$\hat{u}(\tilde{\lambda}) = u_b(\tilde{\lambda})$$

which completes the proof. $\square$

Theorem 3.11 shows that the decentralized control method with full coordination is in general no worse than IBMF, and simple examples can be constructed in which it is much better. However, it can be applied only to weakly coupled systems. This is implied by the convergence properties of the lower-level iterative scheme, which is convergent if the subsystems are in some sense weakly coupled. This limitation will be discussed in more detail in the next section. We note also that functional Q_* in (3.66) is in general nonconvex and nondifferentiable, which makes the coordinator problem rather difficult to solve.

Theorem 3.11 enables us to apply directly the suboptimality results which were derived for IBMF in section 3.3.3. to the suboptimality analysis being done here. If the assumptions of Theorems 3.10, 3.11, and 3.6 are satisfied, then $\|F(c, u) - F_*(c, u)\| \to 0$ uniformly on CU implies that performance loss $Q(c_b(\hat{\lambda}), u_b(\lambda)) - Q(\hat{c}_*, HK_*(\hat{c}_*))$ decreases to zero. Furthermore, if the assumptions of Theorem 3.11 are satisfied then a rough upper bound on performance loss can be obtained from Eq. (3.37) if $\hat{c}(\tilde{\lambda}), \hat{u}_*(\tilde{\lambda})$ is replaced by $c_b(\hat{\lambda})$ and $u_b(\hat{\lambda})$. A more precise bound has not been developed.

3.4.3. PROPERTIES OF THE INFIMAL PROBLEM

In this section the conditions under which the infimal problem solution exists and the lower-level iterative scheme converges to this solution will be derived. To simplify the analysis, $\mathcal{U}$, $\mathcal{C}$, and $\mathcal{Y}$ are assumed to be finite-dimensional Hilbert spaces.

Existence of the solution

It was pointed out in section 3.4.1. that there is an equivalence between the infimal problem solution and the fixed point of mapping $HK_*(\hat{c}(\cdot, \lambda))$. This mapping is the composition of mappings $\hat{c}(\cdot, \lambda)$, K_*, and H. Mapping H is simply a matrix. The interesting properties of mapping K_* can be easily investigated using an implicit function theorem. Hence, it will be assumed that K_* is continuously differentiable. For mapping $\hat{c}(\cdot, \lambda)$, the problem is more complicated and requires a special discussion. We will start with the formulation of the conditions under which $\hat{c}(\cdot, \lambda)$ is continuous. Let λ be fixed in Λ_b.

THEOREM 3.12 (*continuity of* $\hat{c}(\cdot, \lambda)$). *If we assume that for a given u and for any* $i \in \overline{1, N}$,

1. *Functional* $Q_{\mathrm{mod}i}$ *is continuous on* $\mathcal{C}_i \times \mathcal{U}_i$,
2. *Set* CU_i *is expressed by a finite number of functional inequalities, i.e.*

$$CU_i = \{(c_i, u_i) \in \mathcal{C}_i \times \mathcal{U}_i : h_j(c_i, u_i) \leq 0, j \in \mathcal{J}_i\}$$

and is compact,

3. $\forall j \in \mathcal{J}_i$ h_j *is continuously differentiable on* $\mathcal{C}_i \times \mathcal{U}_i$,
4. $(\forall u_i \in U_i^0)(\forall j \in \mathcal{J}_i)$ $h_j(\cdot, u_i)$ *is convex on* $\mathcal{C}_i$,
5. $(\exists \delta > 0)(\forall u_i^k \to u_i, u_i^k \neq u_i)(\exists$ *subsequence* $\{u_i^{k_n}\}$ *of* $\{u_i^k\}$ *such that* $(\forall c_i$ *from the boundary of* $C_i(u_i^{k_n}))$

$$\inf \{\|c_i^*\| : c_i^* \in \partial_{c_i} h_{max}^i(c_i, u^{k_n})\} \geq \delta)$$

where $\partial_{c_i} h^i_{max}$ denotes a subgradient of a functional

$$h^i_{max}(c_i, u_i) \triangleq \max_{j \in \mathscr{I}_i} \{h_j(c_i, u_i)\},$$

with respect to c_i,

then mapping $\hat{c}(\cdot, \lambda)$ is continuous in u.

The proof of this theorem is long and can be found in Ulanicki (1978). Other conditions under which $\hat{c}(\cdot, \lambda)$ is continuous can be derived using the results given in Fiodorow (1977).

Based on Schauder's theorem (Kantorovich and Akilov 1964), we can formulate sufficient conditions for the existence of the fixed point (or the infimal problem solution) of $HK_*(\hat{c}(\cdot, \lambda))$ as follows:

1. Set U^0 is compact and convex,
2. Mapping $HK_*(\hat{c}(\cdot, \lambda))$ is continuous on U^0,
3. $HK_*(\hat{c}(\cdot, \lambda))(U^0) \subseteq U^0$.

Since $U^0 = \Pi_{\mathscr{U}}(CU)$, assumption 1 is satisfied if CU is compact and convex. The conditions under which assumption 2 is satisfied were given in Theorem 3.12. Assumption 3 is difficult to check except when $U^0 = \mathscr{U}$. We can derive other sufficient conditions for the existence of the infimal problem solution by looking at the existence problem from the point of view of perturbation theory. First we will formulate the following general lemma.

LEMMA 3.5. *If continuous mappings $T: A \to \mathscr{U}$ and $T_*: A \to \mathscr{U}$ are given, where $A \subset \mathscr{U}$ such that*

1. *$int\, A \neq \varnothing$,*
2. *There exists a point $\bar{u} \in int\, A$ such that $T(\bar{u}) = 0$,*
3. *There exists neighborhood $\mathscr{V}(\bar{u})$ of $\bar{u}$ such that mapping T is a homeomorphism on $\mathscr{V}(\bar{u})$,*

then the following holds:

$$(\exists \varepsilon_1 > 0)(\forall 0 < \varepsilon < \varepsilon_1)(\exists \delta > 0)$$

$$(\forall u \in \mathscr{V}(\bar{u})\ \|T(u) - T_*(u)\| < \delta) \Rightarrow (\exists \bar{u}_* \in A\ (T_*(\bar{u}_*) = 0) \wedge (\|\bar{u} - \bar{u}_*\| < \varepsilon)).$$

The proof of this lemma is given in Appendix B, section B.4.

We can formulate the following sufficient conditions using Lemma 3.5.

THEOREM 3.13. *If we assume that*

1. *$int\ U^0 \neq \varnothing$,*
2. *For a given λ, a model-based infimal problem has a solution $u_b^m(\lambda)$, i.e.,*

$$u_b^m(\lambda) = HK(\hat{c}(u_b^m, \lambda))$$

and

$$u_b^m(\lambda) \in int\ U^0,$$

3. *There exists neighborhood $\mathscr{V}(u_b^m(\lambda))$ of $u_b^m(\lambda)$ such that mapping $I\text{-}HK(\hat{c}(\cdot, \lambda))$ is a homeomorphism on $\mathscr{V}(u_b^m(\lambda))$,*
4. *Mapping $I\text{-}HK_*(\hat{c}(\cdot, \lambda))$ is continuous on $\mathscr{V}(u_b^m(\lambda))$,*

then the following holds:

$$(\exists \varepsilon_1 > 0)(\forall 0 < \varepsilon < \varepsilon_1)(\exists \delta > 0)$$

$$(\forall c \in \hat{c}(\cdot, \lambda)(\mathscr{V}(u_b^m(\lambda))))\ \|K_*(c) - K(c)\| < \delta) \Rightarrow$$

$$((\exists u_b(\lambda) \in U^0) \wedge (\|u_b(\lambda) - u_b^m(\lambda)\| < \varepsilon)).$$

Proof. Let us note that for any $u \in U^0$

$$\|u - HK_*(\hat{c}(u, \lambda)) - (u - HK(\hat{c}(u, \lambda)))\| \leq \|H\| \|K(\hat{c}(u, \lambda)) - K_*(\hat{c}(u, \lambda))\|.$$

Hence, if we set

$$T(u) = u - HK(\hat{c}(u, \lambda)),$$

$$T_*(u) = u - HK_*(\hat{c}(u, \lambda)),$$

$$u = u_b^m(\lambda),$$

and

$$A = \mathscr{V}(u_b^m(\lambda))$$

then the proof of the theorem will follow directly from Lemma 3.5. $\square$

Theorem 3.13 gives the sufficient conditions for the existence of the infimal problem solution in terms of the properties of the model-based infimal problem. Thus, roughly speaking, if the model-based infimal problem has a solution and the operator $HK(\hat{c}(\cdot, \lambda))$ is a local homeomorphism, then the infimal problem solution exists and it continuously depends on the difference between model and reality expressed by $\|K_*(c) - K(c)\|$. It is very important that we can check whether assumptions 2 and 3 of Theorem 3.13 are satisfied, since they concern the mathematical model of the system. To verify assumption 2 we can use Schauder's theorem. Condition $HK(\hat{c}(\cdot, \lambda))(U^0) \subseteq U^0$, in

Schauder's theorem, can be checked, since mapping K is known. The continuity of $HK\,\hat{c}(\cdot, \lambda))$ may be checked with Theorem 3.12. The conditions under which assumption 3 is satisfied are given in Appendix B, section B.7. To derive such conditions we cannot use an implicit function theorem because operator $\hat{c}(\cdot, \lambda)$ is in general not differentiable. On the other hand, application of Theorem 3.13 is limited to situations where the infimal problem solution belongs to the interior of U^o.

The convergence conditions

The sufficient conditions for convergence of the lower-level iterative scheme will be derived using Banach's contraction mapping theorem (Kantorovich and Akilov 1964). Hence, our main aim is to formulate conditions under which $\hat{c}(\cdot, \lambda)$ has the Lipschitz property and to find the Lipschitz constant of the contraction mapping.

It is easy to find the Lipschitz constant of a continuously differentiable mapping since the mean value theorem can be applied. Unfortunately, mapping $\hat{c}(\cdot, \lambda)$ is usually not differentiable, which can be explained in the following way. For a given value of u, a value of $\hat{c}(\cdot, \lambda)$ is defined as the solution of some optimization problem (see (3.60)). An index set of all active constraints at the solution can change when u is changing in U^o. There exist, however, subsets of U^o where the index set of the active constraints is constant. In the interior of these subsets, $\hat{c}(\cdot, \lambda)$ is differentiable under reasonable assumptions; but it may not be differentiable at the boundary points. Such partial differentiability is the basic property of $\hat{c}(\cdot, \lambda)$ needed to find its Lipschitz constant.

Using the Lipschitz constants of $\hat{c}(\cdot, \lambda)$, found on the differentiability subsets of this mapping by using the mean value theorem, we will find the Lipschitz constant of $\hat{c}(\cdot, \lambda)$ on U^o. We will start with the formulation of certain topological properties of the subsets mentioned above.

Let λ be fixed in Λ_b. Let us assume that the topology of U^o is generated from the topology of $\mathcal{U}$. It will also be assumed that set CU is expressed by a finite number of functional inequalities, i.e.,

$$CU = CU_1 \times \ldots \times CU_N$$

and

$$CU_i = \{(c_i, u_i) \in \mathscr{C}_i \times \mathscr{U}_i : h_j(c_i, u_i) \leq 0, j \in \mathscr{J}_i\}$$

for $i \in \overline{1, N}$.

LEMMA 3.6. *If we assume that*

1. *Mapping $\hat{c}(\cdot, \lambda)$ is continuous on U^o,*
2. *$(\forall i \in \overline{1, N})(\forall j \in \mathscr{J}_i)$ function h_j is continuous on $\mathscr{C}_i \times \mathscr{U}_i$,*

then there exist separable subsets of U^0, $U_1, \ldots, U_L$, that are open in U^0, such that

$$(\forall i \in \overline{1, L})(\forall u', u'' \in U_i)I(\hat{c}(u', \lambda)) = I(\hat{c}(u'', \lambda))$$

where $I(\hat{c}(u, \lambda))$ denotes an index set of all active constraints at point $\hat{c}(u, \lambda)$, and

$$U^0 = \bigcup_{i=1}^{L} \bar{U}_i.$$

The proof of the lemma is given in Appendix B, section B.5.

In the next lemma we will formulate the sufficient conditions for the extension of mapping $\hat{c}(\cdot, \lambda)$ from $\bar{U}_i$ on U^0, $i \in \overline{1, L}$. Let us denote by I_i, $i \in \overline{1, L}$, the index set of all active constraints for $u \in U_i$.

LEMMA 3.7. *Let us suppose that for any $u \in U^0$ and any $i \in \overline{1, L}$ the following assumptions are satisfied:*

1. *Mapping $\hat{c}(\cdot, \lambda)$ is continuous on U^0,*
2. *Functions $Q_{\text{mod}}(\cdot, u, \lambda)$, $h_j(\cdot, u)_N$ are twice continuously differentiable on $\mathscr{C}$ for any $j \in \cup_{i=1}^N \mathscr{J}_i$,*
3. *The following minimization problem has a solution, say, $c^i(u, \lambda)$:*

$$\min_c Q_{\text{mod}}(c, u, \lambda)$$

with respect to

(1)
$$h_{j_1}(c, u) = 0,$$
$$\vdots$$
$$h_{j_s}(c, u) = 0$$

where

$$\{j_1, \ldots, j_s\} = I_i,$$

4. *Vectors $h'_{j\mathscr{C}}(c^i(u, \lambda)u)$ are linearly independent for all $j \in I_i$,*
5. *The following matrix is reversible:*

$$\left[\begin{array}{c:c} (Q_{\text{mod}})''_{cc}(c^i(u, \lambda), u\lambda) + (\langle h(c^i(u, \lambda), u), \mu\rangle)''_{cc} & h_c^{\mathrm{T}\prime}(c^i(u, \lambda), u) \\ \hdashline h'_c(c^i(u, \lambda), u) & 0 \end{array}\right]$$

where

$$h = (h_{j_1}, \ldots, h_{j_s})$$

and μ denotes a vector of the Lagrangian multipliers corresponding to problem (1).

Then,

1. $(\forall i \in \overline{1, L})$ *there exist an open set* $\tilde{U}^0 \supset U^0$ *and a continuously differentiable mapping* $T^i : \tilde{U}^0 \to \mathscr{C}$ *such that*

$$\forall u \in \bar{U}_i \quad T^i(u) = \hat{c}(u, \lambda).$$

If, additionally, $(\forall u \in U^0)$ *the matrices*

$$A = (Q_{mod})''_{cc} - h_{cc}^{T''}(h'_c h_c^{T'})^{-1} h'_c (Q_{mod})'_c \quad and \quad h'_c A^{-1} h_c^{T'}$$

computed at $(\hat{c}(u, \lambda), u)$ *are reversible, then*

2. $(\forall i \in \overline{1, L})(\forall u \in \tilde{U}^0)$ *the derivative of* T^i *is expressed by the following formula*

$$T_u^{i'}(u) = [A^{-1} - A^{-1} h_c^{T'}(h'_c A^{-1} h_c^{T'}) \vdots A^{-1} h_c^{T'}(h'_c A^{-1} h_c^{T'})^{-1}]$$

$$\left[\frac{(Q_{mod})''_{c,u} - h_{c,u}^{T''}(h'_c h_c^{T'})^{-1} h'_c (Q_{mod})'_c}{h'_u} \right]$$

The proof of this lemma is given in Appendix B, section B.6.
We formulate Lemma 3.8 based on Lemmas 3.6 and 3.7.

LEMMA 3.8. *If we assume that*

1. *The assumptions of Lemmas 3.6 and 3.7 are satisfied for a given* λ,
2. *Set* U^0 *is convex and compact,*

then mapping $\hat{c}(\cdot, \lambda)$ *has the Lipschitz property on* U^0, *i.e., there exists a constant* $K(\lambda)$ *such that*

$$\forall u', u'' \in U^0 \quad \|\hat{c}(u'', \lambda) - \hat{c}(u', \lambda)\| \le K(\lambda)\|u'' - u'\|.$$

Proof. Let u' and $u'' \in U^0$ be given. Because of the convexity of U^0, an interval

$$[u', u''] = \{u \in U^0 : u = (1 - t)u' + tu'', 0 \le t \le 1\}$$

is contained in U^0. Let us divide the interval $[u', u'']$ into $[u^0, u^1]$, $[u^1, u^2], \ldots, [u^{S-1}, u^S]$, where $u^0 = u'$ and $u^S = u''$ by using the following procedure:

1. Set $t_0 = 0$,
2. For $s > 0$ determine

$$t_s = \sup_{t} \{t \in [t_{s-1}, 1] : (1 - t)u' + tu'' \in \bar{U}_{i_{s-1}}\},$$

and a corresponding point

$$u^s = (1 - t_s)u' + t_s u'',$$

3.   If $u^s = u''$ then stop; otherwise set $s = s+1$ and go to step 2.

Notice that if $u_s \neq u''$ then $u_s \in \bar{U}_{i_s}$ such that $\bar{U}_{i_s} \neq \bar{U}_{i_m}$ for any $m < s$. Hence the above procedure is well defined and finite since the number of sets $\bar{U}_i$ is finite.

For any value of s the following holds:

$$u^{s-1}, u^s \in \bar{U}_{i_{s-1}}. \tag{3.68}$$

Because U^0 is compact and $T^{i\prime}_{\mathcal{U}}$ is continuous on U^0 for any $i \in \overline{1, L}$ (see Lemma 3.6), the number

$$K^i(\lambda) = \sup_{u \in U^0} \|T^{i\prime}_{\mathcal{U}}(u)\| \tag{3.69}$$

is finite for $i \in \overline{1, L}$. Thus, from part 1 of Lemma 3.7 and the mean value theorem, we conclude that

$$\|\hat{c}(u^s, \lambda) - \hat{c}(u^{s-1}, \lambda)\| \leq K^{i_{s-1}}(\lambda)\|u^s - u^{s-1}\| \quad \text{for any} \quad s \in \overline{1, S}. \tag{3.70}$$

Using the triangle inequality, we obtain:

$$\|\hat{c}(u'', \lambda) - \hat{c}(u', \lambda)\| = \|\hat{c}(u^1, \lambda) - \hat{c}(u', \lambda) + \ldots + \hat{c}(u^S, \lambda) - \hat{c}(u^{S-1}, \lambda)\|$$

$$\leq K^{i_0}(\lambda)\|u' - u'\| + \ldots + K^{i_{s-1}}\|u^S - u^{S-1}\| \leq K(\lambda)\|u'' - u'\|$$

where

$$K(\lambda) = \max_{i \in 1,L} \{K^{i_s}(\lambda)\} \tag{3.71}$$

Hence, the proof of the lemma is complete. $\square$

Notice that formulae (3.69) and (3.71) enable us to compute the Lipschitz constant of $\hat{c}(\cdot, \lambda)$ for a given value of λ. To compute this constant, we may also use the results obtained by Hager (1979).

Let us return to the lower-level iterative scheme and notice the following pitfall. For some k it may be that $u^k \in \mathcal{U}^0$ and u^{k+1} cannot be determined since set $C(u^k)$ is empty. This can happen even when the infimal problem solution exists. If this solution lies in an interior of U^0, then this pitfall can be eliminated by an appropriate choice of the starting point. If not, then the iterative scheme may fail. To eliminate this pitfall, we will modify the scheme. Instead of (3.63), let us consider:

$$u^{k+1} = \Pi_{U^0}(HK_*(\hat{c}(u^k, \lambda))) \tag{3.72}$$

where Π_{U^0} denotes a projection operator from $\mathcal{U}$ on U^0. From the well-known properties of Π_{U^0} it follows that if $HK_*(\hat{c}(\cdot, \lambda))$ satisfies the Lipschitz condition then $\Pi_{U^0}(HK_*(\hat{c}(\cdot, \lambda)))$ also satisfies this condition with a constant no greater than the Lipschitz constant of $HK_*(\hat{c}(\cdot, \lambda))$. Since $U^0 = \mathsf{X}_{i=1}^N U_i^0$

where $U_i^0 \in \mathcal{U}_i$ for $i \in \overline{1, N}$ (see (3.62)), then

$$\Pi_{U^0} = (\Pi_{U^0}, \ldots, \Pi_{U^0}),$$

which implies that the lower-level iterative scheme (3.72) remains completely decentralized. If we introduce a current projection on set U^0, then all points generated according to (3.72) are in U^0. We now proceed to the convergence conditions.

THEOREM 3.14. *If we assume that for a given* $\lambda \in \Lambda_b$

1. *There exists an infimal problem solution* $u_b(\lambda)$,
2. *There exists an open ball* $K(u_b(\lambda); r)$ *such that mapping* $HK_*(\hat{c}(\cdot, \lambda))$ *satisfies the Lipschitz condition on set* $K(u_b(\lambda); r) \cap U^0$ *with a constant less than one,*

then for any initial point from $K(u_b(\lambda); r) \cap U^0$ *the sequence* $\{u^k\}$ *generated by lower-level iterative scheme (3.72) converges to the infimal-level solution* $u_b(\lambda)$.

Proof. Let $u^0 \in K(u_b(\lambda); r) \cap U^0$ be given. From the assumptions it follows that

$$\|u^1 - u_b(\lambda)\| = \|\Pi_{U^0}(HK_*(\hat{c}(u^0, \lambda))) - HK_*(\hat{c}(u_b(\lambda), \lambda))\|$$

$$= \|\Pi_{U^0}(HK_*(\hat{c}(u^0, \lambda))) - \Pi_{U^0}(HK_*(\hat{c}(u_b(\lambda), \lambda)))\| \leq \alpha \|u^0 - u_b(\lambda)\|$$

where $0 < \alpha < 1$.
Hence, $u^1 \in K(u_b(\lambda); r) \cap U^0$. Using mathematical induction we obtain that

$$\forall k \geq 1 \quad u^k \in K(u_b(\lambda); r) \cap U^0 \quad \text{and} \quad \|u^k - u_b(\lambda)\| \leq \alpha^k \|u^0 - u_b(\lambda)\|$$

which implies that

$$\lim_{k \to \infty} u^k = u_b(\lambda),$$

and the proof is completed. $\square$

The conditions under which assumption 1 of the theorem is satisfied were discussed at the beginning of this section. The sufficient conditions for the existence of the Lipschitz constant of mapping $\hat{c}(\cdot, \lambda)$ were given in Lemma 3.8.

To be sure that the lower-level iterative scheme converges by use of Theorem 3.14, the Lipschitz's constant of $HK_*(\hat{c}(\cdot, \lambda))$ must be less than one. If it is, then we say that the control system consists of interconnected subsystems that are weakly coupled. A more precise characterization of weakly coupled systems can be given only in the linear-quadratic case. The properties of the set $K(u_b(\lambda); r) \cap U^o$ are such that we can call it the convergence ball of the lower-level iterative scheme. In many practical cases we know that such a convergence ball exists, but there is no way to find it. Hence, finding a starting point for the lower-level iterative scheme is often a problem. One may solve the model-based infimal problem for a given λ and take the solution as the starting point. The conditions under which such a choice is proper are given in Brdyś and Ulanicki (1978).

3.4.4. EXISTENCE OF THE COORDINATOR PROBLEM SOLUTION

In this section we will derive the sufficient conditions for the existence of the coordinator problem solution. For a given $u \in \mathcal{U}$ let us introduce the following optimization problem:

$$\min Q(c, u)$$

subject to $c \in C(u)$ and

$$u = HF(c, u) + s \tag{3.73}$$

where s is fixed in $S(u) \triangleq \{s \in \mathcal{U} : \exists c \in \mathcal{C} \ (c, u) \in CU \wedge u = HF(c, u) + s\}$. Let us denote by $\hat{c}(u, s)$ the solution of the problem. The following lemma characterizes the infimal problem solutions in terms of the solutions of problem (3.73).

LEMMA 3.9. *If u_b is the infimal problem solution for a certain $\lambda \in \Lambda_b$, then*

1. *Problem (3.73) is coordinable by* IBM *and the following holds*:

$$c_b = \hat{c}(u_b, s)$$

where

$$s = HF_*(c_b, u_b) - HF(c_b, u_b).$$

If the pair $(\bar{c}, \bar{u})$ is such that

$$\bar{c} = \hat{c}(\bar{u}, \bar{s})$$

where

$$\bar{s} = HF_*(\bar{c}, \bar{u}) - HF(\bar{c}, \bar{u}),$$

and if problem (3.73) with $u = \bar{u}$ and $s = \bar{s}$ is coordinable by IBM, *then*

2. *$\bar{u}$ is the infimal problem solution for a certain $\lambda \in \Lambda_b$.*

Proof. *Part* 1: From the assumption, there exists $\lambda \in \Lambda_b$ such that

$$c_b = \arg \min_{C(u_b)} [Q(c, u_b) + \langle \lambda, u_b - HF(c, u_b) \rangle]$$

and

$$u_b = HF_*(c_b, u_b).$$

Let us set

$$s = HF_*(c_b, u_b) - HF(c_b, u_b).$$

Then

$$c_b = \arg \min_{C(u_b)} [Q(c, u_b) + \langle \lambda, u_b - HF(c, u_b) - s \rangle]$$

and

$$u_b - HF(c_b, u_b) - s = 0.$$

This means that problem (3.73) with

$$u = u_b$$

and

$$s = HF_*(c_b, u_b) - HF(c_b, u_b)$$

is coordinable by IBM and c_b is its solution.

Part 2: Let a pair $(\bar{c}, \bar{u})$ satisfy the assumptions of part 2. Therefore, there exists $\bar{\lambda} \in \mathcal{U}$ such that

$$\bar{c} = \arg \min_{C(\bar{u})} [Q(c, \bar{u}) + \langle \bar{\lambda}, \bar{u} - HF(c, \bar{u}) - \bar{s} \rangle]$$

and

$$\bar{u} = HF(\bar{c}, \bar{u}) + \bar{s}.$$

This implies that

$$\bar{c} = \arg \min_{C(\bar{u})} [Q(c, \bar{u}) + \langle \bar{\lambda}, \bar{u} - HF(c, \bar{u}) \rangle]$$

and

$$\bar{u} = HF_*(\bar{c}, \bar{u}).$$

Hence, $\bar{u}$ is the infimal problem solution with $\lambda = \bar{\lambda}$, and the proof of the lemma is finished. $\square$

It is now easy to derive the conditions under which the set of all infimal problem solutions CU_b is compact, where

$$CU_b \triangleq \bigcup_{\lambda \in \Lambda_b} (c_b(\lambda), u_b(\lambda)) \tag{3.74}$$

240

Lemma 3.10   *If we assume that*

1.   *Mappings Q, F, and F_* are continuous on $\mathscr{C} \times \mathscr{U}$,*
2.   *Set CU is compact,*
3.   *Mapping $\mathscr{D} \ni (u, s) \to \hat{c}(u, s) \in \mathscr{C}$ where $\mathscr{D}$ is its domain, is continuous,*
4.   *$\forall (u, s) \in \mathscr{D}$ problem (3.73) is coordinable by* IBM.

then set CU_b is compact.

Proof.   Since $CU_b \subseteq CU$ and CU is compact, set CU_b is bounded. It will be shown that it is also closed.

Let us take a sequence $\{c^n, u^n\}$ which converges to a certain point, say $(\bar{c}, \bar{u})$, such that $\forall n = 1, 2, \ldots,$ $(c^n, u^n) \in CU_b$. Owing to Lemma 3.9 there exists a sequence $\{s^n\}$ such that

$$c^n = \hat{c}(u^n, s^n)$$

where

$$s^n = HF_*(c^n, u^n) - HF(c^n, u^n)$$

and

$$u^n = HF(c^n, u^n) + s^n.$$

Taking into account assumptions (1) and (2), we conclude that

$$\forall n = 1, 2, \ldots, \quad (u_n, s_n) \in \mathscr{D}.$$

Hence, we can use assumption (3) and obtain

$$\lim_{n \to \infty} s^n = \bar{s} = HF_*(\bar{c}, \bar{u}) - HF(\bar{c}, \bar{u}) \tag{3.75}$$

and

$$\bar{u} = HF(\bar{c}, \bar{u}) + \bar{s}. \tag{3.76}$$

Since CU is compact, point $\bar{c}$ belongs to set $C(\bar{u})$. Therefore, relation (3.76) implies that $(\bar{u}, \bar{s}) \in \mathscr{D}$.

Since we know from assumption (1) that mapping $\hat{c}(\,\cdot\,, \,\cdot\,)$ is continuous on its domain, we also know that

$$\hat{c}(\bar{u}, \bar{s}) = \lim_{n \to \infty} \hat{c}(u^n, s^n) = \lim_{n \to \infty} c^n = \bar{c}.$$

This equality, equality (3.75), and assumption (4) allow us to use Lemma 3.9, which tells us that $(\bar{c}, \bar{u}) \in CU_b$. So, set CU_b is compact since it is closed and bounded, and the proof is complete. $\square$

Now we can formulate an essential point of this section.

THEOREM 3.15 (*existence conditions*). *If we assume that* (a) *the assumptions of Lemma 3.10 are satisfied, and* (b) *set* CU_b *is nonempty, then a solution of coordinator problem* CP *of the decentralized control structure with full coordination exists.*

Proof. Let us note that CP (see (3.66)) is equivalent to the following

$$\min_{(c,\,u)\in CU_b} Q(c, u).$$

By assumption (a) the set CU_b is compact. Since it is nonempty and since functional Q is continuous, a solution of this problem exists. $\square$

3.5. ITERATIVE COORDINATION WITH DISTURBANCES

3.5.1. INTRODUCTION

In the previous section it was assumed that disturbances were constant during coordination. It was also assumed that the optimal control problem for the real system consisted in finding a control that optimized the steady state of the system. In this section it is assumed that the disturbances are time-varying and that the system dynamics are fast compared with the changes of the disturbances. Therefore, we deal with a system that operates in the steady but time-varying state. The behavior of the system is described by the following static but nonstationary subsystem equation:

$$y_i(t) = F_{*i}(c_i(t), u_i(t), z_i(t)), \qquad i \in \overline{1, N}, \tag{3.77}$$

where $z_i(t)$ denotes the time-varying disturbance. We assume that $\mathcal{Y}_i$, $\mathcal{C}_i$, $\mathcal{U}_i$, and $\mathcal{Z}_i$ are finite-dimensional Hilbert spaces. Interactions between subsystems are described by the equations

$$u_i(t) = \sum_{j=1}^{N} H_{ij} y_j(t), \qquad i \in \overline{1, N}, \tag{3.78}$$

in which H_{ij} are the interconnection matrixes composed of zeros and ones.

Though the system considered is not truly dynamic, there is a large class of industrial steady-state processes which may be described in the above manner (Findeisen 1974, Pliskin 1975).

Using compact notations, $y = (y_1, \ldots, y_N)$, $c = (c_1, \ldots, c_N)$, $u = (u_1, \ldots, u_N)$, $z = (z_1, \ldots, z_N)$, $F = (F_1, \ldots, F_N)$, the real system equations may be written as

$$y(t) = F_*(c(t), u(t), z(t)),$$

$$u(t) = Hy(t),$$

where

$$y \in \mathcal{Y} = \mathop{\times}_{i=1}^{N} \mathcal{Y}_i, \qquad c \in \mathcal{C} = \mathop{\times}_{i=1}^{N} \mathcal{C}_i, \qquad u \in \mathcal{U} = \mathop{\times}_{i=1}^{N} \mathcal{U}_i,$$

$$z \in \mathcal{Z} = \mathop{\times}_{i=1}^{N} \mathcal{Z}_i, \qquad F_* : \mathcal{C} \times \mathcal{U} \times \mathcal{Z} \to \mathcal{Y}, \qquad H : \mathcal{Y} \to \mathcal{U}.$$

This set of equations is supposed to represent a real physical system and it is postulated that there exists a well-defined implicit mapping $K_* : \mathcal{C} \times \mathcal{Z} \to \mathcal{Y}$, such that the outputs of the real system are uniquely determined by controls c and disturbances z. Unfortunately, in most applications Eq. (3.77) is not known exactly. Unknown and time-varying disturbances make the task of constructing precise models of subsystems very difficult. It is a common procedure in engineering to base the design of a controller on a simple mathematical model that roughly describes the actual system. Let us assume then that we have simplified mathematical models of subsystems of the form

$$y_i(t) = F_i(c_i(t), u_i(t), \bar{z}_i(t)), \qquad i \in \overline{1, N}, \tag{3.79}$$

which, combined, give

$$y(t) = F(c(t), u(t), \bar{z}(t)),$$

where $\bar{z}(t) = (\bar{z}_1(t), \ldots, \bar{z}_N(t))$ denotes the estimation of the unknown function $z(t)$. As before, $F = (F_1, \ldots, F_N)$, $F : \mathcal{C} \times \mathcal{U} \times \bar{\mathcal{Z}} \to \mathcal{Y}$ where $\bar{\mathcal{Z}} \triangleq \mathcal{R}(\bar{z}(\cdot))$. It is postulated that the set of equations (3.78) and (3.79) defines an implicit function $K : \mathcal{C} \times \bar{\mathcal{Z}} \to \mathcal{Y}$ such that outputs $K(c, \bar{z})$ in the model are uniquely determined by control c and disturbance estimate $\bar{z}$. Let us assume that there is a performance index of the system $Q : \mathcal{C} \to \mathbb{R}$ which has the following form

$$Q(c(t), u(t)) = \sum_{i=1}^{N} Q_i(c_i(t), u_i(t)) \tag{3.80}$$

where $Q_i : \mathcal{C}_i \times \mathcal{Y}_i \to \mathbb{R}$. Finally, let us require that the inputs and controls of each subsystem satisfy the relation

$$(c_i(t), u_i(t)) \in CU_i, \qquad i \in \overline{1, N},$$

which, combined, give

$$(c(t), u(t)) \in CU$$

where $CU = \times_{i=1}^{N} CU_i$.

The optimal control problem consists in finding control $\hat{c}(t)$ such that the following Lebesgue integral

$$\int_{t_0}^{t_f} Q(c(t), u(t)) \, dt$$

is minimized subject to the constraints

$$y(t) = F_*(c(t), u(t), z(t)),$$

$$u(t) = Hy(t),$$

$$(3.81)$$

$$(c(t), u(t)) \in CU, \text{ almost all } t \in [t_0, t_f],$$

$$\hat{c}(\cdot) \in \{c(\cdot): c(\cdot) \text{ measurable and essentially}$$

$$\text{bounded on } [t_0, t_f]\}.$$

Note that the optimal control problem formulated as above is neither a static nor a dynamic one. In terms of mathematical programming, our task is to minimize some functional on a certain subset contained in an infinite-dimensional space. However, in terms of optimal control theory, we are dealing with a static but nonstationary optimal control problem. It seems that the latter phrasing is the best one. To show this, let us consider the following problem:

$$\min Q(c, u)$$

subject to

$$y = F_*(c, u, z(t)),$$

$$u = Hy,$$

$$(3.82)$$

$$(c, u) \in CU,$$

where t is fixed in $[t_0, t_f]$. Let us assume that for every $t \in [t_0, t_f]$ a solution of the problem exists and let us denote it by $\tilde{c}(t)$. What is the relation between $\hat{c}(t)$ and $\tilde{c}(t)$?

PROPOSITION 3.16. *If we assume that the integral $\int_{t_0}^{t_f} Q(\tilde{c}(t), \tilde{u}(t))\, dt$ exists, then $\tilde{c}(t)$ solves the optimal control problem.*

Proof. Let us take into account any pair $(c(t), u(t))$ that satisfies constraints (3.81) such that integral $\int_{t_0}^{t_f} Q(c(t), u(t))\, dt$ exists. From the definition of $\tilde{c}(t)$ it follows, that

$$\forall t \in [t_0, t_f] \quad Q(\tilde{c}(t), \tilde{u}(t)) \leq Q(c(t), u(t)).$$

Therefore, using the fundamental property of the Lebesgue integral, we conclude that

$$\int_{t_0}^{t_f} Q(\tilde{c}(t), \tilde{u}(t))\, dt \leq \int_{t_0}^{t_f} Q(c(t), u(t))\, dt.$$

Hence, the proposition is proved. $\square$

244

The above proposition shows that under reasonable assumptions our optimal control problem can be replaced by an infinite number of static problems defined by (3.82) and solved in each instant of time on the interval $[t_0, t_f]$. These static problems are just the same as those considered in the previous sections of this chapter. It will be further assumed that the assumptions of Proposition 3.16 are satisfied. If the disturbance $z(t)$ is a step function and the intervals of time on which $z(t)$ is constant are long enough, then the static problems can be solved by one of the methods from the previous sections. If not, then the situation is much more complicated. We have two possibilities in general. The first is to solve the static problems (3.82) in an open-loop manner, that is, to solve the following model-based optimal static problem in each instant of time:

$$\min Q(c, u)$$

subject to

$$y = F(c, u, \bar{z}(t)),$$

$$u = Hy, \tag{3.83}$$

$$(c, u) \in CU.$$

This problem can be solved by using one of the methods from Chapter 2. Solving (3.83) requires some computation time and it can thus not always be used. When it can be used, it gives a suboptimal approximation of $\hat{c}(t)$ in the form of a step function. In some cases, however, especially if the difference between F and F_* is large and the estimate $\bar{z}(t)$ of the real disturbance $z(t)$ is not good, the open-loop optimal control can be very far from $\hat{c}(t)$.

The second possibility is to measure some real system variables and use these measurements in an algorithm generating an approximation of $\hat{c}(t)$ in the form of a step function. This leads to a kind of closed-loop control and probably gives a better approximation of $\hat{c}(t)$ than that obtained using the open-loop manner. On the other hand, the construction of such an algorithm is not so clear as before. In general, the difficulties are caused by the fact that in solving problem (3.82) for a given t we can use only the measurements taken at the moment t; at the next instant we have another optimization problem in the form of (3.82) because there is another value of the disturbance variable. Thus, it is quite obvious that the direct application of the algorithms from the previous sections leads to satisfactory results only then when the system disturbance changes slowly enough and the algorithm can track the moving solution $\hat{c}(t)$ with satisfactory accuracy. The detailed analysis of the accuracy of the tracking will be given for the IBMF in section 3.5.3. We will now consider the possibilities for the design of a hierarchical decentralized control algorithm in the nonstationary case.

3.5.2. DECENTRALIZED CONTROL OF NONSTATIONARY SYSTEMS

In this section we will try to adapt the decentralized control structure which was considered in section 3.4.1 to the case when some disturbances affect the real system. The presentation will be based on the results given in Brdyś and Michalak (1978). For given values of the disturbance, the disturbance estimate, and the price, z, $\bar{z}$, and λ, let us denote an infimal problem solution (see (3.65)) as $u_b(\lambda, \bar{z}, z)$ and $c_b(\lambda, \bar{z}, z)$. So, the *coordination problem* $\mathrm{CP}^{z,\bar{z}}$ (see (3.66)) tends to minimize the real performance function of the system subject to price, i.e., it performs the task:

Find a coordination variable $\hat{\lambda}(\bar{z}, z)$ such that

$$Q(c_b(\hat{\lambda}, \bar{z}, z), u_b(\hat{\lambda}, \bar{z}, z)) = \min_{\lambda \in \Lambda_b(\bar{z},z)} Q(c_b(\lambda, \bar{z}, z), u_b(\lambda, \bar{z}, z)) \quad (3.84)$$

where for given z and $\bar{z}$, $\Lambda_b(\bar{z}, z)$ denotes a set of all values of price λ for which the lower-level problem solution exists.

Finding the coordinator problem solution at each instant $t \in [t_0, t_f]$ we obtain a control function $c_b(\hat{\lambda}(\bar{z}(t), z(t)), \bar{z}(t), z(t))$ which can be considered as an approximation of $\hat{c}(t)$.

During the solution of the infimal problem a certain number of iterations on the real system must be performed and we assume that during these iterations the disturbances are constant. Hence, the disturbances are time-varying only from the coordinator's point of view. Let us also assume that the disturbance is a step function and can be represented as the sequence $\{z^k\}$. We introduce the following notation:

$$Q_*(\lambda, \bar{z}, z) \triangleq Q(c_b(\lambda, \bar{z}, z), u_b(\lambda, \bar{z}, z)). \quad (3.85)$$

Therefore, the best coordination strategy is to generate a sequence $\{\lambda^k\}$ such that

$$\lambda^k = \arg \min_{\lambda \in \Lambda_b(\bar{z}^k, z^k)} Q_*(\lambda, \bar{z}^k, z^k). \quad (3.86)$$

Because an exact form of Q_* is not known (F can differ from F_* and z^k can differ from $\bar{z}^k$), it is impossible to generate the sequence $\{\lambda^k\}$ in accordance with formula (3.86). An application of optimization methods without knowing the derivatives used in the stationary case can give unsatisfactory results owing to changes in z^k.

It follows from the above that to construct the coordination strategy more realistic criterion for generating λ^k must be formulated. Let us consider the following one:

$$Q_*(\lambda^k, \bar{z}^{k+1}, z^{k+1}) > Q_*(\lambda^{k+1}, \bar{z}^{k+1}, z^{k+1}). \quad (3.87)$$

This condition means that we will gain by changing λ^k to λ^{k+1} in relation to the case when $\lambda^{k+1} = \lambda^k$.

Now we present an algorithm that generates prices λ^k that are, under suitable assumptions, very close to those satisfying criterion (3.87). This algorithm is designed assuming that Q_* is a quadratic. The selection of prices λ^k is done in the following way:

$$\lambda^{k+1} = -Q'_{m\lambda}(\lambda^k, \bar{z}^k)\tau_*^k + \lambda^k \tag{3.88}$$

where $Q_m(\lambda, z)$ is defined by (3.85) taking $F = F_*$ and $z = \bar{z}$, and τ_*^k denotes a step coefficient in iteration k. Notice that the direction $-Q'_{m\lambda}(\lambda^k, \bar{z}^k)$ in iteration k is defined only on the basis of the model. To find the step coefficient τ_*^k we have to make some measurements of the real system.

Let us introduce the following notations. The *intermediate* iterations after the *main* iteration k will be denoted by k_i where $i = 1, 2$. Disturbances and prices will be also denoted with the adequate indexes. The step coefficient τ_*^k in iteration k is determined in the following way. Let us consider the performance $Q_*(\cdot, \bar{z}, z)$ along a line determined in $\mathcal{U}$ by direction $Q'_{m\lambda}(\lambda, \bar{z})$ and a point λ. Owing to the quadratic form of $Q_*(\cdot, \bar{z}, z)$, the following holds:

$$q(\varepsilon, \bar{z}, z) \triangleq Q_*(\lambda - \varepsilon Q'_{m\lambda}(\lambda, \bar{z}), \bar{z}, z) = a\varepsilon^2 + b\varepsilon + c \tag{3.89}$$

where $\varepsilon \in (-\infty, +\infty)$ and a, b, and c are real numbers. Notice that if

$$q(\varepsilon, \bar{z}^k, z^k) = a_1\varepsilon^2 + b_1\varepsilon + c_1 \tag{3.90}$$

then

$$c_1 = q(0, \bar{z}^k, z^k) = Q_*(\lambda^k, \bar{z}^k, z^k). \tag{3.91}$$

The first intermediate iteration consists in finding price λ^{k_1} according to the rule

$$\lambda^{k_1} = -Q'_{m\lambda}(\lambda^k, \bar{z}^k)\tau_1 + \lambda^k, \tag{3.92}$$

where τ_1 is a given step coefficient.
Next, one finds the lower-level problem solution $c_b(\lambda^{k_1}, \bar{z}^{k_1}, z^{k_1})$. Of course, there exist real numbers a_2, b_2, and c_2 such that

$$q(\varepsilon, \bar{z}^{k_1}, z^{k_1}) = a_2\varepsilon^2 + b_2\varepsilon + c_2 \tag{3.93}$$

and

$$q(\tau_1, \bar{z}^{k_1}, z^{k_1}) = a_2\tau_1^2 + b_2\tau_1 + c_2 = Q_*(\lambda^{k_1}, \bar{z}^{k_1}, z^{k_1}). \tag{3.94}$$

In the second intermediate iteration, price λ^{k_2} is found from the formula

$$\lambda^{k_2} = -Q'_{m\lambda}(\lambda^k, \bar{z}^k)\tau_2 + \lambda^k \tag{3.95}$$

where τ_2 is a given step coefficient such that $\tau_2 \neq \tau_1$, and then the lower-level problem solution $c_b(\lambda^{k_2}, \bar{z}^{k_2}, z^{k_2})$ is determined. As above, there exist real numbers $a_3, b_3,$ and c_3 such that

$$q(\varepsilon, \bar{z}^{k_2}, z^{k_2}) = a_3\varepsilon^2 + b_3\varepsilon + c_3 \qquad (3.96)$$

and

$$q(\tau_2, \bar{z}^{k_2}, z^{k_2}) = a_3\tau_2^2 + b_3\tau_2 + c_3 = Q_*(\lambda^{k_2}, \bar{z}^{k_2}, z^{k_2}). \qquad (3.97)$$

The step coefficient τ_*^k is found from the relations (3.91), (3.94), and (3.97), according to the formula:

$$\tau_*^k = -\frac{\tilde{b}}{2\tilde{a}}, \qquad (3.98)$$

where

$$\tilde{a} = \frac{(q(\tau_1, \bar{z}^{k_1}, z^{k_1}) - q(0, \bar{z}^k, z^k))\tau_2 - (q(\tau_2, \bar{z}^{k_2}, z^{k_2}) - q(0, \bar{z}^k, z^k))\tau_1}{\tau_1^2\tau_2 - \tau_2^2\tau_1}$$

$$(3.99)$$

and

$$\tilde{b} = \frac{q(\tau_1, \bar{z}^{k_1}, z^{k_1}) - q(0, \bar{z}^k, z^k)}{\tau_1} - \tau_1\tilde{a}. \qquad (3.100)$$

Hence, iteration k of the coordination strategy consists of two intermediate iterations given by formulae (3.92) and (3.95), on the basis of which the step coefficient τ_*^k is determined; the main iteration defined by formula (3.88) is then found, which gives a new price λ^{k+1}. The real performance value at this point equals $Q_*(\lambda^{k+1}, \bar{z}^{k+1}, z^{k+1})$. What is the difference between λ^{k+1} and a point minimizing $Q_*(\cdot, \bar{z}^{k+1}, z^{k+1})$ on the set $\{\lambda \in \mathcal{U} : \lambda = \lambda^k - \varepsilon Q'_{m\lambda}(\lambda^k, \bar{z}^k), \ \varepsilon \in (-\infty, +\infty)\}$? To answer the question, let us note that there exist real numbers $a_4, b_4,$ and c_4 such that

$$q(\varepsilon, \bar{z}^{k+1}, z^{k+1}) = a_4\varepsilon^2 + b_4\varepsilon + c_4. \qquad (3.101)$$

Hence, this minimizing point equals $\lambda^k - Q'_{m\lambda}(\lambda^k, \bar{z}^k) \cdot \tilde{\tau}$, where

$$\tilde{\tau} = -\frac{b_4}{2a_4}. \qquad (3.102)$$

The relation between the step used in (3.88) and the step $\tilde{\tau}$ which is really optimal is given by the following lemma.

248

LEMMA 3.11. *Let us denote:*

$$a_3 = a_2 + \delta a_2, \qquad a_4 = a_2 + \delta a_2 + \delta a_3,$$

$$b_3 = b_2 + \delta b_2, \qquad b_4 = b_2 + \delta b_2 + \delta b_3,$$

$$c_2 = c_1 + \delta c_1, \qquad c_3 = c_1 + \delta c_1 + \delta c_2,$$

where δa_i, δb_i, *and* δc_i, $i = 1, 2, 3, 4$, *represent changes of coefficients* a_i, b_i, c_i, $i = 1, 2, 3, 4$ *in* (3.90), (3.93), (3.96), *and* (3.101) *caused by changes of the disturbance and the disturbance estimation. Let us assume that for* $i = 1, 2, 3, 4$,

$$|\delta a_i| \leq \Delta a$$

$$|\delta b_i| \leq \Delta b \qquad (3.103)$$

$$|\delta c_i| \leq \Delta c.$$

The following then holds:

$$|\tilde{\tau} - \tau_*^k| \leq \frac{2|\tilde{b}|(|\tilde{a} - a_2|_{max} + 2\Delta a + 2|\tilde{a}|\left(\tau_1|\tilde{a} - a_2|_{max} + \dfrac{\Delta c}{\tau_1} + 2\Delta b\right)}{|2\tilde{a}| - |\tilde{a}|(|\tilde{a} - a_2|_{max} + 2\Delta a)}$$

$$(3.104)$$

where

$$|\tilde{a} - a_2|_{max} = \frac{\tau_2 \Delta c + \tau_1 \tau_2 \Delta b + 2\tau_1 \Delta c + \tau_2^2 \tau_1 \Delta a}{\tau_1^2 \tau_2 - \tau_2^2 \tau_1}.$$

Proof.   From (3.91), (3.94), and (3.96) we have:

$$c_1 = q(0, \bar{z}^k, z^k),$$

$$a_2\tau_1^2 + b_2\tau_1 + c_1 + \delta c_1 = q(\tau_1, \bar{z}^{k_1}, z^{k_1}),$$

and

$$(a_2 + \delta a_2)\tau_2^2 + (b_2 + \delta b_2)\tau_2 + c_1 + \delta c_1 + \delta c_2 = q(\tau_2, \bar{z}^{k_2}, z^{k_2}).$$

Hence

$$a_2 = \frac{\begin{aligned}(q(\tau_1, \bar{z}^{k_1}, z^{k_1}) - q(0, \bar{z}^k, z^k) - \delta c_1)\tau_2 - (q(\tau_2, \bar{z}^{k_2}, z^{k_2}) \\ - q(0, \bar{z}^k, z^k) - \delta c_1 - \delta c_2 - \delta a_2\tau_2^2 - \delta b_2\tau_2)\tau_1\end{aligned}}{\tau_1^2 \tau_2 - \tau_2^2 \tau_1}$$

From the above and formula (3.99) it follows that

$$|\tilde{a}_1 - a_2| \leq \frac{\tau_2 \Delta c + \tau_1 \tau_2 \Delta b + 2\tau_1 \Delta c + \tau_2^2 \tau_1 \Delta a}{\tau_1^2 \tau_2 - \tau_2^2 \tau_1}. \qquad (3.105)$$

The relations (3.94), (3.100), and (3.105) imply that

$$|\tilde{b} - b_2| \le \tau_1 \, |\tilde{a} - a_2|_{\max} + \frac{\Delta c}{\tau_1}$$

and

$$|\tilde{b} - b_4| \le \tau_1 \, |\tilde{a} - a_2|_{\max} + \frac{\Delta c}{\tau_1} + 2\Delta b,$$

$$|\tilde{a} - a_4| \le |\tilde{a} - a_2|_{\max} + 2\Delta a.$$

Hence, using formulae (3.98) and (3.102), we can easily derive inequality (3.104). $\square$

Note that if the disturbance and the disturbance estimation are constant during iteration k, then $\Delta a = \Delta b = \Delta c = 0$ and, from Lemma 3.11, $\tau_*^k = \tilde{\tau}$ regardless of the difference between the system model and its mathematical description. Note also that if $\Delta a \to 0$, $\Delta b \to 0$, and $\Delta c \to 0$, then $\tau_*^k \to \tilde{\tau}$. A fundamental property of the decentralized control algorithm is formulated in the following theorem.

THEOREM 3.17.   *If we assume that*

1.   *$\bar{z}^k$ and z^k are constant for all $k = 0, 1, \ldots,$ and equal to $\bar{z}$ and z,*
2.   *$\Lambda_b = \mathcal{U}$,*
3.   *$(\exists \rho > 0)(\forall \lambda \in \Lambda_b) \; \|Q'_{*\lambda}(\lambda, \bar{z}, z) - Q'_{m\lambda}(\lambda, \bar{z})\| \le \rho$,*
4.   *$\exists A > 0 \; \langle (\lambda - \hat{\lambda}), Q''_{*\lambda\lambda}(\lambda - \hat{\lambda}) \rangle \ge A \, \|\lambda - \hat{\lambda}\|^2$*

where $\hat{\lambda}$ minimizes $Q_*(\,\cdot\,, \bar{z}, z)$ on $\mathcal{U}$, i.e., the functional $Q_*(\,\cdot\,, \bar{z}, z)$ is strictly positive definite on $\mathcal{U}$, then

1.   *For any $\varepsilon > 0$ there exists k_ε such that $\lambda^{k_\varepsilon} \in K(\hat{\lambda}; (\rho/A) + \varepsilon)$, where $K(\hat{\lambda}; (\rho/A) + \varepsilon)$ denotes an open ball with center at $\hat{\lambda}$ and radius equal to $(\rho/A) + \varepsilon$.*
2.   *Sequence $\{Q_*(\lambda^k, \bar{z}, z)$ converges monotonically.*

Proof.   Consider λ and $\varepsilon > 0$ such that

$$\|\lambda - \hat{\lambda}\| \ge \frac{\rho}{A} + \varepsilon.$$

From assumption (4) the following holds:

$$\|Q'_{*\lambda}(\lambda, \bar{z}, z)\| = \|Q''_{*\lambda\lambda}(\lambda - \hat{\lambda})\| \ge A \, \|\lambda - \hat{\lambda}\| \ge \rho + A\varepsilon. \tag{3.106}$$

Let us consider an angle between $Q'_{*\lambda}(\lambda, \bar{z}, z)$ and $Q'_{m\lambda}(\lambda, \bar{z})$. From the scalar product properties it follows that for any $a, b \in \mathcal{U}$

$$\|a - b\|^2 = \langle a - b, a - b \rangle = \|a\|^2 - 2\langle a, b \rangle + \|b\|^2.$$

Thus, taking into account assumption (3) and inequality (3.106), we obtain

$$\langle Q'_{*\lambda}(\lambda, \bar{z}, z), Q'_{m\lambda}(\lambda, \bar{z}) \rangle \geq \alpha$$

where $\alpha = (\rho + A\varepsilon)A\varepsilon$. From the above relation it follows that on the outside of the ball $K(\hat{\lambda}, (\rho/A) + \varepsilon)$, the directions $Q'_{m\lambda}(\lambda, \bar{z})$ are those of uniform improvement. Also, from assumption 1 and consequently from Lemma 3.11, $\tau_*^k = \tilde{\tau}$. Thus, the step coefficients τ_*^k are converging step coefficients, and part 1 follows from the convergence theorem for an algorithm that uses the directions of uniform improvement and the converging step coefficients (Goldstein 1967).

To prove part 2, we note that from assumption 4 the sequence $\{Q_*(\lambda^k, \bar{z}, z)\}$ is bounded from below and by equality $\tau_*^k = \tilde{\tau}$ it is not increasing. The proof of the theorem is completed. $\square$

The theorem gives an asymptotic upper bound on performance loss if a disturbance jumps from one value to another and is constant for a long time. This bound is given by the radius of the accuracy ball $\bar{K}(\hat{\lambda}; \rho/A)$. It seems that Theorem 3.17 describes the behavior of the algorithm during coordination when functional $Q_*(\cdot, \bar{z}, z)$ is quadratic and the disturbance has the form of a step function that does not change frequently. The situation is much more complicated when the disturbance changes frequently. Only partial solutions have been found in this case; one of them is given below.

THEOREM 3.18.   *If we assume that*

1. $\forall (c, u) \in \mathcal{C} \times \mathcal{U} \quad F(c, u) = F_*(c, u),$
2.   *For any* $k = 0, 1, \ldots,$ $z^k \in Z \subset \mathcal{Z}$ *where* Z *is compact,* $\bar{z}^k = z^k,$ *and* $z^k \to z$ *if* $k \to \infty,$
3.   *Functional* $Q_m(\cdot, \bar{z})$ *is convex for any* $\bar{z}$ *from* Z *and a functional* $Q_m(\lambda, \cdot)$ *satisfies the Lipschitz condition with a constant independent of* $\lambda,$
4.   *Sequence* $\{\lambda^k\}$ *is generated according to the rule*

$$\lambda^{k+1} = \lambda^k - \rho^k Q'_{m\lambda}(\lambda^k, \bar{z}^k), \qquad k = 0, 1, \ldots,$$

where ρ^k *denotes a positive step coefficient in iteration* $k,$
5.   *The step coefficients are such that*

$$|z^{k+1} - z^k| \leq \delta^k \to 0 \quad \text{if} \quad k \to \infty$$

and

$$\rho^k \to 0, \sum_{k=0}^{\infty} \rho^k = \infty, \frac{\delta^k}{\rho^k} \to 0 \quad \text{if} \quad k \to \infty,$$

6. *The sequence $\{\lambda^k\}$ is bounded,*
7. $\forall \bar{z} \in Z \quad \Lambda_b(\bar{z}, \bar{z}) = \mathcal{U}$

then the following holds

$$\lim_{k \to \infty} [Q_*(\lambda^k, \bar{z}^k, z^k) - \min_{\lambda \in \Lambda_b(\bar{z}^k, z^k)} Q_*(\cdot, \bar{z}^k, z^k)] = 0.$$

Proof. The proof follows immediately from a general theorem given in Appendix B.8. Let us note that assumption (2) means that we can measure the current values of the disturbance exactly. The theorem seems to be a good starting point for further research.

3.5.3. APPLICATION OF THE INTERACTION BALANCE METHOD WITH FEEDBACK TO NONSTATIONARY SYSTEM COORDINATION

To simplify the analysis, it is assumed that a disturbance estimation $\bar{z}(t)$ is constant an $[t_0, t_f]$. Hence, $\bar{z}$ will be omitted as an argument of the operators F and K. We also assume that $t_f = +\infty$. Note that the analysis here is based on Ruszczyński (1979).

Let us consider the following itcrative, hierarchical control scheme. In iteration k local decision unit i solves its local problem LP_i^k of the form:

For a given value $\lambda^k \in \mathcal{U}$ of the coordination variable find both control and interaction

$$(\hat{c}_i(\lambda^k), \hat{u}_i(\lambda^k)) = \arg \min_{CU_i} Q_{\text{modi}}(\cdot, \cdot, \lambda^k) \tag{3.107}$$

where

$$\forall (c_i, u_i) \in \mathscr{C}_i \times \mathcal{U}_i \quad Q_{\text{modi}}(c_i, u_i) \triangleq Q_i(c_i, u_i) + \langle \lambda_i^k, u_i \rangle - \sum_{j=1}^{N} \langle \lambda_j^k, H_{ji} F_i(c_i, u_i) \rangle$$

and

$$\lambda^k = (\lambda_1^k, \ldots, \lambda_N^k).$$

The *coordinator* generates at times $t_0, t_1, \ldots, t_k, \ldots,$ the prices $\lambda^0, \lambda^1, \ldots, \lambda^k, \ldots,$ according to the rule

$$\lambda^{k+1} = \lambda^k + \varepsilon E R_*(\hat{c}(\lambda^k), \hat{u}(\lambda^k), z(t_{k+1})) \quad \text{for } k = 1, 2, \ldots, \tag{3.108}$$

where ε is a sufficiently small positive number, E is the appropriately chosen linear bounded operator, and $R_*(\hat{c}(\lambda^k), \hat{u}(\lambda^k), z(t_{k+1}))$ denotes the discoordination at time t_{k+1}, that is,

$$R_*(\hat{c}(\lambda^k), \hat{u}(\lambda^k), z(t_{k+1})) = \hat{u}(\lambda^k) - u_*(\hat{c}(\lambda^k), z(t_{k+1})) \tag{3.109}$$

where

$$u_*(c, z(t)) \triangleq HK_*(c, z(t)).$$

The initial price λ^0 is found from the coordination model of the system at $t = t_0$.

Owing to the time needed for data transmission and computation, the control applied to the real system equals

$$c(t) = \hat{c}(\lambda^0) \quad \text{for} \quad t \in [t_0, t_1]$$

and

$$c(t) = \begin{cases} \hat{c}(\lambda^{k-1}) & \text{for} \quad t_k \leq t \leq t_k + \tau, \\ \hat{c}(\lambda^k) & \text{for} \quad t_k + \tau < t \leq t_{k+1} \end{cases} \qquad k = 1, 2, \ldots, \quad (3.110)$$

where $\tau > 0$ denotes delays due to on-line computation and communication between levels. Thus, the operation times $t_1, t_2, \ldots$ should satisfy the inequalities

$$t_{k+1} \geq t_k + T, \qquad k = 1, 2, \ldots, \tag{3.111}$$

where $T \geq \tau$. These inequalities include all delays (among which delays caused by transient processes play the fundamental role) and technological requirements, which limit the frequency of control changes.

If the disturbance is constant over $[t_0, t_f]$ and if t_f is sufficiently large, then the discoordination norm can be made as small as desired. This follows from the convergence analysis of the coordination strategies for IBMF given in section 3.3.5, where suboptimal control has also been discussed. If the disturbance is time-varying, then all we can do is try to keep the current discoordination norm as small as possible by choosing the times for coordinator intervention appropriately. In other words, we try to satisfy the coordination condition of IBMF at each moment t in the interval $[t_0, t_f]$.

To examine this problem, some assumptions limiting the class of possible disturbances must be made. We will assume that we know the range and/or a bound on the rate of change of disturbances. Therefore, we shall assume that there exist a function $\varphi : \mathbb{R} \to \mathbb{R}_+$ and a set $Z \subset \mathcal{Z}$ such that the class Φ of all possible disturbances may be described as follows:

$$\Phi = \{z(\,\cdot\,) \in Z^{\mathbb{R}_+} : \rho(z(t'), z(t'')) \leq \varphi(t'') - \varphi(t') \text{ for all}$$
$$t', t'' \in [t_0, t_f], \text{ such that } t' \leq t''\} \quad (3.112)$$

where $\rho(\cdot, \cdot)$ denotes a metric in $\mathcal{Z}$.

Let us assume that the operator E has been chosen in such a way that there exist in $\mathcal{U}$ a norm, say $\|\cdot\|_0$, equivalent to $\|\cdot\|$ and a positive real number $q < 1$, such that the following inequality is satisfied:

$$(\forall k = 1, 2, \ldots,)(\forall z \in Z) \ \|ER_*(\hat{c}(\lambda^{k+1}), \hat{u}(\lambda^{k+1}), z\|_0$$
$$\leq q \, \|ER_*(\hat{c}(\lambda^k), \hat{u}(\lambda^k), z\|_0. \quad (3.113)$$

This means that for any disturbance that is constant during coordination, the imbalance norm is reduced uniformly on the disturbance value. We show how to construct an operator E in section 3.6.

Continuing with the design of the algorithm, we introduce an operator $\tilde{D}$, $\tilde{D} : \mathscr{C} \times \mathscr{U} \times \mathscr{Z} \to \mathscr{U}$ as follows:

$$\tilde{D}(c, u, z) \triangleq HK(c) - HK_*(c, z). \tag{3.114}$$

Notice that operator $\tilde{D}$ describes a difference between the mathematical model of the system and its mathematical description. Let us assume that with respect to a disturbance this difference satisfies the Lipschitz condition uniformly for a control and an interaction. A positive real number must then exist such that the following inequality holds:

$$(\forall z_1, z_2 \in Z)(\forall (c, u) \in \mathscr{C} \times \mathscr{U}) \quad \|E(\tilde{D}(c, u, z_2) - D(c, u, z_1))\|_0 \le \beta \rho(z_1, z_2).$$

$$\tag{3.115}$$

The simplest way to choose the intervention times is to choose a certain interval $T_c \ge T$ and update the price at times $t_k = t_0 + kT_c$, $k = 1, 2, \ldots$. The following theorem gives the bounds on the discoordination norm when the intervention times are chosen in this a way.

THEOREM 3.19.   *Let there exist constants $d_1 > 0$, $d_2 > 0$ such that for $t \ge t_0$ (a) $\varphi(t + T_c) - \varphi(t) \le d_1$ and (b) $\varphi(t + \tau) - \varphi(t) \le d_2$. Then for any $z(\cdot) \in \Phi$*

(1)
$$\lim_{t \to \infty} \|\eta(t)\|_0 \le \frac{\beta d_1}{1 - q} + \beta d_2.$$

(2)
$$If \ \|\eta(t_{k_0})\|_0 \le \beta d_1/(1-q) \ for \ some \ k_0, \ then$$
$$\|\eta(t)\|_0 \le \beta d_1/(1-q) + \beta d_2 \ for \ all \ t \ge t_{k_0},$$

where $\eta(t)$ denotes a discoordination at time t.

Proof.   Let us define a function $\Delta : [t_0, \infty) \to \mathbb{R}$ as follows:

$$\Delta(t_0) \triangleq \|ER_*(\hat{c}(\lambda^0), \hat{u}(\lambda^0), z(t))\|_0 = \eta(t_0)$$

and for $k = 1, 2, \ldots$,

$$\Delta(t) \triangleq \begin{cases} \Delta(t_k) + \beta(\varphi(t) - \varphi(t_k)) & \text{for} \quad t \in (t_k, t_k + \tau], \\ q\Delta(t_k) + \beta(\varphi(t) - \varphi(t_k)) & \text{for} \quad t \in (t_k + \tau, t_{k+1}]. \end{cases} \tag{3.116}$$

It may be easily checked that for all $t \ge t_0$

$$\|\eta(t)\|_0 \le \Delta(t). \tag{3.117}$$

Indeed, for some $k \ge 1$ let

$$\|\eta(t_k)\|_0 \le \Delta(t_k).$$

254

Then, from (3.110), (3.112), and (3.115), the following inequalities hold for all $t \in (t_k, t_k + \tau]$:

$$\|\eta(t)\|_0 = \|ER_*(\hat{c}(\lambda^{k-1}), \hat{u}(\lambda^{k-1}), z(t))\|_0 = \|ER_*(\hat{c}(\lambda^{k-1}), \hat{u}(\lambda^{k-1}),$$
$$z(t)) - ER_*(\hat{c}(\lambda^{k-1}), \hat{u}(\lambda^{k-1}), z(t_k)) + ER_*(\hat{c}(\lambda^{k-1}), \hat{u}(\lambda^{k-1}),$$
$$z(t_k))\|_0 \le \beta(\varphi(t) - \varphi(t_k)) + \|\eta(t_k)\|_0 \le \Delta(t_k) + \beta(\varphi(t) - \varphi(t_k)) \le \Delta(t).$$

If $t \in (t_k + \tau, t_{k+1}]$, then according to (3.110), (3.112), (3.113), and (3.115), the following is true:

$$\|\eta(t)\|_0 = \|ER_*(\hat{c}(\lambda^k), \hat{u}(\lambda^k), z(t))\|_0 = \|ER_*(\hat{c}(\lambda^k), \hat{u}(\lambda^k), z(t))$$
$$- ER_*(\hat{c}(\lambda^k), \hat{u}(\lambda^k), z(t_k)) + ER_*(\hat{c}(\lambda^k), \hat{u}(\lambda^k), z(t_k))\|_0$$
$$\le \beta(\varphi(t) - \varphi(t_k)) + q \|ER_*(\hat{c}(\lambda^{k-1}), \hat{u}(\lambda^{k-1}), z(t_k))\|_0$$
$$\le q\|\eta(t_k)\|_0 + \beta(\varphi(t) - \varphi(t_k)) \le q\Delta(t_k) + \beta(\varphi(t) - \varphi(t_k)) = \Delta(t).$$

Hence, inequality (3.117) holds for all $t \ge t_0$. Let us note that for $t \in (t_{k-1} + \tau, t_k + \tau]$

$$\Delta(t) \le \Delta(t_k + \tau).$$

Consequently,

$$\overline{\lim_{t \to \infty}} \, \Delta(t) = \overline{\lim_{k \to \infty}} \, \Delta(t_k + \tau). \tag{3.118}$$

It follows from (3.116) that

$$\Delta(t_{k+1}) = q\Delta(t_k) + \beta(\varphi(t_{k+1}) - \varphi(t_k)).$$

Hence

$$\overline{\lim_{k \to \infty}} \, \Delta(t_k) \le \frac{\beta}{1-q} \, \overline{\lim_{t \to \infty}} \, (\varphi(t + T_c) - \varphi(t)) \le \frac{\beta d_1}{1-q}. \tag{3.119}$$

According to (3.116) and assumption (b) of the theorem

$$\Delta(t_k + \tau) \le \Delta(t_k) + \beta d_2. \tag{3.120}$$

Thus, we obtain from (3.118), (3.120), and (3.119) the inequality

$$\overline{\lim_{t \to \infty}} \, \Delta(t) \le \frac{\beta d_1}{1-q} + \beta d_2, \tag{3.121}$$

which, together with inequality (3.117), proves part 1 of the theorem. In order to prove part 2, let us observe that if for some $k = k_0$

$$\|\eta(t_k)\|_0 \le \frac{\beta d_1}{1-q} \tag{3.122}$$

then, according to (3.110), (3.112), (3.113), (3.115), and assumption (a)

$$\begin{aligned}
\|\eta(t_{k+1})\|_0 &= \|ER_*(\hat{c}(\lambda^k), \hat{u}(\lambda^k), z(t_{k+1}))\|_0 \\
&= \|ER_*(\hat{c}(\lambda^k), \hat{u}(\lambda^k)z(t_{k+1})) \\
&\quad - ER_*(\hat{c}(\lambda^k), \hat{u}(\lambda^k), z(t_k)) + ER_*(\hat{c}(\lambda^k), \hat{u}(\lambda^k), z(t_k))\|_0 \\
&\leq \beta d_1 + q \|ER_*(\hat{c}(\lambda^{k-1}), \hat{u}(\lambda^{k-1}), z(t_k))\|_0 \\
&\leq \beta d_1 + q \|\eta(t_k)\|_0 \leq \beta d_1 + q \frac{\beta d_1}{1-q} \\
&= \frac{\beta d_1}{1-q}
\end{aligned}$$

and (3.122) holds for all $k \geq k_0$.

Therefore, to prove part 2 it is enough to prove that

$$\|\eta(t)\|_0 \leq \frac{\beta d_1}{1-q} + \beta d_2 \quad \text{for} \quad t \in (t_k, t_{k+1}) \tag{3.123}$$

where k is fixed and $k \geq k_0$.

Let $t \in [t_k, t_k + \tau]$. Then, according to (3.110), (3.112), (3.115), and assumption (b)

$$\begin{aligned}
\|\eta(t)\|_0 &= \|ER_*(\hat{c}(\lambda^{k-1}), \hat{u}(\lambda^{k-1}), z(t))\|_0 \\
&= \|ER_*(\hat{c}(\lambda^{k-1}), \hat{u}(\lambda^{k-1}), z(t)) \\
&\quad - ER_*(\hat{c}(\lambda^{k-1}), \hat{u}(\lambda^{k-1}), z(t_k)) + ER_*(\hat{c}(\lambda^{k-1}), \hat{u}(\lambda^{k-1}), z(t_k))\|_0 \\
&\leq \beta d_2 + \frac{\beta d_1}{1-q}
\end{aligned}$$

and (3.123) holds. Let $t \in (t_k + \tau, t_{k+1})$. Then, according to (3.110), (3.112), (3.113), (3.115), assumption (a) and the above inequality taken at $t = t_k + \tau$, the following hold

$$\begin{aligned}
\|\eta(t)\|_0 &= \|ER_*(\hat{c}(\lambda^k), \hat{u}(\lambda^k), z(t))\|_0 \\
&= \|ER_*(\hat{c}(\lambda^k), \hat{u}(\lambda^k), z(t)) \\
&\quad - ER_*(\hat{c}(\lambda^k), \hat{u}(\lambda^k), z(t_k + \tau)) + ER_*(\hat{c}(\lambda^k), \hat{u}(\lambda^k), z(t_k + \tau))\|_0 \\
&\leq \beta d_1 + q \|ER_*(\hat{c}(\lambda^{k-1}), \hat{u}(\lambda^{k-1}), z(t_k + \tau))\|_0 \\
&\leq \beta d_1 + q\left(\beta d_2 + \frac{\beta d_1}{1-q}\right) \\
&= \frac{\beta d_1}{1-q} + q\beta d_2 \\
&< \beta d_2 + \frac{\beta d_1}{1-q}
\end{aligned}$$

and (3.123) is also true for this t. Hence, (3.123) holds for all $t \in (t_k, t_{k+1})$, and part 2 is proved. $\square$

The theorem gives an upper bound for discoordination that may occur in the nonstationary system when it is coordinated at a constant frequency by the interaction algorithm. This bound will be called the tracing accuracy. It follows from part 1 that the tracing accuracy depends on the intervals T_c and τ. If φ is uniformly continuous, then the tracing accuracy may be improved by decreasing T_c. Still, inequality (3.111) must hold and consequently the tracing accuracy cannot be brought to zero. However, by simultaneously decreasing T_c, T, and τ, any desired tracing accuracy may be obtained. In contrast, if noncontinuous disturbances occur, decreasing T_c does not necessarily improve performance; the tracing accuracy is limited by jumps of the function φ. Finally, let us observe that the method is less efficient as the bound on the rate of change of the disturbance becomes large. In order to obtain the desired tracing accuracy one should take a very small T_c and update the price frequently, irrespective of the actual disturbances. The efficiency loss can be overcome if the times at which the coordinator intervenes are determined on-line on the basis of current observations of discoordination. Of course, the cost of on-line measurements and data transmission is large compared with operation based on fixed intervention times. On the other hand, the global control cost, which is not considered here, may be less.

Since the total discoordination is the product of local differences $\hat{u}_i(\lambda) - u_{*i}(\hat{c}(\lambda), z(t))$, the observations may be carried out in a decentralized way by taking advantage of the special structure of the control system. Interactions in the real system are observed by the local decision units, and the coordinator comes into operation when one or more local discoordinations become excessive. The control algorithm organized on this idea will be called the *discoordination stabilization method*.

Let each local decision unit have its defined tolerance $\gamma_i \geq 0$, $i \in \overline{1, N}$. The system is supposed to work well if

$$\|\hat{u}_i(\lambda) - u_{*i}(\hat{c}(\lambda), z(t))\|_i \leq \gamma_i \quad \text{for} \quad i \in \overline{1, N},$$

where $\| \cdot \|_i$ denotes a norm on $\mathcal{U}_i$.

Next, let T_m denote the distance between successive times at which measurements of local inputs are made by the local decision units.

We assume that $k - 1$ iterations have been executed and the current price and control equal λ^{k-1}, $\hat{c}(\lambda^{k-1})$. The outline of the discoordination stabilization method is as follows:

Step 0   Index j is set equal to 0.
Step 1   Local decision units at $t_{k-1,j} = t_{k-1} + T + j \cdot T_m$ measure inputs $u_{*i}(\hat{c}(\lambda^{k-1}), z(t_{k-1,j}))$ of their subsystems.

Step 2   If for every $i \in \overline{1, N}$

$$\|\hat{u}_i(\lambda^{k-1}) - u_{*i}(\hat{c}(\lambda^{k-1}), z(t_{k-1,j}))\|_i \leq \gamma_i \tag{3.124}$$

then Step 1 is repeated with $j = j + 1$. Otherwise, the coordinator is notified and Step 3 is executed.

Step 3   The coordinator defines $t_k = t_{k-1,j}$, gathers from all local decision units the discoordination $\hat{u}_i(\lambda^{k-1}) - u_{*i}(\hat{c}(\lambda^{k-1}), z(t_k))$ and updates the price according to (3.108). The new price λ^k is sent to the local decision units.

Step 4   Local decision units solve their problems (LP_i^k) defined by (3.107), save $\hat{u}_i(\lambda^k)$, and apply controls $\hat{c}_i(\lambda^k)$ to the real system. Step 1 is repeated with $k = k + 1$ and $j = 0$.

Before proceeding to the detailed analysis, let us observe that there exists a constant $\bar{\gamma}$ such that

$$(\|\hat{u}_i - u_{*i}\|_i \leq \gamma_i \quad \text{for every} \quad i \in \overline{1, N}) \Rightarrow (\|E(\hat{u} - u_*)\|_0 \leq \bar{\gamma}) \tag{3.125}$$

with $\hat{u} = (\hat{u}_1, \ldots, \hat{u}_N)$, $u_* = (u_{*1}, \ldots, u_{*N})$.

Briefly, the existence of $\bar{\gamma}$ results from the equivalence of the norm $\|\cdot\|_0$ and $\|\cdot\|$ on $\mathcal{U}$ (recall that $\|u\| = \sum_{i=1}^{N} \|u_i\|_i$, for $u = (u_1, \ldots, u_N)$).

Taking advantage of the constant $\bar{\gamma}$, we present the following theorem.

THEOREM 3.20.   *If we assume that*

1.   *There exist constants $d_1 \geq 0$, $d_2 \geq 0$, and $d_3 \geq 0$ such that for all $t \geq t_0$*

$$\varphi(t + T) - \varphi(t) \leq d_1,$$

$$\varphi(t + \tau) - \varphi(t) \leq d_2,$$

$$\varphi(t + T_m) - \varphi(t) \leq d_3.$$

2. $\bar{\gamma} \geq (\beta/(1-q))(d_1 + q d_3),$

then for any $z(\cdot) \in \Phi$

1. $\overline{\lim_{t \to \infty}} \|\eta(t)\|_0 \leq \bar{\gamma} + \beta(d_2 + d_3)$ $\hspace{2cm}$ (3.126)

where $\eta(t)$ denotes the discoordination at time t.

2.   *If for some $k \geq 0$*

$$\|\hat{u}_i(\lambda(t_k + T)) - u_{*i}(\hat{c}(\lambda(t_k + T)), z(t_k + T))\|_i \leq \gamma_i \quad \text{for all} \quad i \in \overline{1, N},$$

then for all $t \geq t_k + T$

$$\|\eta(t)\|_0 \leq \bar{\gamma} + \beta(d_2 + d_3).$$

The proof of this theorem is technically the same as the proof of Theorem 3.19, and is left to the reader. It can be found in Ruszczyński (1979).

258

Theorem 3.20 gives a bound on the tracing accuracy attainable by the discoordination stabilization method. In order to obtain the desired tracing accuracy Δ, one should find $\bar{\gamma} \leq \Delta - \beta(d_2 + d_3)$ according to (3.126). Moreover, $\bar{\gamma}$ should satisfy assumption (2). Next, the γ_i, $i \in \overline{1, N}$, satisfying (3.125) should be chosen. The discoordination stabilization method with these γ_i provides the desired accuracy Δ. It does not follow from this theorem that each local discoordination is kept below its γ_i. According to assumption 2 and (3.126), the tracing accuracy is limited by $\Delta_{\min} = \beta(d_1 + d_3)/(1-q) + \beta d_2$ no matter how small the γ_i have been chosen.

As a matter of fact, the coordination algorithm described in this paper can be used with any intelligent procedure for determining intervals of operation. Once an algorithm having contraction mapping properties (see relation (3.113)) has been evaluated, the problem of tracing the moving solution of the coordination problem is no longer difficult if the rate of change of the disturbances is bounded. If disturbance estimation $\bar{z}(t)$ used in the model (see Eq. (3.79)) is time-varying, then the analysis is more complicated but does not involve any new concepts and is therefore omitted.

The description of disturbances, used in this section seems to be quite close to the description of disturbances in engineering. To avoid confusion, noises due to inaccurate computations and measurements have been ignored. There is every indication, however, that both measurement and computational noises may be taken into account, and ideas suggested by Kheysin (1976) may be useful in further studies.

Simulation results

Let us consider the system described in Figure 3.5. The subsystems are described by the following equations.

Subsystem 1

$$y_1(t) = c_{11}(t) + c_{12}(t) + 2u_1(t) + z_{11}(t)(c_{11}(t))^2$$
$$+ z_{12}(t)(c_{11}(t) - c_{12}(t) - 2)u_1(t) + z_{13}(t).$$

Subsystem 2

$$y_{21}(t) = c_{21}(t) - c_{22}(t) + u_{21}(t) - 3u_{22}(t) + z_{21}(t),$$
$$y_{22}(t) = 2c_{22}(t) - c_{23}(t) - u_{21}(t) + 2u_{22}(t) + z_{22}(t).$$

Subsystem 3

$$y_3(t) = c_3(t) + 4u_3(t) + z_{31}(t)c_3(t) + z_{32}(t).$$

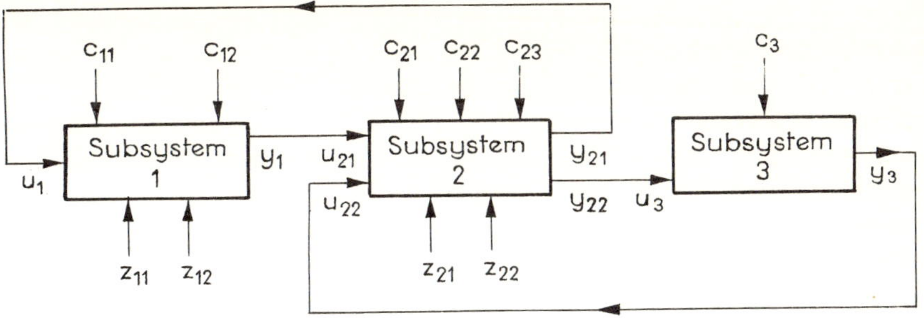

FIGURE 3.5   The structure of the simulated system.

The structure matrix H has the form

$$H = \begin{bmatrix} 0 & 1 & 0 & 0 \\ 1 & 0 & 0 & 0 \\ 0 & 0 & 0 & 1 \\ 0 & 0 & 1 & 0 \end{bmatrix}.$$

We have access to an approximate linear model of the form

Subsystem 1

$$y_1(t) = c_{11}(t) + c_{12}(t) + 2u_1(t).$$

Subsystem 2

$$y_{21}(t) = c_{21}(t) - c_{22}(t) + u_{21}(t) - 3u_{22}(t),$$

$$y_{22}(t) = 2c_{22}(t) - c_{23}(t) - u_{21}(t) + 2u_{22}(t).$$

Subsystem 3

$$y_3(t) = c_3(t) + 4u_2(t).$$

The performance function is defined by

$$Q(c, u) = Q_1(c_1, u_1) + Q_2(c_2, u_2) + Q_3(c_3, u_3)$$

where

$$Q_1(c_1, u_1) = (u_1 - 1)^2 + (c_{11})^2 + (c_{12} - 2)^2,$$

$$Q_2(c_2, u_2) = 2(c_{12} - 2)^2 + (c_{22})^2 + 3(c_{23})^2 + 4(u_{21})^2 + (u_{22})^2,$$

$$Q_3(c_3, u_3) = (c_3 + 1)^2 + (u_3 - 1)^2.$$

Feasible sets for the subsystems are described as follows.

$$CU_1 = \{(c_{11}, c_{12}, c_{13}, u_1) : c_{11} + u_1 \leq 1.006\},$$

$$CU_2 = \mathbb{R}^5,$$

$$CU_3 = \{(c_3, u_3) : c_3 + u_3 \geq -0.5\}.$$

The behavior of the real plant was simulated on the interval $0 \leq t \leq 150$. The disturbances $z_1(t)$, $z_2(t)$, $z_3(t)$ were obtained in a random number generator such that for all $t = 1, 2, \ldots, 150$ and all i, j

$$z_{ij}(0) = 0,$$

$$|z_{ij}(t+1) - z_{ij}(t)| \leq v,$$

$$|z_{ij}(t)| \leq w,$$

where v is the limit of the rate of change and w is the range of disturbances.

Examples of the disturbances obtained with $v = 1$ and $w = 10$ are shown in Figure 3.6.

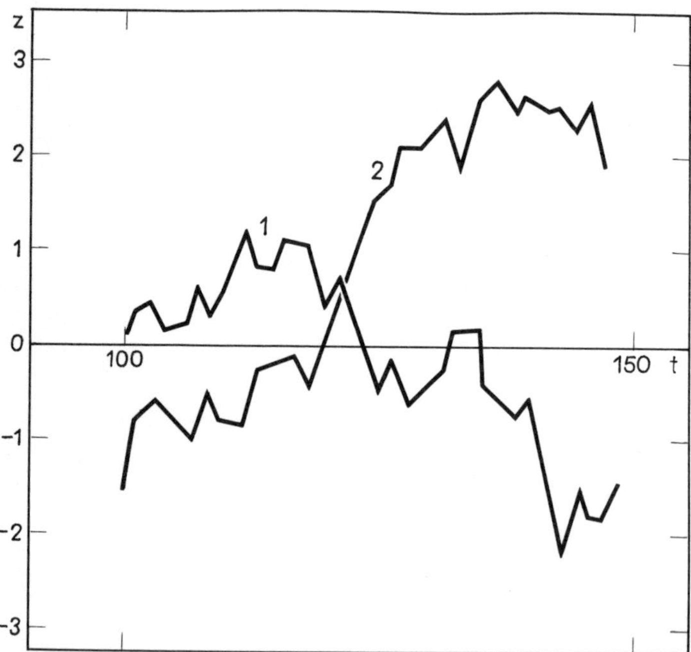

FIGURE 3.6   Examples of disturbances used in the simulation.

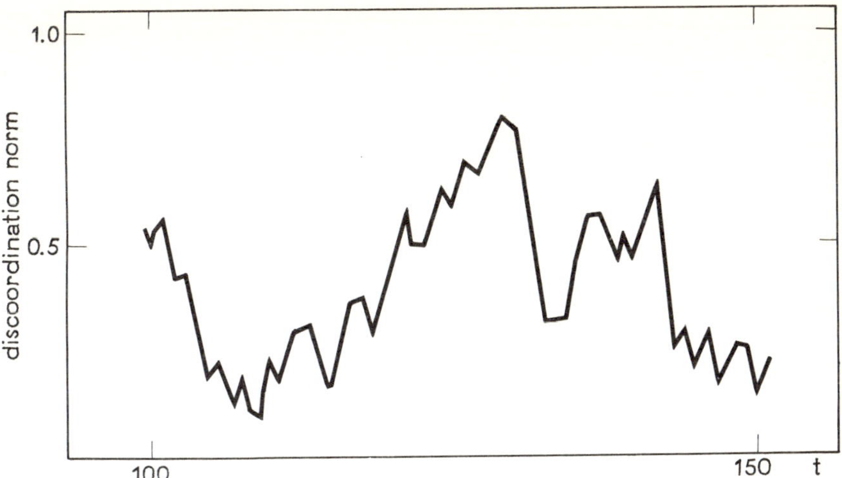

FIGURE 3.7   Behavior of the discoordination norm in the interconnected system.

In the simulation study, the range and rate of change of the disturbances were fixed at $v = 0.015$ and $w = 1.0$. The discoordination norm was computed from the formula

$$[(\hat{u}_1 - u_{*i})^2 + (\hat{u}_{21} - u_{*21})^2 + (\hat{u}_{22} - u_{*22})^2 + (u_3 - u_{*2})^2]^{1/2}$$

The step size ε in (3.108) was fixed at 0.8.

A real system, when controlled by a constant, model-based optimal control $\hat{c}(\lambda^0)$, exhibits the discoordinations indicated in Figure 3.7. When the coordinator intervenes at fixed intervals, the tracing accuracy is much better, as shown in Figures 3.8 and 3.9. However, the method becomes less efficient when the frequency of intervention of the coordinator decreases, which can be seen by the difference between Figures 3.8 and 3.9. The discoordination stabilization method overcomes this difficulty, as indicated in Figure 3.10.

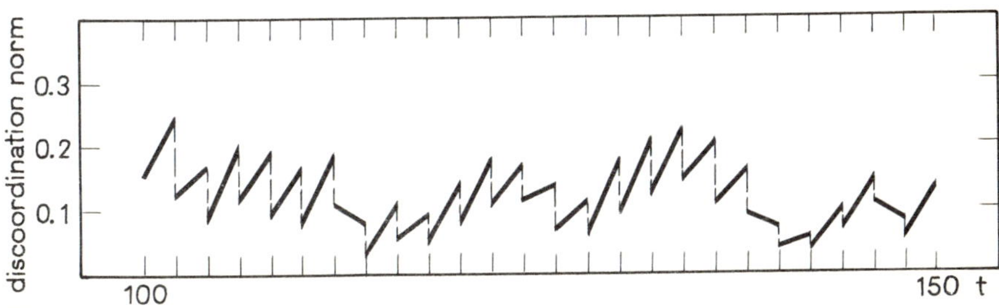

FIGURE 3.8   The discoordination norm when the coordinator intervenes at fixed intervals.

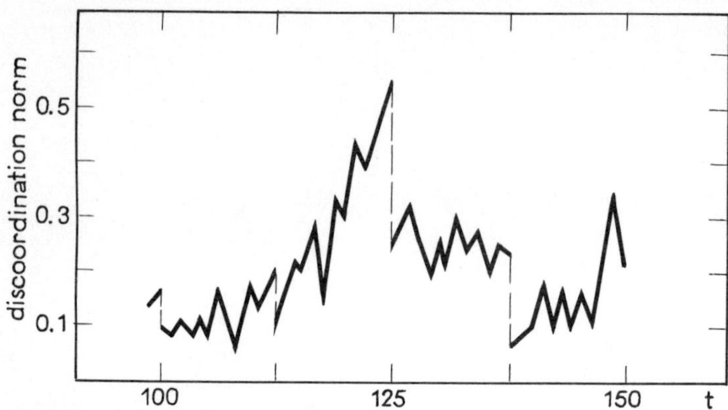

FIGURE 3.9 The discoordination norm with less frequent interventions by the coordinator than in Figure 3.8.

Finally, the correlation between discoordination and real system performance in all the above experiments was studied. The value of the performance on the interval $[0, 150]$ was computed according to the formula

$$\bar{Q} = \sum_{t=0}^{150} Q(\hat{c}(\lambda(t)), u_*(\hat{c}(\lambda(t)), z(t))).$$

The squared discoordination in the coordination process was calculated from the formula

$$\bar{\eta} = \sum_{t=0}^{150} \|\hat{u}(\lambda(t)) - u_*(\hat{c}(\lambda(t)), z(t))\|^2.$$

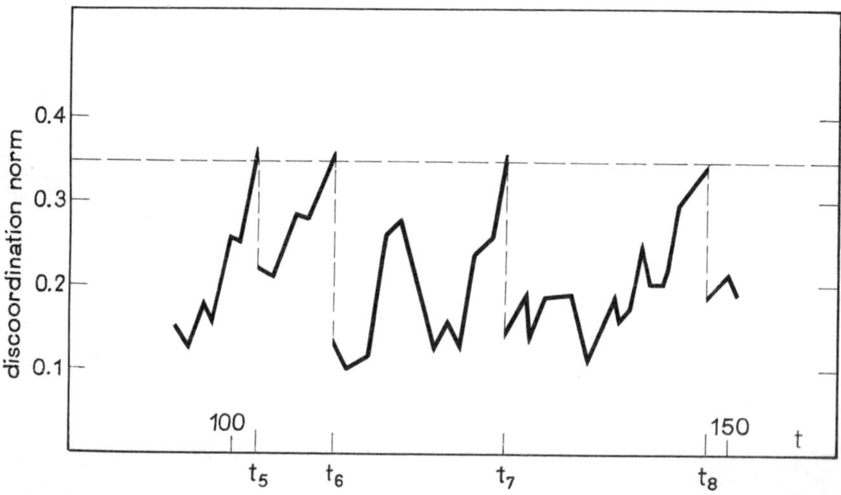

FIGURE 3.10 Discoordination stabilization.

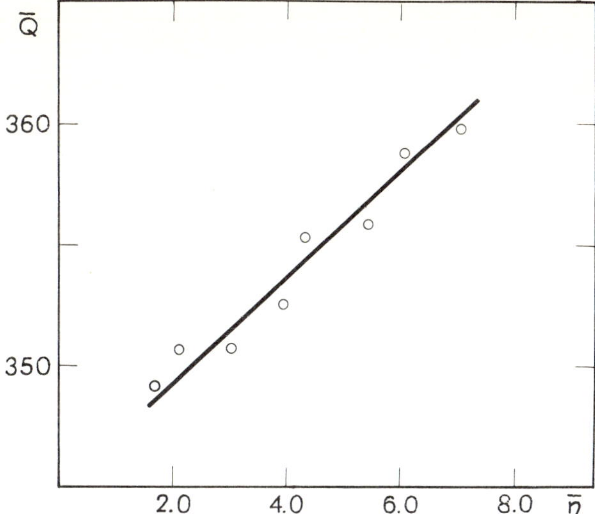

FIGURE 3.11   Correlation between discoordination and performance.

The results are collected in Figure 3.11. It follows from the figure that the coordination improves as the discoordination decreases. Though the coordination condition $\hat{u}(\lambda) - u_*(\lambda, z) = 0$ yields a control that is not necessarily optimal, it still is a good basis for an efficient algorithm.

3.6. FEASIBLE CONTROLS

3.6.1. THE SAFE CONTROL CONCEPT

The coordination procedures presented in the previous sections ensure the feasibility of the control for the real system only after the iterations. Can we avoid an excessive violation of the real system constraints at any stage of the control iterations if we know the limit of the difference between the model and reality? We can if we are able to generate a control that is feasible for the real system on the basis of this limit.

Let us assume that the subsystem model equations contain parameters $\alpha_i \in \mathscr{A}_i$:

$$y_i = F_i(c_i, u_i, \alpha_i), \quad i \in \overline{1, N},$$

and that the models of the constraint relations contain parameters $\beta_i \in \mathscr{B}_i$:

$$G_i(c_i, u_i, \beta_i) \in S_i \subseteq \mathscr{S}_i, \qquad i \in \overline{1, N}.$$

We assume that the real system relations are the same, whereby the system parameters have some fixed values α_{*i}, β_{*i}. If we do not know α_{*i}, β_{*i}, we

264

cannot properly adjust α_i, β_i in the models. We shall, however, assume that we know the sets $\mathscr{A}_i$, $\mathscr{B}_i$, in which α_{*i}, β_{*i} are contained.

DEFINITION 1.   *We say that control $\bar{c}$ is a $(\bar{\alpha}, \bar{\beta})$-feasible control for the model if and only if the following relations are fulfilled:*

$$G(\bar{c}, \bar{u}, \bar{\beta}) \in S$$
$$(\bar{c}, \bar{u}) \in \mathscr{C}\mathscr{U}_0 \subset \mathscr{C} \times \mathscr{U} \tag{3.127}$$

where

$$\bar{\alpha} = (\bar{\alpha}_1, \ldots, \bar{\alpha}_N) \in \mathscr{A} = \mathscr{A}_1 \times \ldots \times \mathscr{A}_N,$$

$$\bar{\beta} = (\bar{\beta}_1, \ldots, \bar{\beta}_N) \in \mathscr{B} = \mathscr{B}_1 \times \ldots \times \mathscr{B}_N,$$

$$S = S_1 \times \ldots \times S_N, \quad \mathscr{C}\mathscr{U}_0 = \mathscr{C}\mathscr{U}_{01} \times \ldots \times \mathscr{C}\mathscr{U}_{0N}, \tag{3.128}$$

$$G(\bar{c}, \bar{u}, \bar{\beta}) = (G_1(\bar{c}_1, \bar{u}_1, \bar{\beta}_1), \ldots, G_N(\bar{c}_N, \bar{u}_N, \bar{\beta}_N))$$

and where $\bar{u}$ is the result of control $\bar{c}$ in the model of the interconnected system when variable α takes the value $\bar{\alpha}$ in set $\mathscr{A}$, that is, it satisfies the equation

$$\bar{u} = HF(\bar{c}, \bar{u}, \bar{\alpha}) \quad \text{giving} \quad \bar{u} = HK(\bar{c}, \bar{\alpha}). \tag{3.129}$$

Set $\mathscr{C}\mathscr{U}_0$ in the above definition may be a cube in which the values of (c, u) are bounded. From this definition it follows that control $\bar{c}$ satisfies all the local constraints, the subsystem equations, and the structure equation in the model when variables α, β take the values $\bar{\alpha}, \bar{\beta}$ in sets $\mathscr{A}$ and $\mathscr{B}$.

DEFINITION 2.   *We say that control $\bar{c}$ is a feasible control for the real system if and only if the relations (3.127), (3.128), and (3.129) are fulfilled with $\bar{\alpha}, \bar{\beta}$ equal to α_*, β_*.*

DEFINITION 3.   *We say that control $c_{\mathscr{A} \times \mathscr{B}}$ is a safe control if and only if it is (α, β)-feasible for the model for every $(\alpha, \beta) \in \mathscr{A} \times \mathscr{B}$.*

It is evident that safe control $c_{\mathscr{A} \times \mathscr{B}}$ is feasible for the real system because $(\alpha_*, \beta_*) \in \mathscr{A} \times \mathscr{B}$. So the feasible control for the real system could also be the safe control and can be found using only the mathematical model of the interconnected system. The following discussion of the existence of a safe control and methods for finding it is based on Brdyś (1975a, 1975b). We assume that $\mathscr{S}_i, \mathscr{U}_i, \mathscr{C}_i$ are Hilbert spaces; S_i is a closed, convex, positive cone; $\mathscr{C}\mathscr{U}_{0i}$ is a closed, convex, bounded set; $\mathscr{A}_i, \mathscr{B}_i$ are compact, topological Hausdorff spaces, and $i \in \overline{1, N}$.

Existence of a safe control

Let us introduce a scalar measure $\psi : \Pi_{\mathscr{C}}(\mathscr{C}\mathscr{U}_0) \times (\mathscr{A} \times \mathscr{B}) \times Z \to \mathbb{R}$, of the value of the constraint function in the model of the interconnected system:

$$\psi(c, \gamma, z) \triangleq \langle z, \bar{G}(c, \gamma) \rangle \tag{3.130}$$

where

$$\gamma = (\alpha, \beta),$$

$$Z \triangleq \{z \in \mathscr{S} : (z \in S^*) \wedge \|z\| \le 1\}, \qquad (3.131)$$

$$\bar{G}(c, \gamma) \triangleq G(c, HK(c, \alpha), \beta)$$

and where $\Pi_{\mathscr{C}}(\mathscr{C}\mathscr{U}_0)$ is a projection of set $\mathscr{C}\mathscr{U}_0$ on space $\mathscr{C}$.

THEOREM 3.21.   (*existence of a safe control*). *If we assume that*

1.   *Mapping $\bar{G}$ is convexlike (see Appendix B.9) on $\mathscr{A} \times \mathscr{B}$ and concavelike on $\Pi_{\mathscr{C}}(\mathscr{C}\mathscr{U}_0)$ with respect to S,*
2. $\forall(\gamma, z) \in (\mathscr{A} \times \mathscr{B}) \times Z$ *functional $\psi(\cdot, \gamma, z)$ is weakly upper semicontinuous on $\Pi_{\mathscr{C}}(\mathscr{C}\mathscr{U})$,*
3. $(\forall z_1, z_2 \in Z)(\forall c \in \Pi_{\mathscr{C}}(\mathscr{C}\mathscr{U}_0))(\forall t \in [0, 1])$

$$\inf_{\gamma \in \mathscr{A} \times \mathscr{B}} [t\psi(c, \gamma, z_1) + (1-t)\psi(c, \gamma, z_2)]$$

$$= t \inf_{\gamma \in \mathscr{A} \times \mathscr{B}} \psi(c, \gamma, z_1) + (1-t) \inf_{\gamma \in \mathscr{A} \times \mathscr{B}} \psi(c, \gamma, z_2),$$

4. $(\forall c \in \Pi_{\mathscr{C}}(\mathscr{C}\mathscr{U}_0))(\forall \alpha \in \mathscr{A})\ (c, HK(c, \alpha)) \in \mathscr{C}\mathscr{U}_0,$
5. $\forall(z, c) \in Z \times \Pi_{\mathscr{C}}(\mathscr{C}\mathscr{U}_0)$ *functional $\psi(c, \cdot, z)$ is lower semicontinuous on $\mathscr{A} \times \mathscr{B}$,*
6. $\forall(\alpha, \beta) \in \mathscr{A} \times \mathscr{B}$ *there exists a control that is (α, β)-feasible for the model*

then a safe control exists.

Proof. To simplify the notation we denote:

$$\inf_{z \in Z} = \inf_z, \qquad \inf_{\gamma \in \mathscr{A} \times \mathscr{B}} = \inf_\gamma, \qquad \sup_{c \in \Pi_{\mathscr{C}}(\mathscr{C}\mathscr{U}_0)} = \sup_c$$

From assumption 6 it follows that:

$$(\forall \gamma \in \mathscr{A} \times \mathscr{B})(\exists c \in \Pi_{\mathscr{C}}(\mathscr{C}\mathscr{U}_0))\ \bar{G}(c, \gamma) \in S,$$

which is equivalent to

$$(\forall \gamma \in \mathscr{A} \times \mathscr{B})(\exists c \in \Pi_{\mathscr{C}}(\mathscr{C}\mathscr{U}_0))(\forall z \in Z)\psi(c, \gamma, z) \ge 0$$

This implies that

$$\inf_\gamma \sup_c \inf_z \psi(c, \gamma, z) \ge 0.$$

266

Thus, taking into account the inequality

$$\forall \gamma \in \mathscr{A} \times \mathscr{B} \quad \inf_{z} \sup_{c} \psi(c, \gamma, z) \geq \sup_{c} \inf_{z} \psi(c, \gamma, z),$$

which is always true, and changing $\inf_\gamma \inf_z$ to $\inf_z \inf_\gamma$, we have

$$\inf_{z} \inf_{\gamma} \sup_{c} \psi(c, \gamma, z) \geq 0. \tag{3.132}$$

From assumptions 1 and 2, the functional $\psi(\cdot, \cdot, z)$ is convexlike on $\mathscr{A} \times \mathscr{B}$ and concavelike on $\Pi_{\mathscr{C}}(\mathscr{C}\mathscr{U}_0)$ for each $z \in Z$. $(\forall (\gamma, z) \in (\mathscr{A} \times \mathscr{B}) \times Z)$, $\psi(\cdot, \gamma, z)$ is weakly upper semicontinuous on $\Pi_{\mathscr{C}}(\mathscr{C}\mathscr{U}_0)$ and the set $\mathscr{C}\mathscr{U}_0$ is weakly compact. Therefore, by the minimax theorem (see Appendix B.9), the following inequality holds:

$$\forall z \in Z \quad \inf_{\gamma} \sup_{c} \psi(c, \gamma, z) = \sup_{c} \inf_{\gamma} \psi(c, \gamma, z). \tag{3.133}$$

Combining (3.132) and (3.133) we have

$$\inf_{z} \sup_{c} \inf_{\gamma} \psi(c, \gamma, z) \geq 0.$$

From assumptions 1, 2, and 3, the function $\inf_\gamma \psi(c, \gamma, z)$ is concavelike on $\Pi_{\mathscr{C}}(\mathscr{C}\mathscr{U}_0)$ and convexlike on Z, and for all $z \in Z$ the function $\Pi_{\mathscr{C}}(\mathscr{C}\mathscr{U}_0) \ni c \to \inf_\gamma \psi(c, \gamma, z) \in \mathbb{R}$ is weakly upper semicontinuous. Therefore, the minimax theorem can be applied again. By this theorem we have

$$\inf_{z} \sup_{c} \inf_{\gamma} \psi(c, \gamma, z) = \sup_{c} \inf_{z} \inf_{\gamma} \psi(c, \gamma, z). \tag{3.134}$$

Relations (3.133) and (3.134) imply that:

$$\sup_{c} \inf_{\gamma} \inf_{z} \psi(c, \gamma, z) \geq 0. \tag{3.135}$$

The function

$$\Pi_{\mathscr{C}}(\mathscr{C}\mathscr{U}_o) \ni c \to \inf_{\gamma} \inf_{z} \psi(c, \gamma, z) \in \mathbb{R}$$

is by assumption 2 weakly upper semicontinuous, and set $\Pi_{\mathscr{C}}(\mathscr{C}\mathscr{U}_0)$ is weakly compact. Therefore, on the basis of the Weierstrass theorem, we conclude that $\sup_c$ in (3.135) is achieved. Hence, we can write

$$(\exists c \in \Pi_{\mathscr{C}}(\mathscr{C}\mathscr{U}_0))(\forall \gamma \in \mathscr{A} \times \mathscr{B})(\forall z \in Z) \quad \psi(c, \gamma, z) \geq 0. \tag{3.136}$$

Thus, due to the definition of ψ and Z, there exists control $\bar{c}$ such that $\bar{G}(\bar{c}, \gamma) \in S$ for all $\gamma \in \mathscr{A} \times \mathscr{B}$. It then follows from assumption 4 that control $\bar{c}$ is safe. $\square$

Assumption 6 seems to be reasonable. It is intimately connected with the quality of the mathematical model. One mathematical model is better than another one when its sets $\mathscr{A}$ and $\mathscr{B}$ approximate more exactly the unknown constant values of the parameters in the process. Assumptions 1, 2, and 5 are commonly used in minimax problems. Satisfying assumption 3 mainly depends on the character of mapping $\bar{G}(c, \cdot)$, where $c \in \mathscr{C}\mathscr{U}_0$. We now give an example in which this assumption is satisfied.

Assume that $Z = \mathbb{R}^n$, $S = \{z \in \mathbb{R}^n : z \leq 0\}$, and $(\forall z \in Z)(\forall c \in \Pi_{\mathscr{C}}(\mathscr{C}\mathscr{U}_0))$ the number $\inf_\gamma \psi(c, \gamma, z)$ is finite. Let us also assume that mapping $\bar{G}$ has the following structure:

$$(\forall c \in \Pi_{\mathscr{C}}(\mathscr{C}\mathscr{U}_0))(\forall \gamma \in \mathscr{A} \times \mathscr{B}) \quad \bar{G}(c, \gamma) = (\bar{G}_1(c, \gamma_1), \ldots, \bar{G}_N(c, \gamma_N))$$

where

$$\mathscr{A} \times \mathscr{B} = (\mathscr{A}_1 \times \mathscr{B}_1) \times \ldots \times (\mathscr{A}_N \times \mathscr{B}_N), \qquad \gamma = ((\alpha_1, \beta_1), \ldots, (\alpha_N, \beta_N)),$$

$$\forall i \in \overline{1, N} \quad (\alpha_i, \beta_i) \in \mathscr{A}_i \times \mathscr{B}_i.$$

By simple computation we can show that assumption 3 is satisfied.

Assumption 2 in Theorem 3.21 is restrictive when $\mathscr{C} \times \mathscr{U}$ is an infinite-dimensional space; sometimes it can be replaced by a less restrictive assumption. Suppose that mapping $\bar{G}$ has the form:

$$(\forall c \in \Pi_{\mathscr{C}}(\mathscr{C}\mathscr{U}_0))(\forall \gamma \in \mathscr{A} \times \mathscr{B}) \quad \bar{G}(c, \gamma) = \Phi(\bar{G}_1(c), \bar{G}_2(\gamma)) \qquad (3.137)$$

where $\bar{G}_1 : \mathscr{C} \to \mathscr{S}$, $\bar{G}_2 : \mathscr{A} \times \mathscr{B} \to \mathscr{S}$, and $\Phi : \mathscr{S} \times \mathscr{S} \to \mathscr{S}$.

For the mappings $\bar{G}$ that belong to the class defined by (3.137), we can formulate the following theorem.

THEOREM 3.22. *If we assume that*

1. *Mapping $\bar{G}$ has the form given by (3.137), where the mappings $\Phi(z, \cdot)$ and $\Phi(\cdot, z)$ are weakly continuous on $\mathscr{S}$ for each $z \in \mathscr{S}$,*
2. *Assumptions 1, 3, 4, 5, and 6 of Theorem 3.21 are satisfied,*
3. *Sets $\bar{G}_1(\Pi_{\mathscr{C}}(\mathscr{C}\mathscr{U}_0))$ and $\bar{G}_2(\mathscr{A} \times \mathscr{B})$ are weakly compact,*

then a safe control exists.

The proof of this theorem is quite similar to the proof of Theorem 3.21 and is left to the reader.

From the above existence theorems it follows that the class of problems for which a safe control $c_{\mathscr{A} \times \mathscr{B}}$ exists is quite wide. We can see from these theorems (see assumptions 3 and 6 in Theorem 3.21) the intuitively obvious fact, that the safe control may not exist if sets $\mathscr{A}$, $\mathscr{B}$ are large, i.e., if the real parameters are uncertain. This means that our approach to the problem of finding a feasible control for a real process using only the model of the process makes sense.

Procedures for generating a safe control

The procedures for finding a safe control involve a search in the set $\mathcal{A} \times \mathcal{B}$, for safe values of model parameters α, β. The first of the two procedures that we will discuss follows from Theorem 3.23.

THEOREM 3.23 (*finding a safe control*). *If we assume that*

1. $\forall c \in \Pi_{\mathscr{C}}(\mathscr{C}\mathscr{U}_0)$ *the functional* $\psi(c, \cdot, \cdot)$ *is weakly lower semi-continuous,*
2. $\forall (\gamma, z) \in (\mathcal{A} \times \mathcal{B}) \times Z$ *the functional* $\psi(\cdot, \gamma, z)$ *is weakly upper continuous,*
3. $(\forall c \in \Pi_{\mathscr{C}}(\mathscr{C}\mathscr{U}_0))(\forall \alpha \in \mathcal{A})$ $(c, HK(c, \alpha)) \in \mathscr{C}\mathscr{U}_0,$
4. *Safe control* $c_{\mathcal{A} \times \mathcal{B}}$ *exists,*

then a solution of the problem

$$\max_{c \in \Pi_{\mathscr{C}}(\mathscr{C}\mathscr{U}_0)} \min_{(\gamma, z) \in (\mathcal{A} \times \mathcal{B}) \times Z} \psi(c, \gamma, z) \tag{3.138}$$

exists. Moreover if $(\bar{c}, \bar{\gamma}, \bar{z})$ *is the solution of this problem then* $\bar{c}$ *is a safe control.*

Proof. Set $\mathscr{C}\mathscr{U}_0$ is weakly compact and mapping $\Pi_{\mathscr{C}}$ is linear and continuous. Therefore, set $\Pi_{\mathscr{C}}(\mathscr{C}\mathscr{U}_0)$ is weakly compact. From assumption 2, functional

$$(\mathscr{C}\mathscr{U}_0) \ni c \rightarrow \min_{(\gamma, z) \in (\mathcal{A} \times \mathcal{B}) \times Z} \psi(c, \gamma, z) \in \mathbb{R}$$

is well defined and weakly upper semicontinuous on $\Pi_{\mathscr{C}}(\mathscr{C}\mathscr{U}_0)$. Thus, on the base of the Weierstrass theorem it follows that a solution of the problem exists.

Let us suppose that $(\bar{c}, \bar{\gamma}, \bar{z})$ solves the problem. The following inequality then holds:

$$\forall c \in \Pi_{\mathscr{C}}(\mathscr{C}\mathscr{U}_0). \min_{(\gamma, z) \in (\mathcal{A} \times \mathcal{B}) \times Z} \psi(\bar{c}, \gamma, z) \geq \min_{(\gamma, z) \in (\mathcal{A} \times \mathcal{B}) \times Z} \psi(c, \gamma, z) \tag{3.139}$$

On the other hand, by assumption 4, we know that

$$\min_{(\gamma, z) \in (\mathcal{A} \times \mathcal{B}) \times Z} \psi(c_{\mathcal{A} \times \mathcal{B}}, \gamma, z) \geq 0. \tag{3.140}$$

Because $c_{\mathcal{A}\times\mathcal{B}} \in \Pi_{\mathcal{C}}(\mathcal{C}\mathcal{U}_0)$ and as a result of inequalities (3.139) and (3.140), we have

$$\min_{(\gamma,\, z)\in(\mathcal{A}\times\mathcal{B})\times Z} \psi(\bar{c}, \gamma, z) \geq 0.$$

From the properties of Z and ψ, and from the previous inequality it follows that

$$\forall \gamma \in \mathcal{A} \times \mathcal{B} \quad \bar{G}(\bar{c}, \gamma) \in S. \tag{3.141}$$

Note that by assumption 3 $\forall \alpha \in \mathcal{A}$ $(\bar{c}, HK(\bar{c}, \alpha)) \in \mathcal{C}\mathcal{U}_0$. Hence, taking into account relation (3.141), we conclude that $\bar{c}$ is a safe control.

Theorem 3.23 implies that it is enough to solve problem (3.138) to find a feasible control. To solve this problem, a procedure for computing the value of mapping $\bar{G}$ is needed.

Because of the way $\bar{G}$ is defined in (3.13), we must solve the subsystem and structure equations simultaneously. If we make some further assumptions, a partial decomposition of the problem will be possible. As can be seen from (3.131), mapping $\bar{G}$ is the decomposition of mappings G and $H \circ K$. We will eliminate mapping $H \circ K$ by adding a modification component to the functional $z \circ G$ and constructing a functional $\tilde{\psi}$ as follows:

$$\tilde{\psi} : \Pi_{\mathcal{C}}(\mathcal{C}\mathcal{U}_0) \times (\mathcal{A} \times \mathcal{B}) \times Z \times \mathcal{U} \times \mathcal{U} \to \mathbb{R},$$
$$\tilde{\psi}(c, \gamma, z, u, \eta) \triangleq \langle z, G(c, u, \beta) \rangle + \langle \eta, u - HF(c, u, \alpha) \rangle. \tag{3.142}$$

The main property of this approach is given in the following theorem.

THEOREM 3.24 (*finding a safe control*). *If the assumptions of Theorem 3.23 hold and*

1. *$\forall c \in \Pi_{\mathcal{C}}(\mathcal{C}\mathcal{U}_0)$, $\gamma \in \mathcal{A} \times \mathcal{B}$, $z \in Z$, $\eta \in \mathcal{U}$ the functional $\tilde{\psi}(c, \gamma, z, \cdot, \eta)$ is concavelike and weakly upper semicontinuous on the set $\Pi_{\mathcal{U}}(\mathcal{C}\mathcal{U}_0)$ (where $\Pi_{\mathcal{U}}$ is a projection mapping from $\mathcal{C} \times \mathcal{U}$ on $\mathcal{U}$),*
2. *$(\forall c \in \Pi_{\mathcal{C}}(\mathcal{C}\mathcal{U}_0))(\forall \alpha \in \mathcal{A})(\exists r > 0)(\forall e \in \mathcal{U}, \|e\| = r)(\exists u \in \Pi_{\mathcal{U}}(\mathcal{C}\mathcal{U}_0)$ $e = u - HF(c, u, \alpha)$,*
3. *$(\forall \beta \in \mathcal{B})(\forall c \in \Pi_{\mathcal{C}}(\mathcal{C}\mathcal{U}_0))$ mapping $\Pi_{\mathcal{U}}(\mathcal{C}\mathcal{U}_0) \ni u \to G(c, u, \beta) \in \mathcal{S}$ is bounded,*
4. *$(\forall c \in \Pi_{\mathcal{C}}(\mathcal{C}\mathcal{U}_0))(\forall u \in \Pi_{\mathcal{U}}(\mathcal{C}\mathcal{U}_0))(\forall \eta \in \mathcal{U})$ the functional $\tilde{\psi}(c, \cdot, \cdot, u, \eta)$ is weakly lower semicontinuous,*

then a solution of the problem

$$\max_{c \in \Pi_{\mathcal{C}}(\mathcal{C}\mathcal{U}_0)} \min_{\eta \in \mathcal{U}} \min_{(\gamma,\, z)\in(\mathcal{A}\times\mathcal{B})\times Z} \max_{u \in \Pi_{\mathcal{U}}(\mathcal{C}\mathcal{U}_0)} \tilde{\psi}(c, \gamma, z, u, \eta) \tag{3.143}$$

exists. Moreover, if some element $(\bar{c}, \bar{\eta}, \bar{\gamma}, \bar{z}, \bar{u})$ solves this problem, then $\bar{c}$ is a safe control.

Proof. Let c, γ, and z be arbitrary points fixed in sets $\Pi_{\mathscr{C}}(\mathscr{C}\mathscr{U}_0)$, $\mathscr{A} \times \mathscr{B}$, and Z, respectively. Since set $\Pi_{\mathscr{U}}(\mathscr{C}\mathscr{U}_0)$ is weakly compact, from assumption 1 we have

$$\inf_{\eta \in \mathscr{U}} \max_{u \in \Pi_{\mathscr{U}}(\mathscr{C}\mathscr{U}_0)} \tilde{\psi}(c, \gamma, z, u, \eta) = \max_{u \in \Pi_{\mathscr{U}}(\mathscr{C}\mathscr{U}_0)} \inf_{\eta \in \mathscr{U}} \tilde{\psi}(c, \gamma, z, u, \eta) \qquad (3.144)$$

Let us note that if $u - HF(c, u, \alpha) \neq 0$ then $\inf_{\eta \in \mathscr{U}} \tilde{\psi} = -\infty$. From assumption 4 of Theorem 3.23, there exists $u \in \Pi_{\mathscr{U}}(\mathscr{C}\mathscr{U}_0)$ such that $u - HF(c, u, \alpha) = 0$. Thus

$$\max_{u \in \Pi_{\mathscr{U}}(\mathscr{C}\mathscr{U}_0)} \inf_{\eta \in \mathscr{U}} \tilde{\psi}(c, \gamma, z, u, \eta) = \langle z, \bar{G}(c, \gamma) \rangle. \qquad (3.145)$$

Combining (3.144) and (3.145), we have

$$\inf_{\eta \in \mathscr{U}} \max_{u \in \Pi_{\mathscr{U}}(\mathscr{C}\mathscr{U}_0)} \tilde{\psi}(c, \gamma, z, u, \eta) = \langle z, \bar{G}(c, \gamma) \rangle.$$

From assumptions 2 and 3, it can be proved that the infimum in the last equality is achieved. Therefore, we know that

$$\forall c \in \Pi_{\mathscr{C}}(\mathscr{C}\mathscr{U}_0) \quad \min_{(\gamma, z) \in (\mathscr{A} \times \mathscr{B}) \times Z} \min_{\eta \in \mathscr{U}} \max_{u \in \Pi_{\mathscr{U}}(\mathscr{C}\mathscr{U}_0)} \tilde{\psi}(c, \gamma, z, u, \eta)$$
$$= \min_{(\gamma, z) \in (\mathscr{A} \times \mathscr{B}) \times Z} \psi(c, \gamma, z).$$

Assumption 4 allows us to change $\min_{(\gamma, z)} \min_{\eta}$ to $\min_{\eta} \min_{(\gamma, z)}$. To complete the proof we need only apply Theorem 3.23. $\square$

It follows directly from Theorem 3.24 that a safe control can be found by solving problem (3.143).

Incorporation into the coordination process

The safe control concept can be incorporated into coordination structures, for example, those considered in section 3.4. In the following sections some possibilities for modifying these structures to achieve feasible control during the iterations will be presented. It should be noted that the methods of feasible control generation using the safe control concept that have been presented in this section can be used to solve another kind of problem, for example, the system start-up problem.

3.6.2. PRICE COORDINATION WITH PROJECTION ON THE SET OF SAFE CONTROLS

Let us denote by $C_{\mathscr{A} \times \mathscr{B}}$ the set of all safe controls. Consider the structure illustrated in Figure 3.12, which is a modification of the structure of IBMF. A feasible control generation (FCG) unit has been introduced. The local prob-

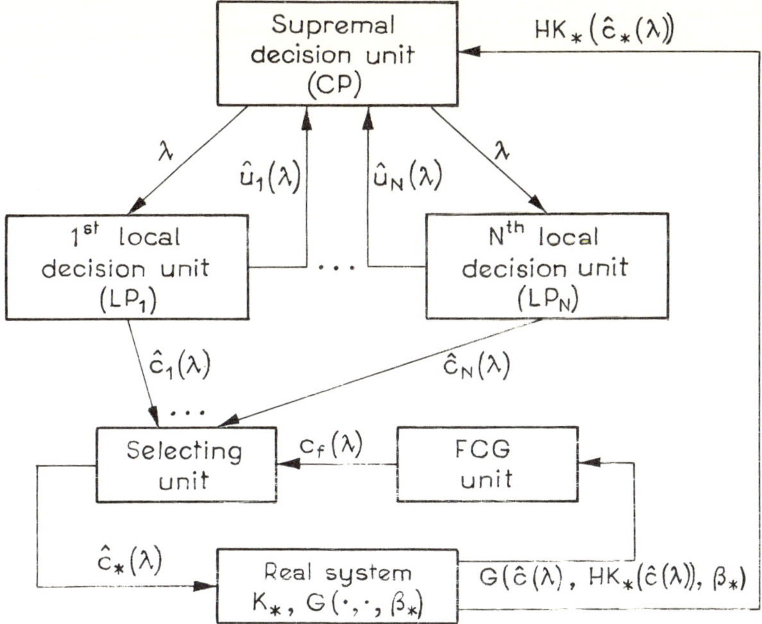

FIGURE 3.12   The structure of coordination by the price mechanism with projection on a set of safe controls.

lems and the coordination task are the same as in IBMF. The control is generated in the following way: first a model-based $\hat{c}(\lambda)$ is applied to the real system and the feasibility of this control is checked by measuring the value of $G(\hat{c}(\lambda), HK_*(\hat{c}(\lambda)), \beta_*)$. If $\hat{c}(\lambda)$ is feasible then the real interaction $HK_*(\hat{c}(\lambda))$ is sent to the coordinator. If not, then on the basis of information about the violated real constraints, the FCG unit generates a new control $c_f(\lambda)$ that is feasible. This $c_f(\lambda)$ is applied to the real system and interaction $HK_*(c_f(\lambda))$ is sent to the coordinator. This task is assigned in Figure 3.12 to the selecting unit that has input $\hat{c}(\lambda)$, $c_f(\lambda)$ and output $\hat{c}_*(\lambda)$, where

$$\hat{c}_*(\lambda) = \begin{cases} \hat{c}(\lambda) & \text{if } \hat{c}(\lambda) \text{ is feasible,} \\ c_f(\lambda) & \text{otherwise.} \end{cases}$$

In order to construct the FCG unit, one must know the set C_f such that $C_f \subseteq C_*$, where C_* is the set of all feasible controls for the real system. Taking into account the properties of the safe controls set $C_{\mathcal{A} \times \mathcal{B}}$, we can choose C_f in the form of $C_{\mathcal{A} \times \mathcal{B}}$.

Let us now consider the following way to implement the FCG unit:

$$c_f(\lambda) = \Pi_{C_{\mathcal{A} \times \mathcal{B}}}(\hat{c}(\lambda)), \tag{3.146}$$

where $\Pi_{C_{\mathscr{A}\times\mathscr{B}}}$ is the projection operator on the set $C_{\mathscr{A}\times\mathscr{B}}$. In this way a new control structure has been obtained. Notice that the coordination task CP has the following form:

Find $\tilde{\lambda} = (\tilde{\lambda}_1, \ldots, \tilde{\lambda}_N)$ such that

$$\hat{u}(\tilde{\lambda}) = HK_*(\hat{c}_*(\tilde{\lambda})). \tag{3.147}$$

The coordinator problem in the considered control structure looks like (3.34) formulated for IBMF. In reality, however, there is an essential difference between these two formulations: the IBMF formulation has no protective mechanism like the selecting and the FCG units. Thanks to this mechanism, only that interaction value that is obtained in the real system when the feasible control is applied is sent to the coordinator. It is obvious, though, that if the real subsystem constraints are known, that is, if $\mathscr{B} = \{\beta_*\}$, then each solution obtained using IBMF is also a solution of the coordinator problem (3.147). However, if the constraint operators G_i, $i \in \overline{1, N}$, are really dependent on the parameters β_i, $i \in \overline{1, N}$, then the controls generated during coordination by IBMF may be infeasible for the real system, and the control obtained at the end of the coordination may also be infeasible. This is mainly the case in which the safe controls with the price mechanism should be applied.

We now consider (a) the existence of $\tilde{\lambda}$, (b) the suboptimality of $\hat{c}_*(\tilde{\lambda})$, and (c) coordination strategies. The sufficient conditions for the existence of $\tilde{\lambda}$ can be derived as they were for IBMF, and the task is left to the reader. With regard to the suboptimality of $\hat{c}_*(\tilde{\lambda})$, quantitative estimates of the difference between the optimal value of the real system performance and the value of this performance taken at $\hat{c}_*(\tilde{\lambda})$ can be obtained, but the detailed analysis is omitted here.

Let us consider the problem of finding suitable coordination strategies for solving the coordinator task (3.147). In terms of the mathematics involved, we must find numerical methods for solving an operator equation of the form

$$R_*^{\mathrm{p}}(\lambda) \triangleq \hat{u}(\lambda) - HK_*(\hat{c}_*(\lambda)) = 0. \tag{3.148}$$

The upper index p is used to stress an existence of the protection mechanism in the control structure. It should be noted that even when functions Q_i, F_i, F_{*i}, G_i, $i \in \overline{1, N}$, have reasonable differentiability properties, the operator R_*^{p} may not be differentiable. Thus, the direct application of Newton's algorithm for solving equation (3.148) is not possible. However, the construction of Newton's algorithm suggests that we can try to apply the following coordination strategy:

$$\lambda^{k+1} = \lambda^k - \varepsilon[R_\lambda'(\lambda^k)]^{-1} R_*^{\mathrm{p}}(\lambda^k), \tag{3.149}$$

where R is a certain differentiable approximation of R_*^p, R_λ' is a Fréchet derivative of R and ε is a positive real number. The following general theorem (Zinchenko 1973) shows that such an approach makes sense.

THEOREM 3.25. *Consider an operator* $P: M \to X_2$, *where* M *is a subset of Banach space* X_1, *and* X_2 *is also a Banach space. Consider an operator* $P_1: M \to X_2$ *that is Fréchet differentiable on* M. *Let us assume that the operators* P_{1x}' *and* $P - P_1$ *satisfy the Lipschitz condition on* M *with appropriate constants* n *and* n_1. *Let us consider the following iterative process:*

$$x^{k+1} = x^k - L^k(x^k) \tag{3.150}$$

where L^k *are operators from* X_1 *into* X_1 *and* $x^0 \in M$. *If we assume that the following conditions are satisfied on* M:

$$\|P(x) - P_{1x}'(x)L^k(x)\| \le \gamma_1 \|P(x)\|,$$

$$\|L^k(x)\| \le \gamma \|P(x)\|,$$

$$\|P(x^0)\| \le \eta \quad \text{and} \quad h \triangleq \gamma_1 + n_1\gamma + \frac{n\gamma^2}{2}\eta < 1,$$

$$\bar{K}(x^0; r) \triangleq \{x \in X_1 : \|x - x^0\| \le r\} \subseteq M$$

where

$$r = \frac{\gamma\eta}{1 - h}$$

then in the closed ball $\bar{K}(x^0; r)$ *there exists a solution* x^* *of the equation* $P(x) = 0$ *and the sequence* $\{x^k\}$ *generated by formula* (3.150) *tends to this solution. The following estimates can also be found:*

$$\|x^k - x^*\| \le \frac{2}{n\gamma} \cdot \frac{h^{2^k}}{1 - h^{2^k}} \quad \text{if} \quad \gamma + n_1\gamma = 0$$

and

$$\|x^k - x^*\| \le rh^k \quad \text{if} \quad \gamma + n_1 n \ne 0.$$

By using Theorem 3.25 we can formulate the satisfactory conditions under which the choice of operator R as the approximation of R_*^p is a proper one. Operator R will be linear and bounded with bounded inverse R^{-1}.

THEOREM 3.26. *If we assume that*

1. *There exists neighborhood* $\Omega(\lambda^0)$ *of* λ^0 *such that control* $\hat{c}(\lambda)$ *is (a) feasible for the real system* $\forall \lambda \in \Omega(\lambda^0)$, *or (b) infeasible for the real system* $\forall \lambda \in \Omega(\lambda^0)$,

2. *Mapping $\hat{c}(\cdot)$ satisfies the Lipschitz condition on $\Omega(\lambda^0)$ with constant k_c,*

3. *Safe feasible set $C_{\mathcal{A}\times\mathcal{B}}$ is convex and closed,*

4. *Operator R is linear, bounded, and has a bounded inverse R^{-1}, R and*

$$0 < \varepsilon < \frac{2(1 - n_1 \|R^{-1}\|)}{\|R\| \|R^{-1}\|^2 \eta}$$

where $\eta = \|R_*^{\mathrm{p}}(\lambda^0)\|$ and n_1 is the Lipschitz constant of $R_*^{\mathrm{p}} - R$,

5. *$\bar{K}(\lambda^0; r(\varepsilon)) \subseteq \Omega(\lambda^0)$*

where

$$r(\varepsilon) = \frac{\|R^{-1}\| \eta}{1 - n_1 \|R^{-1}\| - \frac{1}{2} \|R\| \|R^{-1}\|^2 \varepsilon \eta},$$

then in $\bar{K}(\lambda^0; r(\varepsilon))$ there exists solution $\tilde{\lambda}$ of the coordinator problem and sequence $\{\lambda^k\}$ generated by formula (3.149) tends to $\tilde{\lambda}$.

Proof. Let us suppose that $\hat{c}(\lambda)$ is infeasible. Assumption 1 implies that operator R_*^{p} can be expressed on $\Omega(\lambda^0)$ as the combination of operators $\hat{c}(\cdot)$ and $\Pi_{C_{\mathcal{A}\times\mathcal{B}}}$. Assumption 3 implies that operator $\Pi_{C_{\mathcal{A}\times\mathcal{B}}}$ is well defined. Since $\Pi_{C_{\mathcal{A}\times\mathcal{B}}}$ satisfies the Lipschitz condition with a constant no greater than one, we conclude, taking into account assumption 2, that R_*^{p} satisfies the Lipschitz condition with constant no greater than k_c. This implies that $R_*^{\mathrm{p}} - R$ also satisfies the Lipschitz condition with a constant no greater than $k_c + \|R\|$.

Let us now compute constants γ_1, γ, and n from Theorem 3.25. Since

$$\|R_*^{\mathrm{p}}(\lambda) - \varepsilon R R^{-1} R_*^{\mathrm{p}}(\lambda)\| = (1 - \varepsilon) \|R_*^{\mathrm{p}}(\lambda)\|,$$

then $\gamma_1 = 1 - \varepsilon$. Since

$$\|L^k(\lambda)\| = \|\varepsilon R^{-1} R_*^{\mathrm{p}}(\lambda)\| \leq \varepsilon \|R^{-1}\| \|R_*^{\mathrm{p}}(\lambda)\|,$$

it follows that

$$\gamma = \varepsilon \|R^{-1}\|.$$

Since R is linear and bounded, $n = \|R\|$. The theorem follows from assumptions 4 and 5 and direct application of Theorem 3.25.

The proof for feasible $\hat{c}(\lambda)$ is similar and therefore omitted. $\square$

Notice that if we want to check assumption 4 of Theorem 3.26 for a given operator R we must first compute constant n_1, given constant k_c. The simplest way to do this is to set $n_1 = k_c + \|R\|$, but this is not the best way. To show this, let us consider the number $1 - n_1 \|R^{-1}\|$ (see assumption 4). It follows that

$$1 - n_1 \|R^{-1}\| = 1 - (k_c + \|R\|) \|R^{-1}\| = 1 - k_c \|R^{-1}\| - \|R\| \|R^{-1}\|.$$

Hence, owing to the inequality $\|R\|\,\|R^{-1}\| \geq 1$, we find that $1 - n_1\|R^{-1}\| \leq 0$ and $\varepsilon \leq 0$, which violates assumption 4. Thus, the computation of n_1 must be done very carefully.

Assumption 1 limits us to some special cases. But if the assumption is not satisfied, then R_*^p may even be a discontinuous operator and R cannot be linear. In this case, R must be constructed with great care.

Regarding assumption 2, let as observe that if the performance index is quadratic with a positively defined operator, and mappings F_i, $i \in \overline{1, N}$, are linear, then mapping $\hat{c}(\,\cdot\,)$ satisfies the Lipschitz condition. This follows directly from the projection theorem (Luenberger 1969).

Simulation results

Consider the steady-state system in Figure 3.13. The subsystem models are as follows:

Subsystem 1

$$y_1 = u_{11} + u_{12} + c_1 + \alpha_1(c_1)^2,$$
$$Q_1(c_1, u_1) = (c_1 - 5)^2 + (u_{11} - 4.4)^2 + (u_{12} - 3.65)^2,$$
$$CU_1 = \{(c_1, u_1) \in \mathbb{R}^3 : \beta_{11} \leq c_1 \leq \beta_{12}\},$$
$$\mathscr{A}_1 = \{\alpha_1 \in \mathbb{R}^1 : |\alpha_1| \leq 1\},$$
$$\mathscr{B}_1 = \{(\beta_{11}, \beta_{12}) \in \mathbb{R}^2 : -5 \leq \beta_{11} \leq -1.5,\ 1 \leq \beta_{12} \leq 3\}.$$

Subsystem 2

$$y_{21} = c_{21} - c_{22} + \alpha_{21}(c_{21} - 3.8)^2,$$
$$y_{22} = -u_2 + 2c_{21} + c_{22} + \alpha_{22}(c_{21} - 2.8)^2,$$
$$Q_2(c_2, u_2) = (c_{21} - 10)^2 + 2(c_{22} - 10)^2 + (u_2 - 8.95)^2,$$
$$CU_2 = \{(c_2, u_2) \in \mathbb{R}^3 : c_{21} + c_{22} \leq \beta_2\},$$
$$\mathscr{A}_2 = \{(\alpha_{21}, \alpha_{22}) \in \mathbb{R}^2 : |\alpha_{21}| \leq 0.2,\ |\alpha_{22}| \leq 0.5\},$$
$$\mathscr{B}_2 = \{\beta_2 \in \mathbb{R}^1 : 5 \leq \beta_2 \leq 8\}.$$

The real subsystem equations and constraints have the form:

$$y_1 = u_{11} + u_{12} + c_1 + 0.4(c_1)^2,$$
$$y_{21} = c_{21} - c_{22} + 0.1(c_{21} - 3.8)^2,$$
$$y_{22} = -u_2 + 2c_{21} + c_{22} + 0.25(c_{21} - 2.8)^2,$$
$$-2 \leq c_1 \leq 2,$$
$$c_{21} + c_{22} \leq 6.$$

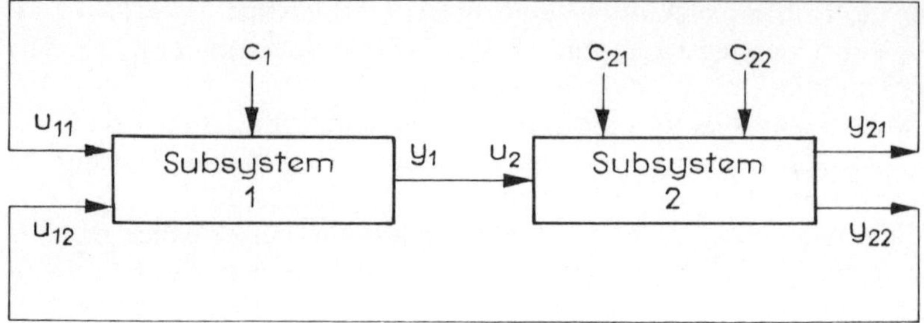

FIGURE 3.13   The structure of the simulated system.

From the above relations it follows that the values of parameters α and β in the real system are:

$$\alpha_{*1}=0.4, \qquad \alpha_{*21}=0.1, \qquad \alpha_{*22}=0.25,$$

$$\beta_{*11}=-2, \qquad \beta_{*12}=2, \qquad \beta_{*2}=6.$$

The set of safe controls has the form:

$$C_{\mathscr{A}\times\mathscr{B}} = \{(c_1, c_2)\in\mathbb{R}^3 : -1.5\leq c_1\leq 1,\ c_{21}+c_{22}\leq 5\}.$$

The operator R has been chosen in the form of

$$R = \begin{bmatrix} -1 & 0 & 0 \\ 0 & -1 & 0 \\ 0 & 0 & -1 \end{bmatrix}.$$

Two possibilities for coefficient ε were tested: $\varepsilon=0.5$ and $\varepsilon=0.2$. In both cases the coordination algorithm was started from $\lambda^0=(0,0,0)$. In all iterations of the algorithm the controls $\hat{c}(\lambda^k)$ generated by the local decisions units were outside the feasible set of the real system. Therefore, the FCG unit intervened in each algorithm iteration. The results of the simulation are given in Table 3.3. k is the iteration number and corresponds to changes in λ. Q^k is the value of the real system performance index after iteration k, and $\|R^p_*(\lambda^k)\|$ is the norm of the discoordination. The final values for the price, controls, and interactions were:

$$\tilde{\lambda} = (7.1987,\ 4.2135,\ 2.1659),$$

$$\hat{c}_1(\tilde{\lambda}) = 2.5000, \qquad \hat{c}_{*1}(\tilde{\lambda}) = c_{f1}(\tilde{\lambda}) = 1.0000,$$

$$\hat{c}_2(\tilde{\lambda}) = (4.4351,\ 2.5648),$$

$$\hat{c}_{*2}(\tilde{\lambda}) = c_{f2}(\tilde{\lambda}) = (3.4351,\ 1.5648),$$

$$u_{*1}(\tilde{\lambda}) = (1.8836,\ 2.6262),$$

$$u_{*2}(\tilde{\lambda}) = 5.7602.$$

TABLE 3.3 Simulation Results for Price Coordination with Projection on the Set of Safe Controls trols

k	Q^k	$\|R_*^p(\lambda^k)\|$
$\varepsilon = 0.5$		
1	257.4	9.5448
2	237.1	7.0618
3	223.4	4.7698
4	218.9	3.3367
5	220.6	3.0532
10	218.95	0.7248
20	218.84	0.0843
40	218.72	0.004
46	218.64	0.00009
$\varepsilon = 0.2$		
1	257.4	9.5448
2	249.0	8.5449
3	241.3	7.5444
4	234.5	6.5534
5	228.8	5.5969
10	218.9	2.4908
20	219.4	1.0185
40	218.95	0.0809
46	218.92	0.0303
95	218.68	0.0009

3.6.3. PRICE COORDINATION WITH FEASIBLE SET IDENTIFICATION

Let us return to the optimal control problem formulation. The problem of providing optimal control is one of finding all $\hat{c}$ such that:

1. There exists $u = HK_*(\hat{c})$ that satisfies the real system equations and the system structure equation,
2. $(\hat{c}, HK_*(\hat{c})) \in CU_* \subset \mathscr{C} \times \mathscr{U}$ where CU_* is the feasible set of the real system,
3. For all c satisfying 1 and 2 we have

$$Q(\hat{c}, HK_*(\hat{c})) \le Q(c, HK_*(c)).$$

This problem cannot be solved because we do not have complete knowledge of CU_* and F_*. Therefore, we will search for suboptimal solutions which, however, must satisfy the real constraints in condition 2. If set CU_*

has the form

$$((c, u) \in CU_*) \Rightarrow (\forall i \in \overline{1, N} \quad (c_i, u_i) \in CU_i)$$

and sets CU_i are completely known, then IBMF would generate a feasible suboptimal control. There are, however, practical control problems that cause the violation of the assumption. For example, as mentioned in section 3.4, the constraint relations may contain y explicitly. To express the real system constraints with only c and u, one should use the approximate model

$$y = F(c, u)$$

that gives an approximation CU to the set CU_*. Sometimes the so-called safe feasible set can be defined off-line using only the model. If the difference between this set and CU_* is too large, then it is not possible to achieve a good enough control. However, by long observation of the process, it is sometimes possible to decrease the uncertainty and, consequently, increase the accuracy of the feasible sets. Especially suited to such a procedure are systems that differ from their models only in the value of certain parameters. We can increase the accuracy of the safe sets by decreasing the sets $\mathcal{A}$ and $\mathcal{B}$.

We can now formulate a method of price coordination with feasible set identification. In particular, we shall assume that a sequence of safe feasible sets is generated by an FSI (Feasible Set Identification) unit and the control iterations based on the information from the FSI unit. First, independent local problems corresponding to the subsystems are formulated. The performance indices in these local problems, as they were with IBMF (see section 3.4), are modified by terms dependent on price λ, which is a coordination variable. The price λ is an element of Hilbert space $\mathcal{U}$, and the sets on which the local indices are minimized may change in the coordination process. These sets are generated by the FSI unit.

Let us assume that the FSI unit produces a family Ξ of the sets X^k, $\Xi = \{X^k\}_{k=0}^{\infty}$, satisfying the following relations (a) $X^0 \subseteq X^1 \subseteq \ldots X^k \subseteq \ldots \subseteq X_{\max} \subseteq CU_*$, and (b) $X^k = \times_{i=1}^{N} X_i^k$ for each $k = 0, 1, \ldots$, where $X_i^k \subset \mathcal{C}_i \times \mathcal{U}_i$. k is the iteration number and i denotes the number of the subsystem. Local problems LP_i^k in iteration k have the following form:

For a given value $\lambda^k \in \mathcal{U}$ of a coordination variable (price) λ and set X_i^k, find both control and interaction

$$(\hat{c}_i(\lambda^k, X_i^k), \hat{u}_i(\lambda^k, X_i^k)) = \arg \min_{X_i^k} Q_{\text{modi}}(\,\cdot\,, \cdot\,, \lambda^k) \qquad (3.151)$$

where

$$\forall (c_i, u_i) \in \mathcal{C}_i \times \mathcal{U}_i \quad Q_{\text{modi}}(c_i, u_i, \lambda) \triangleq Q_i(c_i, u_i) + \langle \lambda_i, u_i \rangle - \sum_{j=1}^{N} \langle \lambda_j, H_{ji} F_i(c_i, u_i) \rangle$$

and $\lambda = (\lambda_1, \ldots, \lambda_N)$.

In compact form they may be written as:

$$\min_{x \in X^k} [Q(x) + \langle \lambda^k, P(x) \rangle] \qquad (3.152)$$

where $x \in X \triangleq \mathscr{C} \times \mathscr{U}$ and $P : X \to \mathscr{U}$, $P(x) \triangleq u - HF(c, u)$.

Let us denote the infimal problem solution of (3.152) by $\hat{x}(\lambda^k, X^k)$, $\hat{x} : \mathscr{U} \times \Xi \to X$.

Coordinator problem CP has the following form:

Generate a sequence $\{\lambda^k\}_{k=0}^{\infty}$ such that

$$\lim_{k \to \infty} [\hat{u}(\lambda^k, X^k) - HK_*(\hat{c}(\lambda^k, X^k))] = 0. \qquad (3.153)$$

Let us define a discoordination in iteration k as

$$R_*(\hat{x}(\lambda^k, X^k)) \triangleq \hat{u}(\lambda^k, X^k) - HK_*(\hat{c}(\lambda^k, X^k)). \qquad (3.154)$$

We can say that the coordinator problem is to generate a sequence of prices $\{\lambda^k\}_{k=0}^{\infty}$ such that the discoordination sequence $\{R_*(\hat{x}(\lambda^k, X^k))\}_{k=0}^{\infty}$ tends to zero. The above method was first formulated by Brdyś, Ruszczyński, and Szymanowski, who also investigated its practical aspects (Szymanowski *et al.* 1976, 1977). The coordination structure of this method is shown in Figure 3.14. Note that due to the iterative nature of the FSI unit, there is no coordination condition in the operator equation form, as there is in IBMF. This is the main difference between this method and other methods using the price mechanism, which where described in Chapter 2 and section 3.4 of this chapter.

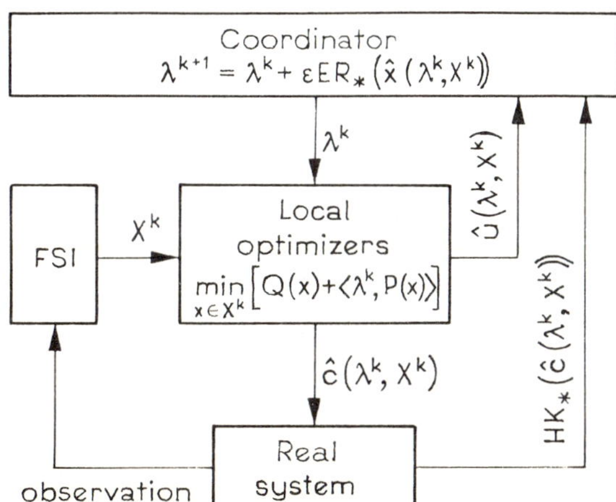

FIGURE 3.14   The structure of coordination by the price mechanism with feasible set identification.

280

Sequence λ^k satisfying (3.153) is generated with the following coordination algorithm or strategy:

Step 1. Based on the mathematical model, solve the equation

$$D(\hat{x}(\lambda, X^0)) = 0 \tag{3.155}$$

where $D: X \to \mathcal{U}$, $D(c, u) \triangleq u - HK(c)$. Use price λ^0 obtained in this way for finding $x^0 = \hat{x}(\lambda^0, X^0)$ and set

$$D_0 \triangleq D'_x(x^0), \qquad P_0 \triangleq P'_x(x^0). \tag{3.155a}$$

Step 2. Select successive prices according to the formula

$$\lambda^{k+1} = \lambda^k + \varepsilon ER_*(\hat{x}(\lambda^k, X^k)) \tag{3.156}$$

where $E = (D_0 A P_0^*)^{-1}$. A is a self-conjugated operator positively defined on $\mathcal{R}(P_0^*)$, i.e.,

$$(\exists m_A > 0, M_A > 0)(\forall x \in \mathcal{R}(P_0^*)) \quad m_A \|x\|^2 \leq \langle x, Ax \rangle \leq M_A \|x\|^2. \tag{3.157}$$

The number ε in the algorithm is assumed to be sufficiently small and positive. Note that step 1 of the above algorithm is equivalent to solving the model-based optimal control problem using IBM (see section 2.3 of Chapter 2). The operator E used in (3.156) in the form of $(D_0 A P_0)^{-1}$ was first proposed by Ruszczyński (1976).

A convergence analysis will now be done under appropriate assumptions related to the mathematical model, the properties of the FSI unit, the properties of the infimal problem solution, and the differences between the model and the mathematical system description represented by the operator $\tilde{D} \triangleq D - R_*$.

Convergence analysis

We have to show first that the algorithm is well defined. Let us assume the following:

(A1) Functions F and Q are Fréchet continuously differentiable twice with respect to both variables and bounded on CU_*.
(A2) For each $(c, u) \in CU_*$, operator $I_u - HF'_u(c, u)$ has a bounded inverse operator.

PROPOSITION 3.27. *If* (A2) *is satisfied, then*

(1) $\quad \mathcal{R}(P_0) = \mathcal{U}$

(2) $\quad \overline{\mathcal{R}(P_0^*)} = \mathcal{R}(P_0^*),$

(3) $\quad \mathcal{N}(P_0^*) = \{0\}.$

Proof. By (A2), $\mathcal{R}(I_u - HF'_u(c^0, u^0)) = \mathcal{U}$. It is also true that $P_0 = [-HF'_c(c^0, u^0), I_u - HF'_u(c^0, u^0]$. Thus, $\mathcal{R}(P_0) = \mathcal{U}$. Part (2) of the thesis follows from (1) and from the closed range theorem (Dunford and Schwartz 1958). Because $\mathcal{N}(P_0^*) = \mathcal{R}(P_0)^\perp$ (Luenberger 1969), by (1) we have that $\mathcal{R}(P_0)^\perp = \{0\}$. $\quad\square$

LEMMA 3.12. *If we assume that assumptions* (A1) *and* (A2) *are satisfied, then operator* $(D_0 A P_0^*)^{-1}$ *exists and is bounded.*

Proof. We will show first that

$$\mathcal{N}(D_0) = \mathcal{N}(P_0). \tag{3.158}$$

By the implicit function theorem we know that

$$\frac{dHK}{dc}(c^0) = (I_u - HF'_u(c^0, HK(c^0)))^{-1} HF'_c(c^0, HK(c^0)).$$

Thus, $D_0 = (I_{\mathcal{U}} - HF'_u(c^0, HK(c^0)))^{-1} P_0$ and relation (3.158) is true. Let us suppose that $D_0 A P_0^* u = 0$ for some $u \in \mathcal{U}$. This means that $A P_0^* u \in \mathcal{N}(D_0)$ and by (3.158), $A P_0^* u \in \mathcal{N}(P_0)$. Thus, $P_0 A P_0^* u = 0$.

From the last equality it follows that

$$\langle P_0^* u, A P_0^* u \rangle = \langle u, P_0 A P_0^* \rangle = 0,$$

which implies that $P_0^* u = 0$ since $A^* = A$. From Proposition 3.27, $u = 0$, so operator $D_0 A P_0^*$ has to have an inverse. It is also linear and continuous since it is a composition of linear and continuous operators.

Now we will prove that $D_0 A P_0^*$ is the operator on $\mathcal{U}$. Let us denote by Π_S the orthogonal projection operator from $\mathcal{C} \times \mathcal{U}$ on linear subspace $S \triangleq \mathcal{R}(P_0^*)$. Let us consider operator $\Pi_S A : S \to S$. If $x \in S$ then

$$\langle x, Ax \rangle = \langle \Pi_S x, Ax \rangle = \langle x, \Pi_S Ax \rangle$$

because Π_S is a self-conjugate operator. From this equality and inequality (3.157) it follows that

$$m_A \|x\|^2 \leq \langle x, \Pi_S Ax \rangle \leq \|x\| \|\Pi_S Ax\|$$

and

$$\|\Pi_S Ax\| \geq m_A \|x\|.$$

Therefore, set $\mathcal{R}(\Pi_S A)$ is closed (Dunford and Schwartz 1958). Let us take $z \in S$ such that $z \perp \mathcal{R}(\Pi_S A)$. Then

$$0 = \langle z, \Pi_S Az \rangle = \langle \Pi_S z, Az \rangle = \langle z, Az \rangle \geq m_A \|z\|^2$$

282

which implies that $z = 0$. Therefore

$$\mathcal{R}(\Pi_S A) = S. \tag{3.159}$$

By Proposition 3.27

$$\mathcal{N}(D_0) = \mathcal{N}(P_0) \quad \text{and} \quad \mathcal{N}(P_0)^\perp = \mathcal{R}(P_0^*),$$

thus

$$\mathcal{N}(D_0)^\perp = \mathcal{R}(P_0^*)$$

and

$$X = \mathcal{N}(D_0) \oplus S. \tag{3.160}$$

From (3.159), (3.160), and Proposition 3.27, it follows that operators

$$P_0^* : \mathcal{U} \to S,$$

$$\Pi_S A : S \to S, \tag{3.161}$$

$$D_0 : S \to \mathcal{U}$$

are onto. Equality (3.160) implies that for all $x \in S$ we have $D_0 A x = D_0 \Pi_S A x$, which, with (3.161), implies that $D_0 A P_0^* : \mathcal{U} \to \mathcal{U}$ is onto.

Summarizing the above results, it has been proved that operator $D_0 A P_0^*$ is linear, bounded, and onto. Hence, from Banach's inverse operator theorem, $(D_0 A P_0^*)^{-1}$ exists and is bounded. $\square$

To guarantee that the algorithm is well defined, we assume that

(A3)   There exist convex neighborhoods $\Omega(\lambda^0)$ of λ^0 and $\omega(x^0)$ of x^0 such that for each $\lambda \in \Omega(\lambda^0)$ and $X^k \in \Xi$ there exist unique infimal problem solutions $\hat{x}(\lambda, X^k)$ and $\hat{x}(\lambda, X^k) \in \omega(x^0)$.

Because of assumption (A3), the following discussion will be limited to prices from the neighborhood $\Omega(\lambda^0)$ of the initial price λ^0. It is obvious that if function $Q_{\mathrm{mod}}(\cdot, \lambda)$ is strictly convex and continuous on CU_* for each λ from a certain neighborhood Ω_1 of λ^0, and the sets X^k are convex, closed, and bounded, we can take $\Omega(\lambda^0) = \Omega_1$ and $\omega(x^0) = \mathscr{C} \times \mathcal{U}$. The following proposition shows that very often there is an $\omega(x^0)$ which is different from the whole space $\mathscr{C} \times \mathcal{U}$.

PROPOSITION 3.28.   *If there exist a set ω and a real positive number d such that*

$$(\forall X^k \in \Xi)(\forall x \in X^k - \omega) \quad Q(x) + \langle \lambda^0, P(x) \rangle \geq Q(x^0) + \langle \lambda^0, P(x^0) \rangle + d.$$

then there exists neighborhood Ω_1 of λ^0 such that

$$(\forall X^k \in \Xi)(\forall \lambda \in \Omega_1(\lambda^0)) \quad \hat{x}(\lambda, X^k) \in \omega$$

if $\hat{x}(\lambda, X^k)$ exists.

The proof is given in Appendix B, section B.10.

To derive the desired properties of the function $\hat{x}(\,\cdot\,, X^k)$, we assume that

$$(A4) \qquad (\exists v > 0)(\forall x \in \omega(x^0))(\forall \lambda \in \Omega(\lambda^0)) \quad Q''_{xx}(x) + \langle \lambda, P(x) \rangle''_{xx} \geq vI.$$

LEMMA 3.13. *If assumptions (A1), (A3), and (A4) are satisfied, then for all $X^k \in \Xi$, the function $\Omega(\lambda^0) \ni \lambda \to \hat{x}(\lambda, X^k) \in \omega(x^0)$ is continuous.*

The proof is given in Appendix B, section B.11.

(A5) Each of the sets X_i^k, $i \in \overline{1, N}$, and $k = 0, 1, \ldots$, is expressed by a finite number of functional inequalities, i.e.,

$$X_i^k = \{(c_i, u_i) \in \mathscr{C}_i \times \mathscr{U}_i : h_j^k(c_i, u_i) \leq 0, j \in J_i^k\}$$

where $h_j^k : \mathscr{C}_i \times \mathscr{U}_i \to \mathbb{R}$ are continuously twice Fréchet differentiable convex functionals and J_i^k is a finite set of indices.

We let λ_1, $\lambda_2 \in \Omega(\lambda^0)$ and set $\lambda_t = t\lambda_2 + (1-t)\lambda_1$ for $0 \leq t \leq 1$. Let us consider the behavior of $\hat{x}(\lambda_t, X^k)$ as a function of real variable t. We define for $\lambda \in \Omega(\lambda^0)$ the set

$$I_0^k(\lambda) \triangleq \{i : h_i^k(\hat{x}(\lambda, X^k)) = 0\}. \tag{3.162}$$

Let $I_0^k = \{i_1, i_2, \ldots, i_l\}$. We define the operator $H^k : \Omega(\lambda^0) \times X \to \mathbb{R}$ as follows:

$$H^k(\lambda, x) \triangleq (h_{i_1}^k(x), h_{i_2}^k(x), \ldots, h_{i_l}^k(x)). \tag{3.163}$$

The next assumption is typical for such a problem.

(A6) For $\lambda \in \Omega(\lambda^0)$, the derivative $H_x^{k'}$ of the operator H^k with respect to x taken at the point $(\lambda, \hat{x}(\lambda, X^k))$ is a surjection, where $k = 0, 1, \ldots,$.

In other words, we assume that at $\hat{x}(\lambda, X^k)$ the gradients of the active constraints are linearly independent. Assumptions (A1)–(A6) prove the existence and uniqueness of Lagrange multipliers for the infimal problem (Luenberger 1969).

Let $\mu \in \mathbb{R}^l$ be these multipliers. We introduce, for fixed $\lambda \in \Omega(\lambda^0)$, and $X^k \in \Xi$, the operator $W(\lambda, X^k) : X \to X$

$$W(\lambda, X^k) \triangleq Q''_{xx}(\hat{x}(\lambda, X_{ii}^k)) + \langle \lambda, P(\hat{x}(\lambda, X^k)) \rangle_{xx} + \langle \mu, H^k(\lambda, \hat{x}(\lambda, X^k)) \rangle''_{xx}.$$

$$\tag{3.164}$$

284

Of course, operator $\langle \mu, H^k(\lambda, x) \rangle''_{xx}(\lambda, \hat{x}(\lambda, X^k))$ is positively semidefined owing to the convexity of the constraints and the inequality $\mu \geq 0$. Therefore, according to (A4), the operator $W(\lambda, X^k)$ with $\lambda \in \Omega(\lambda^0)$, $X^k \in \Xi$ is positively defined, i.e.,

$$W(\lambda, X^k) \geq vI. \tag{3.165}$$

The function $t \to \hat{x}(\lambda_t, X^k)$ with X^k fixed in Ξ is not differentiable in general. It can be shown, however, that at certain points the derivative of this function exists.

LEMMA 3.14.   *If $I_0^k(\lambda_t)$ is a constant set for $t_0 - \tau < t < t_0 + \tau$, then*

1.   *Function $t \to \hat{x}(\lambda_t, X^k)$ is differentiable in t_0 and*

$$\frac{d\hat{x}(\lambda_t, X^k)}{dt}(t_0) = -B(\lambda_{t_0}, X^k)P_x'^*(\hat{x}(\lambda_{t_0}, X^k))(\lambda_2 - \lambda_1) \tag{3.166}$$

where for fixed λ and X^k

$$B(\lambda, X^k): X \to X$$

$$B(\lambda, X^k) = \begin{cases} W^{-1} & \text{for} \quad I_0^k(\lambda_{t_0}) = \varnothing \\ W^{-1} - W^{-1}H_x^*(H_x W^{-1} H_x^*)^{-1} H_x W^{-1} & \text{for} \quad I_0^k(\lambda_{t_0}) \neq \varnothing. \end{cases}$$

$$\tag{3.167}$$

For brevity, we have denoted $W = W(\lambda, X^k)$ and $H_x = H_x^{k'}(\lambda, \hat{x}(\lambda, X^k))$.

2.   *Furthermore, operator $B(\lambda, X^k)$ is nonnegatively defined and self-conjugated, and*

$$\mathcal{N}(B) = \mathcal{R}(H_x^*), \tag{3.168}$$

$$B \leq W^{-1} \tag{3.169}$$

The proof is given in Appendix B, section B.12. Next we assume that

(A7)   For any $\lambda_1, \lambda_2 \in \Omega(\lambda^0)$ and any $X^k \in \Xi$, the section $[\lambda_1, \lambda_2] = \{\lambda_t : \lambda_t = t\lambda_1 + (1-t)\lambda_2, 0 \leq t \leq 1\}$ can be divided into a countable number of sub-intervals so that within each of them the set $I_0^k(\lambda_t)$ is constant.

This assumption means simply that we are dealing with a model whose solution $\hat{x}(\lambda_t, X^k)$ "jumps" a countable number of times from one "wall" of the feasible set to another.

Assumption (A7) is purely technical. It is necessary for some mathematical considerations relating to the formula

$$\hat{x}(\lambda_{t_2}, X^k) - \hat{x}(\lambda_{t_1}, X^k) = \int_{t_1}^{t_2} \frac{d\hat{x}(\lambda_t, X^k)}{dt} \, dt.$$

Assumption (A7) makes no essential practical restrictions on the problem. It is very difficult to set up a model of the system, constraints, and a cost function that do not satisfy (A7). Finally, it should be stressed again that (A7) refers to the mathematical model only.

We shall assume also that in $\Omega(\lambda^0)$ the active constraints and the equations of the model show the property of uniform linear independence:

$$(A8) \quad (\exists \delta > 0)(\forall \lambda \in \Omega(\lambda^0))(\forall X^k \in \Xi)(\forall v \in \mathcal{R}(P_x'^*(\hat{x}(\lambda, X^k))))$$

$$d(v, \mathcal{R}(H_x^{k'*}(\lambda, \hat{x}(\lambda, X^k)))) \geq \delta \|v\|$$

where $d : X \times 2^X \to \mathbb{R}^1$ is the distance between the point and the set involved.

Roughly speaking, this assumption gives sufficient freedom in the manipulation of values $\hat{x}(\lambda, X^k)$ as λ changes. This can be seen from Eqs. (3.166) and (3.168). Assumption (A8) implies that

$$\mathcal{R}(P_x'^*(\hat{x}(\lambda, X^k))) \cap \mathcal{R}(H_x^{k'*}(\lambda, \hat{x}(\lambda, X^k))) = \{0\}.$$

Note that $\mathcal{N}(P_x'^*(\hat{x}(\lambda, X^k))) = \{0\}$ (see Proposition 3.27) and $\mathcal{N}(H_x^{k'*}(\lambda, \hat{x}(\lambda, X^k))) = \{0\}$ (see assumption (A6) and the proof of Proposition 3.27).

It follows that

$$P_x'^*(\hat{x}(\lambda, X^k)) \, du + H_x^{k'*}(\lambda, \hat{x}(\lambda, X^k)) \, dz = 0$$

has the unique solution $du = 0$, $dh = 0$ in $\mathcal{U} \times \mathbb{R}^1$. Since the ranges of $P_x'^*(\hat{x}(\lambda, X^k))$ and $H_x^{k'*}(\lambda, \hat{x}(\lambda, X^k))$ are closed, the equation is equivalent to the solvability of the set of equations

$$P_x'(\hat{x}(\lambda, X^k)) \, dx = du$$

$$H_x^{k'}(\lambda, \hat{x}(\lambda, X^k)) \, dx = dz$$

for any $du \in \mathcal{U}$ and $dz \in \mathbb{R}^l$ (Przeworska-Rolewicz and Rolewicz 1968). The last property in the finite-dimensional case means that the gradients of the system equations, i.e., the rows of matrix $P_x'(\hat{x}(\lambda, X^k))$, and the gradients of the active constraints, the rows of matrix $H_x^{k'}(\lambda, \hat{x}(\lambda, X^k))$, are linearly independent. This is commonly assumed in various works in the field (see Luenberger 1969). The system equations are the equality constraints for the problem. We assume also that the angle between the two subspaces spanned by the two groups of gradients is always greater than a positive angle equal to arc sin δ.

It is convenient to investigate the convergence property of the algorithm in a special norm determined in $\mathcal{U}$ by the new scalar product $\langle \cdot, \cdot \rangle_0$. This scalar product is defined as follows:

$$\forall u', u'' \in \mathcal{U} \quad \langle u', u'' \rangle_0 \triangleq \langle P_0^* u', AP_0^* u'' \rangle. \tag{3.170}$$

LEMMA 3.15 *The bilinear form $\langle \cdot, \cdot \rangle$ defined by (3.170) is a scalar product in $\mathcal{U}$ and the norm $\|\cdot\|_0$ induced by this scalar product is equivalent to the original in $\mathcal{U}$.*

Proof. Operator A is self-conjugated, implying that

$$\langle u', u'' \rangle_0 = \langle P_0^* u', A P_0^* u'' \rangle = \langle A P_0^* u', P_0^* u'' \rangle = \langle u'', u' \rangle_0.$$

On the other hand (see (3.157)),

$$\langle u, u \rangle_0 = \langle P_0^* u, A P_0^* u \rangle \geq m_A \|P_0^* u\|^2$$

which implies that $\langle u, u \rangle_0 = 0$ if and only if $u = 0$ (see Proposition 3.27). The other properties of a scalar product are obvious. So, it has been proved that $\langle \cdot, \cdot \rangle_0$ is a scalar product in $\mathcal{U}$.

According to part 2 of Proposition 3.27 and Banach's inverse operator theorem, there exists a number $\gamma > 0$ such that

$$\forall u \in \mathcal{U} \quad \|P_0^* u\| \geq \gamma \|u\|.$$

Hence

$$\forall u \in \mathcal{U} \quad \|u\|_0^2 = \langle u, u \rangle_0 \geq m_A \gamma^2 \|u^2\|. \tag{3.171}$$

On the other hand

$$(\forall u \in \mathcal{U}) \, \|u\|_0^2 = \langle u, u \rangle_0 \leq \|P_0^*\| \, \|A\| \, \|u\|^2. \tag{3.172}$$

By (3.171) and (3.172) we conclude that the norms $\|\cdot\|$ and $\|\cdot\|_0$ are equivalent, which completes the proof. $\square$

Let us consider now, for fixed $X^k \in \Xi$, the following operator:

$$V_0^k : \Omega(\lambda_0) \times \Omega(\lambda_0) \to X$$

$$V_0^k(\lambda_1, \lambda_2) \triangleq \left(I - \varepsilon E D_0 \int_0^1 B(\lambda_t, X^k) \, dt P_0^* \right)(\lambda_2 - \lambda_1) \tag{3.173}$$

where

$$\lambda_t = t\lambda_2 + (1 - t)\lambda_1, \qquad 0 \leq t \leq 1.$$

The behavior of this operator, which will play a very important role in our considerations, is described by the following lemma.

LEMMA 3.16 *If we assume that*

1. *Assumptions (A1)–(A8) are satisfied.*
2. *There exists a constant $M_w > 0$ such that*

$$(\forall \lambda \in \Omega(\lambda_0))(\forall x \in X) \quad \langle x, W(\lambda, X^k)x \rangle \leq M_w \|x\|^2$$

then

$$(\forall X^k \in \Xi)(\forall \lambda_1, \lambda_2 \in \Omega(\lambda_0)) \quad \|V_0^k(\lambda_1, \lambda_2)\|_0 \le \alpha(\varepsilon)\|\lambda_1 - \lambda_2\|_0 \qquad (3.174)$$

where

$$\alpha(\varepsilon) = \left(1 - \frac{2\varepsilon\delta^2}{M_A M_w} + \frac{\varepsilon^2}{v^2 m_A^2}\right)^{1/2}. \qquad (3.174a)$$

Proof. Relation (3.165) means that $\|W^{-1}(\lambda, X^k)\| \le 1/v$ for all $\lambda \in \Omega(\lambda_0)$ and $X^k \in \Xi$. So, by (3.169), we have

$$(\forall X^k \in \Xi)(\forall \lambda \in \Omega(\lambda_0)) \quad \|B(\lambda, X^k)\| \le \frac{1}{v}. \qquad (3.175)$$

Due to Lemma 3.13, assumption (A7), and the definition (3.167) of $B(\lambda, X^k)$, the function $[0, 1] \ni t \to B(\lambda_t, X^k) \in \mathscr{L}(X, X)$ is continuous almost everywhere for all $X^k \in \Xi$. Hence, on the basis of the above statements, $\int_0^1 B(\lambda_t, X^k)\, dt$ exists (Bourbaki 1961) and for all $X^k \in \Xi$ the operator V_0^k is well defined.

Let us denote by Π_S the projection operator from X on $S \triangleq \mathscr{R}(P_0^*)$. We introduce the following operators:

$$\begin{aligned} A_S &\triangleq \Pi_S A|_S, \\ B_S(\lambda, X^k) &= \Pi_S B(\lambda, X^k)|_S, \\ P_S &\triangleq P_0|_S, \\ D_S &\triangleq D_0|_S. \end{aligned} \qquad (3.176)$$

Let us observe that for all $s_1, s_2 \in S$

$$\begin{aligned} \langle s_1, P_S A|_S s_2 \rangle = \langle s_1, P_S A s_2 \rangle &= \langle A P_S s_1, s_2 \rangle \\ &= \langle A s_1, s_2 \rangle = \langle A s_1, P_S s_2 \rangle = \langle P_S A s_1, s_2 \rangle \\ &= \langle P_S A|_S s_1, s_2 \rangle. \end{aligned}$$

This means that operator A_S is self-conjugated on S.

Additionally, for all $s \in S$

$$\langle s, A_S s \rangle = \langle s, P_S A s \rangle = \langle P_S s, A s \rangle = \langle s, A s \rangle.$$

Hence, from (3.157) we conclude that operator A_S is strongly positively defined on S. This implies that A_S^{-1} exists. Since $S = \bar{S}$ (see Proposition 3.27) and $S = \mathscr{R}(P_0^*)$, the operator P_0^* is equal to the operator P_S^*. By part 3 of Proposition 3.27 and Banach's inverse operator theorem, operator $(P_S^*)^{-1}$ exists.

Relation (3.160) implies that operator D_S^{-1} exists. Hence, the following is true

$$(D_S A_S P_S^*)^{-1} = (P_S^*)^{-1} A_S^{-1} D_S^{-1}. \qquad (3.177)$$

288

According to relations (3.159) and (3.160), $D_0\Pi_S = D_0$ and $D_0AP_0^* = D_0\Pi_S AP_S^* = D_S A_S P_S^*$. The last equality, with relation (3.177), implies that

$$(D_0 AP_0^*)^{-1} = (P_S^*)^{-1} A_S^{-1} D_S^{-1}. \tag{3.178}$$

Let us also observe that the following holds by relation (3.160):

$$D_S^{-1} D_0 = \Pi_S \tag{3.179}$$

Let us now consider the operator $\int_0^1 B_S(\lambda_t, X^k)\, dt$ with X^k fixed in Ξ and λ_1, λ_2 fixed in $\Omega(\lambda_0)$. Note that for all $s \in S$

$$\langle s, B_S(\lambda_t, X^k)s \rangle = \langle s, \Pi_S B(\lambda_t, X^k)s \rangle = \langle \Pi_S s, B(\lambda_t, X^k)s \rangle = \langle s, B(\lambda_t, X^k)s \rangle. \tag{3.180}$$

It was proved in Appendix B, section B.12 (see Eqs. (7) and (8)), that

$$\langle s, B(\lambda_t, X^k)s \rangle = \langle y, W^{-1}(\lambda_t, X^k)y \rangle \tag{3.181}$$

if

$$y = s - H_x^*(H_x W^{-1} H_x^*)^{-1} H_x W^1 y.$$

The simplified notation is used in the last formula. This implies that

$$\|y\| \geq d(s, \mathcal{R}(H_x^*)),$$

which, with assumption (A8), implies that

$$\|y\| \geq \delta \|s\|. \tag{3.182}$$

Relations (3.180), (3.181), (3.182), and assumption 2 allows us to write (see also Appendix B, section B.14)

$$\langle s, B_S(\lambda_t, X^k)s \rangle \geq \frac{1}{M_w} \|y\|^2 \geq \frac{\delta^2}{M_w} \|s\|^2.$$

The last inequality implies (Bourbaki 1961) that

$$(\forall X^k \in \Xi)(\lambda_1, \lambda_2 \in \Omega(\lambda_0))(\forall s \in S) \quad \left\langle s, \int_0^1 B_S(\lambda_t, X^k)\, dt\, s \right\rangle \geq \frac{\delta^2}{M_w} \|s\|^2. \tag{3.183}$$

Let us return now to the definition of V_0^k (see (3.173)).
Relations (3.178) and (3.179) yield

$$V_0^k(\lambda_1, \lambda_2) = \left(I - \varepsilon (P_S^*)^{-1} A_S^{-1} D_S^{-1} D_0 \int_0^1 B_s(\lambda_t, X^k)\, dt P_0^* \right)(\lambda_2 - \lambda_1)$$

$$= \left(I - \varepsilon (P_S^*)^{-1} A_S^{-1} \int_0^1 B_S(\lambda_t, X^k)\, dt P_0^* \right)(\lambda_2 - \lambda_1).$$

Hence

$$\|V_0^k(\lambda_1, \lambda_2)\|_0^2 = \langle \lambda_2 - \lambda_1, \lambda_2 - \lambda_1 \rangle_0 - 2\varepsilon \langle \lambda_2 - \lambda_1, (P_S^*)^{-1} A_S^{-1} RP_S^*(\lambda_2 - \lambda_1) \rangle_0$$

$$+ \varepsilon^2 \langle (P_S^*)^{-1} A_S^{-1} RP_S^*(\lambda_2 - \lambda_1), (P_S^*)^{-1} A_S^{-1} RP_S^*(\lambda_2 - \lambda_1) \rangle_0$$

where, for simplicity, we have denoted $R = \int_0^1 B_S(\lambda_t, X^k)\, dt$. For all $s_1, s_2 \in S$

$$\langle s_1, AA_S^{-1}s_2\rangle = \langle \Pi_S s_1, AA_S^{-1}s_2\rangle = \langle s_1, \Pi_S AA_S^{-1}s_2\rangle = \langle s_1, s_2\rangle. \qquad (3.184)$$

By relations (3.183) and (3.184), we can write

$$\langle \lambda_2 - \lambda_1, (P_S^*)^{-1}A_S^{-1}RP_S^*(\lambda_2 - \lambda_1)\rangle_0$$

$$= \langle P_0^*(\lambda_2 - \lambda_1), AP_0^*(P_S^*)^{-1}A_S^{-1}RP_S^*(\lambda_2 - \lambda_1)\rangle$$

$$= \langle P_0^*(\lambda_2 - \lambda_1), AA_S^{-1}RP_S^*(\lambda_2 - \lambda_1)\rangle = \langle P_0^*(\lambda_2 - \lambda_1), RP_S^*(\lambda_2 - \lambda_1)\rangle$$

$$\geq \frac{\delta^2}{M_W}\|P_0^*(\lambda_2 - \lambda_1)\|^2 \qquad (3.185)$$

and

$$\langle (P_S^*)^{-1}A_S^{-1}RP_S^*(\lambda_2 - \lambda_1), (P_S^*)^{-1}A_S^{-1}RP_S^*(\lambda_2 - \lambda_1)\rangle_0$$

$$= \langle P_0^*(P_S^*)^{-1}A_S^{-1}RP_S^*(\lambda_2 - \lambda_1), AP_0^*(P_S^*)^{-1}A_S^{-1}RP_S^*(\lambda_2 - \lambda_1)\rangle$$

$$= \langle A_S^{-1}RP_S^*(\lambda_2 - \lambda_1), AA_S^{-1}RP_S^*(\lambda_2 - \lambda_1)\rangle$$

$$= \langle A_S^{-1}RP_S^*(\lambda_2 - \lambda_1), RP_S^*(\lambda_2 - \lambda_1)\rangle \leq \frac{1}{m_A}\|RP_S^*(\lambda_2 - \lambda_1)\|^2. \qquad (3.186)$$

Because $B \leq W^{-1}$ (see (3.169)) and $\|W^{-1}\| \leq 1/v$,

$$\|R\| \leq \frac{1}{v}.$$

Thus

$$\|RP_S^*(\lambda_2 - \lambda_1)\|^2 \leq \frac{1}{v}\|P_0^*(\lambda_2 - \lambda_1)\|,$$

which, with (3.185) and (3.186), implies that

$$\|V_0^k(\lambda_1, \lambda_2)\|_0^2 \leq \|\lambda_2 - \lambda_1\|_0^2 - \frac{2\varepsilon\delta^2}{M_w}\|P_0^*(\lambda_2 - \lambda_1)\|^2 + \frac{\varepsilon^2}{v^2 m_A}\|P_0^*(\lambda_2 - \lambda_1)\|^2. \qquad (3.187)$$

It follows from (3.157) that

$$\frac{1}{M_A}\langle P_0^*(\lambda_2 - \lambda_1), AP_0^*(\lambda_2 - \lambda_1)\rangle \leq \|P_0^*(\lambda_2 - \lambda_1)\|^2$$

$$\leq \frac{1}{m_A}\langle P_0^*(\lambda_2 - \lambda_1), AP_0^*(\lambda_2 - \lambda_1)\rangle,$$

290

which implies that

$$\frac{1}{M_A}\|\lambda_2-\lambda_1\|_0^2 \leq \|P_0^*(\lambda_2-\lambda_1)\|^2 \leq \frac{1}{m_A}\|\lambda_2-\lambda_1\|_0^2. \qquad (3.188)$$

Using (3.188) in (3.187), we prove the lemma. $\square$

Now we are ready to consider the contraction properties of an operator $V^k : \Omega(\lambda^0)\to \mathcal{U}$ defined as follows:

$$V^k(\lambda) \triangleq \lambda + \varepsilon ER_*(\hat{x}(\lambda, X^k)) \qquad (3.189)$$

where X^k is fixed in Ξ.

LEMMA 3.17. *If*

1. *The assumptions of Lemma 3.16 are satisfied,*
2. $(\exists L>0)(\forall \lambda_1, \lambda_2 \in \Omega(\lambda^0))(\forall X^k \in \Xi)$

$$\|D_x'(\hat{x}(\lambda_1, X^0))B(\lambda_2, X^k)P_x'^*(\hat{x}(\lambda_1, X^0))$$
$$- D_x'(\hat{x}(\lambda_2, X^0))B(\lambda_2, X^k)P_x'^*(\hat{x}(\lambda_2, X^0))\|_0 \leq L\|\lambda_1-\lambda_2\|_0,$$

3. $(\exists \gamma>0)(\forall \lambda \in \Omega(\lambda^0))(\forall X^k \in \Xi)$

$$\|D_x'(\hat{x}(\lambda, X^k))B(\lambda, X^k)P_x'^*(\hat{x}(\lambda, X^k))$$
$$- D_x'(\hat{x}(\lambda, X^0))B(\lambda, X^k)P_x'^*(\hat{x}(\lambda, X^0))\|_0 \leq \gamma,$$

4. $(\exists \tilde{L}>0)(\forall \lambda_1, \lambda_2 \in \Omega(\lambda^0))(\forall X^k \in \Xi)$

$$\|\tilde{D}(\hat{x}(\lambda_1, X^k))-\tilde{D}(\hat{x}(\lambda_2, X^k))\|_0 \leq \tilde{L}\|\lambda_1-\lambda_2\|_0,$$

5. *There exists continuous function* $\kappa : \mathbb{R}_+ \to \mathbb{R}_+$, $\kappa(0)=0$, *such that for all* $X^k \in \Xi$ *and all* $\lambda \in \Omega(\lambda^0)$

$$\|R_*(\hat{x}(\lambda, X^k))-R_*(\hat{x}(\lambda, X^0))\|_0 \leq \kappa \,(dist\,(X^k, X^0))$$

where

$$dist\,(X^k, X^0) = max\left\{\sup_{x\in X^k}\,\inf_{y\in X^0}\|x-y\|,\ \sup_{y\in X^0}\,\inf_{x\in X^k}\|x-y\|\right\}$$

then the following inequalities are satisfied:

(1) $\|V^k(\lambda_1)-V^k(\lambda_2)\|_0$

$$\leq (\alpha(\varepsilon)+\varepsilon\,\|E\|_0\,L\,max\,\{\|\lambda_2-\lambda^0\|_0, \|\lambda_1-\lambda^0\|_0\}+\varepsilon\|E\|_0\gamma+\varepsilon\|E\|_0\tilde{L})\|\lambda_2-\lambda_1\|_0$$

for all $\lambda_1, \lambda_2 \in \Omega(\lambda^0)$, *and*

(2) $\|V^k(\lambda)-\lambda^0\|_0$

$$\leq \tfrac{1}{2}\varepsilon\,\|E\|_0\,L\|\lambda-\lambda^0\|_0^2+(\alpha(\varepsilon)+\varepsilon\,\|E\|_0\,\gamma+\varepsilon\,\|E\|_0\,\tilde{L})\|\lambda-\lambda^0\|_0+\varepsilon\|E\|_0\kappa_{max}+\varepsilon\eta^0$$

where $\alpha(\varepsilon)$ is given by (3.174), $\eta^0 = \|ER_*(\hat{x}(\lambda^0, X^0))\|_0$, and $\kappa_{max} = \kappa(\text{dist}(X_{max}, X^0))$.

Proof. From the definition of $V^k(\lambda)$ it follows that

$$V^k(\lambda_2) - V^k(\lambda_1) = \lambda_2 - \lambda_1 + \varepsilon E(R_*(\hat{x}(\lambda_2, X^k)) - R_*(\hat{x}(\lambda_1, X^k)))$$
$$= \lambda_2 - \lambda_1 + \varepsilon E(D(\hat{x}(\lambda_2, X^k)) - D(\hat{x}(\lambda_1, X^k)))$$
$$+ \varepsilon E(\tilde{D}(\hat{x}(\lambda_2, X^k)) - \tilde{D}(\hat{x}(\lambda_1, X^k))). \quad (3.190)$$

Let us consider the term $\lambda_2 - \lambda_1 + \varepsilon E(D(\hat{x}(\lambda, X^k)) - D(\hat{x}(\lambda_1, X^k)))$ of the right side of (3.190) and label it by a_1. Taking into account the continuity of function $t \to \hat{x}(\lambda_t, X^k)$ and using the same arguments as in the existence proof for $\int_0^1 B(\lambda_t, X^k)\, dt$ in Lemma 3.16, we can write that $\forall \lambda_1, \lambda_2 \in \Omega(\lambda_0)$ and $\forall X^k \in \Xi$

$$\hat{x}(\lambda_2, X^k) - \hat{x}(\lambda_1, X^k) = \int_0^1 \frac{d\hat{x}(\lambda_t, X^k)}{dt}\, dt. \quad (3.191)$$

It follows from Lemma 3.14 and Eq. (3.191) that

$$\|a_1\|_0 = \|\lambda_2 - \lambda_1 - \varepsilon E \int_0^1 D_x'(\hat{x}(\lambda_t, X^k))B(\lambda_t, X^k)P_x'^*(\hat{x}(\lambda_t, X^k))(\lambda_2 - \lambda_1)\, dt\|_0$$

where $\lambda_t = t\lambda_2 + (1-t)\lambda_1$, $0 \le t \le 1$.

By simple calculations we obtain:

$$\|a_1\|_0 \le \left\| \lambda_2 - \lambda_1 - \varepsilon E \int_0^1 D_x'(\hat{x}(\lambda^0, X^0))B(\lambda_t, X^k)P_x'^*(\hat{x}(\lambda^0, X^0))(\lambda_2 - \lambda_1)\, dt \right\|_0$$

$$+ \varepsilon \|E\|_0 \left\| \int_0^1 (D_x'(\hat{x}(\lambda_t, X^0))B(\lambda_t, X^k)P_x'^*(\hat{x}(\lambda_t, X^0)) \right.$$
$$\left. - D_x'(\hat{x}(\lambda^0, X^0))B(\lambda_t, X^k)P_x'^*(\hat{x}(\lambda^0, X^0)))(\lambda_2 - \lambda_1)\, dt \right\|_0$$

$$+ \varepsilon \|E\|_0 \left\| \int_0^1 (D_x'(\hat{x}(\lambda_t, X^k))B(\lambda_t, X^k)P_x'^*(\hat{x}(\lambda_t, X^k)) \right.$$
$$\left. - D_x'(\hat{x}(\lambda_t, X^0))B(\lambda_t, X^k)P_x'^*(\hat{x}(\lambda_t, X^0)))(\lambda_2 - \lambda_1)\, dt \right\|_0.$$

Let us consider the right side of the above inequality. The first term can be estimated on the basis of Lemma 3.16; the second and third can be estimated using assumptions 2 and 3, respectively. We find then that

$$\|a_1\|_0 \le \alpha(\varepsilon)\|\lambda_2 - \lambda_1\|_0 + \varepsilon \|E\|_0 L \int_0^1 \|\lambda_t - \lambda^0\|_0\, dt \|\lambda_2 - \lambda_1\|_0 + \varepsilon \|E\|_0 \gamma \|\lambda_2 - \lambda_1\|_0$$

$$\le \alpha(\varepsilon)\|\lambda_2 - \lambda_1\|_0 + \varepsilon \|E\|_0 L \max \{\|\lambda_1 - \lambda^0\|_0, \|\lambda_2 - \lambda^0\|_0\} \|\lambda_2 - \lambda_1\|_0$$
$$+ \varepsilon \|E\|_0 \gamma \|\lambda_2 - \lambda_1\|_0. \quad (3.192)$$

Estimating the second term of the right side of (3.190) in accordance with assumption 4 and taking into account relation (3.192), we obtain inequality (1).

Let us turn to the derivation of inequality (2). The following holds:

$$\|V^k(\lambda) - \lambda^0\|_0$$
$$\leq \|V^k(\lambda) - V^k(\lambda^0)\|_0 + \|V^k(\lambda^0) - V^0(\lambda^0)\|_0 + \|V^0(\lambda^0) - \lambda^0\|_0. \quad (3.193)$$

The first term on the right side will be estimated as in the proof of (1). If we take advantage of the fact that in this case $\lambda_2 = \lambda^0$, we obtain

$$\|V^k(\lambda) - V^k(\lambda^0)\|_0$$
$$\leq \tfrac{1}{2}\varepsilon \|E\|_0 L \|\lambda - \lambda^0\|_0^2 + (\alpha(\varepsilon) + \varepsilon \|E\|_0 \gamma + \varepsilon \|E\|_0 \tilde{L}) \|\lambda - \lambda^0\|_0. \quad (3.194)$$

The second term can be estimated using assumption 5:

$$\|V^k(\lambda^0) - V^0(\lambda^0)\| \leq \varepsilon \|E\|_0 \|R_*(\hat{x}(\lambda^0, X^k)) - R_*(\hat{x}(\lambda^0, X^0))\|_0$$
$$\leq \varepsilon \|E\|_0 \kappa(dist\,(X^k, X^0) \leq \varepsilon \|E\|_0 \kappa_{max} \quad (3.195)$$

whereas

$$\|V^0(\lambda^0) - \lambda^0\|_0 = \varepsilon \|ER_*(\hat{x}(\lambda^0, X^0))\|_0 = \varepsilon\eta_0$$

Using this equation and relations (3.194) and (3.195) in (3.193), we obtain inequality (2) of the lemma. $\square$

We know from formula (3.174a) that for sufficiently small ε within a certain bounded interval, the number $\alpha(\varepsilon)$ will be less than 1. So if we demand sufficiently small values of L, $\tilde{L}$, γ, and κ_{max}, we can be sure that operator V^k will have contraction properties. The basic theorem about convergence, Theorem 3.30, is founded on this idea. Before we state the theorem, we will analyze the reasonableness of the assumptions in Lemma 3.17. Assumption 1 invokes assumptions (A1)–(A8), which were discussed in detail when they were presented. The presence of operator $B(\lambda_2, X^k)$ in assumption 2 is not essential because $\|B\| \leq \|W^{-1}\|$. Assumption 2 would be satisfied if it were assumed that functions $D'_x(x)$ and $P'^*_x(x)$ on $\omega(x^0)$ are bounded and satisfy the Lipschitz condition. The same remarks are applicable to assumption 3 if we account for the influence of changes in set X^k on $\hat{x}(\lambda, X^k)$. This influence can be easily estimated using the next proposition.

PROPOSITION 3.29. *Let X be a Hilbert space and X_1 and X_2 its convex closed subsets. Let us assume furthermore that $Q:X \rightarrow \mathbb{R}$ is a functional twice differentiable, and that*:

1. $X_1 \subset X_2$,
2. $\exists \hat{x}_1 \in X_1 \quad Q(\hat{x}_1) = min_{x \in X_1} Q(x)$,
3. $\exists \hat{x}_2 \in X_2 \quad Q(\hat{x}_2) = min_{x \in X_2} Q(x)$,
4. $(\exists m > 0)(\forall x \in X_2) \quad Q_{xx}(x) \geq mI$,
5. $(\exists M > 0)(\forall x \in X_2) \quad \|Q'_x(x)\| \leq M$.

Therefore,

$$\|\hat{x}_1 - \hat{x}_2\| \le \left(\frac{2M}{m} \, dist \, (X_2, X_1)\right)^{1/2}$$

where dist (X_2, X_1) should be understood as it is in Lemma 3.17.

The proof is given in Appendix B, section B.13.

Assumption 4 of Lemma 3.17 limits in some defined way the difference between the model and reality. It is clear that this difference must be limited somehow if we want to control any real process efficiently based on a model. Assumption 5 adds no restrictions since the existence of function $\kappa(\cdot)$ is a result of Proposition 3.29 and the continuity of $R_*(\cdot)$.

We will now give the conditions under which the coordination procedure

$$\lambda^{k+1} = V^k(\lambda^k)$$

leads to the convergent sequence of discoordinations

$$\eta^k = \|ER_*(\hat{x}(\lambda^k, X^k))\|_0. \tag{3.196}$$

THEOREM 3.30 (*convergence theorem*). *Let us introduce the following notation:*

$$\xi \triangleq \frac{1}{\varepsilon} (1 - \alpha(\varepsilon) - \varepsilon \|E\|_0 \gamma - \varepsilon \|E\|_0 \tilde{L}),$$

$$h \triangleq \|E\|_0 L(\eta^0 + \|E\|_0 \kappa_{max})$$

and

$$r = \frac{\xi - \sqrt{\xi^2 - 2h}}{\|E\|_0 L}.$$

If we assume that

1. *The assumptions of Lemma 3.17 are satisfied,*
2. $\sqrt{2h} < \xi,$
3. $\bar{K}(\lambda^0; r) \triangleq \{\lambda \in \mathcal{U} : \|\lambda - \lambda^0\|_0 \le r\} \subset \Omega(\lambda^0),$
4. $X^0 \subseteq X^1 \subseteq \ldots \subseteq X_{max} \subseteq CU_*$ *and* $\lim_{k \to \infty} dist \, (X^{k+1}, X^k) = 0,$

then

1. *Sequence $\{\lambda^k\}$ remains within $\bar{K}(\lambda^0; r)$.*
2. *Sequence $\{\eta^k\}$ defined by (3.196) converges to zero.*
3. *Sequence $\|\lambda^k - \tilde{\lambda}^k\|_0$, where $\tilde{\lambda}^k$ is a fixed point of V^k in $\bar{K}(\lambda^0; r)$,* *converges to zero.*

294

Proof. We set

$$q = \alpha(\varepsilon) + \varepsilon \, \|E\|_0 \, Lr + \varepsilon \, \|E\|_0 \, \gamma + \varepsilon \, \|E\|_0 \, \tilde{L}.$$

Due to assumption 2, $q < 1$. Therefore, according to Lemma 3.17 (part 1) and assumption 3 above, operator V^k is a contraction mapping on $\bar{K}(\lambda^0; r)$:

$$(\forall \lambda_1, \lambda_2 \in \bar{K}(\lambda^0, r)) \, \|V^k(\lambda_1) - V^k(\lambda_2)\|_0 \leq q \, \|\lambda_1 - \lambda_1\|_0. \qquad (3.197)$$

Furthermore, it follows from part 2 of Lemma 3.17 that for $\lambda \in \bar{K}(\lambda^0; r)$ the following is true:

$$\|V^k(\lambda) - \lambda^0\|_0 \leq \tfrac{1}{2}\varepsilon \, \|E\|_0 \, Lr^2 + (\alpha(\varepsilon) + \varepsilon \, \|E\|_0 \, \gamma + \varepsilon \, \|E\|_0 \, \tilde{L})r + \varepsilon \, \|E\|_0 \, \kappa_{max} + \varepsilon \eta^0.$$

Because r is a solution to the quadratic equation of the form

$$\tfrac{1}{2}\varepsilon \, \|E\|_0 \, Lx^2 - (1 - \alpha(\varepsilon) - \varepsilon \, \|E\|_0 \, \gamma - \varepsilon \, \|E\|_0 \, \tilde{L})x + \varepsilon \, \|E\|_0 \, \kappa_{max} + \varepsilon \eta^0 = 0,$$

the right side of the inequality is equal to r. Hence, we know that sequence $\{\lambda^k\}$ remains within $\bar{K}(\lambda^0; r)$ and the proof of part 1 is completed. It follows that operator V^k satisfies the assumptions of Banach's contraction mapping theorem on $\bar{K}(\lambda^0; r)$ and has a fixed point $\tilde{\lambda}^k$ within $\bar{K}(\lambda^0; r)$. Let us examine the behavior of the discoordination sequence $\{\eta^k\}$. We shall estimate the discoordination η^{k+1}:

$$\|\eta^{k+1}\| = \|ER_*(\hat{x}(\lambda^{k+1}, X^{k+1}))\|_0$$
$$\leq \|ER_*(\hat{x}(\lambda^{k+1}, X^{k+1})) - ER_*(\hat{x}(\lambda^{k+1}, X^k))\|_0 + \|ER_*(\hat{x}(\lambda^{k+1}, X^k))\|_0.$$
$$(3.198)$$

The first term on the right side of (3.198) will be estimated using assumption 5 of Lemma 3.17.

$$\|ER_*(\hat{x}(\lambda^{k+1}, X^{k+1})) - ER_*(\hat{x}(\lambda^{k+1}, X^k))\|_0 \leq \|E\|_0 \, \kappa \, (\mathrm{dist}\,(X^{k+1}, X^k)). \qquad (3.199)$$

The second term will be estimated by taking advantage of the contracting properties of operator V^k as follows:

$$\|ER_*(\hat{x}(\lambda^{k+1}, X^k))\|_0 = \frac{1}{\varepsilon} \, \|V^k(\lambda^{k+1}) - V^k(\lambda^k)\|_0$$

$$\leq \frac{1}{\varepsilon} \, q \, \|\lambda^{k+1} - \lambda^k\|_0 = q \, \|ER_*(\hat{x}(\lambda^k, X^k))\|_0 = q\eta^k. \qquad (3.200)$$

From (3.198), (3.199), and (3.200) we obtain

$$\eta^{k+1} \leq \|E\|_0 \, \kappa \, (\mathrm{dist}\,(X^{k+1}, X^k)) + q\eta^k.$$

From assumption 4 and the fact that $q < 1$, we conclude that $\eta^k \to 0$, which completes the proof of part 2.

We know that

$$\|\lambda^k - \tilde{\lambda}^k\|_0 = \|V^{k-1}(\lambda^{k-1}) - V^k(\tilde{\lambda}^k)\|_0$$
$$= \|V^{k-1}(\lambda^{k-1}) - V^k(\lambda^k) + V^k(\lambda^k) - V^k(\tilde{\lambda}^k))\|_0$$
$$\leq \|V^k(\lambda^k) - \lambda^k\|_0 + q\|\lambda^k - \tilde{\lambda}^k\|_0$$

so

$$\|\lambda^k - \tilde{\lambda}^k\|_0 \leq \frac{1}{1-q}\|V^k(\lambda^k) - \lambda^k\|_0 = \frac{\varepsilon}{1-q}\,\eta^k.$$

We have already shown that $\eta^k \to 0$, so it follows that $\|\lambda^k - \tilde{\lambda}^k\|_0 \to 0$, which completes the proof. $\square$

The convergence conditions for Theorem 3.30 are very similar to those for Newton's method (Kantorovich and Akilov 1964). In particular, we did not assume that either of the operators $R_*(\hat{x}(\,\cdot\,, X^k))$ and $D(\hat{x}(\,\cdot\,, X^k))$ is differentiable with respect to λ.

Simulation results

The method presented in this section was tested on two examples of iteratively controlled systems (batch processes). One was a linear dynamic system and the other a nonlinear dynamic system. The systems were described by differential equations, and the control computations were based on simplified mathematical models for which outputs y were found from a three-phase operation made on controls $c(t)$ and inputs $u(t)$. In the first phase, the controls $c(t)$ are transformed into step functions and the inputs $u(t)$ are transformed into piecewise linear functions. Next, the values of controls and inputs at the points of discretization are used as input data for the discrete state equations. From these equations, the values of the outputs at the points of discretization are obtained. In the third phase, a piecewise linear approximation of the output functions is made based on these values.

In order to investigate the influence of model accuracy on the solution of the control problem, one should construct models in which the input and output spaces are the same as for the actual system. For the straightforward discretization of state equations, however, one usually does not have to fulfill this requirement. In the case of system control, this problem takes on a special nature caused by the existence of two kinds of input variables. If, for example, subsystems are described by differential equations and the controls belong to L^2, then the state variables and interaction inputs are usually absolutely continuous functions from W_1^2. For that reason, another type of transformation for c and u was used.

In our investigation, the coordination task is solved on a computer. Therefore, the values of real interactions $HK_*(\hat{c}(\lambda, X^k))$ at each cycle are measured only at fixed intervals, which means that a piecewise linear approximation of $HK_*(\hat{c}(\lambda, X^k))$, based on these measurements, is used in coordination. If we make the measurements in the moments of discretization used in the model, the coordination task, initially formulated in W_1^2, will turn into a finite-dimensional problem. This approach was used in the computations.

In the linear dynamic system there are two subsystems:

Subsystem 1

$$s_1(t) = -u_1(t) + c_1(t), \qquad s_1(0) = 0,$$

$$y_{11}(t) = s_1(t), \tag{3.201}$$

$$y_{12}(t) = \alpha s_1(t).$$

Subsystem 2

$$s_2(t) = u_2(t) + c_2(t), \qquad s_2(0) = 0 \tag{3.202}$$

where s_1, s_2 denote state variables, c_1, c_2 local controls, u_1, u_2 local inputs, and y_{11}, y_{12}, local outputs. The structure of the system is shown in Figure 3.15. Operator H has the form:

$$H = \begin{bmatrix} 0 & I \\ I & 0 \end{bmatrix}, \qquad H : \mathcal{Y}_{11} \times \mathcal{Y}_{12} \rightarrow \mathcal{U}_1 \times \mathcal{U}_2$$

where I denotes the identity operator. The performance index has the form:

$$Q = \int_0^1 \{(c_1(t))^2 + (u_1(t))^2\}\, dt + 10(s_1(1))^2 + \int_0^1 \{(u_2(t))^2 + 10(c_2(t))^2\}\, dt, \tag{3.203}$$

and can be divided into two parts corresponding to the two subsystems. As we mentioned before, the control will be iterative, i.e., after reaching $t = 1$, the initial state $s_1(0) = 0$, $s_2(0) = 0$ is restored and the process starts again.

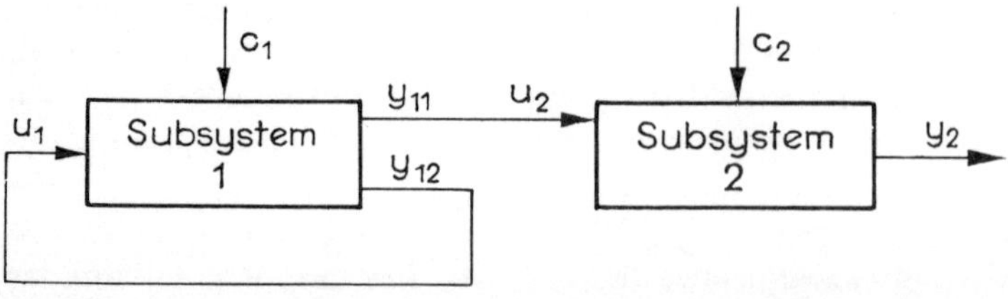

FIGURE 3.15   The structure of the simulated system.

We assume that the feasible set CU_* for the real system is not known.

We assume that Eqs. (3.201) and (3.202) are not known to us precisely and we use the following approximate discrete models:

Subsystem S1

$$s_1(0) = 0,$$

$$s_1(t_{i+1}) = 0.1(c_1(t_i) - u_1 t_i)) + s_1(t_i), \qquad i = 1, \ldots, 9,$$

$$y_{11}(t_i) = s_1(t_i),$$

$$y_{12}(t_i) = s_1(t_i).$$

$$\text{(3.204)}$$

Subsystem S2

$$s_2(t_{i+1}) = 0.1(c_2(t_i) + u_2(t_i)) + s_2(t_i), \qquad \text{(3.205)}$$

$$s_2(0) = 0.$$

where $t_i = (1/10)i$, $i = 0, 1, 2, \ldots, 10$.

There are two sources of differences between the model and the real system. The first is the parameter α, which is set equal to 1 in the model but does not equal 1 in the real system. The second difference is the error introduced into the model by discretization.

We now describe the control algorithm and begin with the design of the FSI unit. We know *a priori* that

$$X^0 = \{(c, u) : s_2(1) \geq z\} \subset CU_* \qquad \text{(3.206)}$$

for $z = z^0 > 1$. We assume that in subsequent iterations the knowledge about set CU_* increases and the FSI unit generates the sequence of sets X^k satisfying the relations:

$$X^k \subset CU_*,$$

$$X^k = \{(c, u) : s_2(1) \geq z^k\}, \qquad \text{(3.207)}$$

$$z^k = \min\left\{z^{k-1}, \frac{v^k}{\gamma \cdot k}\right\}.$$

The numbers v_k are chosen at random from the interval $[0, 1]$, and k is the iteration number. The sets defined in (3.207) are obtained by an identification procedure that will not be discussed here. However, the influence of the identification procedure on the properties of the method should be investigated. For that reason, parameter γ in (3.207) has been introduced. Random properties of the sequence $\{v^k\}$ make the sequence (3.207) close to those obtained in real identification procedures. Let us note that sets X^k defined by (3.207) satisfy assumption 4 of the convergence theorem.

298

Due to the form of the model, discrete versions X_d^k of sets (3.207) will be used in the local problems:

$$X_d^k = X_{1d}^k \times X_{2d}^k,$$

$$X_{1d}^k = \mathscr{C}_1 \times \mathscr{U}_1, \tag{3.208}$$

$$X_{2d}^k = \left\{ (\{c_2(t_i)\}_{i=0}^9, \{u_2(t_i)\}_{i=0}^{10} : 0.1 \sum_{i=0}^9 (c_2(t_i) + u_2(t_i)) \geq z \right\}.$$

Let us denote for simplicity

$$x_1 \triangleq (c_1(t_0), c_1(t_1), \ldots, c_1(t_9), u_1(t_0), \ldots, u_1(t_{10})),$$

$$x_2 \triangleq (c_2(t_0), c_2(t_1), \ldots, c_2(t_9), u_2(t_0), \ldots, u_2(t_{10})).$$

Following the scheme of the algorithm described in this section, we introduce the vector of coordinating variables $\lambda \in R^{22}$ and formulate two independent local problems which in iteration k have the form:

$$\min_{x_1 \in X_{1d}^k} \left[Q_{\mathrm{mod1}}(x_1, \lambda^k) = Q_1(x_1) + 10(s_1(t_{10}))^2 \right.$$

$$\left. + \sum_{i=1}^{10} (\lambda_i^k y_{11}(t_{i-1}) + \lambda_{i+1}^k y_{12}(t_{i-1})) - \sum_{i=1}^{11} \lambda_{i+10}^k x_{1,i+10} \right] \tag{3.209}$$

and

$$\min_{x_2 \in X_{2d}^k} \left[Q_{\mathrm{mod2}}(x_2, \lambda^k) = Q_2(x_2) - \sum_{i=1}^{11} \lambda_i^k x_{2,i+10} \right], \tag{3.210}$$

where λ^k is the price in iteration k, and $y_1(t_i)$ and $s_1(t_{10})$ are calculated on the basis of the model. There are no constraints on the first local problem. The problem is now ready to be solved.

For given λ, the local problems (3.209) and (3.210) together with corresponding model state equations (3.204) and (3.205) form discrete optimal control problems. The discrete form of the maximum principle was used to solve the local problems. The step controls c_1 and c_2 thus obtained were applied to the real system (3.201) and (3.202), which resulted in certain real interactions. The values of these interactions at t_i $(i = 0, 1, \ldots, 10)$ were $H_1 K_*(c(t_i))$ and $H_2 K_*(c(t_i))$ calculated analytically from equations (3.201) and (3.202). Thus, it was possible to obtain discoordinations $u_1(t_i) - H_1 K_*(c(t_i))$ and $u_2(t_i) - H_2 K_*(c(t_i))$ for $i = 0, 1, \ldots, 10$ from the computer. Operators D_0 and P_0 defined in (3.155a) were 20×40 matrices computed directly from the model. Operator A was taken as the inverse of the matrix of second derivatives of $Q_{\mathrm{mod1}} + Q_{\mathrm{mod2}}$ with respect to (x_1, x_2). Operator E, which is used in the algorithm, was a 20×20 matrix. The value of parameter ε was chosen experimentally, as shown below.

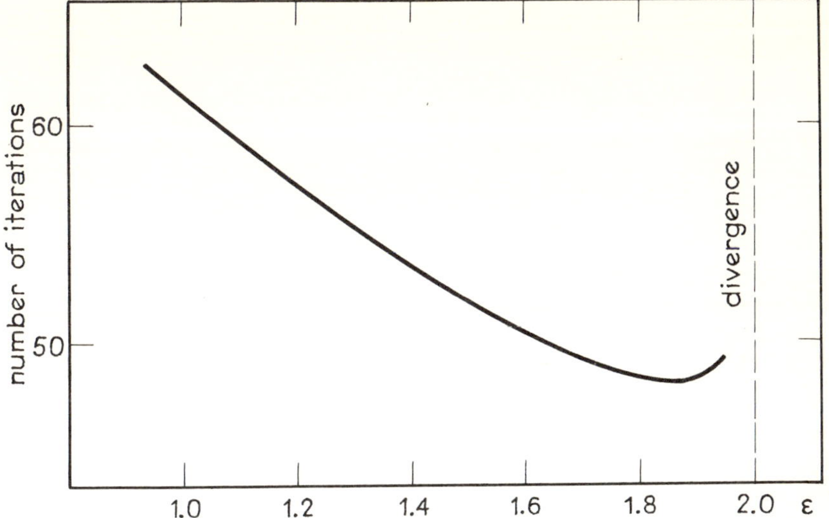

FIGURE 3.16 The influence of parameter ε on convergence properties in the control of a linear dynamic system.

During the numerical simulation, the influence of parameter ε on the convergence properties of the method was investigated, and the results are presented in Figure 3.16. A safe value, $\varepsilon = 1.2$, was chosen. Then, the influence of the FSI unit (parameter γ in (3.207)) on the coordination process was analyzed. This influence is presented graphically by a family of curves of discoordination norm versus iteration number for three values of γ (Figure 3.17). The behavior of z^k in (3.207) is also shown. In all cases the algorithm converges irrespective of the changes in the feasible sets. Coordination with $\gamma = 0.1$ and $\varepsilon = 0.8$ and 1.8 is shown in Figure 3.18.

Finally the influence of discretization on the performance of the method was investigated. Three intervals of discretization were used, $NN = 10$, 15, and 20, and the discrete versions of the local problems and the coordination task were formulated as in (3.209) and (3.210). A plot of the value of the performance at the end of the coordination process versus NN is shown in Figure 3.19.

The second example, a nonlinear dynamic system, is one consisting of subsystems S1 and S2 (Figure 3.20). Subsystem S1 is a periodically loaded chemical reactor in which conversion $A \rightarrow B$ takes place. The work of S1 can be divided into three phases. In the first phase ($0 \leq t \leq t_1$) the reactor is loaded by a constant flow of substance A up to its capacity W_0. The temperature of the substance is $T(0)$. In the second phase ($t_1 \leq t \leq 4$) conversion $A \rightarrow B$ takes place. In the third phase ($4 \leq t \leq 5$) the reactor is

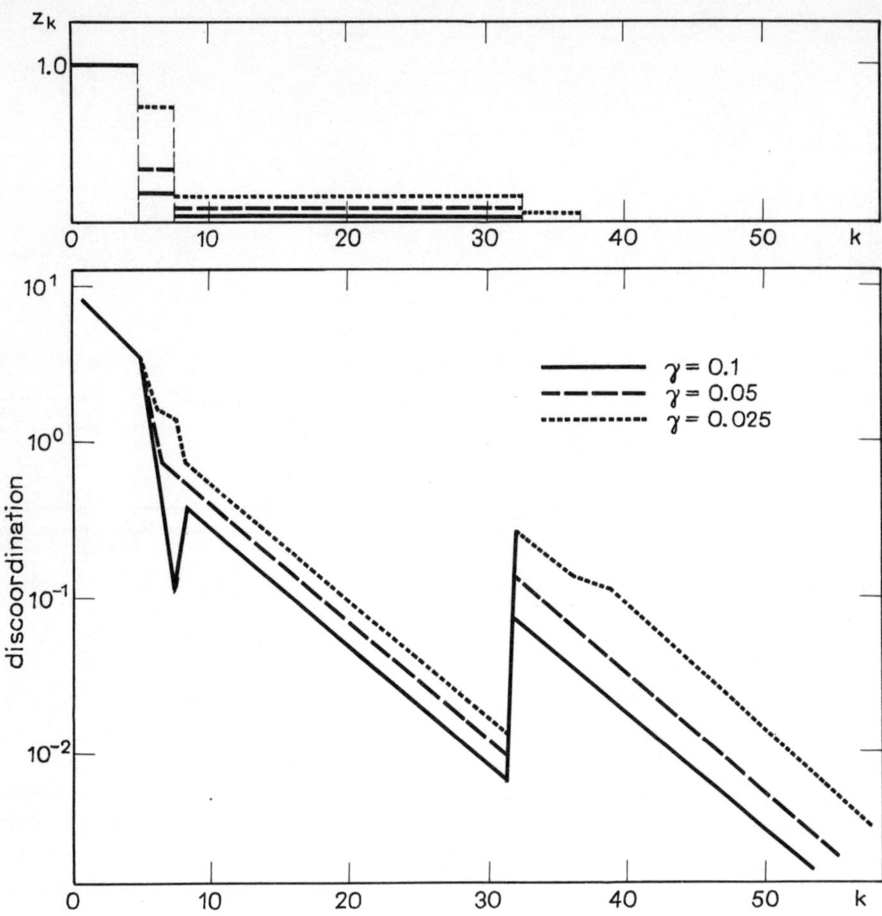

FIGURE 3.17 The influence of the feasible set identification (FSI) unit on the coordination process for a linear dynamic system.

unloaded by a constant flow and the compound flows into S2. The reactor is heated continuously by a flow of heat $H(t)$. There are three state variables in S1: $W(t)$, the volume of liquid in the reactor, $XA(t)$, the concentration of A in the reactor, and $T(t)$, the temperature in the reactor. The controls are t_1, the duration of the filling phase and $H(t)$, the heat flow. Subsystem S2 separates substance B from the compound flowing in from the reactor. S2 is an object without memory. The work of S2 depends on the parameters $XA(t)$ and $T(t)$ of the reactor output flow. Since the separator is supplied by the reactor outflow only during the interval $[4, 5]$, we consider the interactions as functions defined on the interval $[4, 5]$.

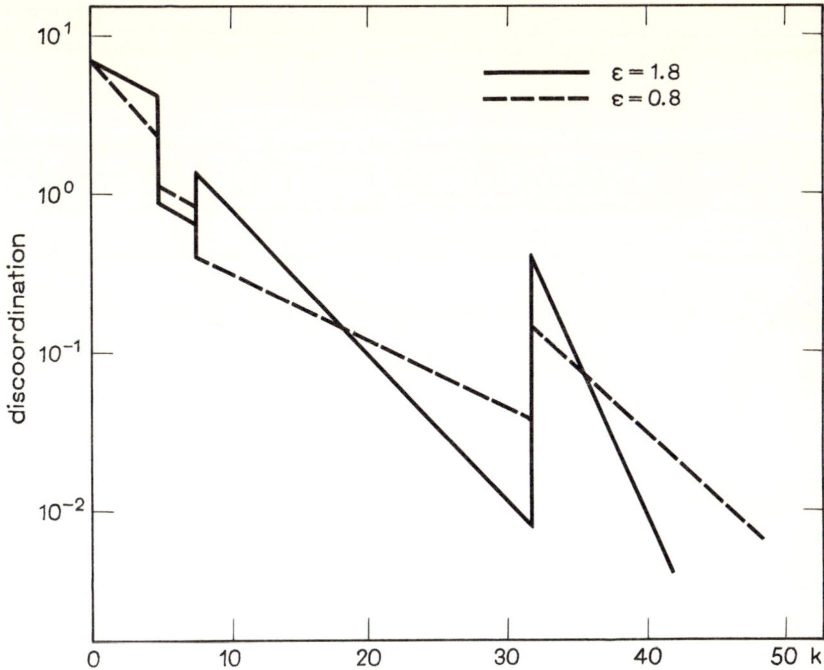

FIGURE 3.18   Coordination with $\gamma = 0.1$ and $\varepsilon = 0.8$ and 1.8.

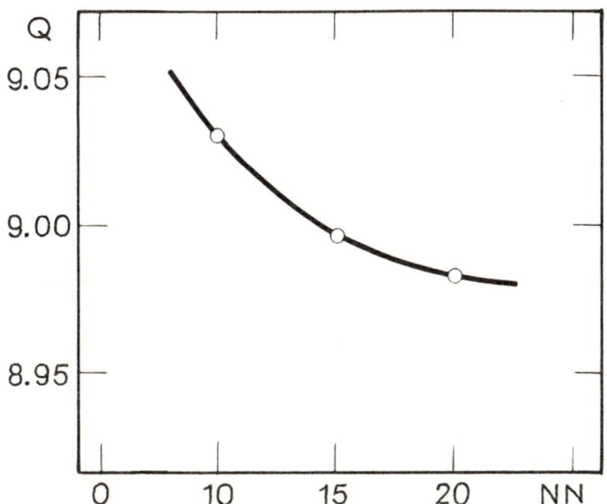

FIGURE 3.19   The influence of discretization on the final value of the performance.

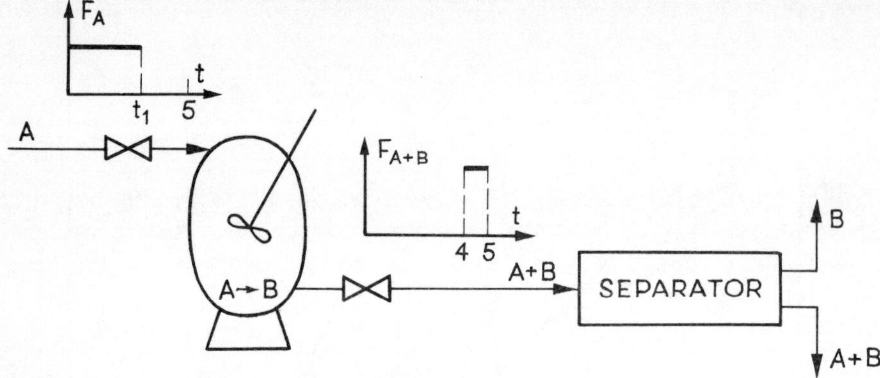

FIGURE 3.20   Example of a nonlinear dynamic system.

The equations of subsystem S1 have the following form:

$$\frac{dW}{dt} = \begin{cases} \dfrac{1}{t_1}, & 0 \le t \le t_1, \\[2mm] 0, & t_1 \le t \le 4, \\[2mm] -1, & 4 \le t \le 5. \end{cases} \tag{3.211}$$

$$\frac{dXA(t)}{dt} = \begin{cases} \dfrac{1-XA(t)}{t_1 \cdot W(t)} - XA(t) \cdot k(T(t)), & 0 < t \le t_1, \\[3mm] -XA(t) \cdot k(T(t)), & t_1 \le t \le 5 \end{cases} \tag{3.212}$$

$$\frac{dT(t)}{dt} = \begin{cases} \dfrac{300-T(t)}{t_1 \cdot W(t)} + \dfrac{H(t)}{W(t)} - XA(t) \cdot k(T(t)), & 0 < t \le t_1, \\[3mm] \dfrac{H(t)}{W(t)} - XA(t)k(T(t)), & t_1 \le t \le 5 \end{cases} \tag{3.213}$$

where $k(T) = (0.00833 \cdot T - 2.5)^3$. The initial state in each cycle was $W(0) = 0$, $XA(0) = 1$, $T(0) = 300\,°K$.

Since the output of S2 does not influence S1, there is no need to introduce the output equations of S2. Still, the performance of the system strongly depends on the work of S2.

The performance index for the reactor is:

$$Q_1 = 0.01 \int_0^5 (H(t))^2 \, dt + \frac{1}{t_1}. \tag{3.214}$$

It represents the cost of control. The performance index for the separator represents the cost of control $R(t)_{[4, 5]}$ and the value of product B:

$$Q_2 = \int_4^5 [-20(1 - \exp(-R(t)))$$
$$\times (1 - \exp(-0.01(1 - XA(t))(500 - T(t))) + 0.1R(t))] \, dt. \quad (3.215)$$

We assumed that the mathematical model was evaluated on the basis of Eqs. (3.211)–(3.213) in the way indicated at the beginning of this section. In the first phase, control $H(t)$ was transformed into a step function constant within intervals $[0, 0.1]$, $[0.1, 1]$, $[1, 2]$, $[2, 3]$, $[3, 4]$, $[4, 5]$. Thus, 6 points of discretization were used. In the second phase, the state equations were discretized at the following points: $0, 0.1, 0.2, 0.4, 0.6, \ldots, 0.2i, \ldots, 5$. In the third phase, the piecewise linear output function was constructed from the values of $XA(t_j)$ and $T(t_j)$ for $t_j = 4 + 0.2j$, and $j = 0, 1, \ldots, 5$. Because of the strong nonlinearity of state equations, a greater number of discretization points were used in the second phase.

Regarding the unknown set CU_*, at the beginning we know that set X^0, defined by the inequalities

$$0 \le t_1 \le 0.1,$$

$$H(t) = 0, \qquad t \in [0, 0.1]$$

$$T(4.0) \le 361 - z^0,$$

for certain $z^0 > 0$, is included in CU_*. We assumed that during the iterations the knowledge about set CU_* increases and the FSI unit generates the sequence of sets X^k defined by the conditions:

$$0 \le t_1 \le 0.1,$$

$$H(t) = 0 \quad \text{for} \quad t \in [0, 0.1],$$

$$T(4.0) \le 361 - z^k, \qquad (3.216)$$

$$z^k = \min \left\{ z^{k-1}, \frac{v^k}{\gamma \cdot k} \right\}.$$

The formula for z^k was the same as in the first example, where the meaning of all variables used was more fully discussed.

Owing to the discrete form of the coordination task and the structure of the model, we can define decision variables for the local problem as follows:

$$c_1 = (H(0.1), H(1), H(2), H(3), H(4), t_1),$$

$$c_2 = (R(4.0), R(4.2), R(4.4), R(4.6), R(4.8)),$$

$$u_2 = (XA(4.0), XA(4.2), XA(4.4), XA(4.6), XA(4.8), XA(5), T(4.0),$$
$$T(4.2), T(4.4), T(4.6), T(4.8), T(5)).$$

304

The local problem for the method is the following for S1:

$$\min_{c_1 \in X_{1d}^k} \left[Q_{\text{mod}1}(c_1, \lambda^k) = Q_1(c_1) - \right.$$

$$\left. \sum_{j=1}^{5} (\lambda_j^k y_{11}(4+(j-1)0.2) + \lambda_{j+5}^k \, y_{12}(4+(j-1)0.2)) \right] \quad (3.217)$$

where $X_{1d}^k = \{c_1 : c_{11} = 0, \, c_{16} \in [0, 0.1]\}$, $\lambda^k = (\lambda_1^k, \ldots, \lambda_{10}^k)$ is a vector of coordination variables, $y_1 = (XA, T)$, and the values of $y_1(4+(j-1)0.2)$ for $j = 1, \ldots, 5$ are obtained in the mathematical model. The local problem for S2 is:

$$\min_{(c_2, \, u_2) \in X_{2d}^k} \left[Q_{\text{mod}2} = Q_2(c_2, u_2) + \sum_{j=1}^{10} \lambda_j^k u_{2,j} \right] \quad (3.218)$$

where $X_{2d}^k = \{(c_2, u_2) : u_{2,7} \leq 361 - z^k\}$. The goal of the coordinator was to bring the discoordinations $u_2(t_j) - H_2 K_*(c(t_j))$ to zero for $t_j = 4 + 0.2(j-1)$, where $j = 1, 2, \ldots, 5$. The values of $H_2 K_*(c(t_j))$ in the real system were computed by a very precise procedure integrating state Eqs. (3.211)–(3.213) with a relative accuracy 0.1 percent. The difference between the model and the real system was about 1.5 percent in $T(t)$ and 10 percent in $XA(t)$.

The local problem for S1 (3.217) was solved by Powell's method. The simple constraint on c_{16} was eliminated by Box's transformation. The model had to be integrated in each evaluation of the cost function in (3.217), so the minimization proved rather time-consuming. Therefore, the problem was not solved with high accuracy for S1.

The solution of problem (3.218), thanks to its static nature, was divided into five independent minimizations for $t = 4.0, 4.2, 4.4, 4.6, 4.8$. These minimization problems were transformed into equations by employing necessary conditions (the problems are convex). The equations were solved by the two-point secant method.

The evaluation of operator E from the model was rather difficult since derivatives D_0 and P_0 could not be calculated analytically. Numerical estimates of the derivatives have been computed. Operator A was taken equal to the unit matrix.

In the calculations, the influence of parameter ε on the convergence properties of the method was investigated first. In this example there were difficulties in choosing an appropriate operator A, since the Hessian of the performance function was not available. Consequently, the identity matrix was used. Unfortunately, it turned out that only very small values of ε guarantee convergence. The results are presented in Figure 3.21. A safe value of $\varepsilon = 0.00022$ was chosen.

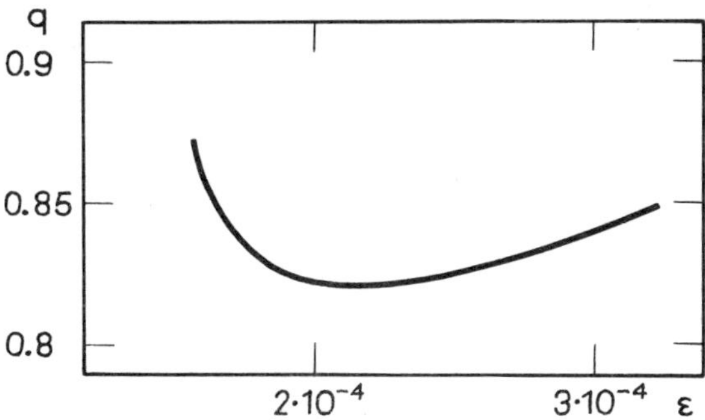

FIGURE 3.21 The influence of parameter ε on convergence properties in the control of a nonlinear dynamic system.

Second, the influence of the FSI unit on the coordination process was analyzed (Figure 3.22). Because of errors inherent in numerical solutions of local problems and numerical simulation of the real system, the coordination condition (3.153) could be satisfied with only limited accuracy. Third, the behavior of real performance during the coordination process was investigated (Figure 3.23). The performance decreases as the coordination proceeds. Small oscillations at the end of the coordination process are caused by random numerical errors.

We draw the following conclusions from the simulation:

- Although full information about the real system equations and feasible sets is not available, the method finds a feasible satisfactory control that is much better than that of open-loop control.
- A convergence proof for the method under very general assumptions about the FSI unit has been given. Such a general approach enables us to construct the FSI unit in a number of different ways, depending on the differences between the model and the real constraints.
- The errors caused by discretization may be taken into account in the framework of this theory.
- In all computational examples the method converged, as predicted by the theory.

The above promising properties indicate that price coordination with feasible set identification may be used to control complex industrial plants. In order to make it more useful, the following research should be done.

First, since real plants are usually nonstationary, the nonstationary case should also be considered. Coordination of a nonstationary system in a similar structure, without the FSI unit, is considered in subsection 3.5.3. Second, the general conditions imposed on the FSI unit allow it to be constructed in various ways. Thus, a choice of specialized identification procedures for some classes of problems should be made. Third, the simulations showed that the quality of the control depends on the discretization step used (see Figure 3.19). It should be determined if the solutions of the discrete versions of the coordination task tend to the solution of the

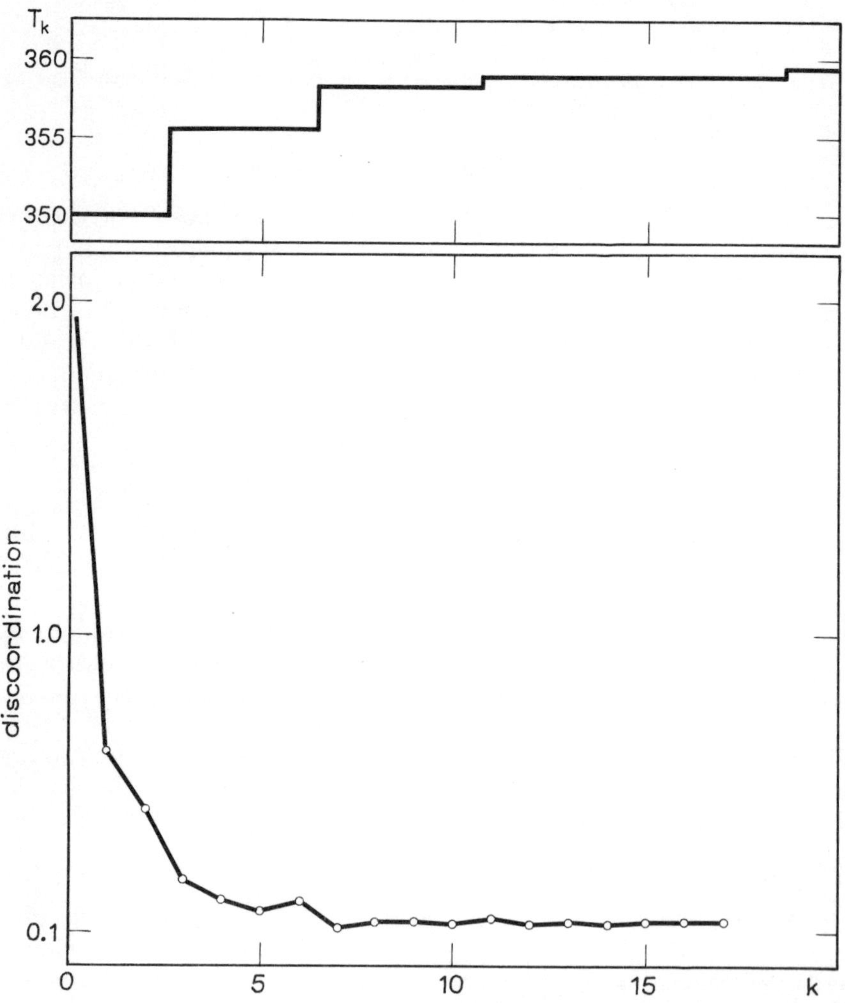

FIGURE 3.22   The influence of the FSI unit on the coordination process.

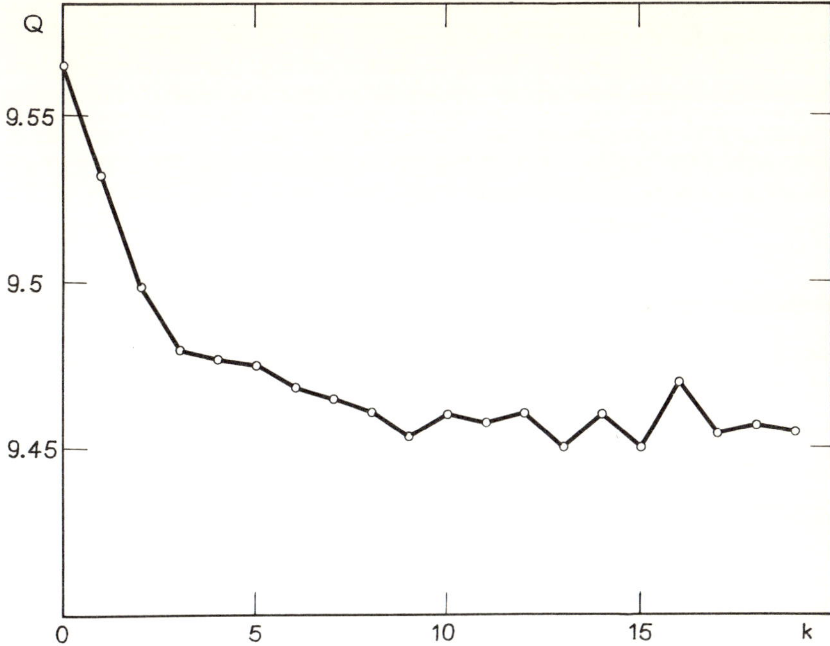

FIGURE 3.23   Behavior of performance during the coordination process.

continuous version as the discretization step tends to zero. Fourth, it was mentioned in the convergence theorem that the rate of convergence is linear, but it seems possible to achieve faster convergence of the coordination algorithm. Moreover, the number of computations may be decreased by an appropriate choice of stop criteria for the lower-level problems. Some research in this direction for two-level methods of mathematical programming has been done by Szymanowski and Ruszczyński (1976).

REFERENCES

Bourbaki, N. (1961). *Fonction d'une Variable Réele*, Elements de Mathematique, Part I, Vol. 4. Hermann, Paris.

Brdyś, M. (1975a). "Methods of feasible control generation for complex systems," *Bull. Pol. Acad. Sci., Ser. Tech. Sci.* 23(12).

Brdyś, M. (1975b). "Finding a feasible control for real process under uncertainty," *Proceedings, 7th IFIP Conference, Nice.*

Brdyś, M., and P. Michalak (1978). "On-line coordination with local feedback for steady-state systems." *Arch. Autom. Telemech.* 23 (4).

Brdyś, M., and B. Ulanicki (1978). "On completely decentralized control with local feedback in large scale systems." *Arch. Autom. Telemech.* 23(1).

Dunford, N., and J. T. Schwartz (1958). *Linear Operators*, Part I. Wiley-Interscience, New York.

Fiacco, A. V., and G. P. McCormick (1968). *Nonlinear Programming—Sequential Unconstrained Minimization.* Wiley, New York.

Findeisen, W. (1974a). "Control and coordination in multilevel systems," *Proceedings, 2nd Polish–Italian Conference on Applications of Systems Theory.* Pugnochiuso, Italy.

Findeisen, W. (1974b). *Wielopoziomowe uklady sterowania [Multilevel Control Systems].* PWN, Warsaw. German translation: *Hierarchische Steuerungssysteme.* Verlag Technik, Berlin, 1977.

Findeisen, W. (1976). "Coordination by price methods," *Proceedings, 3rd Polish–Italian Conference on Applications of Systems Theory.* Bialowieża Poland.

Findeisen, W., M. Brdyś, K. Malinowski, P. Tatjewski, and A. Woźniak (1978). "On-line hierarchical control for steady-state systems." *IEEE Trans. Autom. Control* AC-23: 189–209.

Fiodorow, W. W. (1977). "Regularity conditions and necessary condition for maxmin with connected variables." *Zh. Vychisl. Mat. Mat. Fiz.* 17(1) (in Russian).

Goldstein, A. A. (1967). *Constructive Real Analysis.* Harper and Row, New York.

Hager, W. W. (1979). "Lipschitz continuity for constrained processes." *SIAM J. Control and Optimization* 17(3): 321–328.

Kantorovich, L., and G. Akilov (1964). *Functional Analysis in Normed Spaces.* Pergamon Press, Oxford.

Kheysin, V. E. (1976). "Stability of iterative methods in nonstationary conditions," *Kibernetika*, No. 2 (in Russian).

Kuratowski, K. (1968). *Topology*, Vol. 2. Academic Press, New York.

Lika, D. K. (1975). Functional analysis and approximate solutions of nonlinear equations. in Moldavian. Kishinev, Shtiintsa.

Luenberger, D. G. (1969). *Optimization by Vector Space Methods.* Wiley, New York.

Malinowski, K. (1976). "Lectures on hierarchical optimization and control." *Report*, Center for Control Sciences, University of Minnesota, Minneapolis, Minnesota.

Malinowski, K., and A. Ruszczyński (1975). "Application of interaction balance method to real process coordination." *Control and Cybernetics* 4(2).

Nurminsky, E. A. (1977). "On some problems of nonstationary optimization," *Kibernetika*, No. 2 (in Russian).

Pliskin, L. G. (1975). *Optimization of Continuous Production.* Energia, Moskva (in Russian).

Polak, E. (1971). *Computation Methods in Optimization.* Academic Press, New York.

Przeworska-Rolewicz, D., and S. Rolewicz (1968). *Equations in Linear Spaces*, PWN, Warsaw.

Rice, J. R. (1969). *Approximation of Functions*, Vol. 2. Addison-Wesley, Reading, Massachusetts.

Ruszczyński, A. (1976). "Convergence conditions for the interaction balance algorithm based on an approximate mathematical model." *Control and Cybernetics* 5(4): 29–43.

Ruszczyński, A. (1979). "Coordination of nonstationary systems." *IEEE Trans. Autom. Control* 24(1).

Schwartz, L. (1967). *Analyse mathematique.* Hermann, Paris.

Sion, M. (1958). "On general minimax theorems." *Pacific J. of Mathematics* 8; 171–176.

Stoilov, E. (1977). "Coordination algorithms for nonconvex control problems," *Doctoral dissertation*, Technical University of Warsaw (in Polish).

Szymanowski, J., M. Brdyś, and A. Ruszczyński (1976). "An algorithm for real system coordination" *Proceedings, IFAC Symposium on Large-Scale Systems Theory and Applications*, Udine, Italy.

Szymanowski, J., M. Brdyś, and A. Ruszczyński (1977). "Practical aspects of the feasible price coordination method," *Proceedings, IFAC Workshop*, Toulouse, France.

Szymanowski, J., and A. Ruszczyński (1976). "Convergence analysis for two-level algorithms of mathematical programming," *Proceedings, 9th Symposium on Mathematical Programming,* Budapest.

Ulanicki, B. (1978). "Properties of the completely decentralized control-technical results." *Report,* Institute of Automatic Control, Technical University of Warsaw.

Woźniak, A. (1976). "Parametric method of coordination using feedback from the real process," *Proceedings, IFAC Symposium on Large-Scale Systems Theory and Applications,* Udine, Italy.

Zinchenko, A. I. (1973). "On the approximate solutions of the functional equations with nondifferentiable operators, *Mat. Fiz.,* No. 14 (in Russian).

4.1. PROBLEM DESCRIPTION

4.1.1. THE SYSTEM TO BE CONTROLLED

Throughout this chapter, we assume that the system under control is composed of dynamic subsystems whose inputs and outputs are connected either directly or through intermediate storage elements (inventories). Thus, we consider a complex system that can be described formally as follows:

The *subsystems* are dynamic processes defined on the interval $[t_0, T_0]$ and described by a state transformation mapping and an output mapping. The state transformation mapping is

$$x_i(t) = \phi_{i\Delta}[x_i(t_1), m_{i\Delta}, u_{i\Delta}, z_\Delta], \qquad i \in \overline{1, N}, \tag{4.1}$$

where
$\Delta = [t_1, t]$, $t_0 \le t_1 \le t \le T_0$.
N is the number of system elements.
$x_i(t)$ is the state of subsystem i at time t, $x_i(t) \in X_i$, a real Banach space.
$x_{i\Delta}$ is the state trajectory over interval Δ, $x_{i\Delta} \in C(\Delta, X_i) = \mathscr{X}_{i\Delta}$
$m_i(t)$ is the control input at time t, $m_i(t) \in M_i$, a real Hilbert space.
$m_{i\Delta}$ is the control input over interval Δ, $m_{i\Delta} \in L^2(\Delta. M_i) = \mathscr{M}_{i\Delta}$.
$u_i(t)$ is the interconnection input to subsystem i at time t, $u_i(t) \in \mathbb{R}^{n_{ui}}$.
$u_{i\Delta}$ is the interconnection input over interval Δ, $u_{i\Delta} \in L^2(\Delta, \mathbb{R}^{n_{ui}}) = \mathscr{U}_{i\Delta}$.
$z(t)$ is the disturbance input (or process parameter) at time t, $z(t) \in Z$, e.g., a real Banach space.
z_Δ $z_{i\Delta}$ is the interconnection input over interval Δ, $z_\Delta \in L^2(\Delta, Z) = \mathscr{Z}_\Delta$.

We assume that the initial state plus the inputs until the present determine the current state of each subsystem, i.e., for all t_1, t_2, and t such that $t_0 \leq t_1 \leq t_2 \leq t \leq T_0$ we have

$$\phi_{i\Delta_1}[x_i(t_1), m_{i\Delta_1}, u_{i\Delta_1}, z_{\Delta_1}] = \phi_{i\Delta_2}[x_i(t_2), m_{i\Delta_2}, u_{i\Delta_2}, z_{\Delta_2}], \qquad (4.2)$$

where

$$x_i(t_2) = \phi_{i\Delta_{12}}[x_i(t_1), m_{i\Delta_{12}}, u_{i\Delta_{12}}, z_{\Delta_{12}}]$$

and

$$\Delta_1 = [t_1, t], \qquad \Delta_2 = [t_2, t], \qquad \Delta_{12} = [t_1, t_2].$$

The subsystem output mapping is

$$y_i(t) = F_i^0[x_i(t), m_i(t), u_i(t), z(t)], \qquad i \in \overline{1, N}, \qquad (4.3)$$

where

$$F_i^0 : X_i \times M_i \times \mathbb{R}^{n_{ui}} \times Z \to \mathbb{R}^{n_y}, \qquad y_i(t) \in \mathbb{R}^{n_{yi}}.$$

Combining (4.1) and (4.3), we can define the subsystem input–output relation as follows:

$$y_{i\Delta} = F_{i\Delta}[x_i(t_1), m_{i\Delta}, u_{i\Delta}, z_\Delta], \qquad (4.4)$$

where

$$\Delta = [t_1, t] \quad \text{and} \quad y_{i\Delta} \in L^2(\Delta, \mathbb{R}^{n_{yi}}).$$

For simplicity, we denote

$$x = (x_1, \ldots, x_N), \qquad X = X_1 \times \ldots \times X_N,$$

and we do the same for m, u, y. Thus (4.4) may be written in compact form to describe the whole set of subsystems, the system with its couplings cut, as

$$y_\Delta = F_\Delta[x(t_1), m_\Delta, u_\Delta, z_\Delta]. \qquad (4.5)$$

For the *system couplings*, the interconnection equations, the coupling equation has the following form (for given $\Delta = [t_1, t_2]$):

$$P_\Delta(u_\Delta, y_\Delta) = \sum_{i=1}^{N} P_{i\Delta}(u_{i\Delta}, y_{i\Delta}) = 0, \qquad (4.6)$$

where for given $\Delta = [t_1, t_2]$, $P_{i\Delta}$ is a continuous linear operator with values in some space $\mathscr{P}$ (we assume that $\mathscr{P}$ is a real Hilbert space). Suppose now that for every $i \in \overline{1, N}$ the vectors of the interconnection variables are split into two parts $u_i = (u_{si}, u_{wi})$ and $y_i = (y_{si}, y_{wi})$. We will consider the class of system couplings (4.6) to have the following form:

$$u_{si}(t) = H_i y_s(t) = \sum_{j=1}^{N} H_{ij} y_{sj}(t) \qquad (4.7)$$

for almost every $t \in \Delta$, where H_{ij} are real matrices. Couplings (4.7) are *stiff interconnections*.

312

The remaining components of u and y may be required, for example, to satisfy the following *weak interconnection* relations:

$$\int_{\Delta_l} [\bar{H}_1 u_w(t) - \bar{H}_2 y_w(t)]\, dt = b_l. \tag{4.8}$$

for a given sequence of time intervals Δ_l where $b_l \in \mathbb{R}^{n_w}$, $\bar{H}_1$ and $\bar{H}_2$ are real matrices, and

$$\bar{H}_1 u_w(t) = \sum_{i=1}^{N} \bar{H}_{1i} u_{wi}(t).$$

A similar expression can be written for $\bar{H}_2$. It is easy to see that (4.8) does not relate the input $u_w(t)$ at a particular time to the instantaneous output $y_w(t)$ and therefore this input will be considered as a decision (control) variable.

In most of this chapter, and in particular in sections 4.2 and 4.3, we assume that local variables $m_i(t)$ and $u_i(t)$ may be constrained to a certain set, that is $(m_i(t), u_i(t)) \in MU_i^0 \subset M_i \times \mathbb{R}^{n_{ui}}$. The set MU_i^0 can be defined, for example, as follows:

$$MU_i^0 = \{(m_i(t), u_i(t)) : g_i(m_i(t), u_i(t)) \le 0\},$$

where g_i is a vector-valued mapping. We assume that subsystem state $x_i(t)$ is unconstrained and we consider no global constraints.

4.1.2. THE SYSTEM CONTROL PROBLEM

The overall performance function of the system is specified, for some interval $\Delta_f = [t_0, t_f]$ $(t_f \le T_0)$, in additive form as follows:

$$\begin{aligned}
Q_{\Delta_f}^0(x_{\Delta_f}, m_{\Delta_f}, u_{\Delta_f}, z_{\Delta_f}) &= \sum_{i=1}^{N} Q_{i\Delta_f}^0(x_{i\Delta_f}, m_{i\Delta_f}, u_{i\Delta_f}, z_{\Delta_f}) \\
&= \sum_{i=1}^{N} \left\{ \int_{t_0}^{t_f} q_{0i}(x_i(t), m_i(t), u_i(t), z(t))\, dt + J_i(x_i(t_f)) \right\}, \tag{4.9}
\end{aligned}$$

where q_{0i} and J_i, $i \in \overline{1, N}$, are scalar-valued functions. By making use of (4.1), we can say that performance $Q_{\Delta_f}^0$ implicitly depends on the initial state and on the inputs taken over Δ_f:

$$Q_{\Delta_f}^0 = Q_{\Delta f}(x(t_0), m_{\Delta_f}, u_{\Delta_f}, z_{\Delta_f}). \tag{4.10}$$

The following global off-line system optimization problem (SOP) may now be formulated:

Given prediction $\bar{z}_{\Delta_f}$ of the disturbance trajectory find the model-based, optimal, open-loop controls $\hat{m}_{\Delta_f}(\bar{z}_{\Delta_f})$, $\hat{u}_{w\Delta_f}(\bar{z}_{\Delta_f})$, that minimize $Q^0_{\Delta_f}$ subject to (4.5), (4.6), (4.7), and (4.8) over

$$MU_{\Delta_f} = MU_{1\Delta_f} \times \ldots \times MU_{N\Delta_f}, \tag{4.11}$$

where

$$MU_{i\Delta_f} = \{(m_{i\Delta_f}, u_{i\Delta_f}) : (m_i(t), u_i(t)) \in MU^0_i \quad \text{for } t \in \Delta_f\}.$$

We assume that the initial state $x(t_0)$ is given.

This formulation is not the most general. We treat the disturbance as a random variable and propose to solve the deterministic problem (find a control for the predicted disturbance z) as if it was a given function of time. The reason for this approach was already mentioned in Chapter 1, section 1.1. It is difficult to solve a problem in a stochastic formulation for anything other than a linear, unconstrained system with quadratic performance functions.

SOP is an open-loop formulation, that is, it determines the controls as functions of time. In a closed-loop formulation a decision rule would be determined that relates an instantaneous value of control to the measured value of the system state. The reason is again of a practical nature: many more problems are solvable in the open-loop form than in the closed-loop one and, moreover, we cannot use continuous feedback at the higher levels of a control hierarchy.

4.1.3. THE CONTROL STRUCTURE

When the model-based optimal controls are applied to the real system, the behavior of the system may be far from desirable, because, in general, the disturbances $z^r_{\Delta_f}$ are different from the predictions $\bar{z}_{\Delta_f}$. As a result, coupling equations (4.8) may not be satisfied in the real system; the same could also happen to the local constraints. There is a need to introduce appropriate on-line control structures that would use feedback information from the system. If the optimization problem is formulated and solved as SOP above, the use of feedback is possible in the form of repetitive optimization (see section 4.5), where SOP would be solved at given intervals and the actual state of the system would be used as the new initial value. In most of this chapter we will use predicted disturbances and repetitive optimization.

According to the main theme of this book we are going to discuss in this chapter the hierarchical structures of dynamic control, and we pay most of our attention to dynamic price coordination. We consider this structure to be the most flexible to apply.

4.2. MULTILEVEL STRUCTURES FOR ON-LINE DYNAMIC CONTROL

There are three main kinds of multilevel structures for on-line dynamic control:

- Structures with dynamic price coordination
- Structures based on the state-feedback concept
- Structures using conjugate variables

We shall now describe their main features and properties and draw some comparisons between them. For clarity, we shall make the formulations as simple as possible by omitting the inequality constraints in the formulation given in section 4.1.

4.2.1. DYNAMIC PRICE COORDINATION

The distinctive feature of dynamic price coordination is the use of prices on the inputs u_i and outputs y_i of the subsystems in order to coordinate the local decisions. It has been described in its principal form in section 1.3; we extend it here, although still using rather simple mathematics. A much more comprehensive treatment is given in section 4.3.

The global problem

The control problem for the interconnected system is (compare section 1.3):

$$\text{minimize } Q = \sum_{i=1}^{N} \int_{0}^{t_f} q_{0i}(x_i(t), m_i(t), u_i(t))\, dt \tag{4.12}$$

subject to

$$\dot{x}_i(t) = f_i(x_i(t), m_i(t), u_i(t)), \qquad i \in \overline{1, N} \text{ (state equations)},$$

$$y_i(t) = F_i^0(x_i(t), m_i(t), u_i(t), \qquad i \in \overline{1, N} \text{ (output equations)},$$

$$u(t) = Hy(t) \text{ (interconnections)}$$

the dependence on $z(t)$ is omitted for convenience; $x(0)$ is given and $x(t_f)$ is free or specified. In comparison with section 4.1, our formulation is more specific and simple. We have assumed that the system dynamics can be described by ordinary differential equations and that there are no inequality constraints. The first part of this discussion follows what was said in section 1.3, but in a very abbreviated way.

Decomposition

When the interconnection equation is incorporated into a Lagrangian:

$$L = \sum_{i=1}^{N} \int_0^{t_f} q_{0i}(x_i(t), m_i(t), u_i(t))\, dt + \int_0^{t_f} \langle \lambda(t), u(t) - Hy(t)\rangle\, dt, \quad (4.13)$$

then this Lagrangian can be split into additive parts, which are the basis of local problems:

$$\text{minimize } Q_i = \int_0^{t_f} [q_{0i}(x_i(t), m_i(t), u_i(t))$$

$$+ \langle \hat{\lambda}_i(t), u_i(t)\rangle - \langle \hat{\mu}_i(t), y_i(t)\rangle]\, dt, \quad (4.14)$$

where $y_i(t)$ is determined by the subsystem output equation, the optimization is subject to

$$\dot{x}_i(t) = f_i(x_i(t), m_i(t), u_i(t)),$$

$x_i(0)$ is given, and $x_i(t_f)$ is free or specified as in the original problem. We have to put the optimal values of price vectors $\hat{\lambda}_i$, $\hat{\mu}_i$ into the local problems, which means that the global problem must have been solved ahead of time. As pointed out in section 1.3, there is little sense in the local problems unless we shorten the local horizons and use feedback. When we shorten the horizon from t_f to t_f', (4.14) becomes

$$\text{minimize } Q_i = \int_0^{t_{f'}} [q_{0i}(x_i(t), m_i(t), u_i(t))$$

$$+ \langle \hat{\lambda}_i(t), u_i(t)\rangle - \langle \hat{\mu}_i(t), y_i(t)\rangle]\, dt \quad (4.15)$$

where $x_i(0)$ is given as before, but the target state is taken from the global long-horizon solution, $x_i(t_f') = \hat{x}_i(t_f')$.

The use of feedback at the local level

The short-horizon formulation (4.15) pays off when we repeatedly solve (4.15) rather than solve the global problem only once. Figure 1.21 in Chapter 1 shows the principle of the proposed control structure. In section 1.3 we said that feedback at the local level consists in solving the short-horizon local problems at some intervals $T_1 < t_f$ and in using the actual value of the measured state $x_i^r(kT_1)$ as a new initial value.

More exactly, the operation of the structure is as follows: At $t = 0$ we solve the problem max Q_i for the horizon $[0, t_f']$ with $x_i(0)$. We apply control $\hat{m}_i$ to the real system for an interval $[0, T_1]$, and at $t = T_1$ we again solve max Q_i for horizon $[T_1, t_f]$ with initial state $x_i(T_1) = x_i^r(T_1)$ as measured. Then we apply control $\hat{m}_i$ to the real system on interval $[T_1, 2T_1]$, and so on.

Note that the local problems, which have to be repeated at intervals T_1, are of low dimension and short horizon. We should not forget the disturbances. As spelled out in section 1.3, disturbance prediction would be used while solving (4.12) and (4.15), the global problem and local problems, respectively.

The use of feedback in coordination

Feedback to the coordinator, mentioned in section 1.3 and Figure 1.21, consisted in supplying the actual values x_i^r at time $t_f', 2t_f', \ldots$ so that the global problem could be solved again for each of these new initial values.

Does this feedback to the coordinator make sense when the lower-level problems have to achieve $x_i^r(t_f') = \hat{x}_i(t_f')$ and already use feedback? It does because the model-based target value $\hat{x}_i(t_f')$ is not optimal for the real system and asking the local decision making to achieve exactly $x_i^r(t_f') = \hat{x}_i(t_f')$ may not be advisable or even feasible. Some numerical evidence of this fact can be found in section 4.3.

The feedback to the coordination level need not be at $t_f', 2t_f'$, and so on. It might be advisable to use the feedback and perform the recomputation of the global problem prior to time t_f', and the feedback should then occur prior to time t_f'.

Static elements in the system

We did not mention in section 1.3 that the length of the global problem horizon t_f has to be matched to the slowest system element dynamics and the slowest of the disturbances. The shortened horizon t_f' for the local problems would in fact result from considering repetitive optimization at the coordination level, for example t_f' could be set at $\frac{1}{10}t_f$. The dynamics of a particular system element may then be fast enough to be neglected in its local optimization problem within the horizon t_f'. This means, in other words, that if we take $\hat{m}_i$ and $\hat{u}_i$ from the global optimization solution, the optimal state solution $\hat{x}_i$ follows these with negligible effect on the element dynamics.

To make this assumption more formal, let us consider that the system element has been supplied with first-layer follow-up controls of some appropriately chosen control variables c_i (see section 1.3 of Chapter 1). We are then allowed to assume that c_i determines both x_i and m_i of the original element and the optimization problem becomes

$$\text{minimize } Q_i = \int_0^{t_f} [q_{0i}'(c_i(t), u_i(t))$$

$$+ \langle \hat{\lambda}_i(t), u_i(t) \rangle - \langle \hat{\mu}_i(t), y_i(t) \rangle]\, dt, \tag{4.16}$$

where $q'_{0i}(\cdot)$ is a reformulation of the function q_{0i} with c_i in place of x_i, m_i. Although (4.16) is not a dynamic problem, its results will be functions of time. In particular, $\hat{c}_i$ is a time-varying control because prices $\hat{\lambda}_i$, $\hat{\mu}_i$ are time-varying. The essential assumption under which the dynamic local problem (4.15) reduces to the static problem (4.16) is that dynamic optimal solutions $\hat{m}_i$, $\hat{u}_i$, $\hat{x}_i$ change slowly with respect to the system dynamics.

The use of simplified models

We have made no use till now of the possibility of simplifying the model in the global problem, which is solved at the coordination level at times 0, t'_f, $2t'_f$, ... The global problem may be simplified for at least two reasons: the solution of the full problem may be too expensive, or the data on the real system, in particular the prediction of disturbances, may be too inaccurate to justify computations based on the exact model.

Simplification may concern the dimension of the state vector (aggregated x^c instead of x), the control vector (m^c instead of m), or the inputs and outputs ($u^c = H^c y^c$ instead of $u = Hy$).

The global problem Lagrangian will now be

$$L = \sum_{i=1}^{N} \int_0^{t_f} [q^c_{0i}(x^c_i(t), m^c_i(t), u^c_i(t))\, dt + \int_0^{t_f} \langle \lambda^c(t), u^c(t) - H^c y^c(t) \rangle\, dt.$$

$$(4.17)$$

The simplified solution will yield optimal state trajectory $\hat{x}^c = (\hat{x}^c_1, \hat{x}^c_2, \ldots, \hat{x}^c_N)$ and optimal price function $\hat{\lambda}^c$. The linking of those values to the local problems cannot be done directly because the local problems consider the unaggregated vectors x_i, u_i, and y_i. We have to change the previous requirement $x_i(t'_f) = \hat{x}_i(t'_f)$ into a new one

$$\gamma_i(x_i(t'_f)) = \hat{x}^c_i(t'_f),\qquad\qquad (4.18)$$

which, incidentally, is a more flexible constraint, and we also have to generate a full price vector $\hat{\lambda}$:

$$\hat{\lambda} = R\hat{\lambda}^c \qquad\qquad (4.19)$$

where R is an appropriate price proportion matrix. The prices of the aggregated λ^c may be termed group prices.

We should note that the functions γ_i and matrix R have to be appropriately chosen. The choice could be made by model consideration but at present it is not possible to indicate how this could be accomplished in a general case. We should also note that even with the best possible choice, the optimality of the overall solution will be affected, except in some special cases.

System interconnection through storage elements

We have only considered system interconnections that are stiff, that is, outputs connected to inputs in a permanent way. The full dynamic problem formulation (as given in section 4.1) also considers a weak interconnection of an integral type:

$$\int_{kt_b}^{(k+1)t_b} (u_{ij}(t) - y_{\ell r}(t))\, dt = 0,$$

which corresponds to taking input u_{ij} to subsystem i from a store, with some output $y_{\ell r}$ of subsystem ℓ connected to the same store. To ask that the integral equal zero over $[kt_b, (k+1)t_b]$ means that inflow and outflow have to be in balance over each balancing period t_b (assuming that the capacity of the store is large enough).

A store may be supplied by several outputs and drained by more than one subsystem input. There may also be many stores, for example, for different products. If we assume the same balancing period for all of them, then the integral constraint becomes

$$\int_{kt_b}^{(t+1)t_b} (\bar{H}_1 u_w(t) - \bar{H}_2 y_w(t))\, dt = 0,$$

where u_w, y_w are parts of u, y connected to the stores (the stiffly interconnected parts will be termed u_s, y_s). Matrices $\bar{H}_1, \bar{H}_2$ show the way in which u_w, y_w are connected to the various stores. The number of stores is of course $\dim \bar{H}_1 y_w = \dim \bar{H}_2 u_w$. A state vector w of the inventories can also be introduced

$$w(kt_b + t) = w(kt_b) + \int_{kt_b}^{kt_b + t} (\bar{H}_1 u_w(t) - \bar{H}_2 y_w(t))\, dt. \tag{4.20}$$

At this point it is worthwhile to note that the notion of inventory couplings (4.8) is in general based on an assumption about the existence of inventories in the system. This in turn usually implies the boundedness of stocks at any moment, and we should introduce constraints of the form:

$$\mathscr{C}'_\ell \le - \int_{t_{1\ell}}^{t} [\bar{H}_1 u_w(t) - \bar{H}_2 y_w(t)]\, dt \le \mathscr{C}''_\ell \tag{4.8a}$$

(or $0 \le w(kt_b + t) \le w_{\max}$ for $w(kt_b + t)$ as given by (4.20)). If we drop these constraints from our considerations, we assume implicitly that the fulfillment (or approximate fulfillment) of condition (4.8) implies the fulfillment of (4.8a) owing to the existence of a sufficient capacity reserve. For further discussion of this issue, see Malinowski and Terlikowski (1978).

When both stiff and soft interconnections are present in the system, the global Lagrangian problem becomes

$$L = \sum_{i=1}^{N} \int_0^{t_f} q_{0i}(x_i(t), m_i(t), u_i(t))\, dt + \int_0^{t_f} \langle \lambda(t), u_s(t) - Hy_s(t) \rangle\, dt$$

$$+ \sum_{k=0}^{k=t_f/t_b - 1} \left\langle \eta^k, \int_{kt_b}^{(k+1)t_b} (\bar{H}_1 u_w(t) - \bar{H}_2 y_w(t))\, dt \right\rangle \tag{4.21}$$

and we of course continue to consider

$$\dot{x}_i(t) = f_i(x_i(t), m_i(t), u_i(t)), \qquad i = 1, \ldots, N,$$
$$y_i(t) = F_i^0(x_i(t), m_i(t), u_i(t)), \qquad i = 1, \ldots, N,$$

$x_i(0)$ given, and $x_i(t_f)$ free or specified, $i \in \overline{1, N}$.

In comparison with the previous Lagrangian, a new term has now appeared reflecting the new constraint. Note that prices η^k associated with the integral constraint are constant over period t_b. Note also that if t_b tends to zero, the integral constraint becomes similar to the stiff one and η, which changes in steps, will change continuously, as λ does.

With two kinds of interconnections, the local problems become

$$\text{minimize } Q_i = \int_0^{t_f} [q_{0i}(x_i(t), m_i(t), u_i(t)) + \langle \hat{\lambda}_i(t), u_{si}(t) \rangle - \langle \hat{\mu}_i(t), y_{si}(t) \rangle]\, dt$$

$$+ \sum_{k=0}^{k=t_f/t_b - 1} \left\langle \hat{\eta}^k, \int_{kt_b}^{(k+1)t_b} (\bar{H}_{1i} u_{wi}(t) - \bar{H}_{2i} y_{wi}(t))\, dt \right\rangle, \tag{4.22}$$

where $y_{si}(t) = F_{si}^0(x_i(t), m_i(t), u_i(t))$, $y_{wi}(t) = F_{wi}^0(x_i(t), m_i(t), u_i(t))$ and the optimization is subject to $\dot{x}_i(t) = f_i(x_i(t), m_i(t), u_i(t))$, $x_i(0)$ given, and $x_i(t_f)$ free or specified.

In problem (4.22), inputs u_{wi} taken from the stores are now free control variables and can be shaped by the local decision maker who controlled only m_i in (4.15). The local decisions will be under the influence of prices $\hat{\lambda}$ and $\hat{\eta} = (\hat{\eta}^0, \hat{\eta}^1, \ldots,)$, where both $\hat{\lambda}$ and $\hat{\eta}$ have to be set by the solution of the global problem. Local problem (4.22) has no practical value yet; it will make sense when we introduce local feedback and shorten the horizon. We omit the details and show only the control scheme (see Figure 4.1).

In order to improve the decisions of the coordinator, we made a proposal to feed the actual $x^r(t_f')$ to his level. We have now additional state variables, the inventories w; if price $\hat{\eta}^k$ is wrong, the stores will not balance over $[kt_b, (k+1)t_b]$. We can correct the imbalance by influencing the price η^{k+1} for the next period. The new price should be in line with the difference $\hat{w}((k+1)t_b) - w^r((k+1)t_b)$, where $w^r(\cdot)$ is a value measured in the real

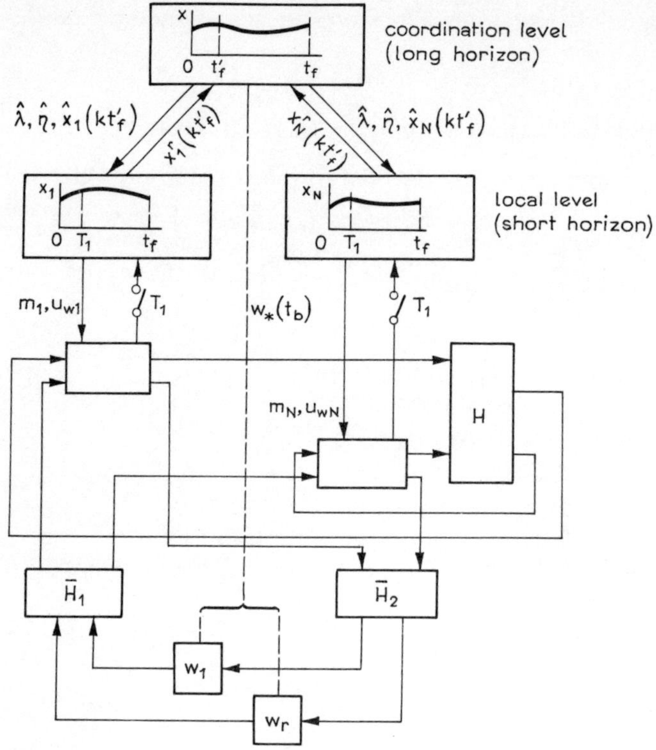

FIGURE 4.1   On-line dynamic price coordination in a system containing inventories in the interconnections.

system. This kind of feedback is also shown in Figure 4.1. Section 4.4 shows some algorithms for improving prices η^k.

Conclusions

We have shown that time-varying prices are a possible coordination instrument that can be used in a multilevel structure of on-line control, if accompanied by prescribed target states. The local problems may be formulated with a short horizon and low dimension. The coordination level must, solve the global problem for the full horizon in order to generate the optimal prices and the target states for the local problems. A simplified global model may be used in appropriate cases.

The price coordination structure applies to systems with stiff interconnections and to system with interconnections through storage elements. The operation of the structure depends on the possibility of solving the optimization problems numerically. Analytical solutions of the dynamic problems involved are not needed, so we are by no means restricted to linear-quadratic systems.

4.2.2. MULTILEVEL CONTROL BASED ON THE STATE FEEDBACK CONCEPT

There has been considerable research devoted to the structure in which the optimal control at time t, $\hat{m}(t)$, is determined as a given function of the current state $x(t)$. Comprehensive solutions exist in this area for the linear system and the quadratic performance case, where the feedback function proved to be linear; that is, we have

$$\hat{m}(t) = K(t)x(t),$$

where $K(t)$ is in general a time-varying matrix. To apply this approach to a complex system, we might use for each local problem

$$\hat{m}_i(t) = K_{ii}(t)x_i(t), \tag{4.23}$$

where K_{ii} is one of the diagonal blocks of matrix K.

The result of such local controls, although all states of the system are measured and used, is not optimal. Note that for $\hat{m}_i(t)$ we should use

$$\hat{m}_i(t) = K_i(t)x(t),$$

that is, we should make $\hat{m}_i(t)$ dependent on the overall state $x(t)$.

We can compensate for the error committed in (4.23) by adding a suitably computed correction signal

$$\hat{m}_i(t) = K_{ii}(t)x_i(t) + \hat{v}_i(t). \tag{4.24}$$

To get $\hat{v}_i(t)$ exactly we should generate it continuously from the overall $x(t)$. This would, however, be equivalent to implementing state feedback for the whole system directly, and we would lose the advantage of having small local problems. Adding $\hat{v}_i(t)$ means, in fact, overriding the local decisions. In particular, $\dim v_i = \dim m_i$. Exactness has to be sacrificed. With this in mind, we propose various solutions, for example (see section 4.6 and Figure 4.2),

* $\hat{v}_i$ will be generated at $t = 0$ for the whole optimization horizon t_f (open-loop compensation)
* $\hat{v}_i$ will be generated at $t = 0$ as before but will be recomputed at $t = t_f' < t_f$, using the actual $x(t_f')$ (repetitive compensation—see subsection 4.6.3)
* $\hat{v}_i$ will not be generated at all; instead we implement in the local problems (see subsection 4.6.2)

$$\hat{m}_i(t) = K_{ii}^0 x_i(t), \tag{4.25}$$

where the feedback gain matrix K_{ii}^0 is adjusted so as to approach optimality.

322

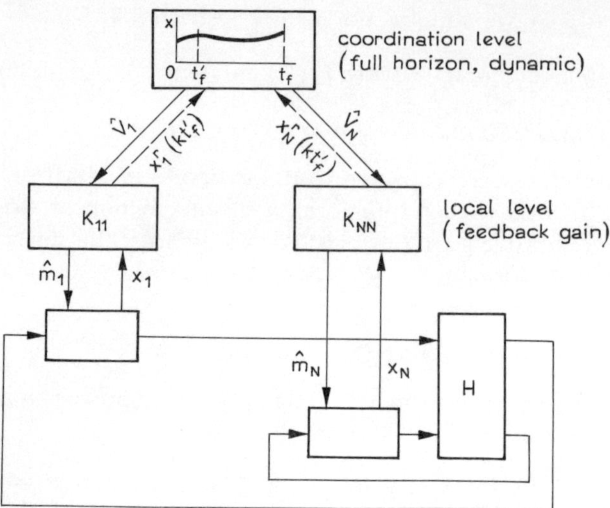

FIGURE 4.2   Dynamic multilevel control based on the feedback gain concept.

This structure may be referred to as decentralized control. We could think of readjusting K_{ii}^0 at some time intervals, which could be looked on as adaptation. This adaptation would present a way of on-line coordination of the local decisions.

Local decision making based on (4.23), (4.24), or (4.25) makes more sense for fully computerized implementation than for a hierarchy of human decision makers. The previous approach based on minimization of the local performance subject to imposed prices seems to describe the workings of the real system better than the state feedback approach. We should also remember that the solutions in the form of feedback gains to the optimization problems are available for a restricted class of these problems only.

4.2.3.   STRUCTURES USING CONJUGATE VARIABLES

It is possible to base the local on-line optimization on the minimization of the simultaneous value of the Hamiltonian, and thus use the trajectories of conjugate variables computed during the global optimization as an additional control decision of the upper-layer controller of the dynamic coordinator. The use of conjugate variables has been described by Findeisen (1977); one can also find some attempts at practical applications (e.g., Foord 1974).

For the complex system optimization problem (4.12), the Hamiltonian would be

$$\mathcal{H} = \sum_{i=1}^{N} q_{0i}(x_i(t), m_i(t), u_i(t)) + \langle \psi(t), f(x(t), m(t), u(t)) \rangle. \qquad (4.26)$$

The interconnection equation

$$u(t) - Hy(t) = u(t) - HF^0(x(t), m(t), u(t)) = 0$$

provides that $u(t)$ is a function of $(x(t), m(t))$ in the interconnected system

$$u(t) = \varphi(x(t), m(t)).$$

Therefore

$$\mathcal{H} = \sum_{i=1}^{N} q_{0i}(x_i(t), m_i(t), \varphi_i(x(t), m(t)))$$
$$+ \langle \psi(t), f(x(t), m(t), \varphi(x(t), m(t))) \rangle. \qquad (4.27)$$

We assume that the global problem has been solved and hence the optimal trajectories of conjugate variables $\hat{\psi}$ can be easily computed. We are going to use the values of $\hat{\psi}$ in the local problems. Having $\hat{\psi}$, we could redetermine the optimal control by performing at the current time t

$$\text{minimize } \mathcal{H} = \sum_{i=1}^{N} q_{0i}(x_i(t), m_i(t), \varphi_i(x_i(x(t), m(t)))$$
$$+ \langle \hat{\psi}(t), f(x(t), m(t), \varphi(x(t), m(t))) \rangle, \qquad (4.28)$$

where the problem is an "instantaneous minimization" and needs no consideration of final state and future disturbances. This information was of course used to solve the global problem and determine $\hat{\psi}$ for the whole time horizon.

To perform (4.28) we need the actual value of state x. We could obtain it by simulating the system behavior starting from time t_1 when the initial condition $x(t_1)$ was given, that is, by using the equation

$$\dot{x}(t) = f(x(t), m(t), \varphi(x(t), m(t))),$$

with $x(t_1)$ given and $m = \hat{m}$ known for $[t_1, t_1]$ from the previous solutions of (4.28). We could also find $x(t)$ by measuring it in the real system, as long as we take into account possible model–reality differences, about which we have incomplete information.

Problem (4.28) is a static optimization, not a dynamic one. We would now like to divide it into subproblems by treating $u(t) - Hy(t) = 0$ as a side condition and solving (4.28) with the Lagrangian

$$L = \sum_{i=1}^{N} q_{0i}(x_i(t), m_i(t), u_i(t))$$
$$+ \langle \hat{\psi}(t), f(x(t), m(t), u(t)) \rangle + \langle \lambda(t), u(t) - Hy(t) \rangle, \qquad (4.29)$$

where $y(t) = F^0(x(t), m(t), u(t))$.

324

Before we get any further with this Lagrangian and its decomposition, let us note how it differs from the Lagrangian for dynamic price coordination. We had there

$$L = \int_0^{t_f} \sum_{i=1}^{N} q_{0i}(x_i(t), m_i(t), u_i(t))\, dt + \int_0^{t_f} \langle \lambda(t), u(t) - Hy(t) \rangle\, dt$$

subject to

$$\dot{x}_i(t) = f_i(x_i(t), m_i(t), u_i(t)), \qquad i \in \overline{1, N}.$$

It was a dynamic problem. In the present case there are no integrals in $L(\cdot)$ and the dynamics are taken care of by the value of conjugate variables $\hat{\psi}$. The differential equations of the system are needed only to compute the current value of x in our new "instantaneous" Lagrangian. No future disturbances have to be known, no optimization horizon is considered—both are included in $\hat{\psi}$.

Assume that we have solved problem (4.29) by using the system model, i.e., by computation, and that we have the current optimal value of price $\hat{\lambda}$, $\hat{\lambda}(t)$. We can then formulate the following static local problems to be solved at time t

$$\underset{m_i(t), u_i(t)}{\text{minimize}}\, L_i = q_{0i}(x_i(t), m_i(t), u_i(t))$$

$$+ \langle \hat{\psi}_i(t), f_i(x_i(t), m_i(t), u_i(t)) \rangle$$
$$+ \langle \hat{\lambda}_i(t), u_i(t) \rangle - \langle \hat{\mu}_i(t), y_i(t) \rangle. \tag{4.30}$$

These local problems could be used in a structure of decentralized control; see Figure 4.3. The local decision makers are asked here to minimize $L_i(\cdot)$ by using the model and to apply control $\hat{m}_i(t)$ to the system elements. The current value $x_i(t)$ is needed to perform the task. The coordination level would supply $\hat{\psi}_i(t)$ and the prices $\hat{\lambda}_i(t)$ and $\hat{\mu}_i(t)$ for the local problem, which would be different for each t. Note that there is no experimental search (hill-climbing) on the system itself.

Figure 4.3 implies that the local model-based problems are solved immediately with no lag or delay. We can therefore assume, conceptually, that the local decision making is more than implementation of a state feedback loop relating control $\hat{m}_i(t)$ to the measured $x_i(t)$. In an appropriate case we could solve problem (4.29) analytically with $\hat{\psi}_i(t)$ and $\hat{\lambda}(t)$ as parameters and the result would be a feedback decision rule. If an analytical solution of (4.29) cannot be done, we have to implement a numerical algorithm of optimization and some time will be needed to perform it. An appropriate version of control with conjugate variables would have to be considered (see section 4.3).

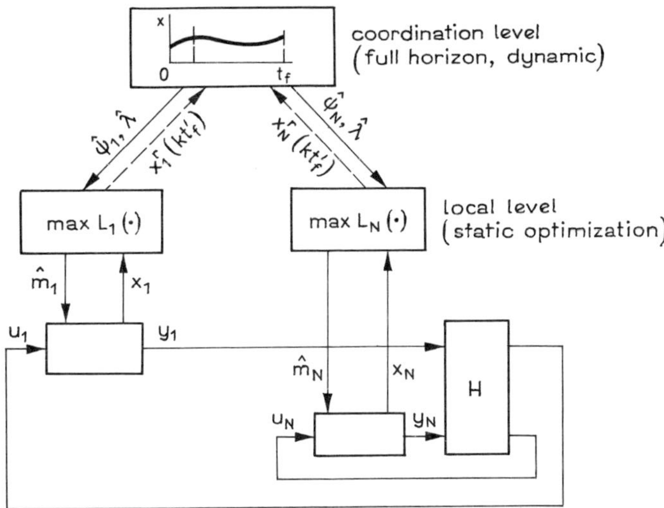

FIGURE 4.3 Dynamic multilevel control using conjugate variables.

Now let us think about feedback to the coordinator. We might decide to let him know the state of the system at some time intervals t_f', that is $x(kt_f')$, on which he could base his solution $\hat{\psi}$ for all $t \geq kt_f'$ and also the prices $\hat{\lambda}$ for the next interval $[kt_f', (k+1)t_f']$. This procedure would be very similar to what we have proposed in dynamic price coordination.

It might be worthwhile to again make some comparisons between dynamic price coordination and the structure using both prices and conjugate variables. Both these cases do not prescribe a state trajectory. In dynamic control, direct coordination with a state trajectory or input and output trajectories would be difficult to perform if model-reality differences are assumed.

In this structure the local problems are static. The local goals are slightly less intuitive, as they involve $\langle \hat{\psi}_i(t), \dot{x}_i(t) \rangle$, that is, the "value of the trend." This would be difficult to explain economically and hence difficult to implement in a human decision-making hierarchy. Since the problem is static, no target state is prescribed.

4.2.4. COMPARISON OF CONTROL STRUCTURES

We have discussed three structures for a dynamic multilevel control system using feedback from the real system in the course of its operation. It is not yet possible to evaluate all advantages and drawbacks of the alternatives However, if the mathematical models do not differ from reality, all the

TABLE 4.1   Comparison of Dynamic Coordination Structures

	Coordinator	Local Problems	Local Goals
Dynamic price coordination	Solves global problem, sets price λ and targets x_i	Dynamic optimization	Maximize performance, achieve target state
State feedback	Solves global problem, supplies compensation signal $\hat{v}_i$	State feedback decision rule	No goal
Conjugate variables	Solves global problem, sets prices λ and conjugate variables $\hat{\psi}_i$	Static optimization	Maximize performance inclusive of $\langle \psi_i(t), \dot{x}_i(t) \rangle$

structures would give the same result, the fully optimal control. The important question is what will happen if the models differ from reality. Quantitative answers do not exist. Some partial results are available, and are presented in later sections.

Another feature of the structures concerns their use in a human decision-making hierarchy. It is then important what the local decision problem will be, the one assigned to the individual decision maker. He may feel uncomfortable, for example, if asked to implement only a feedback decision rule, as in the state feedback structure, or to account for the value of the trend $\langle \hat{\psi}_i(t), \dot{x}_i(t) \rangle$ in his own calculations, as in the structure using conjugate variables. For the human decision maker, the structure with price coordination seems to be the most natural. Table 4.1 shows a comparison of the structures.

Optimizing and regulatory dynamic control

In a hierarchical control system we often decide that it is the control determined at the higher levels that should have the responsibility for optimizing performance, as opposed to the lower-level regulatory control that has to follow the directives specified at higher levels. Usually, the optimizing control is computed for long time horizons and the mathematical models reflect only the dominant ("slow") process dynamics. In this type of control problem

• The process equations are nonlinear and the performance index is neither quadratic nor linear.

- The disturbances to be taken into account are those that have a major effect on the process behavior. They usually change slowly but cannot be considered stationary.
- The interventions of the optimizing controller can only be made at certain intervals (e.g., once a day, once a week, or when an intervention is needed), and only discrete observations of the process behavior (e.g., values of the dominant state variables) are used.

Because of the above characteristic features of the problem, it seems to be very difficult to utilize a stochastic approach for the design of optimizing control algorithms. Moreover, the stochastic control approach may be inappropriate because it would optimize the expected value of the system performance, while we would like to develop a control strategy that would result in satisfactory system behavior for a specific, i.e., predicted, occurrence of dominant disturbances. So it is often more appropriate for us to use a deterministic approach based on prediction of disturbances. This is also the only approach that seems appropriate for an actual application. It will be possible to compensate for errors due to the deterministic treatment when the lower-level regulatory control mechanisms are designed. We should note that the control structure described in subsection 4.2.1, for example, could be considered as an optimizing controller superimposed on some lower-level control systems.

When designing the algorithms of the regulatory control that takes place at the lower levels of a control hierarchy, we usually formulate different objectives and consider different properties of the process from those used in the optimizing control design. The main features of the control problem this time will be as follows:

- The principal task of the regulatory controllers is to cause the process to follow a specified reference trajectory. This trajectory may reduce to stepwise changing set-points if steady-state optimization is used at the higher levels (see Chapters 1 and 3).
- In most cases, the models used to describe the process behavior in some neighborhood of the specified reference trajectory can be linear. Very often, a quadratic performance index describes the regulatory control quality well enough (see, e.g., Athans 1971).
- Usually, it is necessary to take into account fast disturbances (noise) that influence the process. The available measurements will have to be considered incomplete and noise-corrupted.
- Because of the fast disturbances, the regulatory control has to be continuous, at least in principle. Therefore, fast on-line data transmission and processing is required. Heavy demands on the information transfer structure are imposed in that case (see Chapter 5 and section 4.6), and important information constraints may have to be considered.

There are several reasons, the informational aspects among them, for which one tends to consider decentralized rather than centralized control structures for the regulatory control task. If state estimation is required, because of the incomplete and noisy measurements, we have to construct not only the decentralized decision rules, but also the decentralized filtering. The linearity of the models and the quadratic performance indices permit the investigation of the properties of the decentralized structures by means of stochastic control theory.

Some decentralized control structures will be presented in section 4.6. The basic state feedback concepts have been described in subsection 4.2.2. Sometimes, in order to increase the performance of the regulatory control, we introduce an appropriate coordination of the actions of the local controllers. This coordination will be very similar to those in optimizing control, but the linear-quadratic structure of the problem together with the noisy character of the disturbances will make it possible to use stochastic control.

4.3. DYNAMIC PRICE COORDINATION STRUCTURE

4.3.1. TWO-LEVEL SOLUTION OF THE SYSTEM OPTIMIZATION PROBLEM

Let us suppose that the system optimization problem SOP, Eq. (4.11), is being solved in a two-level fashion in the following way. We incorporate the interconnection equation (4.6), that is, both the stiff interconnections (4.7) and the weak interconnections (4.8), into the following Lagrangian:

$$L = Q(x(t_0), m, u, \bar{z}) + \langle \lambda, P(u, y) \rangle, \qquad (4.31)$$

where $\lambda \in \mathcal{P}$, and $\langle \cdot, \cdot \rangle$ denotes a scalar product in $\mathcal{P}$. Equation (4.31) is a more general form of Lagrangian (4.13) in section 4.1. The optimization is considered over the interval Δ_f, but the subscripts Δ_f are omitted for convenience.

For price coordination, it is essential that we be able to solve SOP by the interaction balance method (IBM, see Chapter 2, section 2.4). To do so, we define the infimal problem (IP):

For given $\lambda \in \mathcal{P}$ find $\bar{m}(x(t_0), \lambda, \bar{z})$, $\bar{u}(x(t_0), \lambda, \bar{z})$ minimizing L on MU (see Eq. (4.11)), where for y we substitute output equation (4.5). $\qquad$ (4.32)

We denote by $\hat{L}(x(t_0), \lambda, \bar{z})$ the value of L obtained in the solution to IP. The essential feature of IP is that it can be solved by solving independently N local problems in variables m_i and u_i. Though in general it is not necessary (see section 2.4) we assume that the solution of IP is unique for given $x(t_0)$, $\bar{z} \in \mathcal{Z}_0 \subset \mathcal{Z}$, $\lambda \in \mathcal{P}_0 \subset \mathcal{P}$.

In order to achieve the balance condition, that is, to satisfy (4.6), we have to introduce the supremal problem SP:

Find

$$\hat{\lambda} = \hat{\lambda}(x(t_0), \bar{z}) \in \mathscr{P}_0$$

such that

$$P(\bar{u}(x(t_0), \hat{\lambda}, \bar{z}), \bar{y}(x(t_0), \hat{\lambda}, \bar{z})) = 0, \qquad (4.33)$$

where

$$\bar{y}(x(t_0), \hat{\lambda}, \bar{z}) = F(x(t_0), \bar{m}(x(t_0), \hat{\lambda}, \bar{z}), \bar{u}(x(t_0), \hat{\lambda}, \bar{z}), \bar{z}).$$

If SP has solution $\hat{\lambda}(x(t_0), \bar{z})$, then the resulting controls as defined by IP are optimal in the model and the solution of SOP is found. The conditions under which the solution of SP exists are described in Chapter 2.

4.3.2. OPERATION OF THE CONTROL STRUCTURE

Suppose that we control the real process and that at times t_j (the sequence $\{t_j\}$ may be specified beforehand or may result from on-demand operation) we can obtain the measured or estimated values of the current state in the real system and other information like the current value of integral (4.8) if $t_j \in \Delta_\ell$ for some ℓ. One of the possibilities for system operation has already been briefly described in section 4.2 (see also Figure 4.4). The supremal unit (coordinator) operates as follows.

At times t_{j_k}, where, e.g., $j_1 = 0$, $j_2 = 10$, $j_3 = 20$, the global system optimization problem (SOP) is solved over the interval $\Delta_{fk} = [t_{j_k}, t_f]$, or over $\Delta_{fk} = [t_{j_k}, t_{fk}]$ if we apply the floating horizon approach. The solution is performed by IBM. The measured system state $x^r(t_{j_k})$ as well as the updated prediction $\bar{z}_{\Delta_{fk}}$ are used each time.

The solution provides for model based optimal price trajectory $\hat{\lambda}_{\Delta_{fk}}$ and model-based optimal state trajectory $\hat{x}_{\Delta_{fk}}$ for the whole horizon $[t_{j_k}, t_f]$; segments of these trajectories will be used as the control (decision) variables of the coordinator in interval $[t_{j_k}, t_{j_{k+1}}]$.

As has already been mentioned, final time t_{fk} may be fixed as t_f. This means that we control the system over a finite interval of time, the number of time instants t_{j_k} is finite, and the optimization horizon is shortened after each intervention of the coordinator. If $t_{fk} = t_{j_k} + T_g$ (a constant optimization horizon), then T_g have to be not less than the greatest interval between adjacent time instants t_{j_k} and $t_{j_{k+1}}$. At the same time, t_{fk} has to be a final time instant of some interval Δ_ℓ if integral constraints (4.8) are present. On the other hand, if t_{j_k} belongs to the interior of some Δ_ℓ then we have to modify constraint (4.8) suitably in SOP using the current value of integral (4.8).

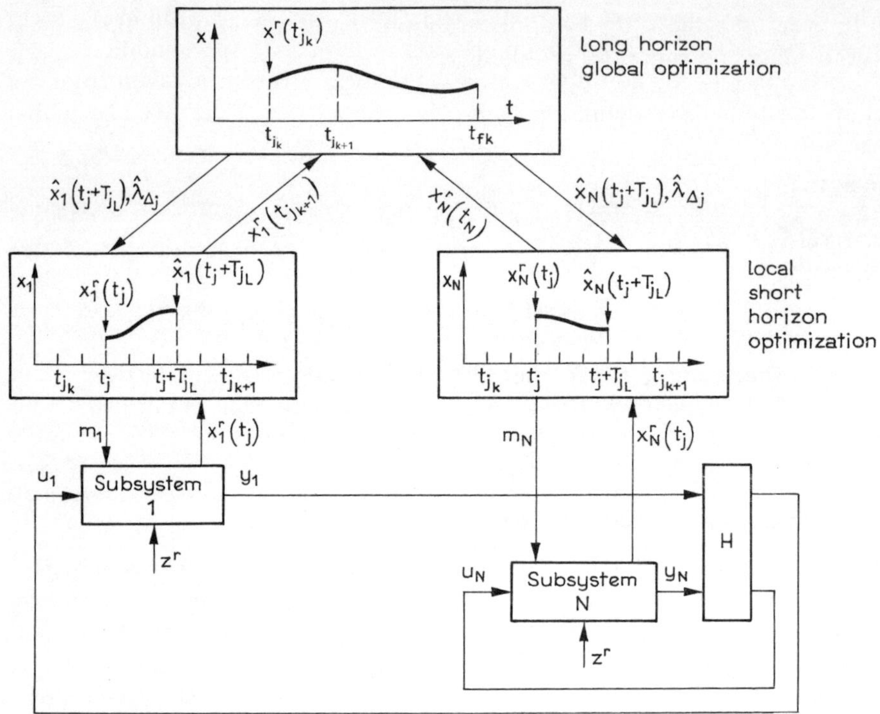

FIGURE 4.4   A two-level structure for dynamic on-line coordination (for simplicity, only stiff interconnections (4.7) are shown).

The local control units operate as follows. At time instants t_j ($t_{j_k} \leq t_j < t_{j_{k+1}}$) the local controllers specify control $\tilde{m}_{i\Delta_j}$, $\tilde{u}_{i\Delta_j}$ for the horizon $\Delta_j = [t_j, t_j + T_{jL}]$, where $\Delta_j \subset \Delta_{fk}$, based on measured state $x_i^r(t_j)$ and updated prediction $\tilde{z}_{\Delta_j}$. An appropriate segment of price trajectory $\hat{\lambda}_{\Delta_j}$ is used here as well as the desired final state $x_i(t_j + T_{jL}) = \hat{x}_{ifk}(t_j + T_{jL})$; both are taken from the supremal unit.

A more formal statement of the local problem (LPS_i) solved at each t_j is the following:

Find

$$\tilde{m}_{i\Delta_j}(x_i^r(t_j), \hat{\lambda}_{\Delta_j}, \tilde{z}_{\Delta_j}, \hat{x}_{ifk}(t_j + T_{jL})), \tilde{u}_{i\Delta_j}(x_i^r(t_j), \hat{\lambda}_{\Delta_j}, \tilde{z}_{\Delta_j}, \hat{x}_{ifk}(t_j + T_{jL})),$$

that minimize the following functional on $MU_{i\Delta_j}$:

$$L_{i\Delta_j} = \int_{\Delta_j} q_{0i}(x_i(t), m_i(t), u_i(t), \tilde{z}(t)) \, dt + \langle \hat{\lambda}_{\Delta_j}, P_{i\Delta_j}(u_{i\Delta_j}, y_{i\Delta_j}) \rangle, \quad (4.34)$$

where $x_i(t)$ and $y_i(t)$ are calculated using (4.1) and (4.3), with $x_i(t_j) = x^r(t_j)$ and $z(t) = \tilde{z}(t) = \tilde{z}_{\Delta j}(t)$, and minimization is subject to the condition $x_i(t_j + T_{jL}) = \hat{x}_{ifk}(t_j + T_{jL})$. The target state for the local problem is taken from the global solution. The definition of $P_{i\Delta_j}$ is obvious and $\hat{\lambda}_{\Delta_j}$ is given by global solution $\hat{\lambda}_{\Delta_j} = \hat{\lambda}_{\Delta_{fk}}|_{\Delta_j}$.

The calculated $\tilde{m}_{i\Delta_j}$ and $\tilde{u}_{i\Delta_j}$ are applied to the real process over time interval $[t_j, t_{j+1}]$ only. At time $t = t_{j+1}$ the next intervention of the on-line control structure takes place.

The interventions of the supremal unit are less frequent than the local ones. Hence, if the sequence $\{t_j\}$ is $t_0, t_1, t_2, \ldots$ spaced at $t_{j+1} - t_j = L$, the sequence $\{t_{j_k}\}$ could be, for example, $t_0, t_{12}, t_{24}, \ldots$, that is, spaced at $l = 12L$, or at similar intervals.

It is evident that the results of the control may heavily depend on the choice of the sequences $\{t_j\}$ and $\{t_{j_k}\}$ and the choice of the local horizons T_{jL}, which may be constant or varying. These problems are rather difficult to assess in a general way. We are able to show only a few sample results in section 4.5.

4.3.3. PROPERTIES OF THE CONTROL STRUCTURE

Optimality

From the construction of a control mechanism it can be easily predicted that when the disturbance predictions in the control scheme coincide with the real disturbance trajectory z^r and $t_{fk} = t_f \forall k$, then the control generated is optimal for the real system. We assume that the solution of SP exists.

If we use floating horizon t_{fk} at the coordination level and $z^r(t) = \bar{z}_{\Delta_{fk}}(t) = \tilde{z}_{\Delta_j}(t) \forall t$, then, as before, the local decision units will repeat the controls generated at the higher layer. The optimality of the behavior of the controlled system will then depend on the proper choice of optimization horizons $[t_{j_k}, t_{fk}]$. In both cases constraints (4.8) are satisfied.

In real cases, when $z^r_{\Delta_{fk}} \neq \bar{z}_{\Delta_{fk}}$ and, in general, when $z^r_{\Delta_j} \neq \tilde{z}_{\Delta_j}$, we have to examine whether it is possible for the control mechanism to operate properly. If it does then we have to know how the interesting variables in the real system (constraint values, performance index, state trajectory $x^r(t)$, and so on) depend on the differences $z^r_{\Delta_{fk}} - \bar{z}_{\Delta_{fk}}$ and $z^r_{\Delta_j} - \tilde{z}_{\Delta_j}$. We can try to obtain some quantitative answers or examine the behavior of the system when the above differences converge to zero. These topics are very difficult to investigate and have not yet been fully explored. However, some results are presented below in the discussions of the coordinator and local-level mechanisms.

332

The coordination level

The coordination level, in order to generate prices $\hat{\lambda}_{\Delta_{fk}}$, has to solve the supremal problem SP at each instant of time t_{j_k}. Since $\hat{\lambda}_{\Delta_{fk}}$ along with $\hat{x}_{\Delta_{fk}}$ are the decision variables of the coordinator used at the local level, we are particularly interested in the properties of $\hat{\lambda}_{\Delta_{fk}}(x(t_{j_k}), \bar{z}_{\Delta_{fk}})$ such as the its uniqueness and its continuity in $x(t_{j_k})$, $\bar{z}_{\Delta_{fk}}$. To simplify the notation, we omit the subscript Δ_{fk} denoting the optimization horizon and treat the initial state value as a component of z.

The solution of SP may be obtained, for example, by application of coordination algorithm (2.68) presented in section 2.4. In the basic formulation of this algorithm we define mapping $W : \mathscr{P}_0 \times \mathscr{Z}_0 \to \mathscr{P}$

$$W(\lambda, z) = -V(\bar{m}(\lambda, z), \bar{u}(\lambda, z) \tag{4.35}$$

where

$$V(m, u, z) = P(u, F(m, u, z)). \tag{4.36}$$

The initial state has been included in z, but we preserve the old symbols to denote the spaces and the various sets of disturbances.

The task of the supremal problem of IBM consists in finding $\hat{\lambda}(\bar{z})$ for given $\bar{z}$, such that

$$W(\hat{\lambda}(\bar{z}), \bar{z}) = 0. \tag{4.37}$$

We assume that IP (see (4.32)) solutions $\bar{m}(\lambda, \bar{z})$, $\bar{u}(\lambda, \bar{z})$ are unique for every $(\lambda, z) \in \mathscr{P}_0 \times \mathscr{Z}_0$, therefore, $W(\cdot, \cdot)$ is well defined on $\mathscr{P}_0 \times \mathscr{Z}_0$. The conditions guaranteeing the existence of $\hat{\lambda}(\bar{z})$ can be found in Chapter 2. In most cases it is difficult to check before the implementation of IBM whether these conditions are satisfied. Nevertheless, they give us some indications concerning the possibility of using IBM for a given system optimization problem.

To find $\hat{\lambda}(\bar{z})$ we can use, among others, algorithm (2.68) from Chapter 2, section 2.4, to generate sequence $\{\lambda^{(n)}\}$ of the values of λ:

$$\lambda^{(n+1)} = \lambda^{(n)} - \varepsilon_n A W(\lambda^{(n)}, \bar{z}). \tag{4.38}$$

The convergence conditions of this algorithm have been thoroughly examined in section 2.4. If the assumptions of Theorem 2.18 or Theorem 2.19 are satisfied, then (4.38) converges to $\hat{\lambda}(\bar{z})$ and $\hat{\lambda}(\bar{z})$ is unique in $\mathscr{P}_1 \subset \mathscr{P}_0$ $\forall \bar{z} \in \mathscr{Z}_0$ (we have assumed that all conditions hold uniformly on $\mathscr{Z}_0$). If $\mathscr{P}_1$ is an open set, then from the uniqueness of $\hat{\lambda}(\bar{z})$ in $\mathscr{P}_1$ it follows that $\hat{\lambda}(\bar{z})$ is a unique point minimizing a convex functional $\hat{\phi}(\cdot, \bar{z})$, where

$$\hat{\phi}(\lambda, \bar{z}) = -L(\bar{m}(\lambda, \bar{z}), \bar{u}(\lambda, \bar{z}), \lambda, \bar{z})$$

and therefore $\hat{\lambda}(\bar{z})$ is a unique solution of SP. See Eq. (4.31).

We now have to consider the continuity in z (on $\mathscr{Z}_0$) of the model-based optimal performance value as well as the continuity of $\hat{\lambda}(\cdot)$. It is evident that the model-based optimal performance for given z is equal to $-\hat{\phi}(\hat{\lambda}(z), z)$.

LEMMA 4.1.   *If we assume that*

1.   *Mappings $Q(w, \cdot)$, $V(w, \cdot)$, where $w = (m, u)$ (see Eq. (4.36) are uniformly continuous in z on $\mathscr{Z}_0$, and this property is uniform in $w \in MU$*

$$(\forall \varepsilon > 0)(\exists \delta > 0)(\forall w \in MU)(\forall z_1, z_2 \in \mathscr{Z}_0) \; \|z_1 - z_2\| \leq \delta:$$

$$|Q(w, z_1) - Q(w, z_2)| < \varepsilon \quad \text{and} \quad \|V(w, z_1) - V(w, z_2)\| < \varepsilon.$$

2.   *For $z \in \mathscr{Z}_0$, $\tilde{\lambda}(z)$ belongs to a bounded set $\mathscr{P}_2 \subset \mathscr{P}$; the conditions sufficient to ensure this are given in Chapter 2, section 2.4.*

then the model-based optimal performance $-\hat{\phi}(\hat{\lambda}(z), z)$ is uniformly continuous on $\mathscr{Z}_0$.

The proof is omitted. The assumptions of Lemma 4.1 are weaker than the assumptions of Lemma 2.13. They are not sufficient to ensure the uniqueness of $\hat{\lambda}(z)$ for $z \in \mathscr{Z}_0$ and the continuity of $\hat{\lambda}(\cdot)$ on $\mathscr{Z}_0$. In order to ensure this continuity, we have to make additional assumptions.

THEOREM 4.1 (*continuity of $\hat{\lambda}(\cdot)$*).   *If the assumptions of Lemma 2.13 and Lemma 4.1 are satisfied and for every $z \in \mathscr{Z}_1 \subset \mathscr{Z}_0$ (where $\mathscr{Z}_1$ is convex) $\hat{\lambda}(z) \in \mathscr{P}_1$, then $\hat{\lambda}(\cdot)$ is continuous on $\mathscr{Z}_1$.*

Proof.   According to the assumptions, $\hat{\phi}(\cdot, z)$ is strongly convex on $\mathscr{P}_1$. Thus $\forall z_1, z_2 \in \mathscr{Z}_1$ we can write

$$\rho\hat{\phi}(\hat{\lambda}(z_1), z_1) + (1-\rho)\hat{\phi}(\hat{\lambda}(z_2), z_1) \geq \hat{\phi}(\rho\hat{\lambda}(z_1)$$
$$+ (1-\rho)\hat{\lambda}(z_2), z_1) + \delta_4\rho(1-\rho)\|\hat{\lambda}(z_1) - \hat{\lambda}(z_2)\|^2$$

or, for $\rho = \frac{1}{2}$,

$$\phi(\hat{\lambda}(z_2), z_1) - \phi(\hat{\lambda}(z_1), z_1) \geq \tfrac{1}{2}\delta_4\|\hat{\lambda}(z_1) - \hat{\lambda}(z_2)\|^2.$$

Now it may be easily shown that under the assumptions of Lemma 4.1 $\hat{\phi}(\lambda, \cdot)$ is uniformly continuous on $\mathscr{Z}_0$, and this property is uniform in $\lambda \in \mathscr{P}_2$. Since $\hat{\phi}(\hat{\lambda}(\cdot), \cdot)$ is also uniformly continuous on $\mathscr{Z}_0$ we can take $\delta > 0$, such that:

$$\forall z_1, z_2 \in \mathscr{Z}_1 \qquad \|z_1 - z_2\| \leq \delta \quad \text{and} \quad \forall \lambda \in \mathscr{P}_1:$$

$$|\hat{\phi}(\lambda, z_1) - \hat{\phi}(\lambda, z_2)| < \frac{\varepsilon^2 \delta_4}{4} \quad \text{and} \quad |\hat{\phi}(\hat{\lambda}(z_1), z_1) - \hat{\phi}(\hat{\lambda}(z_2), z_2)| < \frac{\varepsilon^2 \delta_4}{4}.$$

334

Therefore

$$\forall (z_1, z_2) \in \mathcal{Z}_1 \qquad \|z_1 - z_2\| \le \delta :$$

$$\tfrac{1}{2}\sigma_4 \|\hat{\lambda}(z_1) - \hat{\lambda}(z_2)\|^2 \le \hat{\phi}(\hat{\lambda}(z_2), z_2) - \hat{\phi}(\hat{\lambda}(z_1), z_1)$$

$$+ \hat{\phi}(\hat{\lambda}(z_2), z_1) - \hat{\phi}(\hat{\lambda}(z_2), z_2) < \frac{\varepsilon^2 \sigma_4}{4} + \frac{\varepsilon^2 \sigma_4}{4}$$

and therefore

$$\|\hat{\lambda}(z_1) - \hat{\lambda}(z_2)\| < \varepsilon \quad \square$$

Using a similar technique, we can prove the following lemma:

LEMMA 4.2. *If the assumptions of Theorem 4.1 are satisfied, then* $\bar{w}(\cdot, \cdot)$ *is continuous on* $\mathcal{P}_1 \times \mathcal{Z}_0$ *and* $\bar{w}(\hat{\lambda}(\cdot), \cdot)$ *is uniformly continuous on* $\mathcal{Z}_1$.

The proof is omitted.

Remark. Lemma 4.2 provides for the continuity of the model-based optimal state trajectory in z if the state transformation mappings (4.1) have the appropriate continuity properties.

The results obtained above are of essential importance for the dynamic control structure. The continuity of the model-based optimal price and state trajectories means that if the predicted disturbances are close enough to the real disturbance trajectory, then the data that the supremal controller supplies to the local controllers are close to the actual optimal values. The assumptions that made these results possible are essentially the same as those required to ensure the contraction property and the convergence of the coordination algorithm itself, that is, of the algorithm used for solving the supremal problem.

The continuity of $\hat{w}(\cdot) = \bar{w}(\hat{\lambda}(\cdot), \cdot)$, $\hat{\lambda}(\cdot)$ does not tell us very much about the sensitivity of these control decisions of the coordination level with respect to the prediction $\bar{z}$ of the disturbance trajectory. For the sensitivity analysis the differentiability of $\hat{w}(\cdot)$, $\hat{\lambda}(\cdot)$ is usually essential. However, with the exception of some special cases it is very difficult to obtain global differentiability of $\hat{w}(\cdot)$, $\hat{\lambda}(\cdot)$. Yet, to compute some limits on the decrease in performance or on the trajectory deviation it could be useful if $\hat{\lambda}(\cdot)$ or $\hat{w}(\cdot)$ were Hölder continuous on subset $\mathcal{Z}_1$ of $\mathcal{Z}_0$ such that $\bar{z} \in \mathcal{Z}_1$. To illustrate this possibility we can prove the following theorem.

THEOREM 4.2. *If the assumptions of Theorem 4.1 are satisfied and*

$$(\forall z_1, z_2 \in \mathcal{Z}_1)(\forall w \in MU) \quad |Q(w, z_1) - Q(w, z_2)| \le K_1^0 \|z_1 - z_2\|,$$

$$\|V(w, z_1) - V(w, z_2)\| \le K_2^0 \|z_1 - z_2\|.$$

then the model-based optimal performance $-\phi(\hat{\lambda}(\cdot), \cdot)$ *is Lipschitz continuous and*

$$\forall z_1, z_2 \in \mathscr{Z}_1 \quad \|\hat{\lambda}(z_1) - \hat{\lambda}(z_2)\| \le K_1(\|z_1 - z_2\|)^{1/2}$$

Proof. According to the assumptions, $\forall z \in \mathscr{Z}_1 \; \hat{\lambda}(z) \in \mathscr{P}_2 \subset \mathscr{P}_0$, where $\mathscr{P}_2$ is a bounded set. Therefore $\forall z \in \mathscr{Z}_1 \; \|\hat{\lambda}(z)\| \le L, \; L < +\infty$. Let us take any $\lambda \in \mathscr{P}_2$ and $z_1, z_2 \in \mathscr{Z}_1$. Then

$$\hat{\phi}(\lambda, z_1) - \hat{\phi}(\lambda, z_2) = L(\bar{w}(\lambda, z_2), \lambda, z_2) - L(\bar{w}(\lambda, z_1), \lambda, z_1) \le L(\bar{w}(\lambda, z_1), \lambda, z_2)$$

$$-L(\bar{w}(\lambda, z_1), \lambda, z_1) \le (K_1^0 + LK_2^0)\|z_1 - z_2\|.$$

Similarly, we can show that

$$\hat{\phi}(\lambda, z_1) - \hat{\phi}(\lambda, z_2) \le (K_1^0 + LK_2^0)\|z_1 - z_2\|.$$

Thus, $\hat{\phi}(\lambda, \cdot)$ is Lipschitz continuous on $\mathscr{Z}_1$ uniformly with respect to $\lambda \in \mathscr{P}_2$.
Now

$$\forall z_1, z_2 \in \mathscr{Z}_1 \qquad \hat{\lambda}(z_1), \hat{\lambda}(z_2) \in \mathscr{P}_2.$$

Therefore

$$\hat{\phi}(\hat{\lambda}(z_1), z_1) - \hat{\phi}(\hat{\lambda}(z_2), z_2) \le \hat{\phi}(\hat{\lambda}(z_2), z_1) - \hat{\phi}(\hat{\lambda}(z_2), z_2) \le (K_1^0 + LK_2^0)\|z_1 - z_2\|$$

and

$$\hat{\phi}(\hat{\lambda}(z_2), z_2) - \hat{\phi}(\hat{\lambda}(z_1), z_1) \le (K_1^0 + LK_2^0)\|z_1 - z_2\|.$$

From the above inequalities and the proof of Theorem 4.2 it follows that $\forall z_1, z_2 \in \mathscr{Z}_1$:

$$\tfrac{1}{2}\sigma_4\|\hat{\lambda}(z_1) - \hat{\lambda}(z_2)\|^2 \le 2(K_1^0 + LK_2^0)\|z_1 - z_2\|$$

and so

$$\|\hat{\lambda}(z_1) - \hat{\lambda}(z_2)\| \le \left[\frac{4}{\sigma_4}(K_1^0 + LK_2^0)\right]^{1/2} \cdot (\|z_1 - z_2\|)^{1/2}. \quad \square$$

The continuity of $\hat{\lambda}_{\Delta_{fk}}$ and $\hat{x}_{\Delta_{fk}}$ in $x(t_{j_k})$, $\bar{z}_{\Delta_{fk}}$ should be accompanied by a continuous dependence of the trajectories of the important real variables like $x^r(t)$ or the real performance index over $\Delta_k = [t_{j_k}, t_{j_{k+1}}]$ on $\hat{\lambda}_{\Delta_{fk}}$, $\hat{x}_{\Delta_{fk}}$, $\tilde{z}_{\Delta_j}$, for $j_k \le j < j_{k+1}$, and on $z^r_{\Delta_k}$. This can be considered of course under the assumption that the local-level short-horizon optimization can be performed, i.e., that the LPS$_i$ are solvable, and the continuity of, among others, $x^r_{\Delta_k}$ in $\hat{\lambda}_{\Delta_{fk}}$, $\hat{x}_{\Delta_{fk}}$ will depend on the properties of LPS$_i$ and on the real coupled system behavior. After establishing the desired continuity properties, we can easily draw conclusions about, for example, the continuous dependence of the real performance index value computed over some interval $\bigcup_{k=1}^{k_0} \Delta_k$ on $\bar{z}_{\Delta_{fk}}$, $\tilde{z}_{\Delta_j}$, and $z^r_{\Delta_k}$. When constraints (4.8) are present or local constraints depend on $u_i(t)$, or both, then we may be more interested in estimating the degree to which these constraints may be violated than in examining the real performance value. If we know that $y^r_{\Delta_k}$ (the real output trajectory) is continuous in $\hat{\lambda}_{\Delta_{fk}}$, $\hat{x}_{\Delta_{fk}}$, $\tilde{z}_{\Delta_j}$, $z^r_{\Delta_k}$ ($k = 1, 2, \ldots$) and $\hat{\lambda}_{\Delta_{fk}}$, $\hat{x}_{\Delta_{fk}}$ are continuous in $\bar{z}_{\Delta_{fk}}$, $x^r(t_{j_k})$, then we will have a guarantee that constraints (4.8), for example, will be approximately satisfied for sufficiently small differences $\tilde{z}_{\Delta_j} - z^r_{\Delta_j}$ and $\bar{z}_{\Delta_{fk}} - z^r_{\Delta_{fk}}$.

The on-line local problems

The results presented in the previous section describe the properties of the global system optimization problem, that is, of the long horizon supremal problem. It is also necessary to investigate the properties of the on-line local control problem LPS$_i$ and the properties of their solutions. It can be seen that LPS$_i$ depends on the data from the supremal controller $(\hat{\lambda}_{\Delta_j}, \hat{x}_{ifk}(t_j + T_{jL}))$, on the locally updated prediction of disturbances $\tilde{z}_{\Delta_j}$, and on the real state value at time t_j, that is, $x^r_i(t_j)$. There are two essential questions concerning the solution $\tilde{m}_{i\Delta_j}$, $\tilde{u}_{i\Delta_j}$ of LPS$_i$: When does this solution exist? (In most real cases this is the same as asking if the feasible set of LPS$_i$ is nonempty.) If this solution exists is it unique and continuous in the data supplied to the local controller? We now discuss the questions.

The first is extremely difficult to answer in the general case. The feasible set of LPS$_i$ can often be empty, that is, the final state condition at $t_j + T_{jL}$ cannot be met. Then it is necessary to modify LPS$_i$.

The feasible set of LPS$_i$ (which we denote by $MU^\alpha_{i\Delta_j}$, where α represents all the data on which LPS$_i$ depends) can be expressed as

$$MU^\alpha_{i\Delta_j} = \{(m_{i\Delta_j}, u_{i\Delta_j}) \in MU_{i\Delta_j} : \phi_{i\Delta}(x^r_i(t_j), m_{i\Delta_j}, u_{i\Delta_j}, \tilde{z}_{\Delta_j})$$
$$= \hat{x}_{ifk}(t_j + T_{jk})\}. \tag{4.39}$$

Since $MU^\alpha_{i\Delta_j}$ may prove to be empty if $\tilde{z}_{\Delta_j} \neq \bar{z}_{\Delta_j}$ and $x^r_i(t_j) \neq \hat{x}_{ifk}(t_j)$, which means that $x^r_i(t_j)$ does not lie on the optimal trajectory calculated from the model at the supremal level, the following two modifications of this set can be considered:

(1) $\quad MU^{1\alpha}_{i\Delta_j} = \left\{ (m_{i\Delta_j}, u_{i\Delta_j}) \in MU_{i\Delta_j} : (m_{i\Delta_j}, u_{i\Delta_j}) \right.$

$$\left. = \arg \min_{MU_{i\Delta_j}} \| \phi_{i\Delta_j}(x^r_i(t_j), m_{i\Delta_j}, u_{i\Delta_j}, \tilde{z}_{\Delta_j}) - \hat{x}_{ifk}(t_j + T_{jL}) \| \right\},$$

(2) $\phi_{i\Delta_i}$ satisfies the following Lipschitz condition on $XZ_{\Delta_i} \subset X_i \times \mathscr{Z}_{\Delta_i}$,

$$(\forall (x^1, z^1_{\Delta_i}), (x^2, z^2_{\Delta_i}) \in XZ_{\Delta_i})(\forall (m_{i\Delta_i}, u_{i\Delta_i}) \in MU_{i\Delta_i})$$

$$\|\phi_{i\Delta_i}(x^1, m_{i\Delta_i}, u_{i\Delta_i}, z^1_{\Delta_i}) - \phi_{i\Delta_i}(x^2, m_{i\Delta_i}, u_{i\Delta_i}, z^2_{\Delta_i})\| \le K_{ij}(\|x^1 - x^2\| + \|z^1 - z^2\|).$$

It is now possible to define $MU^{2\alpha}_{i\Delta_i}$ as follows (if we assume that $(x^r_i(t_j), \bar{z}_{\Delta_{fk}|\Delta_i}), (\hat{X}_{ifk}(t_j), \tilde{z}_{\Delta_i}) \in XZ_{\Delta_i})$:

$$MU^{2\alpha}_{i\Delta_i} = \{(m_{i\Delta_i}, u_{i\Delta_i}) \in MU_{i\Delta_i} : \|\phi_{i\Delta_i}(x^r_i(t_j), m_{i\Delta_i}, u_{i\Delta_i}, \tilde{x}_{\Delta_i})$$

$$- \hat{x}_{ifk}(t_j + T_{jL})\| \le K_{ij}(\|\hat{x}_{ifk}(t_j) - x^r_i(t_j)\| + \|\bar{z}_{\Delta_i} - \tilde{z}_{\Delta_i}\|)\},$$

where $\bar{z}_{\Delta_i} = \bar{z}_{\Delta_{fk}}|_{\Delta_i}$.

The obvious feature of the set $MU^{1\alpha}_{i\Delta_i}$ is that if $MU^{\alpha}_{i\Delta_i} \ne \varnothing$ then $MU^{1\alpha}_{i\Delta_i} = MU^{\alpha}_{i\Delta_i}$. However, it is clear that sometimes it is very uneconomical to concentrate effort on reaching target state $\hat{x}_{ifk}(t_j + T_{jL})$. Then the second modification of the feasible set may appear to be more reasonable and convenient, especially if constants K_{ij} can be computed ahead of time. Modification (2), however, may lead to increasing differences between the model-based optimal state trajectory and the real system trajectory. This situation may become dangerous when the supremal controller only rarely intervenes. It is clear that if $\forall j, \bar{z}_{\Delta_i} = \tilde{z}_{\Delta_i} = z^r_{\Delta_i}$, then $MU^{\alpha}_{i\Delta_i} = MU^{1\alpha}_{i\Delta_i} = MU^{2\alpha}_{i\Delta_i}$ and the real optimal control is obtained.

We can also modify LPS$_i$ by adding a penalty term to $L_{i\Delta_i}$:

$$\mathscr{J}^{\mathrm{pen}}_i(\hat{x}_{ifk}(t_j + T_{jL}) - x_i(t_j + T_{jL})),$$

where $\mathscr{J}^{\mathrm{pen}}_i$ is a positively defined functional on X_i. $\mathscr{J}^{\mathrm{pen}}_i$ should be chosen on the basis of simulations done for each individual case.

If set $MU^{\alpha}_{i\Delta_i}$, $MU^{1\alpha}_{i\Delta_i}$, or $MU^{2\alpha}_{i\Delta_i}$ is nonempty and there exists a solution of LPS$_i$, then

$$\tilde{m}_{i\Delta_i}(x^r_i(t_j), \hat{\lambda}_{\Delta_i}, \bar{z}_{\Delta_i}, \hat{x}_{ifk}(t_j + T_{jL})) = \tilde{m}_{i\Delta_i}(\alpha),$$

$$\tilde{u}_{i\Delta_i}(x^r_i(t_j), \hat{\lambda}_{\Delta_i}, \bar{z}_{\Delta_i}, \hat{x}_{ifk}(t_j + T_{jL})) = \tilde{u}_{i\Delta_i}(\alpha)$$

where $\alpha \in \mathbb{A} = X_i \times \mathscr{P}_{\Delta_i} \times \mathscr{Z}_{\Delta_i} \times X_i$ (the norm in $\mathbb{A}$ may be introduced as the sum of norms in X_i, $\mathscr{P}_{\Delta_i}$, $\mathscr{Z}_{\Delta_i}$, X_i). Let us denote $w = (m, u)$ and omit subscripts i and Δ_j. Problem LPS$_i$ may now be written as follows:

For given $\alpha \in \mathbb{A}_0 \subset \mathbb{A}$, find $\tilde{w}(\alpha)$ such that

$$L(\tilde{w}(\alpha), \alpha) = \min_{w \in MU} L(w, \alpha),$$

where L is defined in (4.34).

For the second question concerning the uniqueness and continuity of $\tilde{w}(\cdot)$, the solution of the above problem, on $\mathbb{A}_0$, we can get many results by specifying sufficient conditions guaranteeing this uniqueness and continuity (see Chapter 2). We present only two lemmas stating the sufficient conditions. The proofs are not unusual and therefore are omitted.

LEMMA 4.3.   *If we assume that*

1. $\forall \alpha \in \mathbb{A}_0$, *$\tilde{w}(\alpha)$ exists, is unique, and belongs to a bounded set $W_0 \in \mathcal{M} \times \mathcal{U}$,*

2.   *$\forall \{\alpha^n\}_{n=1}^{\infty}$ such that $\alpha^n \to \alpha^0$, $\alpha^n, \alpha^0 \in \mathbb{A}_0$ and $\forall w^0 \in MU^{\alpha^0}$, there exists $\{w^n\}_{n=1}^{\infty}$ such that $w^n \in MU^{\alpha^n}$ and $w^n \to w^0$,*

3.   *If $w^n \underset{s}{\to} w^0$, $\alpha^n \to \alpha^0$, $w^n \in MU^{\alpha^n}$, $\alpha^n, \alpha^0 \in \mathbb{A}_0$ then $w^0 \in MU^{\alpha^0}$, where $\underset{s}{\to}$ denotes weak convergence. The sufficient conditions for this are: $MU_{i\Delta_i}$ is a weakly closed set and mapping (4.1) satisfies the following*

$$(X_i \ni x_i^n \to x_i^0, (m_{i\Delta_j}^n, u_{i\Delta_j}^n) \underset{s}{\to} (m_{i\Delta_j}^0, u_{i\Delta_j}^0), z_{\Delta_j}^n \to z_{\Delta_j}^0) \Rightarrow$$

$$\phi_{i\Delta_j}(x_i^n, m_{i\Delta_j}^n, u_{i\Delta_j}^n, z_{\Delta_j}^n) \to \phi_{i\Delta_j}(x_i^0, m_{i\Delta_j}^0, u_{i\Delta_j}^0, z_{i\Delta_j}^0),$$

4.   *Functional $L(\cdot, \cdot)$ is continuous on $MU \times \mathbb{A}_0$ and*

$$(\alpha^n \to \alpha^0, \alpha^n, \alpha^0 \in \mathbb{A}_0 \quad and \quad w^n \underset{s}{\to} w^0, w^n, w^0 \in MU) \Rightarrow$$

$$L(w^0, \alpha^0) \leq \underline{\lim} L(w^n, \alpha^n).$$

Then $\forall \{\alpha^n\}_{n=1}^{\infty}$, $\alpha^n \to \alpha^0$, $\alpha^n, \alpha^0 \in \mathbb{A}_0$ we have $\tilde{w}(\alpha^n) \underset{s}{\to} w(\alpha^0)$.

LEMMA 4.4.   *If assumption 1 of Lemma 4.3 is satisfied, set W_0 is convex, and*

1. $(\forall (\alpha^1, \alpha^2) \in \mathbb{A}_0)(\forall w^1 \in W_0 \cap MU^{\alpha^1})(\exists w^2 \in W_0 \cap MU^{\alpha^2})$

such that
$$\|w^1 - w^2\| \leq k \|\alpha^1 - \alpha^2\|$$

and

$$\forall \alpha \in \mathbb{A}_0, \qquad \forall w^1, w^2 \in W_0 \cap MU^\alpha, \qquad \forall \rho \in [0, 1] \exists w^\rho \in W_0 \cap MU^\alpha$$

such that

$$\|\rho w^1 + (1 - \rho) w^2 - w^\rho\| \leq k_0 \rho (1 - \rho) \|w^1 - w^2\|,$$

2. $\forall (\alpha^1, \alpha^2) \in \mathbb{A}_0$ *and $\forall (w^1, w^2) \in W_0$ we have*
$$|L(w^1, \alpha^1) - L(w^2, \alpha^2)| \leq k_1 \|w^1 - w^2\| + k_2 \|\alpha^1 - \alpha^2\|$$

and

$$\rho L(w^1, \alpha^1) + (1-\rho)L(w^2, \alpha^1) \geq L(\rho w^1 + (1-\rho)w^2, \alpha^1)$$
$$+ \sigma\rho(1-\rho)\|w^1 - w^2\|^2 \qquad \forall \rho \in (0, 1],$$

where $\sigma > k_1 k_0$.

then $\forall \alpha^1, \alpha^2 \in \mathbb{A}_0$, *the following inequality is satisfied*

$$\|\tilde{w}(\alpha^1) - \tilde{w}(\alpha^2)\| \leq \left[k^2\|\alpha^1 - \alpha^2\|^2 + \frac{4(k_1 k + k_2)}{\sigma - k_1 k_0}\|\alpha^1 - \alpha^2\| \right]^{1/2}.$$

In Lemma 4.4 the essential role is played by assumption 1 concerning the continuity of MU^α with respect to α.

The next questions concerning the operation of the lower-level control are connected with the behavior of the controlled process. The dependence of $x^r_\Delta(t)$ on the local control decisions and the continuity of $x^r_{\Delta_{fk}}$ and the other trajectories (e.g., $y^r_{\Delta_i}$) with respect to $x^r(t_j)$ and the local-level control decisions are especially important. From (4.1), (4.2), and (4.3) we can see that

$$x^r_{i\Delta_i} = \Phi_{i\Delta_i}(x^r(t_j), \tilde{m}_{\Delta_i}, \tilde{u}_{w\Delta_i}, z^r_{\Delta_i}), \tag{4.40}$$

where $\Phi_{i\Delta_i}$ transforms the initial state, control, and disturbance trajectories into the state trajectory over Δ_i of subsystem i, that is,

$$\Phi_{i\Delta_i} : X \times \mathcal{M}_{\Delta_i} \times \mathcal{U}_{w\Delta_i} \times \mathcal{Z}_{\Delta_i} \to \mathcal{X}_{i\Delta_i}.$$

We consider, of course, a coupled system for $\tilde{m}_{\Delta_i} = (\tilde{m}_{1\Delta_i}, \ldots, \tilde{m}_{N\Delta_i})$, and for $\tilde{u}_{w\Delta_i}$. Except for linear systems, it is usually difficult to establish sufficient conditions guaranteeing the continuity of mapping (4.40), especially when the direct description of the system elements is given by differential equations. Some results of this type may be found in Ioffe and Tikhomirov (1974) and Pontriyagin (1961).

When mapping (4.40) is Lipschitz continuous on some subset of $X \times \mathcal{M}_{\Delta_i} \times \mathcal{U}_{w\Delta_i} \times \mathcal{Z}_{\Delta_i}$ and for upper-level decisions and lower-level decisions some proper inequalities are satisfied (like those given by Theorem 4.2 and Lemma 4.4), we can estimate the upper bounds on the performance and the state (or interaction) trajectory deviation for given $z^r_{\Delta_{fk}} - \bar{z}_{\Delta_{fk}}$ and $z^r_{\Delta_i} - \tilde{z}_{\Delta_i}$. However, in general, such bounds will be much too large.

Final remarks

It is rather obvious that the results of an application of the dynamic price control structure may depend heavily on the choice of intervention times t_j, t_{j_k} and on the choice of T_{iL}, the local optimization horizon. It is extremely difficult to give qualitative or quantitative estimates of the influence of t_j and

340

t_{j_k} on the system's behavior. It should be noted that changing the number of interventions increases the operational cost connected with the implementation of the structure. Some results for the basic repetitive control problems are given in section 4.5.

Sometimes it may be preferable or even necessary to use $\hat{u}_{i\Delta_{fk}}(x^r(t_{jk}), \bar{z}_{\Delta_{fk}})$, the model-based optimal interaction input trajectory computed at $t = t_{j_k}$, or some part of it as the upper-level control decisions. In that case we would take $u_{i\Delta_i}$ in LPS$_i$ as $\hat{u}_{i\Delta_{fk}}(x^r(t_{j_k}), \bar{z}_{\Delta_{fk}})|_{\Delta_i}$. Such an approach would decrease the complexity of the local problems and sometimes even improve system performance. If we use one of the methods described in section 2.5 to solve SOP then it may be necessary to fix $u_{i\Delta_i}$ in LPS$_i$.

Findeisen and Malinowski (1976) introduced a simple algorithm between the lower and the global level that corrects prices λ_w associated with couplings (4.8). In the next section we will consider the possibility of constructing such a correction mechanism in a simplified case where the subsystems can be treated as static and nonstationary.

In section 4.2.1. the possibility of using simplified models at the upper level was discussed. Unfortunately, the methods that could be used to simplify nonlinear models do not yet exist, except possibly for the singular perturbation method; we also do not know what the influence of model simplification is on the quality of control decisions generated at the upper level. Therefore, we will have to choose the proper simplified model in each case through experiments and numerical simulation.

4.3.4. SIMULATION RESULTS

A dynamic system with stiff interconnections and fixed time horizon

Assume that we are given the dynamic system in Figure 4.5 described by:
Subsystem equations

$$\dot{x}_1(t) = z_a(t)m_1(t) + z_b(t)u_1(t)$$
$$y_1(t) = x_1(t)$$
$$\dot{x}_2(t) = -\tfrac{1}{2}x_2(t) + z_a(t)m_2(t) + z_a(t)u_2(t)$$
$$y_2(t) = x_2(t) + m_2(t)$$
$$\dot{x}_3(t) = z_a(t)m_3(t) + z_a(t)u_3(t)$$
$$y_3(t) = x_3(t)$$

Interconnection equations

$$u_1 = y_3$$
$$u_2 = y_1$$
$$u_3 = y_2$$

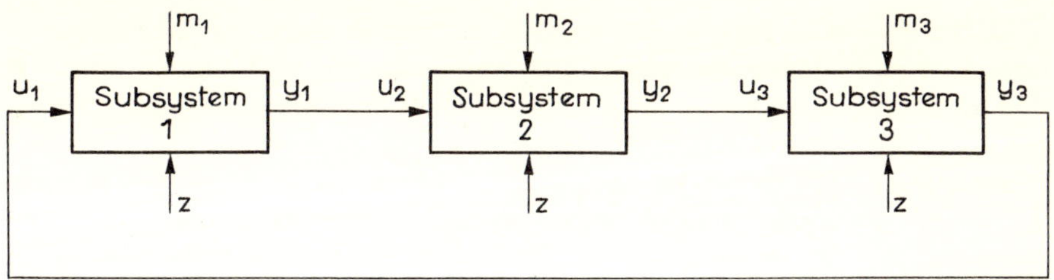

FIGURE 4.5 The structure of the simulated dynamic system with a fixed time horizon.

Performance index

$$Q_\Delta = Q_{1\Delta} + Q_{2\Delta} + Q_{3\Delta},$$

where

$$Q_{1\Delta} = \int_\Delta \left(m_1^2(t) + 4(u_1(t) - 2)^2 \right) dt,$$

$$Q_{2\Delta} = \int_\Delta \left(m_2^2(t) + (u_2(t) - \tfrac{1}{2})^2 \right) dt,$$

$$Q_{3\Delta} = \int_\Delta \left(m_3^2(t) + (u_3(t) - \tfrac{1}{3})^2 \right) dt.$$

The off-line global optimization has been performed over interval $\Delta_f = [0, 5]$; the prediction of disturbances was assumed to be $\bar{z}_a(t) = 1$ and $\bar{z}_b(t) = 0.5$.

The optimization was done by applying the interaction balance method. Interval Δ_f was discretized into 20 parts, $\bar{\Delta} = 0.25$. The supremal problem solution supplied model-based optimal prices $\hat{\lambda}_{\Delta_f}$, controls $\hat{m}_{\Delta_f}$, and state trajectory $\hat{x}_{\Delta_f}$.

The dual function $\hat{L}(x(0), \lambda_{\Delta_f}, \bar{z}_{\Delta_f})$ discussed in section 4.3.1 attained the value 8.90 at $\hat{\lambda}_{\Delta_f}$. After applying $\hat{m}_{\Delta_f}$ to the simulated coupled system (but still with $z_{\Delta_f} = \bar{z}_{\Delta_f}$), we find that the value of the performance function equals 9.11. This value differs slightly from $\hat{L}(x(0), \lambda_{\Delta_f}, \bar{z}_{\Delta_f})$ as a result of the discretization and the finite accuracy of the solution of the supremal problem.

We assumed that constant trend disturbances and sinusoidal disturbances were present in the real system:

$$z_a^r(t) = 1 + \beta t, \qquad z_b^r(t) = 0.5 + \alpha t,$$

$$z_a^r(t) = 1 + \beta_1 \sin \frac{2\pi}{5} t, \qquad z_b^r(t) = 0.5 + \alpha_1 \sin \frac{2\pi}{5} t$$

TABLE 4.2 Comparison of Open-Loop and Repetitive Control with Constant Trend Disturbances and a Fixed Time Horizon

α	β	Open-Loop Control	Real Optimal Performance	Repetitive Control $k=1$ $T_{jL}=0.5$	$T_{jL}=1.5$	$k=1,2$ $T_{jL}=0.5$	$T_{jL}=1.5$
0	0	9.111	9.111	9.111	9.111	9.111	9.111
0.12	0.24	505.825	9.589	10.58	23.31	10.486	11.557
−0.12	−0.24	11.958	6.978	8,77	10.00	8.82	9.26

For various values of α and β, the open-loop control ($\hat{m}_{\Delta_f}$ applied to the real system over Δ_f) was compared with a two-level repetitive control scheme without updating the disturbance predication. The interval $[0, 5]$ was split into 10 parts with intervals $t_j = 0.5j$ $(j=0,\ldots,9)$. We assumed that the local optimization horizon T_{jL} was 0.5 or 1.5. Global optimization was either not repeated ($k=1$; $t_{j_1}=0$) or repeated once ($k=1,2$; $t_{j_1}=0$, $t_{j_2}=2.5$). The results of the simulation are displayed in Tables 4.2 and 4.3. Real optimal performance is attainable when z_a^r and z_b^r can be predicted exactly.

Table 4.4 displays more results of the comparison between the open-loop control and the repetitive scheme with $T_{jL}=0.5$ and without global recoordination. Tables 4.2–4.4 show that in most cases local repetitive control significantly improves system performance; yet, it may also perform poorly (see case 1 in Table 4.4). The requirement that the final state value in the local, short-horizon, optimization problems be that prescribed by the long-horizon global solution may be the reason. Feedback to the upper level and increasing the local horizons may improve performance, but updating the prediction of disturbances at times t_j (especially at times t_{j_k}) could improve performance even more.

It is significant that even though the differences between $(\bar{z}_a, \bar{z}_b)$ and (z_a^r, z_b^r) for sinusoidal and constant trend disturbances were very large, the

TABLE 4.3 Comparison of Open-Loop and Repetitive Control with Sinusoidal Disturbances and a Fixed Time Horizon

α_1	β_1	Open-Loop Control	Real Optimal Performances	Repetitive Control $k=1$ $T_{jL}=0.5$	$T_{jL}=1.5$	$k=1,2$ $T_{jL}=0.5$	$T_{jL}=1.5$
0	0	9.111	9.111	9.111	9.111	9.111	9.111
0.2	0.4	88.757	6.895	8,524	16.368	8.523	11.927
−0.2	−0.4	97.385	9.6401	10.789	15.546	10.742	18.843

TABLE 4.4 Comparison of Open-Loop and Repetitive Constant Trend Disturbances Control with $T_{jL} = 0.5$, and a Fixed Time Horizon

	α	β	Open-Loop Control	Repetitive Control
1	−0.20	−0.40	11.91	16.28
2	−0.16	−0.32	11.89	10.24
3	−0.12	−0.24	11.96	8.77
4	−0.08	−0.16	11.71	8.56
5	−0.04	−0.08	10.62	8.75
6	0.04	0.08	15.89	9.55
7	0.08	0.16	82.04	10.05
8	0.12	0.24	505.82	10.58
9	0.20	0.40	5,114.43	11.55
10	−0.08	−0.10	14.27	8.21
11	−0.16	−0.20	16.12	10.06

repetitive control scheme with price coordination was able to provide good results.

A dynamic system with stiff interconnections and floating time horizon

Assume that we have the same dynamic system as above except that the optimization horizon at the coordination level is always of the same length. The basic control policy then consists in performing the model-based optimization at time instants t_{j_k} $(k = 1, 2, \ldots)$ over time horizon $[t_{j_k}, t_{j_k} + T_g]$. We can compare this basic control policy with the two-level repetitive control scheme in which price coordination is performed at t_{j_k} over interval $\Delta_{fk} = [t_{j_k}, t_{j_k} + T_g]$, and the local control problems are solved at times t_j. Assume that $k = 1, 2, \ldots, j = 0, 1, \ldots, j_k = 0, 5, 10, \ldots$, and $T_{jL} = 0.5, T_g = 5$. We apply this control scheme without updating the disturbance prediction. The results of the simulation are displayed in Tables 4.5 and 4.6 where

TABLE 4.5 Comparison of Control Results with Constant Trend Disturbances and a Floating Time Horizon

			Basic Control Policy	Two-Level Repetitive Control with Price Coordination	
α		β		$T_{jL} = 0.5$	$T_{jL} = 1.5$
0.12		0.24	25.717	10.240	10.607
−0.12		−0.24	12.550	9.851	10.71

TABLE 4.6 Comparison of Control Results with Sinusoidal Disturbances and a Floating Time Horizon

		Basic Control Policy	Two-Level Repetitive Control with Price Coordination	
α_1	β_1		$T_{jL} = 0.5$	$T_{jL} = 1.5$
0.2	0.4	23.367	9.775	13.215
−0.2	−0.4	16.242	11.453	12.844

all the performance functions are evaluated over time horizon $[0, 5]$. The two-level repetitive control is performed at times $t_j \neq t_{j_k}$.

It can be seen that the application of the two-level control structure brings a considerable improvement when compared with the basic control policy. The results also show that a better performance value was achieved with the short local optimization horizons T_{jL}. Updating the prediction of disturbances at time instants t_{j_k} and t_j should bring even more improvement in the performance value.

4.3.5. THE CONTROL STRUCTURE WITH CONJUGATE VARIABLES

We discussed conjugate variables in section 4.2.3. Here, we assume that the state transformation (4.1) results from the differential equation:

$$\dot{x}_i(t) = f_i(x_i(t), m_i(t), u_i(t), z(t)), \tag{4.41}$$

given $x_i(t_1)$, where $f_i : \mathbb{R}^{n_{xi}} \times \mathbb{R}^{n_{mi}} \times \mathbb{R}^{n_{ui}} \times Z \to \mathbb{R}^{n_{xi}}$; we assume also that $\forall t_1, t, t_0 \leq t_1 < t \leq T_0 \ x_{i\Delta} \in W^{2,1}[\Delta; \mathbb{R}^{n_{xi}}](\Delta = [t_1, t])$ and that mappings $f_i, q_{oi}, F_i^0, \mathcal{J}_i$ are continuous and have continuous derivatives with respect to $x_i(t)$.

We consider a dynamic system described by Eqs. (4.3), (4.7), and (4.41). In SOP we have performance function (4.9) and constraints on $m_i(t), u_i(t)$ in the form: $m_i(t), u_i(t)) \in MU_i^0$ (see section 4.1).

Supremal unit—periodic optimization

As in the control structure discussed in the previous section, we introduce the periodic (repetitive) global optimization at the higher level. SOP will be solved at times t_{j_k}. However, we want to use $\hat{\lambda}_{\Delta_{fk}}$ and the trajectories of the conjugate variables $\hat{\psi}_{\Delta_{fk}} = \hat{\psi}_{\Delta_{fk}}(x^r(t_{j_k}), \bar{z}_{\Delta_{fk}})$ as the control decisions of the supremal unit. Trajectories $\hat{\psi}_{\Delta_{fk}}$ may be easily computed when we find $\hat{\lambda}_{\Delta_{fk}}, \hat{x}_{\Delta_{fk}}, \hat{m}_{\Delta_{fk}}$ from the solution of SOP by the interaction balance method. Of course, the components of $\hat{\psi}_{\Delta_{fk}} = (\hat{\psi}_{1\Delta_{fk}}, \dots, \hat{\psi}_{N\Delta_{fk}})$ can be computed independently in the local decision units of the optimization (dynamic coordinator) level.

It is also possible to treat Eq. (4.41) together with Eq. (4.7) as the coupling Eq. (4.6). In this case we can form a Lagrangian:

$$L^d(x_{\Delta_{fk}}, m_{\Delta_{fk}}, u_{\Delta_{fk}}, \bar{z}_{\Delta_{fk}}, \lambda_{\Delta_{fk}}, \psi_{\Delta_{fk}}) = Q^0_{\Delta_{fk}}(x_{\Delta_{fk}}, m_{\Delta_{fk}}, u_{\Delta_{fk}}, \bar{z}_{\Delta_{fk}})$$

$$+ \int_{\Delta_{fk}} \langle \lambda(t), u(t) - HF^0(x(t), m(t), u(t), \bar{z}(t)) \rangle \, dt$$

$$\int_{\Delta_{fk}} \langle \psi(t), f(x(t), m(t), u(t), \bar{z}(t)) - \dot{x}(t) \rangle \, dt \quad (4.42)$$

Using the Lagrangian we can apply IBM. The local problems of IBM will be solved by minimization of (4.42) (for given $\bar{z}_{\Delta_f}$, $\lambda_{\Delta_{fk}}$, $\psi_{\Delta_{fk}}$) with respect to $m_{\Delta_{fk}}$, $u_{\Delta_{fk}}$, $x_{\Delta_{fk}}$ over the set $(MU_{\Delta_{fk}} = MU_{1\Delta_{fk}} \times \ldots \times MU_{N\Delta_{fk}}) \times \mathcal{X}_{\Delta_{fk}}$. The coordination problem will be to find $\hat{\lambda}_{\Delta_{fk}}$, $\hat{\psi}_{\Delta_{fk}}$ for which the local problem solutions satisfy Eqs. (4.3) and (4.41). The properties of this decision mechanism can be investigated by using the general results presented in Chapter 2. In this situation the local problems of IBM are not constrained by differential equations, but they cannot be separated into independent instantaneous static optimization problems since $x_{i\Delta_{fk}} \in W^{2,1}[\Delta_{fk}, \mathbb{R}^{n_{xi}}]$. It is also easy to introduce the constraints on the instantaneous state values when using this approach. For these reasons, the use of conjugate variables as coordination instruments has recently become more popular (Mahmoud 1977 and Singh *et al.* 1975). It is especially popular for discrete time dynamic problems. If the coordinating instruments $\lambda_{\Delta_{fk}}$, $\psi_{\Delta_{fk}}$ are iterated with different frequencies, the use of conjugate variables can lead easily to a three-level optimization method. Unfortunately, there is an obvious drawback connected with using $\psi_{\Delta_{fk}}$ as a coordinating instrument of IBM. The applicability conditions of IBM become more strenuous than in the case when the system dynamics are treated as the equality constraints in the local problems of IBM (see section 4.3.1).

Local control

We will now describe the use of $\hat{\psi}_{fk}$ (and other trajectories obtained from the upper level) and the measurements (or estimates) of $x^r(t)$, $z^r(t)$ at the local decentralized control layer. We assume here that the conditions under which the maximum principle holds are satisfied (see Athans and Falb 1966 and Boltyanski 1969) and the Hamiltonian (see section 4.2)

$$\mathcal{H} = \sum_{i=1}^{N} q_{0i}(x_i(t), m_i(t), u_i(t), z(t)) + \langle \hat{\psi}_{\Delta_{fk}}(t), f(x(t), m(t), u(t), z(t)) \rangle$$

$$+ \langle \hat{\lambda}_{\Delta_{fk}}(t), u(t) - HF^0(x(t), m(t), u(t), z(t)) \rangle$$

attains its minimum with respect to $m(t)$ (or $(m(t), u(t))$ when needed) at a unique point for every $x_{\Delta_{fk}}$, $z_{\Delta_{fk}}$ from some neighborhood of $\hat{x}_{\Delta_{fk}}$, $\bar{z}_{\Delta_{fk}}$.

The formulation of local control problems requires some consideration. Let us assume for a moment that the operation of the lower-layer control mechanism is continuous. Then we can write the local problems (LOP_i) in the form:

Find $\tilde{m}(t)$, $\tilde{u}(t)$, such that

$$\mathcal{H}_i(x_i(t), \tilde{m}_i(t), \tilde{u}_i(t), z(t), \hat{\lambda}_{\Delta_{fk}}(t), \hat{\psi}_{i\Delta_{fk}}(t))$$

$$= \min_{(m_i(t), u_i(t)) \in MU_i^0} \mathcal{H}_i(x_i(t), m_i(t), u_i(t), z(t), \hat{\lambda}_{\Delta_{fk}}(t), \hat{\psi}_{i\Delta_{fk}}(t)) \quad (4.43)$$

where

$$\mathcal{H}_i(x_i(t), m_i(t), u_i(t), z(t), \lambda(t), \psi(t))$$

$$= q_{0i}(x_i(t), m_i(t), u_i(t), z(t)) + \langle \psi_i(t), f_i(x_i(t), m_i(t), u_i(t), z(t)) \rangle$$

$$+ \langle \lambda_i(t), u_i(t) \rangle - \sum_{j=1}^{N} \langle \lambda_j(t), H_{ji}F_i^0(x_i(t), m_i(t), u_i(t), z(t)) \rangle.$$

The values of $x_i(t)$, $z(t)$ in problem (4.43) can be chosen in the following ways:

1. $x_i(t) = x_i^r(t)$, $z(t) = \bar{z}(t)$—we use the current measurement (or estimate) of the state and the prediction $\bar{z}(t)$ of the disturbance
2. $x_i(t) = \hat{x}_{i\Delta_{fk}}(t)$, $z(t) = z^r(t)$—the current state is not measured but we have available the current value of the disturbance acting on the system
3. $x_i(t) = x_i^r(t)$, $z(t) = z^r(t)$—the current state and disturbance values are available

If $z(t) = z^r(t) \forall t$ then our control structure (where $\tilde{m}(t)$ is applied to the real system) will have the optimality property; if $z(t) \neq z^r(t)$ then in case 3 we use the greatest amount of information about the real system behavior. Of course in case 3 we can expect the highest operational costs. It should be noted that (4.43) is a static optimization problem and we do not have to consider future values of the state and disturbances. The future of the system is implicitly taken into account by introducing into (4.43) the value of trend $\langle \hat{\psi}_{i\Delta_{fk}}(t), \dot{x}(t) \rangle$.

In most cases it is either not reasonable or simply not possible to compute and use the current value of $\tilde{m}(t)$ in solving problem (4.43) because

- We design our structure for the optimizing control that operates periodically and usually we will have a regulatory control layer (or layers) in the control structure (see section 4.2)

- It is not possible to find an analytical solution rule for problem (4.43) that is easy to program—we need some time to perform the numerical optimization
- To measure (or estimate) the real state and disturbance values and continuously send the information to the controllers is impossible or, at the least, extremely uneconomical

Assuming that the interventions of the lower decentralized control layer are made at times t_j as in repetitive control with price coordination, we can define the local problems (LOP_i^I) in the following way:

Find $\tilde{m}_{i[t_j,\,t_{j+1}]}$, $\tilde{u}_{i[t_j,\,t_{j+1}]}$ minimizing on $MU_{i[t_j,\,t_{j+1}]}$

$$((m_{i[t_j,\,t_{j+1}]},\, u_{i[t_j,\,t_{j+1}]}) \in MU_{i[t_j,\,t_{j+1}]} \quad \text{if} \quad \forall t \in [t_j,\, t_{j+1}] : (m_i(t),\, u_i(t)) \in MU_i^0)$$

$$(4.44)$$

the integral performance

$$\int_{t_j}^{t_{j+1}} \mathcal{H}_i(\tilde{x}_i(t),\, m_i(t),\, u_i(t),\, \tilde{z}(t),\, \hat{\lambda}_{\Delta_{fk}}(t),\, \hat{\psi}_{\Delta_{fk}}(t)\, dt,$$

where $\tilde{z}_{[t_j,\,t_{j+1}]}$, $\tilde{x}_{i[t_j,\,t_{j+1}]}$ are the prediction of the disturbance and the prediction of the state trajectory. We can take for example $\tilde{z}(t_j) = z^r(t_j)$ and

$$\tilde{x}_i(t_j) = x_i^r(t_j), \qquad \tilde{x}(t) = \hat{x}_{i\Delta_{fk}}(t) - \hat{x}_{i\Delta_{fk}}(t_j) + x_i^r(t_j).$$

The control trajectories $\tilde{m}_{i[t_j,\,t_{j+1}]}$ are used as the control decisions of the local control units. If $z^r(t) = \tilde{z}(t)\forall t$ then the control structure is optimal.

To solve problem LOP_i^I we usually have to discretize it; in the simplest case LOP_i^I could be replaced by the following optimization problem (LOP_i^{II}):

Find $\tilde{m}_i(t_j)$, $\tilde{u}_j(t_j)$ minimizing over MU_i^0 the Hamiltonian

$$\mathcal{H}_i(x_i^r(t_j),\, m_i(t_j),\, u_i(t_j),\, z^r(t),\, \hat{\lambda}_{\Delta_{fk}}(t_j),\, \hat{\psi}_{\Delta_{fk}}(t_j)). \qquad (4.45)$$

We define the control trajectory over $[t_j,\, t_{j+1}]$ as $m_i(t) = \tilde{m}_i(t_j)\ \forall t \in [t_j,\, t_{j+1}]$. The optimality property is not satisfied even when $z^r(t) = \tilde{z}(t)$. If the frequency of interventions of the local controllers is too low then it may be necessary to use a more fine discretization of LOP_i^I. Figure 4.6 illustrates the operation of the conjugate variable control structure.

Properties of the conjugate variable control structure

With conjugate variables, the optimization problems solved at the local control layer are static. At the same time, disturbance predictions are only for a

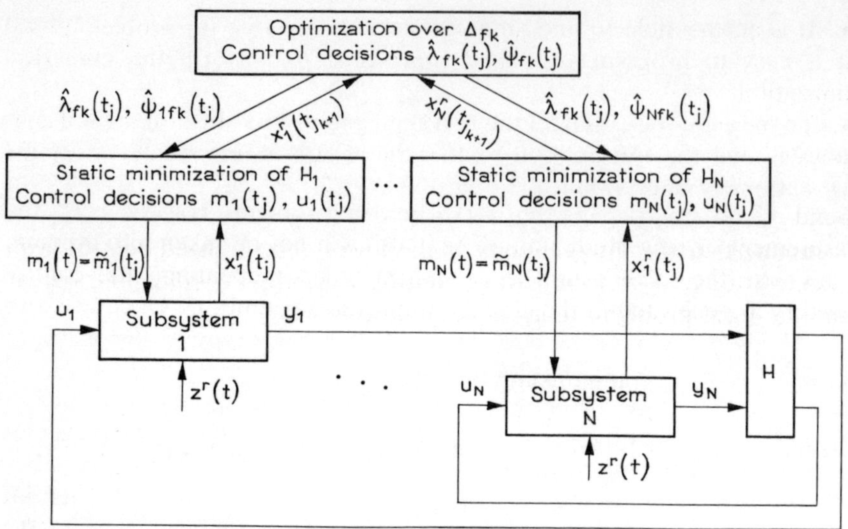

FIGURE 4.6 Control structure using conjugate variables.

short horizon (over $[t_j, t_{j+1}]$) for local problems. From this, we can deduce the following advantages of the control structure using conjugate variables:

- Frequent interventions of the local controllers are possible if needed, and we can use the available information to improve the control. The local problems are more complicated in dynamic price coordination and it may be difficult to solve them as frequently as, for example, LOP_i^{II}.
- There is no danger that the empty feasible sets will appear in the local problems, as they can in local problems LPS_i.
- If functions (4.3) depend only on $x_i(t)$, $z_i(t)$ and we can measure the real interaction values $u_i^r(t)$, then it is reasonable to set $u_i(t) = u_i^r(t_j)$ in problem LOP_i^{II}. With frequent interventions of the local controllers it should be possible then to protect the system constraints from being seriously violated.

We can also see very easily the obvious drawbacks of the conjugate variables scheme:

- When solving their control problems with $x_i(t) = x_i^r(t)$, the local controllers do not take into account the difference between $\hat{x}_{\Delta_{fk}}(t)$ and $x^r(t)$, which can become large when $z^r(t) \neq \bar{z}(t)$.
- Let us assume that the integrand in the performance index has the form:

$$q_{0i}^1(x_i(t), z(t)) + q_{0i}^2(m_i(t), z(t)) + q_{0i}^3(u_i(t), z(t))$$

and the differential Eqs. (4.41) are the following

$$\dot{x}_i(t) = f_i^1(x_i(t), z(t)) + f_i^2(m_i(t), z(t)) + f_i^3(u_i(t), z(t)) + f_i^4(z(t))$$

The functions defining a typical linear-quadratic control problem are in this form. With the expressions in this form, any information about the real state value $x_i^r(t)$ will not be effectively used in problems LOP_i^I or LOP_i^{II}. If functions q_{0i}^2 and f_i^2 do not depend on $z(t)$, then in problems LOP_i^I or LOP_i^{II} the measurement or new prediction of $z^r(t)$ also will not be taken into account. In this case the local controllers will simply be "repeating" the control trajectory as generated in the global optimization at times t_{jk}. Of course, this will not happen in the general case; nevertheless, this example demonstrates that when controlling the system with conjugate variables, we cannot effectively exploit the available on-line information about the system behavior.

- In the "maximum principle" structure the local goals are less acceptable intuitively than in the dynamic price coordination structure since they involve $\langle \hat{\psi}_{i\Delta_{fk}}(t), \dot{x}(t) \rangle$, that is, the value of the trend. This would be difficult to explain in economic terms and hence difficult to implement in a human decision-making hierarchy. For the human decision maker the structure with dynamic local problems seems to be more appealing.

The conditions guaranteeing proper functioning of both layers of the conjugate variable structure are mainly connected with the applicability of IBM and the convergence of suitable coordination strategies (Chapter 2). It is important but difficult to formulate these conditions in a form depending explicitly on the properties of Eqs. (4.41). The conditions given in Chapter 2 concerned the properties of Eqs. (4.4). The investigation of these conditions requires rather detailed analysis of the properties of differential equations and is beyond the scope of this book.

When formulating the global and local problems, we discussed the optimality property of the structure when $z^r(t) = \bar{z}(t)$. For obvious reasons, we are more interested in qualitative and quantitative analysis of the effects that will be produced by an application of the structure when $z^r(t) \neq \bar{z}(t)$ and when we use discretized local problems such as LOP_i^{II}. As in the discussion of dynamic price coordination, we can investigate the continuity of important decision variables like $\hat{\lambda}_{\Delta_{fk}}$, $\hat{\psi}_{\Delta_{fk}}$ with respect to the initial state and the real and predicted disturbance trajectories. We can also try to establish bounds on the deviations of the actual trajectories from the real optimal trajectories, depending on the given difference $z^r(t) - \bar{z}(t)(\forall t)$. Since the detailed considerations are similar to what has been presented in section 4.3.3, we will not present them here. It should be noted, though, that if $\hat{\lambda}_{\Delta_{fk}}$, $\hat{\psi}_{\Delta_{fk}}$, $\bar{m}_{i[t_j, t_{j+1}]}$ are differentiable functions of their arguments (initial state, disturbance trajectory) then we may try to derive the formulae that would allow us to compute, for a given system, the sensitivities of the state trajectories and the performance index. These formulae would be the second variation of the performance (Wierzbicki 1977). In some case it may be more convenient to use these formulae to obtain the quantitative estimates beforehand instead of performing direct numerical simulations. Detailed investigations are also required to identify the effects of the discretization of LOP_i^I. The considerations concerning the "maximum principle" structure are not very advanced so far. Few results are available concerning the properties of control structures for continuous processes with discrete observations and control actions (see section 4.5).

Possible modifications and final remarks

As we mentioned before, local problem LOP_i^{II} may be modified by setting $u_i(t) = u_i^r(t)$ or $u_i(t) = \hat{u}_{i \Delta_{fk}}(t)$, where $\hat{u}_{i \Delta_{fk}}$ is the interaction input trajectory computed at the global layer. The latter approach will be necessary if instead of IBM we apply some other multilevel methods, for example, the augmented Lagrangians discussed in section 2.5 for the global optimization.

It is also possible to update the values of ψ and λ with different frequencies. For example, we can recompute ψ less frequently than λ. The use of simplified models for the global dynamic optimization is also possible.

4.4. PRICE COORDINATION FOR A STATIC NONSTATIONARY SYSTEM WITH DYNAMIC INVENTORY COUPLINGS

4.4.1. PROBLEM DESCRIPTION AND CONTROL STRUCTURE

The theory of systems whose elements are static although time varying and whose couplings are exclusively in the form of (4.8) (or something similar), that is, with $y(t) \triangleq y_w(t)$ and $u(t) \triangleq u_w(t)$, may be useful in many applications. We assume that the dynamics of static system elements are very fast compared with the dynamics of system couplings (4.8) and the frequency of intervention of the on-line controllers.

In this case, the system elements are:

$$y_i(t) = F_i(m_i(t), u_i(t), z(t)) \tag{4.46}$$

where

$$F_i : M_i \times \mathbb{R}^{n_{ui}} \times Z \rightarrow \mathbb{R}^{n_{yi}}.$$

We will not consider updating the disturbance predictions except in the global optimization. We thus demonstrate exclusively the opportunities offered by direct use of feedback information about the real system behavior. It is, of course, possible to consider different off-line and on-line models.

At the global optimization layer, the system optimization problem is to be solved with the performance function that is defined for any interval $\Delta_{fk} = [t_0, t_{fk}]$ as follows (compare Eq. (4.9)):

$$Q(m_{\Delta_{fk}}, u_{\Delta_{fk}}, z_{\Delta_{fk}}) = \sum_{i=1}^{N} \int_{\Delta_{fk}} q_i(m_i(t), u_i(t), \bar{z}(t))\, dt \qquad (4.47)$$

where $\bar{z}(t) = \bar{z}_{\Delta_{fk}}(t)$ is the disturbance prediction.

The Lagrangian (4.11) will now have the form:

$$L_{\Delta_{fk}}(m_{\Delta_{fk}}, u_{\Delta_{fk}}, \lambda_{\Delta_{fk}}, z_{\Delta_{fk}}) = Q(m_{\Delta_{fk}}, u_{\Delta_{fk}}, z_{\Delta_{fk}}) + \int_{\Delta_{fk}} \langle \lambda(t), \bar{H}_1 u(t) - \bar{H}_2 y(t) \rangle\, dt,$$

where $\lambda(t)$ is of constant value on each time interval Δ_ℓ (see Eq. (4.8)) and the intervals Δ_ℓ add up to Δ_{fk}.

If we use IBM to solve the system optimization problem then it is clear that for nondynamic elements the solutions to IP (see Eq. (2.55)) will be $\bar{m}(\lambda(t), z(t))$ and $\bar{u}(\lambda(t), z(t))$, i.e., they will result from the static problem:

$$\underset{MU^0}{\text{minimize}} \; \mathscr{L}(m(t), u(t), \lambda(t), \bar{z}(t)), \qquad (4.48)$$

where

$$\mathscr{L}(m(t), u(t), \lambda(t), \bar{z}(t)) = q(m(t), u(t), \bar{z}(t))$$
$$+ \langle \lambda(t), \bar{H}_1 u(t) - \bar{H}_2 F(m(t), u(t), z(t)) \rangle.$$

The task of the supremal decision unit (coordinator) is to solve the following problem (SPI):

Find price vector function $\hat{\lambda}_{\Delta_{fk}}$ (where $\Delta_{fk} = [t_{j_k}, t_{f_k}]$) constant on every time interval $\Delta_\ell \subset \Delta_{fk}$, such that

$$\int_{\Delta_\ell} [\bar{H}_1 \bar{u}(\hat{\lambda}(t), \bar{z}(t)) - \bar{H}_2 \bar{y}(\hat{\lambda}(t), \bar{z}(t))]\, dt = b_\ell \qquad (4.49)$$

where

$$\bar{y}(\hat{\lambda}(t), \bar{z}(t)) = F(\bar{m}(\hat{\lambda}(t), \bar{z}(t)), \bar{u}(\hat{\lambda}(t), \bar{z}(t)), \bar{z}(t)).$$

After solving the off-line model-based optimization problem formulated as (4.48) and (4.49), we obtain $\bar{m}(\hat{\lambda}(t), \bar{z}(t))$, $\bar{u}(\hat{\lambda}(t), \bar{z}(t))$. This means that we have, in fact, outlined the system inventory policy over any given sequence of time intervals $\{\Delta_j\}$:

$$-\int_{\Delta_j} [\bar{H}_1 u(t) - \bar{H}_2 y(t)]\, dt = \bar{R}_j, \qquad (4.50)$$

352

where

$$\bar{R}_j = -\int_{\Delta_j} [\bar{H}_1 \bar{u}(\hat{\lambda}(t), \bar{z}(t)) - \bar{H}_2 \bar{y}(\hat{\lambda}(t), \bar{z}(t))] \, dt.$$

$\bar{R}_j$ is the desired net inventory increase or decrease (inventory level increment) over a given Δ_j. We will assume that $\forall j \, \exists \ell \, \Delta_i \subset \Delta_\ell$ for some ℓ; in this case $\hat{\lambda}(t) = \hat{\lambda}_j = $ a constant for $t \in \Delta_j$.

When we apply the model-based optimal controls to the real system, Eqs. (4.50) may not be satisfied when we substitute for $y(t)$ the real values of the subsystem outputs $y^r(\hat{\lambda}(t), z^r(t))$, i.e., the outputs obtained with real disturbances.

Now we can consider the following on-line control mechanism (Figure 4.7) that differs from the one presented in section 4.3:

1.   At times t_{j_k} (see the discussion of the supremal unit in section 4.3.2) the global optimization over time interval Δ_{fk} is solved for an updated prediction of disturbances $\bar{z}_{\Delta_{fk}}$ and the real value of the inventory levels at $t = t_{j_k}$. The solution, that is, the model-based optimal price function $\hat{\lambda}_{\Delta_{fk}}$

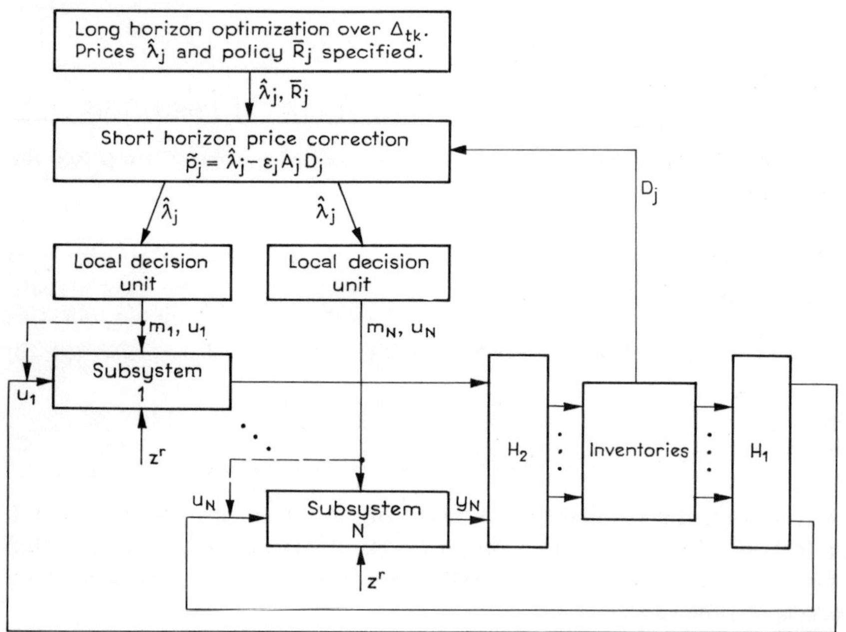

FIGURE 4.7   The structure of on-line inventory price coordination for a static time-varying system.

and the sequence $\{\bar{R}\}$—the inventory policy—is supplied to the local controllers;

2. The task of the control mechanism used at times t_j, located between consecutive times t_{j_k}, is to cause the real inventory level at $t = t_j$, that is, $\sum_{s=1}^{j} R_s^r(\tilde{\lambda}_s)$, to follow the level desired, $\sum_{s=1}^{j} \bar{R}_s$, where

$$R_j^r(\tilde{\lambda}_j) = -\int_{\Delta_j} [\bar{H}_1 \bar{u}(\tilde{\lambda}_j, \bar{z}(t)) - H_2 y^r(\tilde{\lambda}_j, z^r(t))]\, dt$$

and $\tilde{\lambda}_j$ is updated at times t_j.

The control interventions made at times t_j are described and analyzed in the next section. Note that step 2 does not involve optimization; it is a problem of achieving levels already set.

4.4.2. CONTROL OF THE INVENTORY LEVEL

The task of the "high-frequency" control mechanism, step 2 above, should be to generate at each time t_j a price $\tilde{\lambda}_j$, such that the following condition is satisfied:

$$\sum_{s=1}^{j} (R_s^r(\tilde{\lambda}_s) - \bar{R}_s) = 0. \tag{4.51}$$

This means that we would like to follow exactly at each consecutive $t = t_j$ the model-based optimal inventory levels as determined by the "low frequency," upper-layer controller (step 1 above). Condition (4.51) is similar to the coordination condition of the interaction balance method with feedback described in Chapter 3.

Since the model and the real system descriptions are different $(\bar{z}(t) \neq z^r(t))$, the strict fulfillment of condition (4.51) cannot be achieved and we have to set a more realistic goal. In this section we consider the possibility of developing an on-line control strategy for solving the following control problem:

Over a sequence $\{\Delta_j\}$ $(j_k \leq j < j_{k+1})$ of time intervals, adjust prices $\tilde{\lambda}_j$ during the system operation so as to satisfy the following condition at all $t = t_j$:

$$\left\| \sum_{s=1}^{j} (R_s^r(\tilde{\lambda}_s) - \bar{R}_s) \right\| \leq C_0 \tag{4.52}$$

where C_0 is a given positive number.

We would like to make C_0 as small as possible, but we can expect that some compromise has to be made between the value of C_0 and the simplicity of the control strategy. It is also clear that we cannot expect C_0 to be arbitrarily small. However, we may consider the possibility of increasing the number of time intervals Δ_j, that is, of splitting the control horizon into more parts, to make C_0 as small as we need it to be.

To simplify the notation we introduce:

$$D_j = \sum_{s=1}^{j-1} [R_s^r(\tilde{\lambda}_s) - \bar{R}_s], \qquad j = j_k, j_k + 1, \ldots, \tag{4.53}$$

D_j is the deviation of the real inventory level from the model-based optimal level at the beginning of time interval j (D_j would be zero for $j = j_k$). Also,

$$G_j(\lambda_j) = R_j(\lambda_j) - \bar{R}_j + D_j, \tag{4.54}$$

where

$$R_j(\lambda_i) = \int_{\Delta_j} r(\lambda_j, z(t)) \, dt,$$

and

$$r(\lambda_j, \bar{z}(t)) = -[\bar{H}_1 \bar{u}(\lambda_j, \bar{z}(t)) - \bar{H}_2 \bar{y}(\lambda_j, \bar{z}(t))].$$

$G_j(\lambda_j)$ is the value of deviation D_{j+1} predicted at the beginning of Δ_j on the basis of the desired increment $\bar{R}_j$, the known deviation D_j, and the assumed price λ_j.

From (4.53) we know that real deviation D_{j+1} will be:

$$D_{j+1} = D_j + R_j^r(\tilde{\lambda}_j) - \bar{R}_j = G_j(\tilde{\lambda}_j) + [R_j^r(\tilde{\lambda}_j) - R_j(\tilde{\lambda}_j)]. \tag{4.55}$$

The term in brackets in the above formula satisfies the inequality:

$$\|R_j^r(\tilde{\lambda}_j) - R_j(\tilde{\lambda}_j)\| \le \int_{\Delta_j} \|\bar{H}_2[y^r(\tilde{\lambda}_j, z^r(t)) - \bar{y}(\tilde{\lambda}_j, \bar{z}(t))]\| \, dt. \tag{4.56}$$

Suppose now that at time t_j we measure D_j and choose $\tilde{\lambda}_j$ so as to obtain

$$G_j(\tilde{\lambda}_j) = 0. \tag{4.57}$$

This can obviously be done by using the model and solving the global system optimization problem with (4.57) as a constraint. If (4.57) is satisfied, the bound on D_{j+1} is given by (4.56). Therefore, if the following condition is fulfilled:

$$(\forall t)(\forall (m(t), u(t)) \in MU^0) \, \|\bar{H}_2(F(m(t), u(t), z(t)) - F(m(t), u(t), z^r(t)))\| \le \beta, \tag{4.58}$$

then from Eqs. (4.55)–(4.57) we obtain

$$\|D_{j+1}\| \le \beta |\Delta_j|, \forall j.$$

Condition (4.58) can be easily satisfied if set MU^0 is bounded.

The choice of $\tilde{\lambda}_j$ proposed by (4.57) provides the solution of (4.52) if $C_0 \geq \beta |\Delta_j|$ Note that we can reduce $\|D_{j+1}\|$ if we shorten the intervals Δ_j. However, this strategy requires that the full-horizon system optimization problem be solved at the beginning of each time interval Δ_j in order to find $\tilde{\lambda}_j$. The computational effort and the amount of information that has to be exchanged between the coordinator and the local decision units can render this approach either impossible or highly uneconomical. It seems, therefore, that we need a more realistic compromise between the degree of complication of the on-line control strategy and the reduction of D_{j+1}. A possibility is offered by the coordination algorithm (2.68) in section 2.4 and its contraction property. We can use this algorithm to find a solution of Eq. (4.57) at the supremal level; the price vector values would be adjusted using the formula:

$$\lambda_j^{(n+1)} = \lambda_j^{(n)} - \varepsilon_{j_n} A_j G_j(\lambda_j^{(n)}), \qquad n = 1, 2, \ldots \tag{4.59}$$

where $\varepsilon_{jn} > 0$ and A_j is a symmetric, positively defined matrix. Under proper assumptions, the above algorithm has the contraction property, that is,

$$(\exists \delta_j < 1)(\forall n) \ \|G_j(\lambda_j^{(n+1)})\|_{A_j} \leq \delta_j \ \|G_j(\lambda_j^{(n)})\|_{A_j}.$$

Since we have $G_j(\hat{\lambda}_j) = D_j$, the control strategy using the above algorithm can be based on the following decision rule:

Adjust the values of λ_j according to (4.59), with $\lambda_j^{(1)} = \hat{\lambda}_j$, until

$$(\delta_j)^{n_i} \sqrt{\frac{\mu_2^i}{\mu_1^i}} \leq \delta < 1, \tag{4.60}$$

where μ_1^i, μ_2^i are, respectively, the smallest and the largest eigenvalues of A_j. Then take $\tilde{\lambda}_j = \lambda_j^{(n_j+1)}$ and apply it to the system over Δ_j.

If algorithm (4.59) provides contraction in norm $\|\cdot\|$ in a single step, then we may also use the following price correction:

$$\tilde{\lambda}_j = \hat{\lambda}_j - \varepsilon_j A_j D_j. \tag{4.61}$$

Algorithm (4.61) can be used if we are sure that $\forall j, \delta_j \leq \delta < 1$. When applying it, we have to exchange information between the coordinator and the local units only once.

In both cases we can estimate the norm of deviation D_{j+1} as follows:

$$\|D_{j+1}\| \leq \delta \|D_j\| + \beta a,$$

where condition (4.58) is assumed to hold and $\forall j, |\Delta_j| \leq a$. Therefore

$$\forall j \ \|D_{j+1}\| \leq \beta a \sum_{s=j_k}^{j-1} (\delta)^{s-j_k} \leq \beta a \frac{1}{1-\delta}, \qquad (D_{j_k} = 0). \tag{4.62}$$

356

Thus, the proposed approach solves the control problem (4.52) with $C_0 = \beta a(1/1 - \delta)$. However, since the values of δ_1 and δ may vary with time intervals Δ_j, we cannot conclude from (4.62) that $\|D_j\| \to 0^+$ when we apply price correction (4.60) with $n_j = n^0$ for all j and when $a \to 0^+$. On the other hand, the application of decision rule (4.60) may require an increasing number n of iterations done at the beginning of each Δ_j as $|\Delta_j| \to 0^+$. However, it will be shown in the next section that if we fix the number n_j in (4.60) and choose, under certain assumptions, the proper length a of equal time intervals Δ_j, then it will be possible to follow with desired accuracy the model-based optimal level of inventories in the real system.

4.4.3. PROPERTIES OF DECISION RULE (4.60)

We will now formulate the conditions under which decision rule (4.60) with fixed $n^0 = n_j \forall j$ allows us, with a proper division of Δ_{fk} into Δ_j, to solve control problem (4.52) with any desired $C_0 > 0$. Let us consider sequences $\{\Delta_j\}$ of time intervals such that $|\Delta_j| = |\Delta_{fk}|/J$ $(j = j_k, \ldots, j_k + J)$. The requirement that $n^0 = n_j \forall j$ will be fulfilled if the following condition is satisfied:

For given n^0 $\exists \delta_0 < 1$, $\exists J_0$ such that

$$\forall J \geq J_0 \quad \|G_j(\tilde{\lambda}_j)\| \leq \delta_0 \|G_j(\hat{\lambda}_j)\| = \delta_0 \|D_j\|,$$

for $j = j_k, \ldots, j_k + J$ and $\tilde{\lambda}_j = \lambda_j^{(n^0+1)}$, where

$$\lambda_j^{(n+1)} = \lambda_j^{(n)} - \varepsilon A_j G_j(\lambda_j^{(n)}), \qquad \lambda_j^{(1)} = \hat{\lambda}_j, \qquad n = 1, 2, \ldots, n^0. \quad (4.63)$$

Let us assume that $A_j = V_j^{-1}$. V_j can be defined using a hypothetical decision rule defined $\forall t \in \Delta_j$ in the form $\tilde{\lambda}_j(t) = \lambda^{(n^0+1)}(t)$, where

$$\lambda_j^{(n+1)}(t) = \lambda_j^{(n)}(t) - \varepsilon [V(t)]^{-1} g_j(\lambda_j^{(n)}(t), \bar{z}(t)) \qquad (4.64)$$

and

$$g_j(\lambda_j(t), \bar{z}(t)) = r(\lambda_j(t), \bar{z}(t)) - r(\hat{\lambda}_j, \bar{z}(t)) + \frac{D_j}{|\Delta_j|},$$

$$\lambda_j^{(1)}(t) = \hat{\lambda}_j, \qquad n = 1, 2, \ldots, n^0.$$

From the definition of $g_i(\lambda_j(t), \bar{z}(t))$ we have

$$g_i(\hat{\lambda}_j, \bar{z}(t)) = d_j = \frac{D_j}{|\Delta_j|} \qquad \forall t \in \Delta_j. \qquad (4.65)$$

Let us now define an operator V_j as follows:

$$V_j = \int_{\Delta_j} V(t) \, dt. \qquad (4.66)$$

Operator $V(t)$ can be constructed by using, for example, the approach described in Appendix A.

Suppose that the above algorithm has the contraction property described in Chapter 2, section 2.4, that is,

$$\|g_j(\tilde{\lambda}_j(t), \bar{z}(t))\| \le \delta(t) \quad \|g_j(\hat{\lambda}_j, \bar{z}(t))\| = \delta(t)\|d_j\|$$

$\forall t \in \Delta_j$ and $\forall \bar{z}_{\Delta_{fk}} \in \mathcal{Z}_{\Delta_{fk}}$, where $\delta(t) < 1$. The following theorem specifies the conditions under which condition (4.63) is satisfied.

THEOREM 4.3. *If, $\forall \bar{z}_{\Delta_{fk}} \in \mathcal{Z}^0_{\Delta_{fk}}$, the following conditions are fulfilled:*

1. *$\exists c_1 > 0$ such that $\forall t \in \Delta_{fk}$ $\varepsilon\|[V(t)]^{-1}\| \le c_1$. ($V(t)$ has to be self-conjugated and positive definite.)*
2. *$\exists c_2 > 0$ such that $\forall t, \tau \in \Delta_{fk}$ $\|V(t) - V(\tau)\|/\varepsilon \le c_2|t - \tau|$,*
3. *$\exists c_3 > 0$ such that $\forall t \in \Delta_{fk}$ and $\forall \lambda^1(t), \lambda^2(t) \in \mathbb{P}_0 \subset \mathbb{R}^{n_w}$*

$$\|r(\lambda^1(t), \bar{z}(t)) - r(\lambda^2(t), \bar{z}(t))\| \le c_3\|\lambda^1(t) - \lambda^2(t)\|,$$

4. *$\exists \delta < 1$ such that $\forall t \in \Delta_{fk}$ $\delta(t) \le \delta$,*
5. *Sequence $\{\lambda_j^{(n)}(t)\}$ generated by algorithm (4.64) belongs to $\mathbb{P}_0$ $\forall t \in \Delta_{fk}$,*

then condition (4.63) is fulfilled for some δ_0 and J_0.

The proof is given in Malinowski and Terlikowski (1978).

4.4.4. SIMULATION RESULTS

The model used for the numerical simulation does not represent any particular system but it has all the important features which could be present in a real example. The results illustrate two important properties of the price control structure: the necessity of choosing a proper operator A and a proper value of ε in Eq. (4.59) or Eq. (4.61), and the possibility of following the upper-level inventory policy by using a proper price correction algorithm with a sufficiently high frequency of intervention.

Consider a static, time-varying system with dynamic inventory couplings as presented in Figure 4.8. The subsystem equations of type (4.46), performance functions (4.47), and local constraints have the following form:

Subsystem 1

$$y_1(t) = z_a(t)[m_{11}(t) - m_{12}(t)] + 2u_1(t) + z_c(m_{11}(t))^2$$
$$+ z_d[m_{11}(t) + m_{12}(t) - 2]u_1(t) + z_b(t),$$
$$q_1 = (u_1(t) - 1)^2 + (m_{11}(t))^2 + (m_{12}(t) - 2)^2,$$
$$MU_1^0 = \{(m_1(t), u_1(t)) \in \mathbb{R}^2 \times \mathbb{R}, m_{11}(t) + u_1(t) \le \alpha\}$$

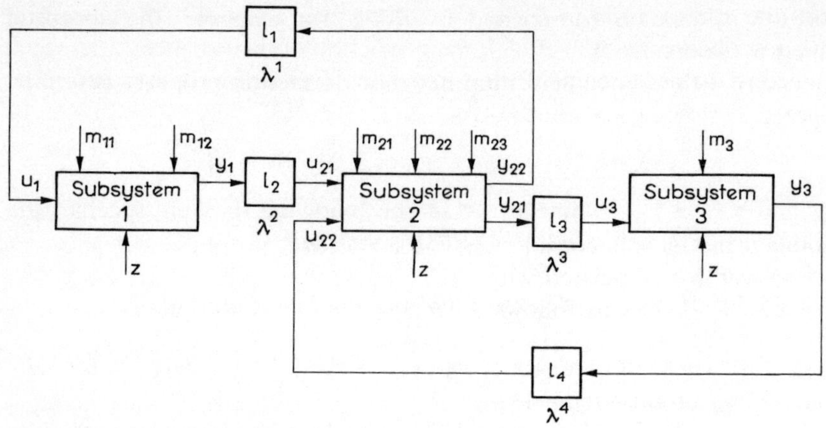

FIGURE 4.8   The structure of a time-varying static system with inventory couplings $I_1, \ldots, I_4$.

Subsystem 2

$$y_{21}(t) = z_a(t)[m_{21}(t) - m_{22}(t)] + u_{21}(t) - 3u_{22}(t) + z_b(t),$$

$$y_{22}(t) = z_a(t)[2m_{22}(t) - m_{23}(t)] - u_{21}(t) + u_{22}(t) + z_b(t),$$

$$q_2 = 2(m_{21}(t) - 2)^2 + (m_{22}(t))^2 + 3(m_{23}(t))^2$$
$$+ 4(u_{21}(t))^2 + (u_{22}(t))^2.$$

$$MU_2^0 = \mathbb{R}^3 \times \mathbb{R}^2.$$

Subsystem 3

$$y_3(t) = z_a(t)m_3(t) - 4u_3(t) + z_e m_3(t)u_3(t) + z_b(t),$$

$$q_3 = (m_3(t) + 1)^2 + (u_3(t) - 1)^2.$$

$$MU_3^0 = \{(m_3(t), u_3(t)) \in \mathbb{R} \times \mathbb{R}, -m_3(t) - u_3(t) \leq \beta\},$$

The dynamic inventory couplings have the form:

$$\int_{\Delta_f} [\bar{H}_1 u(t) - \bar{H}_2 y(t)]\, dt = 0,$$

where

$$\bar{H}_1 = I_{4 \times 4} \qquad \bar{H}_2 = \begin{bmatrix} 0 & 1 & 0 & 0 \\ 1 & 0 & 0 & 0 \\ 0 & 0 & 0 & 1 \\ 0 & 0 & 1 & 0 \end{bmatrix}$$

and

$$\Delta_f = [t_0, t_f] = [0, 10].$$

The off-line optimization problem was solved over Δ_f (step 1 of the control scheme described in section 4.4.2); the prediction of disturbances $z_a(t)$, $z_b(t)$ and the constant parameters of the subsystems z_c, z_d, z_e were assumed to be as follows:

$$\bar{z}_a(t) = 1 + 0.01t, \qquad \bar{z}_b(t) = 0$$
$$\bar{z}_c = \bar{z}_d = \bar{z}_e = 0,$$

i.e., a linear model was considered. The model-based optimal price $\hat{\lambda}_{\Delta_f} = \hat{\lambda}$ (constant over Δ_f) associated with the coupling constraints (see Figure 4.9) are, for $\alpha = \beta = 1{,}000$, as follows:

$$\hat{\lambda}_1 = -0.6577, \qquad \hat{\lambda}^2 = -0.1772, \qquad \hat{\lambda}^3 = -0.5426, \qquad \hat{\lambda}^4 = -0.7568.$$

The model-based optimal performance was $\hat{Q} = 19.9253$. After dividing the time interval Δ_f into n equal parts Δ_j, the control scheme described in section 4.4.2 was applied; for the simulation of the control scheme we have assumed the following real system parameters and disturbances:

$$z_a^r(t) = \bar{z}_a(t), \qquad z_b^r(t) = 0.1t, \qquad z_c^r = z_d^r = z_e^r = 0.02.$$

It should be noted that for these disturbances the real optimal performance $\hat{Q}^r$ would equal 25.5. Price corrections were first made according to formula (4.61) with $A_j = I$; thus, at the beginning of each interval Δ_j, new prices $\tilde{\lambda}_j$ were generated as follows:

$$\tilde{\lambda}_j = \hat{\lambda}_j - \varepsilon D_j.$$

The results obtained for $n = 10$, 20, and 40 for the various values of ε are presented in Table 4.7. In the rows denoted by $D^1, \ldots, D^4$ the deviations of the level of the inventories at the end of the whole interval Δ_f are given. In the first column ($\varepsilon = 0$), the results obtained for open-loop control without price correction are given. The last row of the table displays the values of performance Q obtained in the real system to show that they do not change very much, except when $\varepsilon = 0.50$.

It can be seen that the proper choice of ε is essential. For $n = 10$ and $\varepsilon = 0.5$, the contraction property of the price correction algorithm was not achieved and the application of the control structure was not successful. For $\varepsilon = 0.25$, the contraction was obtained with $\delta < 0.8$ and the final deviation of the inventory levels from the model-based optimal values decreased. For $\varepsilon = 0.25$, we show in Figure 4.9 how the prices $\hat{\lambda}^1, \ldots, \hat{\lambda}^4$ vary over time. For smaller values of ε (and especially for $\varepsilon = 0.0625$) the deviations increased again. In all cases, though, when the contraction was obtained, the performance value, which is not the prime goal in this control scheme, was close to the model-based optimal value. For $n = 20$ and $n = 40$, smaller values of D^i were obtained and the contraction was also achieved for $\varepsilon = 0.5$.

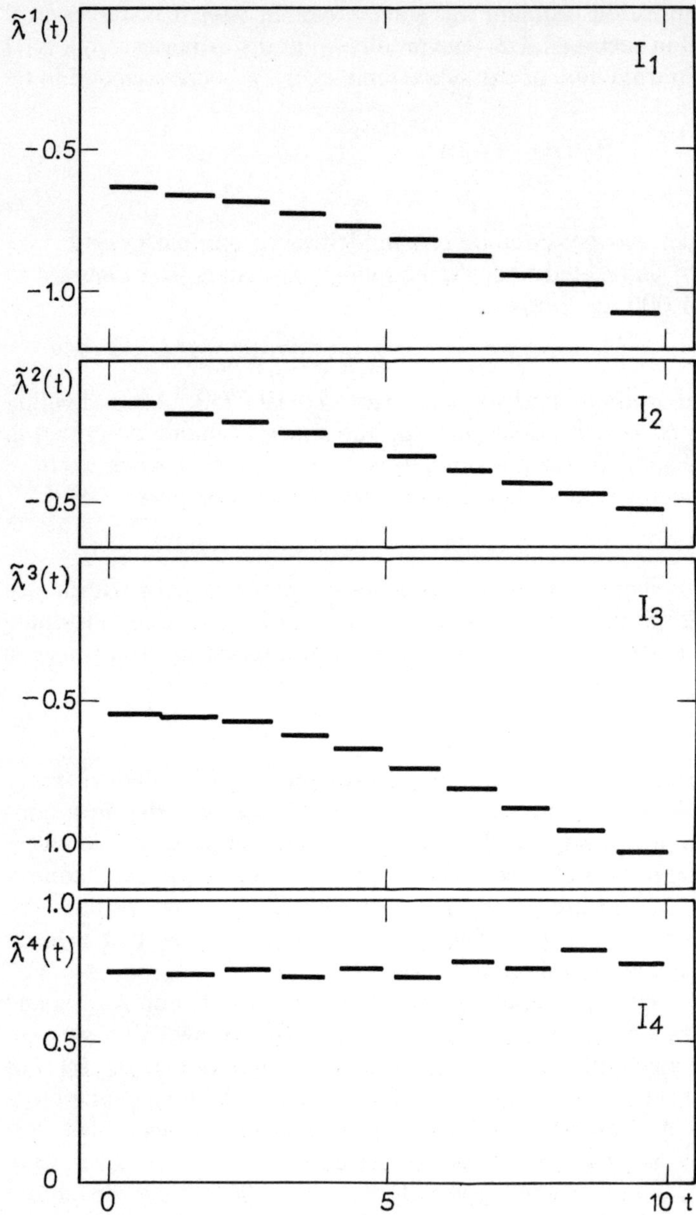

FIGURE 4.9   The prices associated with the coupling constraints.

TABLE 4.7 Simulation Results for Price Coordination for a Static Nonstationary System with Dynamic Inventory Couplings

	0	0.0625	0.125	0.25	0.50	1.0
$n = 10$						
D^1	5	3.41	2.76	1.90	−1374.8	
D^2	5	3.40	2.5	1.62	365.5	
D^3	5	3.76	3.1542	2.30	167.7	
D^4	5.02	0.68	0	0.42	−5,021.7	
Q	19.2	18.98	19.73	21.21	1,066,336	
$n = 20$						
D^1		3.36	2.69	1.91	0.39	
D^2		3.33	2.43	1.56	1.03	
D^3		3.72	3.09	2.26	1.22	
D^4		0.64	0.00	−0.21	−2.37	
Q		19.04	19.84	21.38	25.45	
$n = 40$						
D^1					0.15	−0.83
D^2					0.89	36
D^3					−1.38	71
D^4					−0.19	−189
Q					23.20	165,458

In the simulation, the price corrections were also performed according to formula (4.61) with

$$A_j = V_j^{-1}; \qquad V_j = \int_{\Delta_j} V(t)\, dt,$$

where $V(t)$ is an approximation of $r(\hat{\lambda}_j(t), \bar{z}(t))$ given by

$$V(t) = \varepsilon F_w(\bar{w}(\hat{\lambda}(t), \bar{z}(t)), \bar{z}(t))[F_w(\bar{w}(\hat{\lambda}(t), \bar{z}(t)), \bar{z}(t))]^*,$$

where $w = (m, u)$, and

$$F(w(t), z(t)) = -\bar{H}_1 u(t) + \bar{H}_2 f(m(t), u(t), z(t))$$

See Eq. (4.66) and Appendix A, section A.2. The elements of $V(t)$ for the example can be easily computed:

$$V(t) = \varepsilon \begin{bmatrix} 2\bar{z}_a^2(t) + 11 & -3 & -2\bar{z}_a^2(t) - 4 & 3 \\ -3 & 2\bar{z}_a^2(t) + 5 & -1 & 0 \\ -2\bar{z}_a^2(t) - 4 & -1 & 5\bar{z}_t^2(t) + 3 & 3 \\ 3 & 0 & 3 & \bar{z}_a^2(t) + 17 \end{bmatrix}$$

This $V(t)$ does not have a diagonal character. In Table 4.8 the results of the

TABLE 4.8   Simulation Results with $A_j = V_j^{-1}$ and $n = 10$

	1.0	2.0	2.5	3.0	3.5	4.0	4.5
D^1	1.63	0.91	0.74	0.63	0.52	0.16	−1.92
D^2	0.95	0:46	0.37	0.3	0.19	−0.77	−7.08
D^3	1.74	0.99	0.82	0.69	0.57	0.17	−2.17
D^4	1.7	0.95	0.77	0.65	0.57	0.5	0.41
Q	23.92	24.91	25.14	25.31	25.55	28.58	95.65

computations are displayed for $n = 10$. There is considerable improvement over the simple price correction algorithm with $A = I$. For $\varepsilon = 3.5$ the final deviations of the level of inventories were smaller than those achieved using a range of ε in the simple iteration rule with a fine division of the time horizon ($n = 40$). We can also see that $V(t)$ made the correction algorithm less sensitive to changes of the value of ε; good results were obtained for $1 \le \varepsilon \le 4$).

Tables 4.9 and 4.10 present the results for $\alpha = 1.006$ and $\beta = 0.5$; with these values of α and β the local constraints become active. The results show that the control structure can behave quite well when the local constraints become active. As in the previous case, the more elaborate correction algorithm (with $A_j = V_j^{-1}$) proved to be much more successful.

TABLE 4.9   Simulation Results with $d = 1.006$, $\beta = 0.5$, and $A_j = I$

	0	0.125	0.25	0.5
D^1	5.0	2.7	1.97	−461.0
D^2	4.98	3.11	1.96	13.0
D^3	5.0	3.21	2.36	140.0
D^4	5.01	0.3	−0.05	−726.0
Q	20.77	20.29	21.74	59,961.0

TABLE 4.10   Simulation Results with $\alpha = 1.006$, $\beta = 0.5$, and $A_j = V_j^{-1}$

	0	2.5	2.5	3.5	4	4.5
D^1	5.0	1.38	0.64	0.44	0.19	−0.85
D^2	4.98	2.54	0.61	0.34	−0.28	−1.26
D^3	5.0	1.83	0.88	0.62	0.35	−0.08
D^4	5.01	2.18	1.06	0.78	0.74	0.79
Q	20.77	24.82	25.81	26.20	27.84	35.56

4.5. THE REPETITIVE CONTROL ALGORITHM

The repetitive control scheme was used at the global and local levels of the control structures considered in sections 4.2–4.4. Here we study the most elementary applications of the repetitive control algorithm.

4.5.1. THE BASIC CONTROL STRUCTURE

Here we assume that we are given a dynamic process described by differential equations:

$$\dot{x}(t) = f(x(t), m(t), z(t)) \tag{4.67}$$

where

$$f : \mathbb{R}^{n_x} \times \mathbb{R}^{n_m} \times \mathbb{R}^{n_z} \to \mathbb{R}^{n_x}, \qquad t \in [t_0, T],$$

$x(t)$ is the state of the process, $m(t)$ the control vector, and $z(t)$ the disturbance and/or parameter vector. We introduce the following control scheme for Eq. (4.67): at time t_j $(j = 0, 1, 2, \ldots, \ldots J)$ the control for period $[t_j, t_{j+1}]$ is based on the solution of the following current process optimization problem (OP):

Given the real (measured or estimated) value of the process state $x^r(t_j)$ and the prediction of the disturbance trajectory $\bar{z}_{\Delta_j}$, where $\Delta_j = [t_j, t_{jf}]$, find model-based optimal control $\tilde{m}_{\Delta_j}$, that is, $\tilde{m}(t)$ for all $t \in \Delta_j$, that minimizes performance:

$$Q_{\Delta_j} = \int_{t_j}^{t_{jf}} q(x(t), m(t), \bar{z}(t)) \, dt + J_{jf}(x(t_{jf})) \tag{4.68}$$

subject to $\dot{x}(t) = f(x(t), m(t), z(t))$, $x(t_j) = x^r(t_j)$ and $m(t) \in M^0 \subset \mathbb{R}^n$ $\forall t \in \Delta_j$. The control trajectory from the solution of OP is applied to the process over $[t_j, t_{j+1}]$ (see Figure 4.10). The final time in the optimization horizon may be fixed, i.e., $t_{jf} = t_f \forall j$, or shifted, e.g., $t_{jf} = t_{(j-1)f} + \Delta T_f$. It is also obvious that we can compare the above control scheme with the open-loop control (computed once and forever at $t = t_0$) only if $t_{jf} = t_f$ (see the first simulated application in section 4.3). In this case, the final state may be fixed $(x(t_f) = x^f)$.

4.5.2. THE PROPERTIES OF REPETITIVE CONTROL IN THE LIMIT

We assume now that in the basic repetitive scheme of the previous section the prediction of disturbances $\bar{z}(t)$ is not updated and the final time is fixed; it is $t_{jf} = t_f$ and $\bar{z}_{\Delta_j} = \bar{z}_{\Delta_0}|_{\Delta_j}$. Let $\tilde{m}(x(t), t)$ be an optimal feedback control law

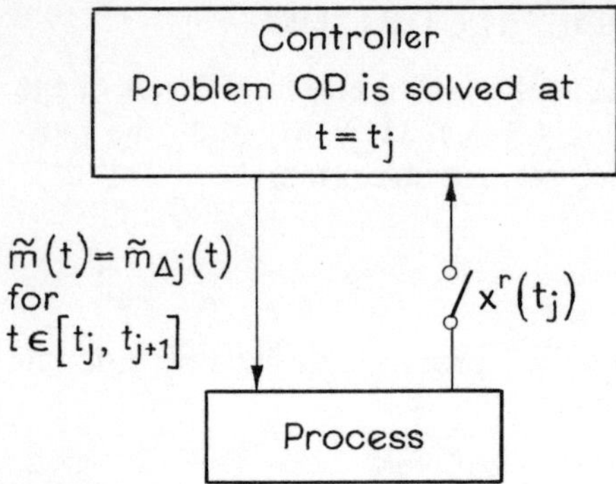

$\tilde{m}(t) = \tilde{m}_{\Delta j}(t)$
for
$t \in [t_j, t_{j+1}]$

$x^r(t_j)$

FIGURE 4.10   The basic repetitive control structure.

computed for given prediction $\bar{z}_{\Delta_0}$ and let $\bar{x}^r(t)$ be the value of the state trajectory in the real system when this control law is applied. Of course, in most cases we cannot obtain the feedback law in an analytical form, but the hypothetical application of the law offers us a good basis for comparisons. By $\tilde{m}(t), x^r(t)(\tilde{m}(t) = \tilde{m}_{\Delta_j}(t)$ for $t \in [t_j, t_{j+1}])$ we denote, respectively, the control and the state trajectories obtained from the application of the repetitive control scheme. We set $t_{j+1} - t_j = h$, $j = 0, 1, \ldots, J$, $t_J = t_f - h$.

To compare repetitive control with the continuous control law $\bar{m}(x(t), t)$, we can use the following theorem (Nowosad 1978).

THEOREM 4.4.   *If we assume that*

1.   *Mapping f in (4.67) is continuous in its arguments and has continuous derivatives with respect to $x(t)$ and $m(t)$*
2.   *The control law $\bar{m}(x(t), t)$ is continuous in t and continuously differentiable with respect to $x(t)$*
3.   *Prediction $\bar{z}(t)$ is continuous on $[t_0, t_f]$, as is $z^r(t)$*
4.   *The real state values $\bar{x}^r(t)$ and $x^r(t)$ belong to some uniformly bounded set*

then the state trajectory resulting from the application of the repetitive control scheme converges uniformly to $\bar{x}^r(t)$ when the frequency of intervention increases to infinity $(h \to 0^+)$.

Assumption 3 may be relaxed; we can take a uniformly bounded piecewise continuous $z^r(t)$ with a finite number of discontinuity points. Assumption 2 is, in fact, a special case of the so-called regular synthesis (Boltyanski 1969), and we can also make it a bit weaker. The control law $\bar{m}(x(t), t)$ may be the regular synthesis law continuous with respect to $x(t)$. The assumptions about the continuity of $\bar{m}(x(t), \bar{z}(t))$ with respect to time may be weakened in the same way as the assumptions concerning $z(t)$. Condition 4 is quite natural and basic. The following lemma gives some conditions that ensure the fulfillment of this condition.

LEMMA 4.5 (*Ioffe and Tikhomirov* 1974). *Suppose that for the process described by* (4.67), *where* $x(t_0) = x_0$, $m(t) \in M^0$, $z(t) \in Z^0$, *the following assumptions are satisfied:*

1. *Mapping f is differentiable with respect to $x(t)$, and mappings f and f_x are continuous in their arguments,*
2. $\forall m(t) \in M^0$, $z(t) \in Z^0$ *and* $\forall x \in \mathbb{R}^{n_x}$, *mapping* $t \to f(x, m(t), z(t))$ *is measurable on* $[t_0, t_f]$ *and*

$$|\langle x, f(x, m(t), z(t)) \rangle| \le c \|x\|^2 + r(t)$$

where $c > 0$, $r(t) \in L^1([t_0, t_f], \mathbb{R})$.

then the set of state trajectories $\mathcal{K}(x_0)$ *realized by all measurable* $m_{\Delta_0}, z_{\Delta_0}(m(t) \in M^0, z(t) \in Z^0)$ *is bounded.*

Theorem 4.4 can be applied to the linear-quadratic problem because under typical assumptions the conditions of this lemma are fulfilled. It is possible to show that Theorem 4.4 is also valid for the linear-quadratic problem when $x(t) \in X$, where X is a Banach space (Bamberger *et al.* 1975).

The above considerations show that under appropriate assumptions repetitive control is similar to closed-loop control. Therefore, if closed-loop control provides for good system behavior, then repetitive control will also do well when the frequency of the repitions is sufficiently high (even without updated disturbance predictions). It should be pointed out here that judgments about system behavior on the basis of performance sensitivity are quite logical in principle; in practice, however, with constraints on state and/or interaction variables we might be more interested in the trajectory sensitivity. It can be shown that in many cases the trajectory sensitivity for closed-loop control is much less than the trajectory sensitivity for open-loop control (see Kreindler 1972). The ability to approach the optimal state trajectory accurately is particularly important in previously described hierarchical control structures. For example, the lower-level optimization in the dynamic price coordination structure depends on the model-based optimal values of the state trajectory. Since it is not possible to use continuous feedback at the higher levels, we can hope to achieve good results by applying the repetitive control scheme with reasonable operating costs and no need to know the optimal feedback control law.

The considerations presented in this section are qualitative; quantitative results concerning the influence of increasing or decreasing the frequency of repetitions are very difficult to obtain and do not exist for the general case.

4.5.3. QUANTITATIVE ANALYSIS OF THE LINEAR-QUADRATIC CONTROL PROBLEM

To evaluate the repetitive scheme quantitatively, we must satisfy the following three conditions:

- We have to know explicitly how the optimal control and state trajectories depend on the disturbances and the initial conditions.
- We have to assume that the disturbance and the disturbance prediction are in a simple form.
- We should know the form of the optimal control law $\bar{m}(x(t), t)$ to investigate the properties of the problem in the limit.

For the above reasons, it is possible to analyze only a linear-quadratic control problem with additive disturbances, where the disturbances and the predictions are assumed to be constant over the control horizon. Therefore, we describe the process by the following equation:

$$\dot{x}(t) = A(t)x(t) + B(t)m(t) + z, \qquad x(t_0) = x_0,$$

and the performance index is

$$Q(x_{\Delta_i}, m_{\Delta_i}, z) = \tfrac{1}{2}\langle x(t_f), Fx(t_f)\rangle + \tfrac{1}{2}\int_t^{t_f} [\langle x(t), P(t)x(t)\rangle + \langle m(t), R(t)m(t)\rangle]\, dt$$

We assume that the endpoint of the time horizon is fixed.

For the model-based optimization (4.68) we set $\bar{z} \neq z^r$, $\bar{z}, z^r$—constant, $\bar{z}, z^r \in \mathbb{R}^{n_x}$. For convenience and without loss of generality we assume that $\bar{z} = 0$. We assume also that $R(t)$ and F are positively defined and $P(t)$ is semi-positively defined. Under these assumptions, the optimal solution of the control problem exists and is unique. The control and state trajectories obtained in the open-loop structure (optimal control computed once and then applied over Δ_0) are given by the formulae:

$$m^0(t) = S(t, t_0)x_0$$

$$x^0(t) = L(t, t_0)x_0 + T(t, t_0)\bar{z}$$

where

$$S(t, t_0) = R^{-1}(t)B^T(t)[\Phi_{21}(t, t_0) + \Phi_{22}(t, t_0)\Omega(t_f, t_0)]$$

$$L(t, t_0) = \Phi_{11}(t, t_0) + \Phi_{12}(t, t_0)\Omega(t_f, t_0)$$

$$T(t, t_0) = \int_{t_0}^t v(t, \tau)\, d\tau$$

and

$$\Omega(t_f, t_0) = -[\Phi_{22}(t_f, t_0) + F\Phi_{12}(t_f, t_0)]^{-1}[\Phi_{21}(t_f, t_0) + F\Phi_{11}(t_f, t_0)],$$

matrices $\Phi(t, t_0)$ and $v(t, t_0)$ satisfy the equations

$$\frac{d}{dt}\Phi(t, t_0) = \begin{bmatrix} A(t) & B(t)R^{-1}(t)B^T(t) \\ P(t) & -A^T(t) \end{bmatrix} \cdot \Phi(t, t_0)$$

$\Phi(t_0, t_0) = I$:

$$\frac{d}{dt}v(t, t_0) = A(t)v(t, t_0) \qquad v(t_0, t_0) = I.$$

Now we can compare open-loop control with the control generated by the repetitive scheme. The following lemma has been proved by Malinowski and Nowosad (1977):

LEMMA 4.6. *The difference between the performance values obtained for open-loop control and those for the repetitive scheme is given by a quadratic formula in disturbance vector z^r and does not depend on the initial condition* $x(t_0) = x_0$.

Therefore we have

$$\Delta Q = Q^0 - Q^r = (z^r)^T V(z^r),$$

where

$$V = \tfrac{1}{2}\sum_{j=1}^{J} V_j.$$

For $J = 0$ we have just one intervention of the controller, which is equivalent to open-loop control.

$$V_j = V_{0j} + (V_{1j} + V_{1j}^T) + (V_{2j} + V_{2j}^T), \tag{4.69}$$

where

$$V_{0j} = \left[\int_{t_{j-1}}^{t_j} v^T(t_j, \tau)\, d\tau\right] W_{0j}\left[\int_{t_{j-1}}^{t_j} v(t_j, \tau)\, d\tau\right],$$

$$V_{1j} = \left[\int_{t_{j-1}}^{t_j} v^T(t_j, \tau)\, d\tau\right] W_{0j}\left[\int_{t_j}^{t_f} v(t_j, \tau)\, d\tau\right],$$

$$V_{2j} = \left[\int_{t_{j-1}}^{t_j} v^T(t_j, \tau)\, d\tau\right]\left[\int_{t_j}^{t_f}\int_{t_j}^{t} [L(\tau, t_j) - v(\tau, t_j)]^T P(\tau)v(\tau, t)\, d\tau\, dt\right],$$

$$W_{0j} = \int_{t_j}^{t_f} v^T(t, t_j)P(t)v(t, t_j)\, dt + v^T(t_f, t_j)Fv(t_f, t_j) + K(t_j)$$

and $K(t_j)$ is the solution of the Riccati equation

$$\dot{K}(t) + K(t)A(t) + A^T(t)K(t) + K(t)B(t)R^{-1}(t)B^T(t)K(t) - P(t) = 0,$$
$$K(t_f) = -F$$

368

for $t = t_j$. The following theorem can now be proved (Malinowski and Nowosad 1977).

THEOREM 4.5. *Assume that $P(t) \equiv 0$ and V_{0j} is positively defined; V_{0j} is always positively defined when $A(t)$ is constant and has real eigenvalues. Then an intervention at time t_j gives a better performance value under repetitive control than under open-loop control if the following condition is satisfied: $\sigma_i > -1$ for all eigenvalues σ_i of matrix $V_{0j}^{-1}(V_{1j} + V_{1j}^T)$. In particular, repetitive control provides for a better preformance value than open-loop control if $A(t) \equiv 0$ for any number J of interventions.*

In formula (4.69) the first term represents essentially the profit we get from using the exact value of the real system state at time t_j. If $z^r(t) = z^r \neq \bar{z} = 0$ for $t \in [t_0, t_s]$ and $z^r = \bar{z} = 0$ for $t \in [t_s, t_f]$, then we surely have an advantage over open-loop control when making one intervention at time t_s; the profit is expressed by $(z^r)^T V_{0s}(z^r)$, where V_{0s} denotes the first term of (4.69) for $t_s = t_j$. The remaining two terms in (4.69) are more difficult to interpret: they reflect the effects of the continuing difference between z^r and $\bar{z}$. When the final state penalty weighting matrix F has sufficiently large eigenvalues then the repetitive control scheme in general ensures a smaller final state deviation than open-loop control.

Now we will demonstrate that the assumptions of Theorem 4.5 are essential and cannot be eliminated. The following example of a linear oscillator shows that if these assumptions are not satisfied, then repetitive control may make Q^r larger than Q^0.

Suppose that

$$\dot{x}_1(t) = \omega x_2(t) + z$$
$$x_2(t) = -\omega x_1(t) + m(t)$$
$$x_1(0) = x_2(0) = 0$$
$$Q = \tfrac{1}{2}(x_1(t_f))^2 + \tfrac{1}{2}(x_2(t_f))^2 \frac{1}{2}\int_0^{t_f}(m(t))^2\, dt.$$

Assume that we take $t_f = 2\pi/\omega$, $J = 1$, and repeat the optimization at $t_1 \in (0, 2\pi/\omega)$. The open-loop optimal control (with $\bar{z} = 0$) is obviously $\hat{m}(t) = 0$. Since $x_1^0(t) = (z/\omega)\sin\omega t$, $x_2^0(t) = -(2z/\omega)\sin^2(\omega/2)t$; thus, $x_1^0(t_f) = x_2^0(t_f) = 0$ and $Q^0 = 0$. When making the repetition we have $x^r(t_1) \neq x^0$; therefore, $\tilde{m}(t) \neq 0$ for $t \in [t_1, t_f]$. This means that $Q^r > 0$ and $\Delta Q < 0$.

Under the assumptions of Theorem 4.5, if we increase the number of interventions without moving the previous intervention times but introducing new $t_{J+1}, t_{J+2}, \ldots$ then ΔQ will increase. Generally, the dependence of

ΔQ on $\{t_j\}_{i=0}^{J}$ is very complex. For fixed J, one can prove only the following result (the simple proof is omitted):

LEMMA 4.7. *For the linear-quadratic control problem and fixed J there exists an optimal distribution of intervention times* $\{t_j\}_{j=0}^{J}$, *where* $t_{j+1} \geq t_j$.

However, at present there is no practical way to determine this optimal distribution. In this section we have concentrated on the differences between the repetitive control scheme and closed-loop (or open-loop) control. It would also be very useful to examine the difference between $\tilde{m}(t)$, $x^r(t)$, Q^r, that is, the trajectories and the performance value in the repetitive control structure, and the real optimal control and performance. In particular, we are interested in the continuity analysis with respect to the difference $z^r(t) - \bar{z}(t)$ or in the bounds on trajectory sensitivity and performance degradation (see also section 4.3). These topics are being investigated. The analysis of the repetitive scheme becomes more difficult when $t_{jf} \neq t_f$ (the floating time horizon), and when the predictions of z are updated. We presume, however, that in these cases the repetitive scheme will be even more advantageous. We note finally that in every application, a numerical simulation will be required to find the proper sequence $\{t_j\}_{j=0}^{J}$ of repetitions.

4.6. MULTILEVEL STATE FEEDBACK STRUCTURES FOR LINEAR SYSTEMS WITH QUADRATIC PERFORMANCE FUNCTIONS

4.6.1. PRELIMINARY DISCUSSION

The part of hierarchical and decentralized control theory for dynamic systems that concerns state feedback structures for linear systems with quadratic performance functions would perhaps be impossible to present in one book. It would have to cover decentralized design, hierarchical and decentralized filtering, stability of composite systems, and so on. We have devoted the previous sections of this chapter to control algorithms of the repetitive type that are recommended for the higher layers of the hierarchy, where nonlinear processes, discrete observations, and discrete coordination interventions must usually be considered. The analysis of these repetitive schemes was rather difficult and it was not possible to obtain too many quantitative results. Nevertheless, we consider the topic to be important for applications. To complete the picture of hierarchical control for dynamic systems, we briefly present in this section some of the concepts and results of linear–quadratic state feedback theory.

The control algorithms considered in this section can, of course, also be used in the steady-state process control structures where they would regulate the process to the specified set points which may be determined by static optimization, for example.

Two-layer control structure

In complex systems, the basic repetitive control scheme of section 4.5 or the more elaborate hierarchical algorithms of section 4.3 for large systems are usually not applied directly to the system elements. They are used rather as the control strategy for the higher-layer controllers in the multilayer structure (see Chapter 1). Typically, a second-layer optimizing controller might adjust its decisions at times $t = t_j$ (sections 4.3–4.5) and specify the model-based optimal control and state trajectories over a chosen time interval Δ_j for the first layer regulatory controller (compare the remarks at the end of section 4.2). If the basic repetitive scheme of section 4.5 is used in the second layer to control indirectly the process described by Eqs. (4.67), then the first-layer direct control strategy can be based on

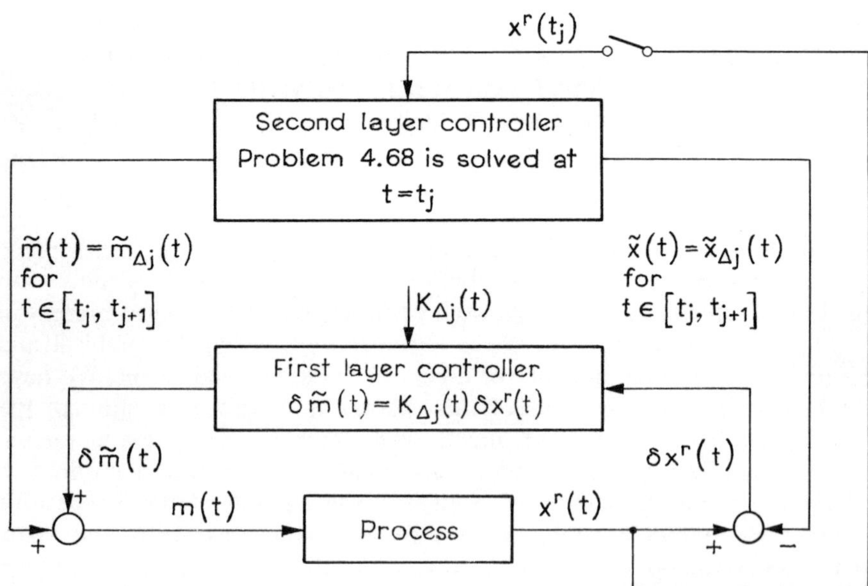

FIGURE 4.11   A two-layer control structure with continuous first-layer control. If $x^r(t)$ cannot be measured then the estimate $\hat{\delta}x^r(t)$ is used instead of $\delta x^r(t)$.

1. Equations (4.67) linearized along $\tilde{m}_{\Delta_j}$, $\tilde{x}_{\Delta_j}$, and $\bar{z}_{\Delta_j}$ (see problem (4.68)), that is,

$$\delta \dot{x}(t) = A(t)\delta x(t) + B(t)\delta m(t) + F(t)\delta z(t), \qquad (4.70)$$

where $A(t) = f_x'(\tilde{x}(t),\ \tilde{m}(t),\ \bar{z}(t))$, $B(t) = f_m'(\tilde{x}(t),\ \tilde{m}(t),\ \bar{z}(t))$, and $F(t) = f_z'(\tilde{x}(t),\ \tilde{m}(t),\ \bar{z}(t))$

2. Properly defined first-layer objectives

The classical state feedback techniques can now be used along with appropriate estimation (filtering) algorithms, if necessary. The algorithms provide us with estimate $\hat{\delta x}^r(t)$ of $\delta x^r(t) = x^r(t) - \tilde{x}(t)$. Suppose that we use $\tilde{\delta}m(t) = K_{\Delta_j}(t) \cdot \hat{\delta x}^r(t)$ as the first-layer control law. Then our control structure will compute at time t_j the new trajectories $\tilde{m}_{\Delta_j}$, $\tilde{x}_{\Delta_j}$, and $K_{\Delta_j}(t)$ for $t \in [t_j, t_{j+1}]$ and apply the control law $\tilde{\delta}m(t) = K_{\Delta_j}(t)\hat{\delta x}^r(t)$ over $[t_j, t_{j+1}]$ (see Figure 4.11). It should be noted that if the process is to operate in the steady state, that is, when the constant state is considered to be optimal or desirable, then the linearized Eq. (4.70) will be stationary (with $A(t) = A$ and so forth). For this reason and because the major part of state feedback control theory is developed for stationary systems, we will consider only such systems in the next sections. However, some of the methods and concepts can be easily extended to the nonstationary linear case.

The linear-quadratic control problem

We consider the following linear-quadratic control problem of a complex system:

$$\dot{x}_i(t) = A_{ii}x_i(t) + G_{0i}u_i(t) + B_i m_i(t) \qquad (4.71a)$$

$$y_i(t) = G_{Ii}x_i(t) \qquad (4.71b)$$

$$u_i(t) = H_i y(t) = \sum_{j=1}^{s} H_{ij}y_j(t), \qquad (4.71c)$$

where $x_i(t) \in \mathbb{R}^{n_{xi}}$, $u_i(t) \in \mathbb{R}^{n_{ui}}$, $m_i(t) \in \mathbb{R}^{m_{mi}}$, $y_i(t) \in \mathbb{R}^{n_{yi}}$, and $i \in \overline{1, s}$. The performance index is

$$J = \frac{1}{2}\sum_{i=1}^{s}\left[\int_0^T (\langle x_i(t), Q_i x_i(t)\rangle + \langle m_i(t), R_i m_i(t)\rangle)\,dt + \langle x_i(T), Q_i^0 x_i(T)\rangle\right]. \qquad (4.72)$$

We use J instead of Q in order to stress the difference between optimizing control and regulatory control design. The linear system description (4.71) may be obtained, as mentioned before, by linearizing the process equations at the optimal steady-state value of the state and control. In this case, $x_i(t)$, $u_i(t)$, $m_i(t)$ will be the deviations from the nominal state, interaction input, and control input values, respectively. It should be noted that decomposition

372

(4.71), which is used for the first layer of control, does not have to be identical to the decomposition used at the higher layers; usually it is much more fine, that is, more subsystems are formed. Also the global dimension of control vector $m(t) = (m_1(t), \ldots, m_s(t))$ can be different from the control vector considered at the higher layers. In many cases it may be impossible to use some of the control inputs as manipulated variables for regulation (for example, when the controls have to be manually adjusted or when a change of their value requires much time and effort). There is no satisfactory theory concerning the important question of the choice of manipulated variables for regulation and the choice of the observed variables if the system state cannot be observed directly. A few aspects of this problem were presented in section 1.2.

It is also by no means clear how to establish the weighting matrices Q_i, Q_i^0, and R_i in performance index (4.72). In fact, a large number of control system design techniques, like the frequency response methods or pole placement techniques, do not exploit this form of the performance index at all. If we accept (4.72), time horizon T has to be appropriately chosen depending on the value of $t_{j+1} - t_j$ (see the first part of this section) and on the process dynamics; if $t_{j+1} - t_j$ is large enough, we can use $T = \infty$, which simplifies the control design. If the task of the first-layer controller is to keep some regulated output variables w_i^r, where $w_i^r = C_i^r x_i$, to the specified set points w_{id}^r from, for example, the steady-state optimization at the higher layer, then matrices Q_i in (4.72) could be $Q_i = C_i^{r^T} Q_i^1 C_i^r$, where Q_i^1 has to "weight" the regulated outputs. Note that in the linearization of the process equations at times t_j, w_i^r will be an increment of the regulated output and w_{id}^r will be zero.

In most cases the process state x_i is not directly accessible and we have to observe it through some measured variables w_i, where

$$w_i(t) = C_i x_i(t), \qquad w_i(t) \in \mathbb{R}^{n_{wi}} \tag{4.73}$$

If the measurements are corrupted by an additive noise $\eta_i(t)$ and if it is possible to describe the fast disturbances acting on any subprocess as another additive factor $\xi_i(t)$, then we can formulate the following control problem:

$$\dot{x}_i(t) = A_{ii} x_i(t) + G_{0i} u_i(t) + B_i m_i(t) + \xi_i(t) \tag{4.74a}$$

$$w_i(t) = C_i x_i(t) + \eta_i(t), \qquad i = 1, \ldots, s, \tag{4.74b}$$

plus Eqs. (4.71b) and (4.71c) and performance index. By $\xi_i(t)$, $\eta_i(t)$ we usually mean white Gaussian noises, but it is possible to consider other disturbances as well. We should write stochastic equation (4.74a) in the form

$$dx = A_{ii} x_i(t)\, dt + G_{0i} u_i(t)\, dt + B_i m_i(t)\, dt + d\Xi$$

where $\Xi(t)$ is a Wiener process, but since notation (4.74a) and (4.74b) in the linear case does not lead to misunderstandings we will use it for convenience.

The first-layer control algorithm will usually involve (a) a procedure for tuning the controllers, and (b) the implementation of the control law. Both the tuning procedure and the control law can be decentralized or centralized. For example, we may think about the centralized tuning of a decentralized controller or the decentralized tuning of a centralized controller, for which a decentralized procedure for solving the overall Riccati equation can be developed (Kokotovič 1972, Laub 1974). In complex dynamic systems, it is especially important to obtain the decentralized control law because in this way we reduce considerably the on-line information requirements (see Chapter 5) and increase the system reliability. Therefore, we will concentrate on decentralized control laws, which may be tuned in either a decentralized or centralized way.

A decentralized control law can be tuned at times t_j only, or updated more frequently. If it is updated more frequently and the tuning procedure is not decentralized, it is sometimes referred to as decentralized control with periodic coordination.

4.6.2. DECENTRALIZED CONTROL STRUCTURES

Simple decentralized control

Consider a linear-quadratic optimal control problem given by Eqs. (4.71) and (4.72), where for design purposes we set $T = \infty$. This problem can be written in compact form as

$$\dot{x}(t) = Ax(t) + Bm(t)$$

$$J = \frac{1}{2} \int_0^\infty (\langle x(t), Qx(t)\rangle + \langle m(t), Rm(t)\rangle)\, dt, \qquad Q \geq 0, \qquad R > 0, \tag{4.75}$$

where the system defined by matrices (A, B) is controllable, B, Q, and R are block diagonal matrices with entries B_i, Q_i, and R_i, $x = (x_1, \ldots, x_s)$, and $m = (m_1, \ldots, m_s)$. Matrix A is of the form

$$A = A_0 + D, \tag{4.76}$$

where A_0 is block diagonal with entries A_{ii} and the block entries of D are as follows:

$$D_{ij} = \begin{cases} 0, & \text{for} \quad i = j. \\ G_{0i}H_{ij}G_{Ij}, & \text{for} \quad i \neq j. \end{cases} \tag{4.77}$$

The optimal control law $m^*(t) = K^*x(t)$, where $K^* = -R^{-1}B^TM$ can be obtained by solving the Riccati matrix equation

$$A^{T}M + MA - MBR^{-1}B^{T}M + Q = 0. \tag{4.78}$$

Since A is not block diagonal, Eqs. (4.78) cannot be presented as s independent sets of equations and moreover, the resulting control law is not decentralized, that is, for each subsystem

$$m_i^*(t) = \sum_{j=1}^{s} K_{ij}^* x_j(t) = K_i^* x(t) \tag{4.79}$$

Our main goal is to obtain a set of controllers for each subsystem, where each controller is allowed to observe and control the variables of his own subsystem only. The simplest way to achieve this is as follows (Bailey and Ramapriyan 1973). Suppose that we neglect the interactions and consider matrix D equal to zero. Then the classical design procedure using the approximation $A = A_0$ will be *decentralized* into s problems:

Find a feedback control law so as to minimize

$$J_i = \frac{1}{2}\int_0^{\infty} (\langle x_i(t), Q_i x_i(t)\rangle + \langle m_i(t), R_i m_i(t)\rangle)\, dt \tag{4.80}$$

subject to $\dot{x}_i(t) = A_{ii}x_i(t) + B_i m_i(t)$.

After solving the Riccati equation for each subsystem, we obtain the *decentralized control law*:

$$m_i(t) = K_{ii}^0 x_i(t), \qquad K_{ii}^0 = -R_i^{-1}B_i^T M_i^0, \qquad i \in \overline{1, s} \tag{4.81}$$

where M_i^0 is the solution of the Riccati equation for (4.80). If, as a whole, the closed-loop system

$$\dot{x}(t) = (A + BK^0)x(t), \tag{4.82}$$

where K^0 is composed of blocks K_{ii}^0, is asymptotically stable, then the value J^1 of the performance index obtained in this system is

$$J^1 = \tfrac{1}{2}x_0^T M^1 x_0, \tag{4.83}$$

where M^1 is the solution of the linear Lyapunov equation

$$M^1(A + BK^0) + (A + BK^0)^T M^1 = -(K^{0T}RK^0 + Q)$$

and x_0 is the initial state. The following theorem (Bailey and Ramapriyan 1973) summarizes the results regarding the stability of the proposed design and gives bounds for the suboptimality.

THEOREM 4.5.   *Let us denote by*

J^* the optimal value of the performance in the system when the centralized control law (4.79) is applied,
J^1 the suboptimal value of performance when the decentralized control law (4.81) is applied (see Eq. 4.83),
J^0 The optimal value of the performance in a hypothetical decoupled system with $A = A_0$

and by $\sigma(\Sigma)$ the set of eigenvalues of the matrix

$$\Sigma = (K^{0\mathrm{T}}RK^0 + Q)^{-1}(M^0 D + D^T M^0), \tag{4.84}$$

where $M^0 = block\ diag\ (M_i^0)$. If the following condition is satisfied

$$\sigma(\Sigma) \subset (-1, 1) \tag{4.85}$$

then the closed-loop system (4.82) is stable and for any initial conditions the following relation holds:

$$(1 + \sigma_m)^{-1} J^0 \leq J^* \leq J^1 \leq (1 - \sigma_M)^{-1} J^0,$$

where σ_m, σ_M are, respectively, the minimum and the maximum eigenvalues of (4.84).

Condition (4.85) may be interpreted as a criterion of weak coupling with respect to stability in the sense that the simple decentralized control $m(t) = K^0 x(t)$ is stable. This condition can always be satisfied when $\|D\|$ is small enough. Yet it will also be satisfied for large $\|D\|$ when the symmetric part of $M^0 D$ is small.

When we use a finite optimization horizon T in problem (4.71), (4.72), the procedure remains the same except that we have to solve a differential Riccati equation. We obtain a time-dependent control law

$$m_i(t) = K_{ii}^0(t) x_i(t), \tag{4.86}$$

where $K_{ii}^0(t) = -R_i^{-1} B_i^T M_i^0(t)$ and

$$\dot{M}_i^0(t) = -M_i^0(t) A_{ii} - A_{ii}^T M_i^0(t) + M_i(t) B_i R_i^{-1} B_i^T M_i(t) - Q,$$

$M_i^0(T) = Q^0$. The same can be done when A_{ii}, B_i, R_i, and Q_i are time dependent.

General cases of decentralized control

Note that matrices K_{ii}^0 and hence K^0 in (4.82) were determined by decentralized calculations (decentralized tuning). It is obvious that there may exist a block diagonal feedback matrix better than K^0. Therefore, if we decide to use a centralized design procedure we may try to find a block diagonal matrix K^D for which the closed-loop system is stable. Its performance, which is expressed by

$$J(K^D) = \tfrac{1}{2}x_0^T M^2 x_0, \tag{4.87}$$

where M^2 is the solution of equation

$$M^2(A + BK^D) + (A + BK^D)^T M^2 = -(K^{D^T} RK^D + Q), \tag{4.88}$$

reaches at K^D its minimum value over the set of all block-diagonal matrices that produce a stable system. However, in this way we obtain a matrix K^D that depends on the initial condition x_0.

Since we would like to design a control law independent of the initial conditions we assume either that all components of x_0 are independent random variables with $E[x_{iko}] = 0$, $E[x_{ko}^2] = r$, $k = 1, \ldots, n_{xi}$ or that x_0 is uniformly distributed over an n-dimensional sphere. Then from (4.87) we have

$$E[J(K^D)] = \tfrac{1}{2} tr M^2. \tag{4.89}$$

In order to find K^D we can minimize (4.89), subject to the conditions that K^D is a block diagonal and $M^2 > 0$. It is difficult to specify the conditions that guarantee the existence of a solution to this problem.

This parametric optimization problem is not decentralized and not even easily decomposable (see Bailey and Wang 1972).

Once we have decided that the on-line decentralized feedback control laws can be designed (tuned) by the centralized procedure, we may also use the following indirect approach. Assume that the parameters of the centralized control law (4.79) have been calculated and we know that

$$m_i^*(t) = K_i^* x(t),$$

where $K_i^* = -R_i^{-1} B_i^T M$ and M satisfies Eq. (4.78). It is now possible to minimize the norm of the difference between $m_i^*(t)$ and the decentralized state feedback control

$$m_i(t) = K_{ii} x_i(t).$$

The basic design objective is not to minimize the norm but to minimize the performance loss. Nevertheless, since J in Eq. (4.75) is continuous in m, then the proximity of $m_i(t)$ to $m_i^*(t)$ should imply the proximity of the suboptimal and optimal performance values. The best $\hat{K}_{ii}$, which minimizes $\|m_i(t) - m_i^*t)\|$, results from

$$\min_{K_{ii}} \|K_{ii} I_{n_{xi}, n_x} - K_i^*\|$$

and is given by

$$\hat{K}_{ii} = K_i^* I_{n_{xi},n_x}^T [I_{n_{xi},n_x} \cdot I_{n_{xi},n_x}^T]^{-1} = K_{ii}^*,$$

where $I_{n_{xi},n_x} = [0 \mid \ldots \mid 0 \mid I \mid 0 \mid \ldots \mid 0].$

We should note that the above continuity argument can be misleading and it can be easily demonstrated by examples that the above design might result in an unstable closed-loop system. If the composite system (4.71) has no unstable modes fixed with respect to class $\mathcal{K}^D$ of block-diagonal state feedback matrices, then we can always find a stabilizing matrix $K^D \in \mathcal{K}^D$ (Wang and Davison 1973; see also Theorem 4.6, which follows). Note that a complex number p is called a fixed mode of $\dot{x} = Ax + Bm$ with respect to class $\mathcal{K}$ of state feedback matrices if $p \in \sigma(A + BK) \, \forall K \in \mathcal{K}$.

When only output variables w_i (see Eq. (4.73)) can be measured, we have to design a decentralized output feedback controller. This may be done by constructing an observer of the state of the system described by Eqs. (4.71) and (4.73). If this system is observable, the state may be reconstructed with any desired accuracy for $t \geq t_0 > 0$. With an observer, we could use the state estimate $\hat{x} = (\hat{x}_1, \ldots, \hat{x}_s)$ in the decentralized control laws. This will make sense, however, only if we are able to design a decentralized state observer. It is also possible to use another approach based on decentralized local output feedback with dynamic components in the feedback laws. The set of local dynamic feedback laws will then have the following form:

$$\dot{x}_{si}(t) = S_i x_{si}(t) + T_i w_i(t) \tag{4.90a}$$

$$m_i(t) = V_{ii} x_{si}(t) + K_{ii} w_i(t), \tag{4.90b}$$

where $x_{si}(t) \in \mathbb{R}^{n_{zi}}$ is the state of feedback controller i and S_i, T_i, V_{ii}, and K_{ii} are real constant matrices of appropriate size. After applying feedback laws (4.90) to the complex system (4.71), (4.73), we obtain a closed-loop system that can be described in compact notation as follows (see (4.75), (4.76)):

$$\begin{bmatrix} \dot{x}(t) \\ \dot{x}_s(t) \end{bmatrix} = \begin{bmatrix} A + BKC & BV \\ \hline TC & S \end{bmatrix} \cdot \begin{bmatrix} x(t) \\ x_s(t) \end{bmatrix}, \tag{4.91}$$

where $C = \text{block diag}(C_i)$, $S = \text{block diag}(S_i)$ and so on for T, V, and K.

THEOREM 4.6 (Wang and Davison 1973). *There exists a set of local feedback control laws (4.90) such that the closed-loop system (4.91) is asymptotically stable if and only if the system*

$$\dot{x}(t) = Ax(t) + Bm(t)$$

$$w(t) = Cx(t)$$

does not have any unstable mode fixed with respect to the class $\mathcal{K}^D$ of block-diagonal output feedback matrices K^D, that is, there exists no complex

number p such that $Re\, p \geq 0$ *and* $p \in \sigma(A + BK^D C)$ $\forall K^D \in \mathcal{K}^D$. $K^D = block$
diag (K_i) *and* K_i *is of size* $n_{mi} \times n_{wi}$.

In the proof of the above theorem, Wang and Davison describe the construction of control law (4.90). There is, of course, some freedom left in the design and we can try to find among all decentralized feedback controllers that produce a stable system the ones for which the given performance index (for example, Eq. (4.89)) is minimized.

Decentralized filtering

Consider now a stochastic system that can be modelled by Eqs. (4.74a), (4.71b), and (4.71c), that is,

$$\dot{x}_i(t) = A_{ii}x_i(t) + G_{0i}u_i(t) + B_i m_i(t) + \xi_i(t)$$
$$y_i(t) = G_{Ii}x_i(t) \tag{4.92}$$
$$u_i(t) = \sum_{j=1}^{s} H_{ij}y_j(t)$$

and observed (measured) outputs

$$w_i(t) = C_i x_i(t) + \eta_i(t), \tag{4.93}$$

where $\xi_i(t)$ is the local disturbance noise and $\eta_i(t)$ is the local observation noise. We assume that the initial state $x_i(0)$ (or, more precisely, $x_i(t_j)$—see section 4.6.1) is a Gaussian random vector with mean x_i^0 and covariance P_i^0, and the noise processes ξ_i and η_i are white Gaussian processes with zero mean and covariance matrices Ξ_i and N_i, respectively. We also assume that ξ and η are independent on $x(0)$, that $\xi_i(t)$ and $\eta_j(t)$ are stochastically independent $\forall i \neq j$, $\forall t$, and also that $x_i(0)$ is independent of $x_j(0)$, $\xi_i(t)$ of $\xi_j(t)$, and $\eta_i(t)$ of $\eta_j(t)$ $\forall i \neq j$, $\forall t$.

Consider now a stochastic equivalent of the deterministic control problem (4.75), with performance index

$$J = \frac{1}{2} \lim_{T \to \infty} E\left\{ \frac{1}{T} \int_0^T (\langle x(t), Qx(t) \rangle + \langle m(t), Rm(t) \rangle)\, dt \right\}. \tag{4.94}$$

If pair (A, B) is controllable and $(A, Q^{1/2})$, (A, C) are observable then there exists a unique control law that stabilizes system (4.91) and minimizes (4.94); see Athans (1971). A', B, and C are defined in relations (4.75)–(4.77). This control is given in the form of feedback control law (4.79), where in place of $x(t)$ we have to use the optimal estimate $\hat{x}(t)$ of state $x(t)$ generated by a centralized Kalman filter in which one has to use all available information $m(t) = (m_1(t), \dots, m_s(t))$ and $w(t) = (w_1(t), \dots, w_s(t))$.

If we want to use decentralized control law (4.81) then in order to preserve a completely decentralized *a priori* and *a posteriori* information pattern which is characteristic of (4.81) (that is, if we want to use decentralized control as well as decentralized design or tuning), we will have to construct a decentralized filter compatible with this information pattern. A reduced-order local Kalman filter can be designed by neglecting the interactions in (4.92) and solving the stochastic control problem with $D = 0$ (see (4.76)). The local filters are described then by the equation

$$\dot{x}_i(t) = A_{ii}\hat{x}_i(t) + B_i m_i(t) + K_{ii}^F(w_i(t) - C_i\hat{x}_i(t)), \qquad (4.95)$$

where gains $K_{ii}^F = P_i C_i^T N_i^{-1}$ and P_i result from the Riccati filter equations (see, e.g., Jazwinski 1970), which depend only on the local data. It should be noted that (4.95) does not give an unbiased estimate of $x_i(t)$, yet it is possible to formulate the conditions under which the closed-loop system resulting from the application of the decentralized control law $m_i(t) = K_{ii}^0\hat{x}_i(t)$ to (4.92) and (4.95) is stable. Teneketzis (1976) has proved the stability for sufficiently small $\|D\|$ in (4.76) and has investigated bounds on the estimate error covariance matrix.

The local filter design and implementation can be improved if one can also observe interaction inputs u_i, that is, if the following measurements are available:

$$w_{i1}(t) = u_i(t) + \eta_{i1}(t) \qquad (4.96a)$$
$$w_{i2}(t) = C_i x_i(t) + \eta_{i2}(t). \qquad (4.96b)$$

Let us assume that $\eta_{i1}(t)$, $\eta_{i2}(t)$ are stochastically independent. Then we have a special case of the problem considered by Sanders *et al.* (1974, 1975), whose results allow us to design a decentralized filter giving an *unbiased estimate* $\hat{x}_i(t)$ of $x_i(t)$.

Each of the s local filters has the form:

$$\dot{x}_i(t) = A_{ii}\hat{x}_i(t) + B_i m_i(t) + G_{0i}w_{i1}(t) + K_{ii}^F(t)[w_{i2}(t) - C_i\hat{x}_i(t)], \quad (4.97a)$$

where $x_i(0) = x_i^0$ and the gain matrices $K_{ii}^F(t)$ are given by

$$K_{ii}^F(t) = P_i(t)C_i^T N_{i2}^{-1} \qquad (4.97b)$$

$$\dot{P}_i(t) = A_{ii}P_i(t) + P_i(t)A_{ii}^T - P_i(t)C_i^T N_{i2}^{-1} C_i P_i(t) + \Xi_i + G_{0i}N_{i1}G_{0i}^T, \qquad P_i(0) = P_i^0, \qquad (4.97c)$$

where covariance matrix N_i of (η_{i1}, η_{i2}) is given by

$$N_i = \begin{bmatrix} N_{i1} & 0 \\ 0 & N_{i2} \end{bmatrix}.$$

In the above construction, the interaction measurement noise is treated as the local process noise. Consider the following important theorem from Tacker *et al.* (1976).

380

Theorem 4.8. *If pairs $(A_i, C_i N^{-1/2})$ and (A_i, C_i) are observable and if system (4.92), (4.96) can be stabilized by decentralized state feedback, then it can also be stabilized when local estimates from (4.97) are used in place of the local states.*

If we use filters for local control that have the dimension of the full state of the system, and if we use a completely centralized design procedure, we can work out more complicated decentralized control schemes for the stochastic control problem; see Chong and Athans (1971).

Decentralized disturbance-accommodating controllers

Consider composite system (4.74), (4.71b,c), where the process and observation disturbances can be modeled as follows:

$$\xi_i(t) = E_i z(t), \qquad \eta_i(t) = F_i z(t), \tag{4.98}$$

and where $z(t)$ is the output of a "disturbance generator:"

$$\dot{\mathfrak{z}} = A_z \mathfrak{z}$$
$$z = C_z \mathfrak{z} \tag{4.99}$$

with random initial conditions. The disturbances of the type $z(t) = a_0 + a_1 t + \ldots + a_{n-1} t^{n-1}$ with unknown coefficients $a_0, \ldots, a_{n-1}$ fall easily into this class since they satisfy the equation $z^{(n)}(t) = 0$.

Suppose also that the regulated outputs $w_i^r(t)$ are measured, that is, they belong to the measured outputs $w_i(t)$. For simplicity, we consider the case when $w_i^r(t)$ have to be brought to some specified constant values w_{id}^r in the presence of disturbances (4.99). We mentioned in section 4.6.1 that the constant values of w_{id}^r may result from the steady-state approach used at the higher control layers—see Bailey and Malinowski (1977).

The control problem is now as follows: we are looking for a regulator that will asymptotically regulate $w_i^r(t)$ to w_{id}^r in the presence of disturbances z, provide for a fast response to changes in w_{id}^r, and use a decentralized information pattern. An attractive approach that has some of these features is Davison's decentralized robust regulator theory (Davison 1976a,b, 1977). This approach yields a set of disturbance-accommodating controllers that:

1. Regulate $w_i^r(t)$ to w_{id}^r;
2. Use feedback to maintain zero steady-state error with respect to set points in spite of the influence of a defined class (4.99) of disturbances;
3. Are robust in the sense that they successfully regulate over a certain range of process parameters;
4. Are maximally decentralized.

The process equations are:

$$\dot{x}_i(t) = A_{ii}x_i(t) + B_i m_i(t) + G_{0i}u_i(t) + E_i z(t)$$
$$y_i(t) = G_{Ti}x_i(t) \qquad (4.100)$$
$$u_i(t) = \sum_{j=1}^{s} H_{ij}y_j(t)$$

and the observation equation:

$$w_i(t) = C_i x_i(t) + F_i z(t),$$

where a part of $w_i(t)$ is the regulated output

$$w_i^r(t) = C_i^r x_i(t) + F_i^r z(t),$$

and $z(t)$ is generated by (4.99). The design of the robust regulator proceeds in two stages:

1. The dynamic servo compensators

$$\dot{x}_{ci}(t) = A_{ci}x_{ci} + B_{ci}(w_i^r(t) - w_{id}^r) \qquad (4.101)$$

are chosen to ensure that the process tracks w_{id}^r in the presence of z. Roughly speaking, the servo compensators are designed to cancel the unstable poles of the disturbance generator (4.99).

2. $w_i(t)$ and $x_{ci}(t)$ are used as the inputs to the decentralized local output feedback controllers with dynamic compensation (stabilizing compensators)—see Eq. (4.90)—which are designed to stabilize the augmented system (4.100), (4.101) with E_i, $F = 0$. The design procedure given by Wang and Davison (1973) requires some specification of the required closed-loop pole assignment, which can depend on the control objectives (e.g., on some performance index) as well as on the sensitivity of the design with respect to process parameter variations and process nonlinearities. The decentralized controller will in the end have the following structure:

$$m_i(t) = K_{ii}^{01}x_{ci}(t) + K_{ii}^{02}x_{si}(t), \qquad (4.102)$$

where $x_{si}(t)$ is the output of the stabilizing compensator.

Davison (1976a) showed that, roughly speaking, when each servo compensator (4.101) is controllable, the transmission zeros of the process (4.100) do not cancel the poles of any servo compensator, and some other conditions are met, the robust regulator can provide arbitrary assignment of all poles of the process. The robust regulator regulates the process for any changes in plant parameters A_i, B_i, C_i for which the control structure remains stable. The tracking property is however lost when the unstable poles of generator (4.99) change their values. Therefore, in practice this approach seems to be restricted to the cases in which dynamic system (4.99) models the structural properties of disturbances, for example, polynomial disturbances that are modeled by the equation $z^{(n)}(t) = 0$.

Whenever the set point values change significantly (at times t_j) the tuning procedure should, at least in theory, recompute the new servo and/or stabilizing compensator parameters. However, owing to the robust properties of the regulator, it will continue to regulate as long as the control structure remains stable. The question of when is it appropriate to recompute the regulator parameters is complex. Some aspects of this question, including the possibilities for sequential tuning of the local controllers, have been considered by Davison (1976b) and Davison and Gesing (1977).

In some cases it may not be possible to stabilize the complex system by using the completely decentralized information pattern (see Theorem 4.7). Even a stable, completely decentralized control may produce unsatisfactory system performance. This does not necessarily mean that we should entirely abandon the idea of decentralized control since in many cases one can significantly improve the decentralized design by introducing a few chances for communication between the controllers. For example, in the designing of control law (4.102) one may find that some unstable poles of the process are fixed for the class of control laws (4.102), it is then often possible to move them to the desired positions by adding appropriate off-diagonal blocks. This means that m_i will have the form:

$$m_i(t) = K_{ii}^{01} x_{ci}(t) + K_{ii}^{02} \hat{x}_{si}(t) + K_{ij}^2 \hat{x}_{sj}(t) + \cdots, \tag{4.103}$$

which indicates the need for some communication among controllers.

A similar thing could happen if one wanted to improve the performance of a decentralized filter, for example, by sending estimates $\hat{x}_i(t)$ from local filter i to other local filters. The benefits of this method should be compared with increased communication costs.

4.6.3. DECENTRALIZED CONTROL WITH PERIODIC COORDINATION

The central controller and its functions

In this section we will consider some possibilities of introducing a central controller into the state feedback control structures for the processes described by Eqs. (4.71), (4.73), or Eqs. (4.74a,b), (4.71b,c). We will refer to this central controller as the on-line coordinator or—simply—the coordinator, although the concept and role of this coordinator differs considerably from the coordinator concept in a hierarchical decision-making (hierarchical optimization) mechanism (see Chapter 2).

To specify the role of the on-line coordinator we have to proceed very carefully since, as noted by Sandell *et al.* (1976), there are many pitfalls that may virtually cancel decentralization. For example, we may allow the coordinator to have too many capabilities with respect to information and decisions. Obviously, the information pattern is a decisive factor here and if we want to maintain a reasonable decentralization then, roughly speaking, we will have to specify carefully what the coordinator knows and in what manner he will influence the local controllers.

We assume, as in the previous section, that each local controller knows the model data (the dynamics (4.71a) or (4.74a), performance index, probabilistic information) associated with his own subsystem and that he knows his own past measurements and control decisions. It should be noted that in many cases it may be more reasonable to assume that the local controllers have zero memory and at each instant of time they can use only the current observation.

Consider first the case in which the coordinator operates on the same time scale as the local controllers, that is, that his interventions can be as frequent as the local actions. What prior information is available to the coordinator and what current information will be exchanged between the coordinator and the local controllers? We could assume that the coordinator does not know the complete model of the controlled system (no prior information) and cannot gather this kind of data from on-line communication with the local controllers. For control of dynamic systems, such an assumption does not seem to be very productive since, unlike the coordinator in the static case considered in Chapter 3, the coordinator here cannot make iterations on the system with which to work out a proper control decision. Nevertheless, this assumption may be correct in some situations, for example, if the local controllers do not want to reveal the parameters or even the structure of their subsystems. However, this might be more relevant for the theory of management systems (involving human decision makers) than for process control problems. Let us, therefore, assume that the coordinator has all of the model data.

As far as the on-line information is concerned, we might assume that the coordinator has full and error-free access to all measurements taken by the local controllers and that he also knows all local control decisions. In this case, however, as pointed out by Chong and Athans (1975), the coordinator will tend to override the local decisions and to virtually take over the control of local subsystems by continuous intervention. It would be unreasonable to prevent the coordinator from doing so since we have assumed that he communicates with the local controllers at all times and knows the best possible control decision to be made at any moment. We could as well eliminate the local control units.

It makes more sense to assume that the coordinator does not know the measurements taken by each local controller but that he can obtain instantaneous and error-free information about the local control decisions; see Aoki (1973). It turns out then (Sandell *et al.* 1976) that the coordinator can again calculate control decisions that are almost optimal, so he knows how to override the local control decisions and take better local actions. This happens because information about the local measurements can be indirectly included in the local control decisions transmitted to the coordinator. If the coordinator operates on the same time scale as the local controllers, then he cannot be involved in complicated numerical tasks since he would have to work out his decisions very fast, that is, as fast as the local controllers. Note that the local controls may be continuous or almost continuous.

The above considerations make it rather clear that instantaneous and continuous two-way communication between the coordinator and the local controllers destroys the decentralized nature of the control, especially if the exchange of information is error-free, and has to be abandoned. It is also usually not acceptable because of the large costs of data transmission. Therefore, it is reasonable to assume that the coordinator will influence the local units less frequently than the local controllers operate on the process. In other words, the coordinator will become a higher-layer controller. This will also allow him more time to perform the computations on the overall system model.

We consider two possibilities for on-line information transmission to the coordinator:

1.   The coordinator obtains all the information that is available to the local controllers and has a perfect memory, but he is permitted to make control actions only at times t_ℓ

2.   The coordinator obtains only some aggregated or discrete data from the local units (like the coordinator described in sections 4.2 and 4.3)

We will briefly discuss some control structures in which the coordinator takes control actions at times t_ℓ and the local controllers use state feedback decision rules. We will refer to this control scheme as decentralized state feedback control with periodic coordination.

Periodic price coordination for a deterministic control problem

Assume that we are given the linear-quadratic optimal control problem described by Eqs. (4.71), (4.72) with finite control horizon T, and that at times $t_\ell \in [0, T)$ the dynamic coordinator receives information about the actual value $x^r(t_\ell)$ of the system state. At each t_ℓ the coordinator has two tasks with respect to the local controllers:

- The coordinator has to change the local incentives by modifying the local performance indices
- The coordinator may have to provide for some time trajectories required by the local-level controllers

Suppose that at time t_e coupling equation (4.71b) is incorporated into the following Lagrangian

$$L = \frac{1}{2} \sum_{i=1}^{s} \left[\int_{t_e}^{T} (\langle x_i(t), Q_i x_i(t) \rangle + \langle m_i(t), R_i m_i(t) \rangle)\, dt + \langle x_i(T), Q_i^0 x_i(T) \rangle \right]$$

$$+ \sum_{i=1}^{s} \int_{t_*}^{T} \left\langle \lambda_i(t), u_i(t) - \sum_{j=1}^{S} H_{ij} y_j(t) \right\rangle dt. \quad (4.104)$$

This Lagrangian can be split into s parts for given $\lambda_i(t)$, $\mu_i(t) = \sum_{j=1}^{s} H_{ji}^T \lambda_j(t)$. We can consider the following problem LP^I $(i \in \overline{1, s})$:

Minimize with respect to $m_i(t)$, $u_i(t)$ $(t \in [t_*, T])$ the functional

$$L_i = \frac{1}{2} \int_{t_e}^{T} [\langle x_i(t), Q_i x_i(t) \rangle + \langle m_i(t), R_i m_i(t) \rangle + \langle \lambda_i(t), u_i(t) \rangle - \langle \mu_i(t), G_{Ii} x_i(t) \rangle]\, dt$$

$$+ \langle x_i(T), Q_i^0 x_i(T) \rangle \quad (4.105)$$

subject to

$$\dot{x}_i(t) = A_{ii} x_i(t) + B_i m_i(t) + G_{0i} u_i(t)$$
$$x_i(t_e) = x_i^r(t_e).$$

Suppose that problem LP^i has a unique solution $\bar{m}_i(\lambda_i, \mu_i)(t)$, $\bar{u}_i(\lambda_i, \mu_i)(t)$ $(t \in [0, T])$. We can use the interaction balance method to solve the overall system optimization problem at $t = t_e$ (see sections 2.4 and 4.3.1) and find the balance prices $\hat{\lambda}_i(t)$. The coordinator transmits $\hat{\lambda}_i(t)$, $\hat{\mu}_i(t)$ to the local controllers and then, in order to develop the local feedback control, we can use the following two-stage procedure:

Stage 1. For given $\hat{\lambda}_i(t)$, $\hat{\mu}_i(t)$ optimization problem (4.105) is solved by each local controller and $\bar{u}_i(\hat{\lambda}_i, \hat{\mu}_i)(t) = \hat{u}_i(t)$ is found.

Stage 2. For given $\hat{\mu}_i(t)$ and $\hat{u}_i(t)$, the following control problem CP^i

$$\min_{m_i} \hat{L}_i = \int_{t_e}^{T} [\langle x_i(t), Q_i x_i(t) \rangle + \langle m_i(t), R_i m_i(t) \rangle - \langle \hat{\mu}_i(t), G_{Ii} x_i(t) \rangle]\, dt$$

$$+ \langle x_i(T), Q_i^0 x_i(T) \rangle \quad (4.106)$$

subject to

$$\dot{x}_i(t) = A_{ii}x_i(t) + B_i m_i(t) + G_{0i}\hat{u}_i(t) \tag{4.106}$$

is used to yield the following decentralized local feedback law with compensation signal

$$m_i(t) = K_{ii}^0(t)x_i^r(t) + v_i(t) \tag{4.107}$$

where

$$K_{ii}^0(t) = -R_i^{-1}B_i^T M_i^0(t), \qquad v_i(t) = -R_i^{-1}B_i^T v_i^1(t) \tag{4.108}$$

and $M_i^0(t)$, $v_i^1(t)$ satisfy the equations

$$\begin{aligned}
\dot{M}_i^0(t) &= -M_i^0(t)A_{ii} - A_{ii}^T M_i^0(t) + M_i^0(t)B_i R_i^{-1}B_i^T M_i^0(t) - Q_i \\
M_i(T) &= Q_i^0
\end{aligned} \tag{4.109}$$

$$\begin{aligned}
\dot{v}_i^1(t) &= M_i^0(t)B_i R_i^{-1}B_i^T v_i^1(t) - A_{ii}^T v_i^1(t) - M_i^0(t)G_{0i}\hat{u}_i(t) \\
&\quad - G_{Ii}^T\hat{\mu}_i(t), \qquad v_i^1(T) = 0
\end{aligned} \tag{4.110}$$

Now suppose that problem LP^i does not have a unique solution. We cannot use IBM to solve the overall optimization problem (though the interaction balance method with input prediction could be used—see section 2.5), and therefore the coordinator has to transmit $\hat{\mu}_i(t)$ and $\hat{u}_i(t)$, the predicted interaction inputs, to the local controllers. The design of local procedures is along the lines of stage 2 above.

Problem LP^i can have a unique solution if $Q_i > 0$, $R_i > 0$, and $Q_i^0 > 0$; there are other conditions as well. However, it is clear that we cannot solve this problem with the maximum principle; the solution may be obtained by using a standard technique of dynamic programming. If we add to performance J (see 4.72) the following term:

$$\sum_{i=1}^s \int_0^T \langle u_i(t), S_i u_i(t)\rangle \, dt,$$

then LP^i, with $\langle u_i(t), S_i u_i(t)\rangle$, has a unique solution when $R_i > 0$, $S_i > 0$, and $Q_i, Q_i^0 \geq 0$. The decentralized control law (4.107) is implemented over time horizon $[t_\ell, t_{\ell+1}]$. We then transmit $x^r(t_{\ell+1})$ to the coordinator and repeat the whole procedure. It should be noted that the most complicated Riccati equation (4.109) has to be solved only once by each local controller at the initial time $t = 0$.

The information pattern of this control structure is similar to the information pattern of the dynamic price coordination structure that was considered in section 4.3; the coordinator receives only part of the on-line information that is used by the local controllers (state values $x^r(t_\ell)$ at times t_ℓ) and provides these controllers with the performance incentives and—if necessary—with the local input prediction $\hat{u}_i(t)$. Unless the dimension of $u_i(t)$ is much smaller than the dimension of $m_i(t)$, the coordinator has to transmit a considerable amount of data to the local units—the amount of information transmitted from the coordinator to the local controllers could, in fact, be more than the amount of information involved in the control signals that would have to be sent to the process actuators by a completely centralized controller. Therefore, as far as the information pattern is concerned, the main benefit of this structure is that we need to transmit only some of the measurements to the coordinator. If the dynamic coordinator is allowed to operate on the same time scale as the local controllers, he will take over all the control activities by overriding the local decisions.

Periodic price coordination for a stochastic control problem with completely decentralized filtering.

Assume now that at the design stage we consider the subprocess equations (4.74a) instead of (4.71), that is, we acknowledge that the disturbance input $\xi_i(t)$ is a white noise. We assume also that the process behavior can be observed through noisy measurements (4.74b). We want to preserve the pattern of information exchange between the local controllers and the dynamic coordinator that was presented above. Witsenhausen (1968) has demonstrated that for the solution of the linear-quadratic Gaussian problem with nonclassical information patterns, the separation theorem does not hold; nevertheless, for technical and computational reasons we will still use the separated filtering algorithm and the linear control law as in the previous section. Therefore, we consider the following control structure:

1. At times t_ℓ, the coordinator receives state estimates $\hat{x}_i(t_r)$ from the local controllers; given this information and prior deterministic information, he specifies as before $\hat{u}_i(t)$ and $\hat{\mu}_i(t)$ (or $\hat{\lambda}_i(t)$, $\hat{\mu}_i(t)$) and transmits them to the local controllers.

2. The local controllers compute the parameters of their local control laws (4.107) as in the previously described case but use in place of $x_i^r(t)$ the local filter output $\hat{x}_i(t)$. The local filter is designed for the process

$$\dot{x}_i(t) = A_{ii}x_i(t) + B_i m_i(t) + G_{0i}\hat{u}_i(t) + \xi_i(t)$$
$$w_i(t) = C_i x_i(t) + \eta_i(t). \tag{4.111}$$

It should be noted that the local filters do not provide for unbiased estimates $\hat{x}_i(t)$ of $x_i^r(t)$, and the coordinator may compute wrong incentives for the local controllers if he uses the biased estimates. If it is possible to measure the interaction inputs (see Eq. (4.96a)) then we may implement filter (4.97a) to get an unbiased estimate of $x_i^r(t)$. This could improve the performance of the control structure. In any case, this control scheme looks interesting and deserves investigation.

Periodic price coordination for a stochastic control problem with local and global filtering

We briefly outline the control scheme presented by Chong and Athans (1975). They considered a control structure with periodic price coordination for a linear discrete-time stochastic system. All prior information about the process dynamics, probabilistic data, and performance index is available to the coordinator. The local controllers have only the information related to their subsystems and can measure local noisy output. The task of the coordinator is

- To correct estimates generated by the local-level Kalman filters;
- To specify new time paths for the local control mechanisms, which depend on the version of the control scheme that is used;
- To specify additional terms that are added to the local performance indices.

For his task, the coordinator receives all the measurements available to the local controllers and all the information about the local control. However, the coordinator is allowed to communicate his decisions to the local controllers only at times nl, $l = 1, \ldots, l_0$; the local controllers apply controls at all times $k = 1, \ldots, N$. A full-order Kalman filter is used by the coordinator to generate the optimal state estimates. There are essentially two different ways for the coordinator to formulate the decision problem at time nl:

- He may neglect the fact that he will act in the future; in that case, he works with a so-called open-loop feedback optimal structure.
- He may take his present as well as his future interventions into consideration. This leads to a closed-loop optimal structure with periodic coordination, which was developed by Chong and Athans. In this case, however, the decoupling of the local control problems is very difficult; the coordinator has to transmit more information to the local controllers than in the open-loop structure.

The local problems are well-defined stochastic control problems. Based on the information transmitted from the coordinator, the local controllers can develop their own feedback decision rules and their local filters. Chong and Athans show that when the coordinator operates on the same time scale as the local controllers, he himself generates stochastically optimal controls for the overall system.

What do we gain by using this decentralized feedback structure? It leaves some autonomy to the local controllers, but more information is exchanged within the control structure than in completely centralized control. The information requirements may be difficult to accept, and the on-line computational requirements are no less than they would be in a completely centralized case. Nevertheless, the theory of the structure is interesting and valuable for further studies connected with partially decentralized control schemes.

Final remarks

It should be noted that all control schemes described in this subsection would remain unchanged for time-dependent matrices $A_{ii}(t)$, $B_i(t)$, and so on. A linear description of the controlled process, a white noise model of the disturbance, and a quadratic performance function describe the first-layer regulatory control problem (see section 4.2) and thus virtually limit the applications of the theory presented in section 4.6 to regulatory control design. Since the first-layer actions are relatively frequent, we should avoid coordination or communication between the first-layer controllers as it will be expensive because of the wide bandwidth required. It seems that, as far as the regulatory control structures are concerned, we should often try to develop completely decentralized control schemes, some of which were presented in section 4.6.2.

REFERENCES

Aoki, H. (1973). "On decentralized linear stochastic control problems with quadratic cost." *IEEE Trans. Autom. Control* AC-18(3): 243–249.

Athans, M. (1971). "The role and use of the stochastic Linear–Quadratic–Gaussian problem in control system design." *IEEE Trans. Autom. Control,* AC-16: 529–541.

Athans, M., and P. L. Falb (1966). *Optimal Control.* McGraw-Hill.

Bailey, F. N., and K. Malinowski (1977). "Problems in the design of multilayer–multiechelon control structures." *Proceedings, 4th IFAC Symposium on Multivariable Technological Systems,* Fredericton, Canada, pp. 31–38.

Bailey, F. N., and H. K. Ramapriyan (1973). "Bounds on suboptimality in the control of linear dynamic systems." *IEEE Trans. Autom. Control.* AC-18: 532–534.

Bailey, F. N., and F. C. Wang (1972). "Decentralized control strategies for linear systems." *Proceedings,* Asilomar Conference.

Bamberger, A., C. Saques, and I. P. Yvon (1975). "Contrôle en boucle ouverte adaptée de systèmes distribués" Iria Laboria, *Rapport de Recherche no.* 128.

Boltyanski, W. G. (1969). *Mathematical Methods of Optimal Control.* Nauka, Moscow (in Russian).

Chong, C., and M. Athans (1971). "On the stochastic control of linear systems with different information sets." *IEEE Trans. Autom. Control* AC-16: 423–430.

Chong, C., and M. Athans (1975). "On the periodic coordination of linear stochastic systems." *Proceedings, 6th IFAC Congress*, Part IIIA, Boston, Massachusetts.

Davison, E. J. (1976a). "The robust decentralized control of a general servomechanism problem." *IEEE Trans. Autom. Control* AC-21: 14–24.

Davison, E. J. (1976b). "Decentralized robust control of unknown systems using tuning regulators." *Proceedings*, 14th Allerton Conference on Circuits and Systems.

Davison, E. J. (1977). "Recent results on decentralized control of large scale multivariable systems." *Proceedings, 4th IFAC Symposium on Multivariable Technological Systems*, Fredericton, Canada, pp. 1–10.

Davison, E. J., and W. Gesing (1977). "Sequential stability and optimization of large scale decentralized systems." *Proceedings, 4th IFAC Symposium on Multivariable Technological Systems*, Fredericton, Canada.

Findeisen, W. (1974). *Wielopoziomowe uklady sterowania* [Multilevel Control Systems]. PWN, Warsaw. German translation: *Hierarchische Steuerungssysteme*. Verlag Technik, Berlin, 1977.

Findeisen, W. (1977). "Multilevel structures for on-line dynamic control." PP–77–6, International Institute for Applied Systems Analysis, Laxenburg, Austria; and *Ricerche di Automatica* 8 (1).

Findeisen, W., and K. Malinowski (1976). "A structure for on-line dynamic coordination." *Proceedings, IFAC Symposium on Large-Scale Systems Theory and Applications*, Udine, Italy.

Findeisen, W., and K. Malinowski (1979). "Two level control and coordination for dynamical systems." *Arch. Autom. Telemech.* 24(1): 3–27.

Foord, A. G. (1974). "On-line optimization of a petrochemical complex." *Doctoral dissertation*, Cambridge University, Cambridge, England.

Ioffe, A. D., and W. M. Tikhomirov (1974). *Theory of Extremal Problems*. Nauka, Moscow (in Russian).

Jazwinski, A. M. (1970). *Stochastic Processes and Filtering Theory*. Academic Press, New York.

Kokotovič, P. V. (1972). "Feedback design of large linear systems." In J. B. Cruz, ed., *Feedback Systems*. McGraw-Hill, New York.

Kreindler, E. (1972). "Comparative sensitivity of optimal control systems". In J. B. Cruz, ed., *Feedback Systems*. McGraw-Hill, New York.

Laub, A. (1974). "Decentralization of optimization problems by iterative coordination." *Doctoral dissertation*, University of Minnesota, Minneapolis, Minnesota.

Mahmoud, M. S. (1977). "Multilevel systems control and applications: a survey." *IEEE Trans. Syst. Man Cybern.* SMC-7(3): 125–143.

Malinowski, K., and K. Nowosad (1977). "Applications and properties of repetitive control structure." 4th International Conference on System Science, Wroclaw, Poland.

Malinowski, K., and T. Terlikowski (1977). "Problems in optimization of complex systems with inventory couplings." *Richerche di Automatica* 8(1): 60–86.

Malinowski, K., and T. Terlikowski (1978). "Coordination of a nonstationary system with dynamic inventory couplings." *Technical Report*, Institute of Automatic Control, Technical University of Warsaw, Poland.

Nowosad, K. (1978). "The properties of repetitive control under high sampling frequency." *Arch. Autom. Telemech* 23(3). (in Polish)

Pontriyagin, L. S. (1961). *Ordinary Differential Equations*. Nauka, Moscow. (in Russian). (1962). Addison-Wesley, Reading, Massachusetts (in English).

Sandell, N. R., Jr., P. Varaiya, and M. Athans (1976). "A survey of decentralized control methods for large scale systems." *Proceedings, Engineering Foundation Conference on Systems Engineering for Power: Status and Prospects*, Henniker, New Hampshire, pp. 334–352.

Sanders, C. W., E. C. Tacker, and T. O. Linton (1974). "A new class of decentralized filters for interconnected systems." *IEEE Trans. Autom. Control* AC-19.

Sanders, C. W., E. C. Tacker, and T. D. Linton (1975). "Decentralized filtering algorithms for interconnected systems." *Proceedings, 6th IFAC Congress*, Boston, Massachusetts.

Singh, M. G. (1977). *Dynamical Hierarchical Control*, North Holland, Amsterdam.

Singh, M. G., S. A. W. Drew, and J. F. Coales (1975). "Comparisons of practical hierarchical control methods for interconnected dynamical systems." *Automatica* 11 (4): 331–350.

Tacker, E. C., C. W. Sanders, and T. D. Linton (1967). Some results in decentralized filtering and control. *Proceedings, IFAC Symposium on Large-Scale Systems Theory and Applications*, Udine, Italy.

Teneketzis, D. (1976). "Perturbation methods in decentralized stochastic control." *Report* ESL-R-664, M.I.T., Cambridge, Massachusetts.

Wang, S. H., and E. J. Davison (1973). "On the stabilization of decentralized control systems." *IEEE Trans. Autom. Control* AC-18: 473–478.

Wierzbicki, A. P. (1977). *Models and Sensitivity of Control Systems*. PWN, Warsaw (in Polish).

Witsenhausen, H. S. (1968). "A counterexample in stochastic optimal control." *SIAM J. Control* 6.

<h1>5 Information Problems in
Hierarchical Systems</h1>

5.1. INTRODUCTION

As noted in Chapter 1, the design of a control system involves the design of an information process and a decision process; the former provides data for the decisions taken by the latter. Most of the work in traditional control theory and hierarchical control theory focuses on the design of a decision process in which the information process is either given or chosen arbitrarily. In this chapter we focus on the information process. We consider information and how we measure it, models for information structures in complex control systems, measures of the amount, value, and cost of information, and the design of information structures.

Unfortunately, the results are much more tentative than those in the rest of this book. Many of the ideas are new, most of the concepts are relatively untested, and many important results are missing. Yet, when we consider that informational factors are central to the whole justification of hierarchical systems, it becomes clear that this is a topic of central importance in the overall theory. We hope that the questions raised will encourage further study of these issues.

5.2. THE INFORMATION PROCESS

We begin with a simple and informal definition of information. *Information is a commodity that improves decisions.* The question of how this improvement occurs will be considered in detail below. For now, we give a simple example. Consider a farmer who has to plant either crop W or crop D. Crop W grows well in a wet season and crop D grows well in a dry season. Thus, the farmer's profit will depend on the future weather and also on the future market prices of W and D, both of which are relatively uncertain quantities. The former depends on the actions of "nature" or "chance" and the latter is a function of the actions taken by his neighbor. In order to choose the best action, the one that gives maximum profit, the farmer would like to know the coming weather and the future market price. Farmers pay for weather and market forecasts and they complain about economic loss from unexpected weather or market changes. On the other hand, a careful identification of the totality of information used by a specific farmer in making his decision would be very difficult. We know what information does but it is often difficult to say precisely what information is.

INFORMATION IN CONTROL

The uses of information in control problems can best be illustrated by some simple examples. Let us first consider a single-controller static problem with outcome (output) $y = g(m)$ where m is a manipulated input. Preferences are represented by the performance index $Q(y, m)$, which is to be maximized. Here the controller must choose m to solve the problem

$$\underset{m}{\text{maximize }} Q(y, m)$$

subject to $y = g(m)$. The only information needed by the controller is *process information* about g and *goal information* about Q.

Now consider a more complex version where there is an external disturbance u affecting the outcome. That is, the process is now $y = g(m, u)$, but the performance index remains the same. The controller now needs additional information about u. This is *coupling information*; it represents the effects of external actions (externalities) that are coupled into the local process. The controller's success in maximizing Q depends on the quality of the coupling information he receives. We can represent the coupling information received by $v = h(u)$. The controller's problem is now: given $v = h(u)$, choose m to solve

$$\underset{m}{\text{maximize }} Q(y, m)$$

subject to $y = g(m, u)$.

The above examples suggest that a controller needs three types of information: goal information, process information, and coupling information. In general, the first two are given in advance and we assume that they are distributed according to some prior arrangement. The focus in this chapter will be on coupling information: what is needed and how it is distributed.

5.3. THE SYSTEM MODEL

The systems considered will be modeled as a collection of simple process elements, or simply elements, with each element controlled by a separate decision agent (decision unit). The agents are assumed to act only once; repeated actions are represented by agents who communicate. In this respect the development is similar in spirit to that used by Witsenhausen (1968, 1974). However, we are primarily interested in the design of information structures while he focuses on classification. In addition we assume that the order of action of the agents is fixed so there will be no questions about causality.

Each process element is described by an outcome function

$$y = g(m, u) \tag{5.1}$$

where y is the outcome resulting from action m and coupling input u. Each agent is described by an information function

$$v = h(u) \tag{5.2}$$

and a decision rule

$$m = d(v). \tag{5.3}$$

y, m, u, and v are assumed to be members of an appropriate space, most commonly, a finite-dimensional vector space.

A *closed system* denoted $\bar{\mathscr{S}}$ will be represented by a set $\bar{\mathscr{P}}$ of process elements and a set $\bar{\mathscr{A}}$ of associated agents. The elements and agents of $\bar{\mathscr{S}} = (\bar{\mathscr{P}}, \bar{\mathscr{A}})$ are assumed to be located at the vertices of a space–time grid. The term *space* is used in a generic sense. It may represent real space, sectors of an economy, divisions of a firm. Space is divided into S elements indexed by $s = 1, \ldots, S$. Time is divided into $T+1$ elements indexed by $t = 0, 1, \ldots, T$. Time elements are assumed to be ordered by the usual ordering of the integers in $[0, T]$. The time order of the agents actions is given by their time index.

We can draw an (S by $T+1$) array of points called an *s–t grid*; it is shown in Figure 5.1. We then think of $\bar{\mathscr{S}}$ as a collection of elements and agents located at these points. The process elements, agents' outcome functions, information functions, and decision rules will be indexed by *st*. Thus, the complete set of outcome functions for $\bar{\mathscr{S}}$ is denoted by

$$y_{st} = g_{st}(m_{st}, u_{st}) \qquad s = 1, \ldots, S; \qquad t = 0, \ldots, T. \tag{5.4}$$

The couplings u_{st} into g_{st} are in general a function of the actions of all the *other* agents. They are represented by

$$u_{st} = k_{st}(m). \tag{5.5}$$

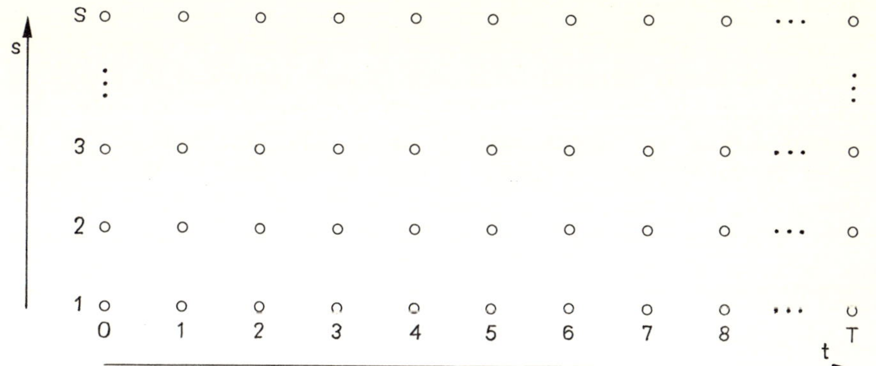

FIGURE 5.1   A system on an s–t grid.

In addition, agent A_{st} receives the functional form of k_{st} as part of his process information. Thus, the part of the coupling information unknown to A_{st} is

$$m = (m_{10}, \ldots, m_{ST}),$$

a vector representing the actions of all of the agents in $\bar{\mathscr{S}}$. In some cases it is convenient to combine (5.4) and (5.5) and write

$$y_{st} = \tilde{g}_{st}(m_{st}, m). \tag{5.6}$$

The fact that u_{st} does not depend on m_{st} will be assumed but not noted explicitly.

The coupling structure of $\bar{\mathscr{S}}$ can be represented by a directed graph on the s–t grid called the *coupling graph*. A branch of the coupling graph is drawn from point ij to point kl if the dependence of u_{kl} on m_{ij} is not trivial.

Exogenous disturbances coupled into the outcomes of other elements are represented by the actions of a subset of $\bar{\mathscr{S}}$ called the *chance subsystem* and denoted $\mathscr{S}_0$. The agents in $\mathscr{S}_0$ are called *chance agents* and denoted $\mathscr{A}_0$. The chance agents receive no information and their actions are independent of the actions of other agents in $\bar{\mathscr{S}}$. Thus, they act first, and the chance agents and their elements are located at $t = 0$. The elements of the chance agents are dummies but can be represented by

$$y_{so} = g_{so}(m_{so}), \qquad s = 1, \ldots, S. \tag{5.7}$$

It is possible to model the actions m_{so} of the chance agents as random variables from an appropriate probability space. The remaining elements in $\bar{\mathscr{S}}$ (i.e., $\bar{\mathscr{S}} - \mathscr{S}_0$) are denoted by $\mathscr{S}$ and called the *system*.

5.3.1. SUBSYSTEMS

A subset of $\bar{\mathcal{S}}$ having either no couplings entering from outside itself or couplings only with $\mathcal{S}_0$ is called a *subsystem*. Subsystems (agents and process elements) will be denoted by $\mathcal{S}_i$ for $i > 0$. The elements and agents in $\mathcal{S}_i$ will be denoted $\mathcal{P}_i$ and $\mathcal{A}_i$, respectively. Several special types of subsystems will be identified based on the topological properties of their coupling graphs. A subsystem is *deterministic* if it has no coupling with $\mathcal{S}_0$. A subsystem is *causal* if none of its coupling branches are directed to the left. A subsystem is *static* if all of its elements have the same time index; otherwise, it is *dynamic*. A subsystem is *lumped* if all of its elements have the same space index; otherwise, it is *distributed*. A subsystem is *sequential* if its coupling graph has no loops; otherwise, it is *cyclic*. Note that the process elements and agents of a sequential subsystem are partially ordered by the interconnections in their coupling graph. This partial ordering will be called the *coupling order*.

5.3.2. GOALS

The agents may each have individual goals such as

$$\text{maximize } Q_{st}(y_{st}, m_{st})$$

with respect to m_{st}, or a subset of agents $\mathcal{A}_i$ may have a common goal such as

$$\text{maximize } Q_{\mathcal{A}_i}(y_{\mathcal{A}_i}, m_{\mathcal{A}_i}) \tag{5.8}$$

with respect to $m_{\mathcal{A}_i}$, where $Y_{\mathcal{A}_i}$, $m_{\mathcal{A}_i}$ are vectors of outcomes and actions for all agents in $\mathcal{A}_i$ and their processes. A group of agents having a common goal is called a *team*. The team goal is *separable* if

$$Q_{\mathcal{A}_i}(y_{\mathcal{A}_i}, m_{\mathcal{A}_i}) = \sum_{st \in \mathcal{A}_i} Q_{st}(y_{st}, m_{st}). \tag{5.9}$$

5.3.3. SYSTEM EXAMPLES

Some simple examples of systems will be given to clarify the concepts.

Example 5.1. *A single-element system* $\mathcal{S}_1$

The simplest system of interest has a single element in $\mathcal{S}_0$ and a single element in $\mathcal{S}$. The coupling graph is shown in Figure 5.2. The element at $(1, 0)$ is a disturbance generator coupling disturbances into the element at

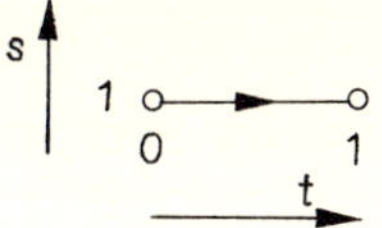

FIGURE 5.2   Coupling graph for the system S_1.

(1, 1). The element at (1, 1) is described by

$$y_{11} = g_{11}(m_{11}, u_{11})$$
$$u_{11} = k_{11}(m_{10}).$$

(5.10)

The problem for agent A_{11} might be to maximize

$$Q_{11}(y_{11}, m_{11}).$$

(5.11)

If m_{10} is known this is a deterministic problem. If m_{10} is random this is a stochastic problem and we assume Q_{11} includes an appropriate expectation operation.

Example 5.2.   An N-element static system $\mathscr{S}_N^s$

The coupling graph for a typical N-element static system may look like the one shown in Figure 5.3. Here each of the elements at $t = 1$ is represented by

$$y_{i1} = g_{i1}(m_{i1}, u_{i1})$$
$$u_{i1} = k_{i1}(m)$$

$$i = 1, \ldots, N.$$

(5.12)

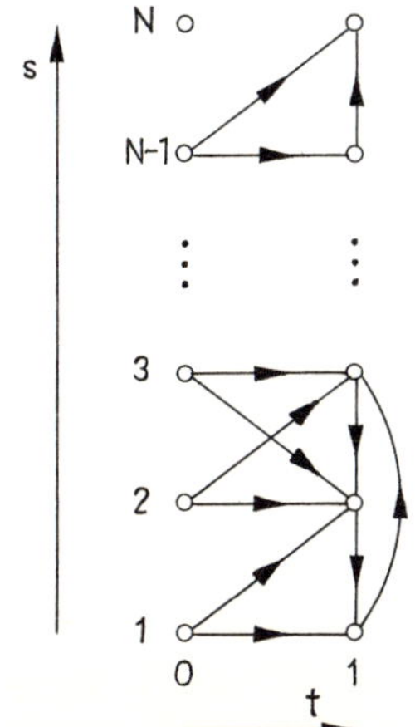

FIGURE 5.3   Coupling graph for a typical S_N^s.

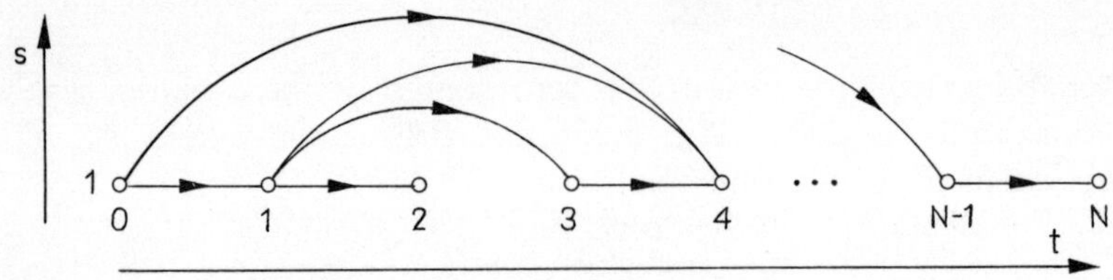

FIGURE 5.4 Coupling graph for a typical S_N^ℓ.

For example, the element at $(3, 1)$ is, according to Figure 5.3, represented by

$$y_{31} = g_{31}(m_{31}, u_{31})$$
$$u_{31} = k_{31}(m_{20}, m_{30}, m_{11}). \tag{5.13}$$

If the agents in this system are a team there will be a single performance index

$$Q(m_{11}, \ldots, m_{N1}, y_{11}, \ldots, y_{N1}). \tag{5.14}$$

Example 5.3.   An N-element, causal, lumped system $\mathscr{S}_N^\ell$

The coupling graph for a given N-element, causal, lumped system is shown in Figure 5.4. Note that causality requires that couplings flow only to the right. The outcome function for element $(1, k)$ is

$$y_{1k} = g_{1k}(m_{1k}, u_{1k})$$
$$u_{1k} = k_{1k}(m) \qquad k = 1, \ldots, N. \tag{5.15}$$

Because of the causal assumption, u_{1k} can be a function only of m_{1j} for $j < k$. Note that every subset of $\mathscr{S}_N^\ell$ to the left of $t = t'$, for all $t' \le N$, is a subsystem.

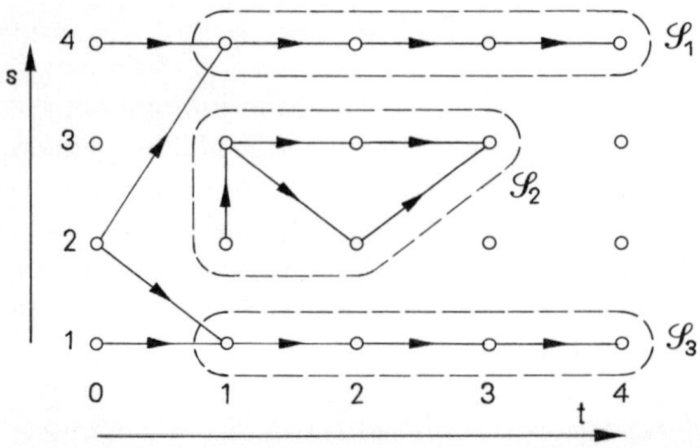

FIGURE 5.5 Coupling graph for a typical general system $\bar{S}$.

Example 5.4. A general space–time system $\bar{\mathscr{S}}$

A typical general space–time system $\bar{\mathscr{S}}$ is represented by the coupling graph shown in Figure 5.5. Note that $\bar{\mathscr{S}}$ is made up of three subsystems $\mathscr{S}_1$, $\mathscr{S}_2$, and $\mathscr{S}_3$. Subsystem $\mathscr{S}_2$ is deterministic.

5.3.4. INFORMATION STRUCTURE

Each agent receives information just prior to his turn to act. All agents act in an order determined by their time index, t, and all agents having the same time index are assumed to act simultaneously. Agent A_{st} receives process information describing g_{st} and k_{st}, goal information describing Q_{st} (or $Q_{\mathscr{A}_i}$ if the agent is a member of a team), and the coupling information. The vehicle for distributing this information is the system information structure: a network of communication links between the agents. In order to focus on coupling information, we assume that process and goal information is distributed in one of two polar cases. Process and goal information is *globally distributed* if each agent knows the process and goal information for all the other agents in $\mathscr{S}$. Process and goal information is *locally distributed* if each agent knows only his own process and goal information. Goal information can be locally distributed in a team only if the team performance index is separable.

The coupling information received by agent A_{st} may include outcomes, actions, observations, or messages from other process elements or agents. Since all of these are eventually a function of m, we describe the coupling information by the information function

$$v_{st} = h_{st}(m). \tag{5.16}$$

As before, m is a vector of actions of all the agents in $\bar{\mathscr{S}}$, but we explicitly assume that v_{st} is independent of m_{st}. The set of information functions (5.16) for an entire closed system $\bar{\mathscr{S}}$, combined with a statement about the distribution of process and goal information, describes the system's *information structure*. The coupling information flows implied by an information structure can be represented by a directed graph called the *information flow graph*. The information flow graph is drawn on the s–t grid in a way similar to the way in which the coupling graph is drawn. That is, a branch is directed from ij to kl if h_{kl} is a nontrivial function of m_{ij}. For clarity, it is often desirable to eliminate $\mathscr{S}_0$ and branches from elements of $\mathscr{S}_0$ from the information flow graph. This *reduced information flow graph* is equivalent to the precedence graph described by Ho and Chu (1974).

Information structures can be classified on the basis of topological properties of their information flow graphs. An information structure is *deterministic* if its information flow graph has no couplings with $\mathscr{S}_0$. An information structure is *causal* if all of the branches of its information flow graph are directed to the right. Note that this causality condition is stronger than that used in section 5.3.1. We assume that in a causal information structure an agent cannot be informed of the actions of other agents acting simultaneously. An information structure is *static* if all of the connected elements of its reduced information flow graph have the same time index; otherwise, it is *dynamic*. An information structure is *sequential* if its information flow graph has no loops; otherwise, it is *cyclic*. We can now assert some simple facts.

ASSERTION 5.1.　*A causal static information structure has no couplings in its reduced information flow graph.*

ASSERTION 5.2.　*A causal dynamic information structure is sequential.*

ASSERTION 5.3.　*A sequential information structure induces a partial ordering of the agents of $\bar{\mathscr{S}}$.*

Note that the ordering described in Assertion 5.3 is different from the coupling order and will be called the *information order*.

ASSERTION 5.4.　*A system with a causal information structure can be "played out" in real time.*

By *played out in real time* we mean that each agent can be presented with his information and then asked to act in a real-time sequence. Thus, a causal information structure describes an extensive game (see Witsenhausen 1974).

Most of the previous discussion of information structures in decision/control problems has focused on systems with causal information structures. However, many important multiperson decision/control problems involve noncausal information structures. That is, the information required by some agents depends on the actions taken by other agents acting at the same time or even in the future. Noncausal information structures are found in models of hierarchical control systems, economic mechanisms (Hurwicz 1971), planning processes (Burton and Obel 1977), and iterative optimization and estimation schemes (Laub and Bailey 1978). A system with a noncausal information structure cannot be played out in real time but may be solved by a *planning/action process*. A planning/action process involves a planning stage during which appropriate plans of action are selected, and an action stage during which the actions are taken. This involves a delay while plans are developed and a special information structure strictly for planning. One could include the planning/action explicitly in the coupling and information flow graphs. Alternatively, if the noncausal information flow is sequential it can be made causal by delaying the actions of some of the agents.

ASSERTION 5.5. *A sequential information structure can be made causal by changing the sequence of action of the agents.*

A more complex situation is an information structure that is nonsequential, and thus noncausal by Assertion 5.2. Such systems can be solved by a special iterative planning/action process called a *tâtonnement process*. Tâtonnement processes have been studied extensively in the economics literature (Hurwicz 1971) and are receiving increasing attention in the literature of hierarchical control theory. As we will see below, the tâtonnement processes have advantages and disadvantages. However, it is important to realize that in most control problems the choice of information structure is made by the control system designer.

5.3.5. EXAMPLES OF INFORMATION STRUCTURE

Some typical information structures will be described to clarify the concepts in the preceding section.

Example 5.5. Three static information structures

The information flow graphs for three typical static information structures are shown in Figure 5.6. The first is causal, the second is not causal but sequential, and the third is cyclic.

Example 5.6. A dynamic information structure

The information flow graph for the dynamic information structure shown in Figure 5.7a has three parts. The upper two are causal. The lower is cyclic. The corresponding reduced information flow graph is shown in Figure 5.7b.

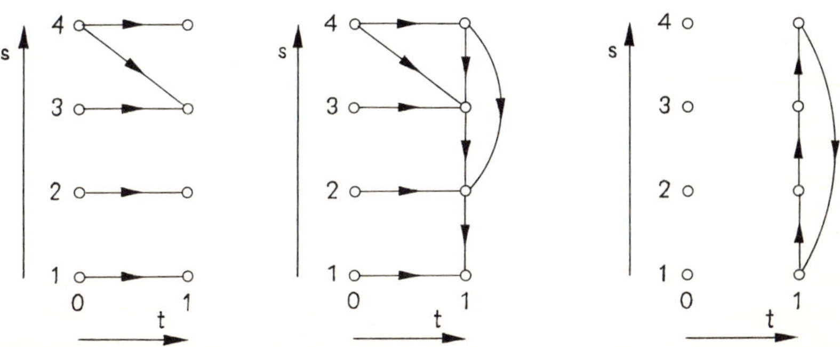

FIGURE 5.6 The information flow graphs for three static information structures.

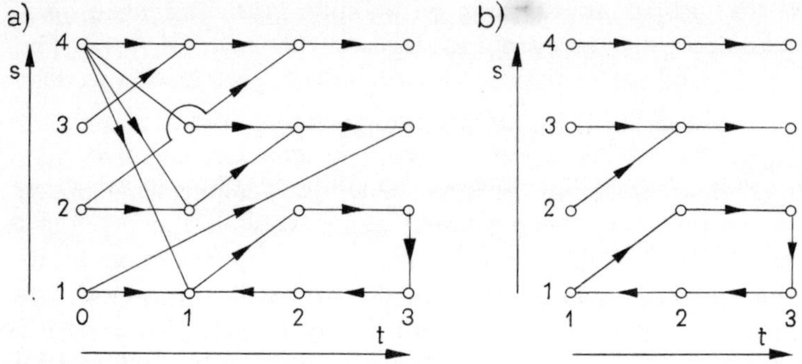

FIGURE 5.7   The information flow graph (a) and reduced information flow graph (b) for a dynamic information structure.

5.3.6.   APPLICATIONS

We now demonstrate the use of the proposed system model to represent some traditional decision and control problems.

Example 5.7.   A single-person decision problem (SPDP)

Since many of the concepts presented in sections 5.4 and 5.5 are based on results obtained for SPDP, this problem will be discussed in detail. SPDP is characterized by (see Raiffa and Schlaifer 1968):

A state space X (The term state space in decision theory refers to the "state of nature." Thus x is a random variable representing uncertainty.)
An action space M
An observation space V
An information function $h : X \rightarrow V$
A performance index $\tilde{Q} : M \times X \rightarrow R$
A stochastic model $p(x, v)$ on $X \times V$. (The notation $p(x, v)$ will be used to represent the joint probability density of x and v when the variables x and v are continuous and the set of joint probabilities when the variables x and v are discrete.)

The traditional decision problem is: given information function $v = h(x)$, find a decision rule $d \in D$, where $m = d(v)$, that achieves

$$\max E\{\tilde{Q}(d(v), x)\} \quad \text{with respect to} \quad d \in D$$

where E denotes the expectation with respect to $p(x, v)$.

Using the models developed above, we see that the system implied by SPDP is $\mathcal{S}_1$ (Example 5.1). The chance agent A_{10} chooses m_{10} (corresponding to $x \in X$) according to an appropriate probability rule. The agent A_{11} observes v_{11} and selects m_{11} according to decision rule $m_{11} = d_{11}(v_{11})$. The outcome function g_{11} and performance index Q_{11} for A_{11} are chosen so that

$$Q_{11}(m_{11}, g_{11}(m_{11}, m_{10})) = \tilde{Q}(m_{11}, m_{10}).$$

The information flow graph is the same as the coupling graph in this case.

Example 5.8.   *A static multiperson decision problem* (MPDP)

A static N-person decision problem is characterized by (see Marschak and Radner 1972):

A state space X
N action spaces M_i
N observation spaces V_i
N information functions $h_i : X \rightarrow V_i$, $i = 1, \ldots, N$
N performance indices $\tilde{Q}_i : M_1 \times \ldots \times M_N \times X \rightarrow R$
A stochastic model $p(x, v_1, v_2, \ldots, v_N)$

The problem for each of the N decision makers is to find a decision rule d_i, where $d_i : V_i \rightarrow M_i$, that achieves

$$\max E\{\tilde{Q}_i(m_1, \ldots, d_i(v_i), \ldots, m_N, x)\}$$

with respect to d_i. If this is a team problem, then there is only one performance index $\tilde{Q}$ and the problem is to find a set of N decision rules $d_i, \ldots, d_n$ that achieve

$$\max E\{\tilde{Q}(d_1(v_1), \ldots, d_N(v_N), x)\}$$

with respect to $d_1, \ldots, d_N$.

A static MPDP with N decision makers can be represented by the static system $\mathscr{S}_N^s$ shown in Figure 5.3. The outcome functions g_{i1} and performance indices Q_{i1}, $i = 1, \ldots, N$, are chosen to be consistent with $\tilde{Q}_i$ as in Example 5.7. The random variables corresponding to $x \in X$ are generated by one or more elements of $\mathscr{S}_0$. The information structure will be static because $v_{i1} = h_{i1}(m_{10}, \ldots, m_{N0})$. In most cases it is assumed that the process and goal information is globally distributed.

Example 5.9.   *A discrete-time dynamic system*

A discrete-time dynamic system on the time interval $t = 1, 2, \ldots, N$ is traditionally characterized by a difference equation model

$$\begin{aligned} x_{t+1} &= f_t(x_t, m_t) \\ y_t &= \tilde{g}_t(x_t, m_t), \end{aligned} \qquad t = 1, 2, \ldots, N \qquad (5.17)$$

an initial state $x_1 = \bar{x}$, and a performance index

$$\tilde{Q}_T(y_T) + \sum_{i=1}^{N} \tilde{Q}_i(y_i, m_i). \tag{5.18}$$

Sequential solution of and substitution in the difference equation shows that at step j the outcome is

$$y_j = \tilde{f}_j(\bar{x}, m_1, \ldots, m_j). \tag{5.19}$$

This suggests that we can represent a discrete-time dynamic system by using the lumped system $\mathscr{S}_N^\ell$. The agents A_{1j} choose actions $m_{1j} = m_j$ for $j = 0, \ldots, N$. The initial state $\bar{x}$ is chosen by A_{10} as $m_{10} = \bar{x}$. The coupling function k_{1j} is chosen so that

$$u_{1j} = k_{1j}(m) = (m_{10}, \ldots, m_{1,j-1})$$

and the outcome function is chosen so that

$$y_{1j} = g_{1j}(m_{1j}, u_{1j}) = \tilde{f}_j(\bar{x}, m_1, \ldots, m_j)$$

for all $1 < j \leq N$. The agents have a team goal

$$Q_{\mathscr{A}_N^1}(m_{11}, \ldots, m_{1,N-1}, y_{11}, \ldots, y_{1N}) = \sum_{j=1}^{N} [Q_{1i}(m_{1i}, y_{1i})]$$

where the Q_{1i} are chosen to correspond to the $\tilde{Q}_i$ in (5.18). Note that this is a team over time rather than space.

The information structure for a discrete-time dynamic system will depend on the observations available to the agents. A common though not necessary assumption is that each agent is really the same controller acting at different times. In this case, the assumption of globally distributed process and goal information is realistic. The coupling information available may vary. In the traditional model of a controller with output feedback, the coupling information available to agent A_{1j} is, from (5.19),

$$v_{1j} = y_{j-1} = \tilde{f}_{j-1}(\bar{x}, m_1, \ldots, m_{j-1}).$$

Thus the information flow graph in this case is the same as the coupling flow graph shown in Figure 5.4.

Example 5.10. *A multilevel control system*

A deterministic static system composed of N processes

$$y_i = g_i(m_i, u_i) \tag{5.20}$$

with couplings

$$u_i = k_i(y) \tag{5.21}$$

local performance indices $Q_i(m_i, y_i)$, and team performance index

$$Q(m, y) = \sum_{i=1}^{N} Q_i(m_i, y_i)$$

can be represented by the system $\mathscr{S}_N^s$ described in Example 5.2 with $\mathscr{S}_0$ omitted. The coupling graph for any specific case depends on the coupling function

$$u_{i1} = k_{i1}(m) \tag{5.22}$$

obtained by solving (5.20) and (5.21) jointly.

A multilevel control system with N infimal units and one coordinator is represented by a similar system $\mathscr{S}_{N+1}^s$ where the additional agent $A_{N+1,1}$ is termed the coordinator. Since the coordinator's function is strictly information distribution, he is not connected to the other elements in the coupling graph.

The traditional assumption in hierarchical control is that the process and goal information is only locally distributed. The traditional information flow has each agent receiving information about the actions of other agents through the coordinator.

5.3.7.   SOME RESULTS

The information structure affects the decisions made via the quality and quantity of the information provided. At this point we can say something about the quantity of information provided in terms of the information flow. In section 5.4 we will also consider the quality of the information flow, its informativeness. For the results obtained below we will limit our attention to the two polar cases of either globally distributed or locally distributed process and goal information.

ASSERTION 5.6.   *A deterministic subsystem with globally distributed process and goal information requires no coupling information.*

Explanation: In such a subsystem each agent can calculate the actions of the other agents and thus compute m locally.

Assertion 5.6 states that open-loop control is optimal when there is no uncertainty. This suggests that when process and goal information is globally distributed the primary function of coupling information is prediction of the actions of the agents in $\mathscr{S}_0$.

An information function $v = h(m)$ is said to be *perfectly informative* about m if observation of v reveals the precise value of m (see section 5.4 for further discussion of informativeness). An information flow is said to be

406

sufficient if it leads to optimum actions by all agents under the assumption that it is perfectly informative.

ASSERTION 5.7. *When process and goal information is globally distributed a sufficient information flow is a feedforward structure providing information about the actions of $\mathscr{A}_0$ to all agents in $\mathscr{S}$.*

Explanation: Once the actions of $\mathscr{A}_0$ are known, the closed system $\bar{\mathscr{S}}$ is equivalent to a deterministic system $\mathscr{S}$.

Of course, such a feedforward of information is generally unrealistic and feedback is used to develop an estimate of the actions of $\mathscr{A}_0$. Assertion 5.7 suggests that when process and goal information is globally distributed, two information structures are equivalent if they convey the same information about the actions of $\mathscr{A}_0$.

ASSERTION 5.8.   *When process and goal information is only locally distributed and the agent's preferences are independent, a sufficient information flow graph is one that follows the coupling graph.*

Explanation: In this case the individual agents face single-person decision problems with uncertainty z. The sufficient information flow will thus have to include information about all actions that affect z. This is indicated by the coupling graph.

ASSERTION 5.9.   *In a deterministic system with locally distributed process and goal information but team preferences, a sufficient information flow provides global distribution of process and goal information.*

Explanation: Global distribution reduces the problem to the case considered in Assertion 5.6.

Such an information structure is of course not very attractive. The coordination mechanisms discussed elsewhere in this book are designed to avoid global distribution of information. In section 5.5.3 we examine some informational questions arising in this case.

Note that the function of the information structure differs considerably between the two cases of globally and locally distributed process and goal information. This leads to a wide difference in attitude toward information in the literature associated with these two cases: the stochastic control literature traditionally assumes global distribution and the hierarchical control literature generally assumes local distribution. Unfortunately, many of the interesting multiperson control/decision problems are those where the process and goal information lies between these two extremes. In such problems, the observation v can include messages containing process and goal information as well as coupling information and the overall problem becomes very complex.

The framework developed provides a structure for considering a broad class of multiperson dynamic control problems. It is not the most general since the order of action is fixed in advance, but it seems appropriate for many complex control problems. The framework includes but is not limited to such problems as classical (discrete) control, decision problems, static and dynamic team problems, and static and dynamic hierarchical control problems.

5.4. PERFORMANCE OF THE INFORMATION STRUCTURE

In the previous section we identified information flows and some properties of systems based on information flow only. In this section we are interested not only in the flow but also in its effectiveness in improving actions.

5.4.1. PARTITIONS AND SUBFIELDS

An information function $v = h(m)$ conveys information to an agent about m. If we assume that m is a member of a set M and v is a member of a set V, then we can say that h induces a partition of M into subsets

$$M_v^h = \{m \in M | h(m) = v\} \quad \text{for all} \quad v \in V.$$

The set of subsets $\mathcal{M}^h = \{M_v^h | v \in V\}$ is called a partition of M and represents the resolution of v in describing m. That is, in viewing M through h we find that all m in each M_v^h are equivalent. Moreover, there is a one-to-one correspondence between information functions h and partitions $\mathcal{M}^h$. Thus, they represent alternative descriptions.

Partitions can be compared by their fineness. A partition $\mathcal{M}^{h_1}$ is said to be finer than $\mathcal{M}^{h_2}$ (denoted $\mathcal{M}^{h_1} \leq \mathcal{M}^{h_2}$) if every element $M_v^{h_1} \in \mathcal{M}^{h_1}$ is a subset of some element of $\mathcal{M}^{h_2}$. An information function h_1 is said to be *more informative* than h_2 if $\mathcal{M}^{h_1} \leq \mathcal{M}^{h_2}$.

Another approach to the comparison of information structures is to define a field $\mathcal{F}$ of subsets of M and a corresponding field $\mathcal{F}_h$ generated by the subsets M_v^h. The field $\mathcal{F}_h$ is simply the set $\mathcal{M}^h$ plus additional subsets of M generated by taking unions, complements, and intersections of the elements of $\mathcal{M}^h$. The elements M_v^h of $\mathcal{M}^h$ are the atoms of the field $\mathcal{F}_h$. For simplicity, we will assume that all sets are finite. The concepts developed can be generalized for infinite sets. If the partition $\mathcal{M}^{h_1}$ is finer than $\mathcal{M}^{h_2}$ then clearly

$\mathscr{F} \supseteq \mathscr{F}_{h_1} \supseteq \mathscr{F}_{h_2}$. The following statements are equivalent:

h_1 is more informative than h_2

$\mathscr{M}^{h_1} \leq \mathscr{M}^{h_2}$

$\mathscr{F}_{h_1} \supseteq \mathscr{F}_{h_2}$

$\mathscr{M}^{h_1}(\mathscr{F}_{h_1})$ is finer than $\mathscr{M}^{h_2}(\mathscr{F}_{h_2})$

$\mathscr{M}^{h_2}(\mathscr{F}_{h_2})$ is coarser than $\mathscr{M}^{h_1}(\mathscr{F}_{h_1})$

Note that if h_1 is more informative than h_2 then $\mathscr{M}^{h_1} \leq \mathscr{M}^{h_2}$ and there is a condensing function $f: V \rightarrow V$ such that $h_2(m) = f(h_1(m))$. Thus we call h_2 a *condensation* of h_1.

Given a set H of information functions h we note that fineness gives a partial ordering of H. That is, for any h_i, $h_j \in H$ we say $h_i \geq h_j$ if $F_{h_i} \supseteq F_{h_j}$. The fact that the ordering is only partial means that the requirement of "finer than" is too strong. In general, two information functions may be incomparable because they are finer in some places and coarser in others. However, even under these circumstances some interesting results can be obtained.

5.4.2. NOISELESS INFORMATION FUNCTIONS

An information function $h: M \rightarrow V$ is said to be *noiseless* if it can be represented by a partition $\mathscr{M}^h$, that is, if for each $m \in M$ there is a unique $v \in V$ such that $h(m) = v$. The finest possible partition resolves M into its elements m. However, in general, a coarser partition may be satisfactory, first, because only a subset of M may be of interest to one or a group of agents, and, second, because even in the part of M of interest there may be certain m's that are equivalent in terms of performance. Thus, given a subsystem $\mathscr{S}_i$ we take M_i as the subset of M of interest to $\mathscr{A}_i$ (the agents in $\mathscr{S}_i$) and $\mathscr{F}^i$ as the finest field of interest to $\mathscr{A}_i$.

Information centralization

If two agents A_{ij} and A_{kl} in $\mathscr{S}_i$ have information functions h_{ij} and h_{kl} then it is generally true that neither $\mathscr{F}^i_{ij} \supseteq \mathscr{F}^i_{kl}$ nor $\mathscr{F}^i_{kl} \supseteq \mathscr{F}^i_{ij}$. In this case we say that the two agents are *informationally distinct*. In the terminology of Hexner and Ho (1977), these two agents each have a private information structure with respect to the other. On the other hand, it may be true that $F^i_{ij} \supseteq F^i_{kl}$. In this case we say that A_{ij} is *informationally superior* to A_{kl}. According to Hexner and Ho (1977), the common information structure of these two agents is $\mathscr{F}^i_{kl}$. Equivalently, we may say that A_{kl} has no private information structure with respect to A_{ij}. It is, of course, possible that agent ij is informationally superior to several other agents in $\mathscr{S}_i$. If some agent in $\mathscr{S}_i$ is informationally superior to all other agents in $\mathscr{S}_i$ then we say that $\mathscr{S}_i$ is *informationally centralized*.

In a causal static information structure the fineness of $\mathscr{F}_{ij}^i$ for each agent is independent of the actions of the other agents and so centralization is a property of the information structure design. In other cases, centralization depends on the actions taken by the agents in $\mathscr{S}_i$ and prior statements about information centralization/decentralization are more difficult.

If $\mathscr{S}_i$ is informationally centralized because A_{ij} is informationally superior to all other agents, then h_{ij} is more informative than all other h_{kl} in $\mathscr{S}_i$ and there are condensation functions f_{kl} such that

$$v_{kl} = h_{kl}(m) = f_{kl}(h_{ij}(m))$$

for all $A_{kl} \subset \mathscr{A}_i$.

Information nesting

If $\mathscr{S}_i$ has a sequential information structure, the agents $\mathscr{A}_i$ are partially ordered according to the information ordering. Thus each agent (except starting agents) has one or more predecessors. A sequential information structure is said to be *partially nested* if each agent is informationally superior to all of his predecessors. Such an information structure provides complete information along its branches of flow. Each agent knows at least as much as his predecessors. Thus, when the information structure is partially nested, the flow arrows in the information flow graph also indicate informativeness. Ho and Chu (1974) indicate this by augmenting their precedence graph with dotted informativeness lines.

Equivalent information structures

In section 5.3 we suggested that when process and goal information is globally distributed, two information structures are equivalent if they provide the same information about the action of the chance agents $\mathscr{A}_0$ (see Assertion 5.7). We now formalize this by stating that if process and goal information is globally distributed, then two information structures are *equivalent* if they are equally informative with respect to $M_0 = \{m_{i0}\}$, $i = 1, \ldots, S$. We can then prove the following:

ASSERTION 5.10. *A partially nested dynamic information structure is equivalent to the static information structure obtained by deleting all flows between elements in $\mathscr{S}$.*

Explanation: Since the information structure is partially nested, each agent knows as much as his predecessor in the information ordering. Thus, the dynamic flows between agents of $\mathscr{S}$ do not increase the agents' knowledge of M_0.

This result is roughly equivalent to a result obtained by Ho and Chu (1972a). It is important because it reduces a class of systems with dynamic information structures to systems with static information structures. Since there are explicit solution techniques available for some static structures they can now be applied to dynamic structures. Unfortunately, the requirement of a partially nested dynamic information structure is overly restrictive. The required information nesting is seldom achieved in practice except for the case of a single decision maker. Ho and Chu (1972b) have also shown how suboptimal controls for systems with non-nested information structures can be studied using systems with nested structures as bounds.) The following example shows the major application.

Example 5.11. *A lumped stochastic dynamic system*

Consider a stochastic version of the lumped dynamic system discussed in Example 5.9. The difference equation model is now

$$x_{t+1} = f_t(x_t, m_t, w_t),$$
$$y_t = \tilde{g}_t(x_t, m_t),$$

where for each $t = 1, \ldots, N$, w_t is a random variable. In an s–t grid model, w_t is generated by the chance subsystem with $w_k = m_{k0}$, for $k = 1, \ldots, N$. If we make the standard assumptions that $v_{1j} = y_{1,j-1}$ and that all agents are controlled by a single controller then each agent "remembers" what his predecessors observed. Thus, the information structure is partially nested and equivalent to the static one obtained by deleting branches in $\mathscr{S}$. The problem can thus be considered equivalent to a static team problem. If it is also one of those static team problems that can be solved, then an optimal strategy is available. Generally, only linear-quadratic static team problems yield simple solutions (Marschak and Radner 1972).

As noted in Ho and Chu (1972a), this is an important result because it gives an extremely simple proof of the linearity of the solution to the linear-quadratic Gaussian (LQG) control problem under very mild assumptions.

5.4.3. NOISY INFORMATION FUNCTIONS

A noiseless information function fails to be completely informative about M only because of the aggregation or condensation of M into subsets M_v^h by h. A noisy information function can also fail by sending the wrong message v.

In addition to the partition approach, the information function h can be represented by the conditional probability density function $p(v|m)$ of sending a correct v for given m. A noiseless information function is such that for each m, $p(\cdot|m) = 1$ for some $v \in V$. If this is not true, the information function h is said to be noisy and is more conveniently described probabilistically, e.g., by $p(v|m)$, $p(m|v)$, or $p(v, m)$. Thus a noisy information function can introduce both aggregation of information and erroneous representation.

In dealing with noisy information functions, the concept of condensation is replaced by the concept of garbling. If h_1 and h_2 represent two noisy information functions with $p(v_2|v_1, m) = p(v_2|v_1)$ then we say that v_2 is a *garbling* of v_1.

Conceptually at least, a noisy information function $h_n : M \rightarrow V$ can always be redefined as a noiseless information function h_ℓ by expanding the space M to include the noise. That is, $h_\ell : M \times N \rightarrow V$ can be defined equivalent to h_n. The advantages of working with noisy information structures appear to lie in the statistical tools available (Marschak and Miyasawa 1968, Wyner 1970).

We can also introduce analogous concepts of informational distinction, superiority, centralization, decentralization, and so on in terms of garbling. For example, A_{ij} is informationally superior to A_{kl} if v_{kl} is a garbling of v_{ij}. At this point, though, the extension does not seem profitable.

5.4.4.   THE VALUE OF INFORMATION

Since we have defined information as a commodity that improves decisions, the simplest definition of the value of information arises from this property. That is, the value of information is equal to the improvement in performance it produces. The important question here is what measure of performance should be used. Up to this point we have only used the performance measure $Q(m, y)$ defined in section 5.3.4. More realistically, we should write our performance measure as $Q(m, y, h)$ where h reflects informational factors. For example, if we think of Q as representing profit, then the h term brings in the cost of information. If this Q is separable as

$$Q(m, y, h) = Q_0(m, y) - Q_c(h), \tag{5.23}$$

we will call Q_0 the *gross performance* and Q_c the *information cost*. While this separable form of the performance index is less general, it is the only case for which significant general results have been obtained (e.g., see Marschak and Miyasawa 1968).

For a single-agent problem the gross value of information can be defined as follows. The single agent faces the problem of selecting an action that achieves

$$\max_{m} Q_0(m, y)$$

subject to $y = g(m, u)$. We will assume that if the agent receives no information he chooses $m = m_0$, and that if he receives information $v = h(u)$ he chooses $m = m_h$. The gross value of information function h (given u) is then

$$V_g(h; u) = Q_0(m_h, g(m_h, u)) - Q_0(m_0, g(m_0, u)).$$

If u is a random variable, the gross expected value of h is

$$\bar{V}_g(h) = E_u\{V(h; u)\}.$$

When $Q(m, y, h)$ can be defined and is separable as shown in (5.23), the net value of information can be obtained in an equivalent fashion (Marschak and Radner 1972). We can now compare two information functions h_1 and h_2 by comparing $\bar{V}_g(h_1)$ and $\bar{V}_g(h_2)$. We say that h_1 is *more valuable than* h_2 if $\bar{V}_g(h_1) \geq \bar{V}_g(h_2)$. Note that this comparison depends on both the performance measure Q_0 and the probability distribution $p(z)$. The fineness theorem below (Marschak and Radner 1972) relates the informativeness of h to the value of h in a broad class of single-agent problems.

THEOREM 5.1. *If h_1 is more informative than h_2 (i.e., $\mathcal{M}^{h_1} \leq \mathcal{M}^{h_2}$), then for all $p(z)$ and Q_0, h_1 is more valuable than h_2.*

An equivalent statement can be made about noisy information functions (Marschak and Radner 1972).

Conceptually, the above approach can be used to compare information structures for entire subsystems having a single team performance index. However, complications can arise when the actions of one agent affect the information available to another agent, as shown in the following well-known example (Witsenhausen 1968).

Example 5.12.   A two-person team with dynamic information structure

We assume that there are two elements in $\mathcal{S}$ coupled as shown in Figure 5.8. The outcome functions of the process elements are

$$y_{11} = m_{11} + m_{10}$$
$$y_{12} = m_{12} + y_{11} = m_{12} + m_{11} + m_{10}.$$

The team performance index is

$$Q(y_{11}, y_{12}, m_{11}, m_{12}) = \tfrac{1}{2}y_{11}^2 + \tfrac{1}{2}m_{11}^2.$$

Process and goal information is assumed to be globally distributed. The coupling information provided is

$$v_{11} = m_{10}$$
$$v_{12} = m_{10} + m_{20} + m_{11}.$$

The coupling information flow is shown in Figure 5.8(b). Note that the information structure is dynamic since v_{12} depends on m_{11} but it is not partially nested since A_{12} does not have as much information about m_{10} as

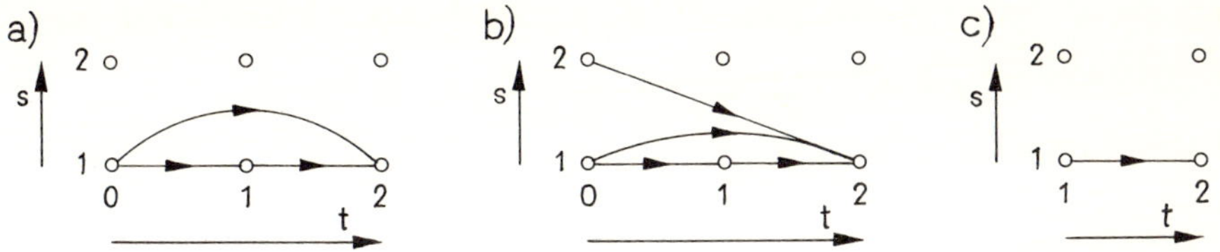

FIGURE 5.8 A two-person team: (a) coupling graph, (b) information flow graph, and (c) reduced information flow graph.

A_{11}. Moreover, the information available to A_{12} depends on the action m_{11} chosen by A_{11}. Thus, the information structure and the actions are interrelated. In this case, the informativeness of h_{12} depends on the action taken by A_{11} and cannot be discussed independently. We might also expect that the information cost q_c would depend on the action of A_{11}. Thus, the separation proposed in (5.23) and the related measures dealing with the net value of information may be problematical. Finally, it may be very difficult to obtain even a gross value of information since the best action for A_{11} is difficult to calculate. For the example given, there is no known best action (Ho *et al.* 1978).

5.4.5. THE COST AND AMOUNT OF INFORMATION

The informativeness of an information function tells us something about the usefulness of the information in certain decision situations. The value of information tells us something about its effect on system performance. If we are to attempt a cost–benefit analysis, some measure of the cost of information is also needed. Unfortunately, useful measures of the cost of information are not easily obtained. In general situations the total cost of information depends on production (information-gathering) costs, transmission costs, and market factors (supply and demand). Since all of these factors depend on a host of exogenous variables, it seems unrealistic to search for a general model for the cost of information. However, there are special cases where the transmission cost dominates and can be modeled. Consider the initial example of the farmer selecting a crop. Here we might argue that, with respect to weather data, the production costs are borne by the government and there is no market because the data is freely distributed. Thus, the only cost to the farmer is the transmission cost, say, the cost of a long-distance phone call. Fortunately, classical information theory (Wyner 1970) provides an attractive approach to modeling the cost of transmitting information in terms of the amount transmitted.

Assume that we have a noisy information function characterized by $p(m_i | v_j)$ providing information $v_j \in V$ about a discrete random signal $m_i \in M$ with probability $p(m_i)$. To simplify the discussion we assume that m and v

414

are discrete. For the continuous case see Gallager (1968). Consider the following concepts:

Information: $I(m_i) = -\log_2 p(m_i)$
Entropy: $H(M) = E\{I(m_i)\} = -\sum_i p(m_i) \log_2 p(m_i)$
Conditional information: $I(m_i | v_j) = -\log_2 p(m_i | v_j)$
Conditional entropy: $H(M | v_j) = E\{I(m_i | v_j)\} = -\sum_i p(m_i | v_j) \log_2 p(m_i | v_j)$
Equivocation: $H(M | V) = E\{H(M | v_j)\} = \sum_j p(v_j) H(M | v_j)$
Mutual information: $I(M, V) = H(M) - H(M | V)$

The information I is a measure of the amount of information conveyed by a symbol m_i from M. Since $\log_2$ is used, the units are bits. The entropy H is the average uncertainty (missing information) about the source M in bits per symbol. The conditional information and entropy have similar meanings given that v_j has been received. The equivocation is the average amount of information about M that is missing when symbols from V are received. The mutual information is thus the amount of information sent by the information function.

We next assume that the information v about m is obtained through a transmission device called a channel. A channel may be a sensor, a computer memory, a human observer, and so on. If the average power $E\{\Phi(M)\}$ at the input to the channel is constrained by $E\{\Phi(M)\} \leq \alpha$ then the capacity of the channel is defined as

$$C(\alpha) = \rho_c \sup_{p(m)} I(M, V) \tag{5.24}$$

subject to $E\{\Phi(M)\} \leq \alpha$, where ρ_c is the channel symbol rate in symbols per second. A fundamental result of information theory is that for any fixed $\alpha = \bar{\alpha}$, $C(\alpha)$ is the maximum rate in amount per unit time at which information can be sent through the channel essentially error free, that is, with errors as small as we want them to be by appropriate coding. Thus, for a given channel, the channel capacity $C(\alpha)$ is a measure of the amount of information that can be sent. Moreover, since the cost of using a channel is often closely correlated with its capacity, the relation (5.24) can be used to relate cost to amount of information when transmission costs dominate. In the next section we will relate the amount of information to system performance.

5.5. INFORMATION COMPRESSION

In sections 5.3 and 5.4 we identified the information structure and investigated some measures of its effectiveness in improving system performance as measured by a performance index. We also noted that in general there is a cost encountered in operating an information structure. Thus, we expect that the design of a satisfactory information structure requires a cost–benefit analysis. While the relevant costs are generally difficult to quantify, in section 5.4.5 we noted that the informativeness of an information structure could be related to the amount of information (mutual information) transmitted and thus to the required capacity of the information channel employed. The fact that channel capacity is often closely related to channel operating costs suggests a direct relation between the amount of information sent and the cost of the information structure. This is, of course, consistent. Experience with communication systems and control systems confirms this expectation (e.g., see the example cited in Chapter 1, section 1.4).

In order to reduce the cost of the information structure, we should consider reducing the amount of information sent, a problem of information compression. Ideally we would like to find compression schemes that reduce the amount of information without producing a deterioration in performance. More realistically, we seek compression schemes that yield an increase in net performance (performance minus cost of information).

While there has been a great deal of work on information compression (also called data compression) in the communications literature (e.g., see Davisson and Gray 1976), the cost–benefit measures and techniques used are seldom appropriate for control problems. Thus, while some of these communication techniques are conceptually useful, there is still a great deal of work required in the development of information compression techniques appropriate to control system problems.

5.5.1. SUFFICIENT STATISTICS AND STATE VARIABLES

The most commonly used data compression schemes are based on sufficient statistics and/or state variables. They are generally costless compression schemes, that is, they are designed to obtain compression with no degradation in system performance.

Sufficient statistics

A sufficient statistic is simply a compression of the information that does not degrade performance. If an agent is provided with information $v = h(m)$, and $\bar{v} = f(v)$ is a compression of v such that $\bar{v}$ is as valuable as v, then $\bar{v}$ is called a *sufficient statistic* for the agents decision problem. Such compressions always exist since $v \equiv \bar{v}$ is a sufficient statistic. The difficulty is in finding a sufficient statistic that reduces the information cost (e.g., reduces the amount of information). The sources of compression are most clearly seen in single-person decision problems.

416

Example 5.13.   *A single-person decision problem*

Consider a single-person decision problem (Example 5.7) with

   Information: $v = h(m_0)$, $m_0 \in M_0$
   Outcome: $y = \tilde{g}(m, m_0)$
   Performance index: $Q(m, y)$.

The agent is to choose m to maximize the expected value of Q.

A completely informative h would partition M_0 into its elements. However, a coarser partition may be satisfactory for several reasons. First, there may be several m_0's producing the same outcome. That is,

$$\tilde{g}(m, m_0^1) = \tilde{g}(m, m_0^2) \quad \text{for all} \quad m \in M. \tag{5.25}$$

Second, different outcomes may have the same performance. That is

$$Q(m, \tilde{g}, (m, m_0^1)) = Q(m, \tilde{g}(m, m_0^2)) \quad \text{for all} \quad m \in M. \tag{5.26}$$

We can say that a partition that resolves M_0 down to those elements that are equivalent in (5.25) is *outcome sufficient.* Similarily, a partition that resolves M_0 down to those elements that are equivalent in (5.26) is *performance sufficient.* Clearly, an outcome-sufficient partition is also performance sufficient. It is also clear that a performance-sufficient partition is as valuable as a completely informative partition since decision rules $m = d(v)$ based on either partition will produce equivalent performances.

The concepts of outcome- and performance-sufficient partitions can be easily extended to multiperson problems with static information structures. Extensions to dynamic information structures encounter the problems noted in Example 5.12.

State variables

State variable compression is an extension of the outcome-sufficient partition concept to systems with dynamic couplings. If we consider a causal, deterministic, lumped dynamic system (e.g., Example 5.9) the coupling graph shows each element (ik) coupled to its predecessors $(i1), \ldots, (ik-1)$. Now assume that each agent has a decision rule that is a function of the actions of all preceding agents. That is, for agent A_{ik}

$$m_{ik} = d_{ik}(m_{i1}, \ldots, m_{ik-1}) \tag{5.27}$$

and the required information is

$$v_{ik} = h_{ik}(m_{i1}, \ldots, m_{ik-1}). \tag{5.28}$$

In some cases, this information flow pattern can be simplified through the introduction of state variables.

A state space is a sequence of sets X_k such that

$$g_{ik}(m_{ik}, u_{ik}) = \tilde{g}_{ik}(m_{ik}, x_{ik}), \qquad (5.29)$$

where $x_{ik} \in X_k$, the function $\tilde{g}_{ik}$ is given, and x_{ik} is given by an updating rule

$$x_{ik} = f(x_{ik-1}, m_{ik-1}). \qquad (5.30)$$

Clearly the trivial solution

$$X_k = M_1 \times \ldots \times M_{k-1},$$

with $\tilde{g}_{ik} = g_{ik}$ and $x_{ik} = z_{ik}$, always exists. Again, the problem is to find a state-space representation that produces a satisfactory data compression.

Conceptually, the sets X_k can be defined as follows (Witsenhausen 1976). Given $(A, <)$ where A is a finite set and $<$ is a total ordering on A, we define a *cut* as a partition of A into mutually exclusive subsets A_α, A_β such that for every $a_i \in A_\alpha$ we have $a_i < a_j$ for every $a_j \in A_\beta$. Since a causal, lumped system is sequential, it is totally ordered by its coupling order. Thus the sets $M = \underset{i}{\times} M_i$ and $Y = \underset{i}{\times} Y_i$ are totally ordered and we can define cuts M_α, M_β and Y_α, Y_β. Now, for any $t \in (1, T)$, we define a cut Y_α, Y_β where $y_\alpha = (y_1, \ldots, y_t)$ and $y_\beta = (y_{t+1}, \ldots, y_T)$. Similar equations can be written for m_α and m_β. By causality, y_α depends only on m_α, but y_β depends on both m_α and m_β:

$$y_\beta = G(m_\alpha, m_\beta).$$

We next define an equivalence relation on M_α by $m_\alpha^1 \sim m_\alpha^2$ when

$$G(m_\alpha^1, m_\beta) = G(m_\alpha^2, m_\beta) \quad \text{for all} \quad m_\beta \in M_\beta.$$

Then X_t is the quotient space $M_\alpha/\sim$. Note that X_t is simply an output-sufficient partition of M_α. When we can, in addition, find a simple relation such as (5.30) for "updating" $x_t \in X_t$ from stage to stage, then the result is a compression of the data sent by the original information structure given by (5.28).

The fact that X_t is an output-sufficient partition suggests an extension of the above approach using a payoff-sufficient partition. That is, we note that when the system has an additive team performance index, then $\{Q_t\}$, where

$$Q_t = \sum_{k=1}^{t} Q_{ik}(m_{ik}, u_{ik}), \qquad t = 1, \ldots, T,$$

is also a totally ordered set, and by causality

$$Q_\beta = Q(m_\alpha, m_\beta).$$

Thus, we can define a new state space X_t^Q as the quotient space $M_\alpha/\approx$, where $m_\alpha^1 \approx m_\alpha^2$ when

$$Q(m_\alpha^1, m_\beta) = Q(m_\alpha^2, m_\beta) \quad \text{for all} \quad m_\beta \in M_\beta.$$

418

The question of whether satisfactory data compression can be obtained in this manner has not been addressed.

The state variables and sufficient statistics can be combined for nondeterministic dynamic systems; see Witsenhausen (1973) and Striebel (1965). The use of state variables in developing output-sufficient partitions can be extended to arbitrary sequential systems. Since sequential coupling induces a partial ordering of the system elements, we can always select a total ordering that is compatible with this partial ordering and proceed as before. Since there may be many possible total orderings consistent with a given partial ordering, there are many possible state spaces. One would hope to select one giving the best information compression. For further comments see Witsenhausen (1976).

5.5.2. INFORMATION COMPRESSION USING RATE DISTORTION THEORY

The discussion of information compression in the preceding section assumed that the compression would be costless. Beyond this, it may be valuable to investigate costly information compression techniques. Such compression schemes produce performance degradation and must be subjected to cost–benefit analysis. That is, it should be shown that the benefit in reduced information cost is larger than the cost in reduced system performance. While the proposed cost–benefit analysis may generally be very difficult, when the cost of information is dominated by transmission cost, classical information theory provides some useful tools.

In section 5.4.5 we noted that the capacity of an information channel could be described by $C(\alpha)$ in (5.24). Recall that the capacity of a channel is the number of bits per second it can process essentially without error. It can also be shown (Wyner 1970) that given a distortion function $D(m, v)$, a general source of signals m can be coded to produce information at a maximum rate

$$R_d(\beta) = \rho_s \inf_{p(v|m)} I(M, V) \tag{5.31}$$

subject to $E\{D(m, v)\} \leq \beta$ so that the average distortion $E\{D(m, v)\}$ does not exceed β (ρ_s is the source rate in symbols per second). Thus, given a channel with capacity $C(\alpha)$, signal m can be coded so that $R_d(\beta) \leq C(\alpha)$ as long as the average distortion is not required to be less than β. Equivalently, if an average distortion less than β is desired, then a channel capacity $C(\alpha) \geq R_d(\beta)$ is required. That is, the minimum average distortion is given by

$$\beta = \inf_{p(v|m)} E\{D(m, v)\} \tag{5.32}$$

subject to $\rho_s I(M, V) \leq C(\alpha)$. If we can now relate the average distortion to degradation in system performance, then the theoretical results presented above suggest a link between system performance and information channel capacity. The following example shows how these results can be used in a single-person decision problem.

Example 5.14.   *Information compression in* SPDP

Consider a single-person decision problem (Example 5.7) with $m_0 \in M_0$ having probability $p(m_0)$. Assume a noisy information function $v = h(m_0)$ modeled by $p(v|m_0)$. We will also assume for simplicity that m_0 and v are discrete. If m_0 were known exactly then the optimal action would be $m^*(m_0)$ and $\tilde{Q}$ would attain

$$Q^*(m^*, m_0) = \max_m \tilde{Q}(m, m_0).$$

Of course, in general m_0 will not be known precisely, Thus the agent may choose $m \neq m^*$, and incur regret

$$r(m, m_0) = \tilde{Q}(m^*, m_0) - \tilde{Q}(m, m_0). \tag{5.33}$$

For a given information function $v = h(m_0)$ and decision rule $m = d(v)$, we can define the expected regret as

$$R(h, d) = E\{r(d(h(m_0))), m_0\}, \tag{5.34}$$

where the expectation is taken with respect to $p(m_0)$.

We now note that the choice of information function h implies a choice of $p(v|m_0)$ and thus mutual information $I(M_0, V)$. At the same time a choice of h and d implies a choice of expected regret $R(h, d)$ as given by (5.34). Thus, if we consider the I–R space shown in Figure 5.9, the choice of (h, d) corresponds to the choice of a point $I(h) \equiv I(M_0, V)$ and $R(h, d)$ in this space. With regard to system design, the optimal I–R point is $(0, 0)$. At this point, we would have zero mutual information (requiring zero channel capacity) and also have zero regret. Unfortunately, not all points in the I–R space are feasible. A point I_1, R_1 is feasible if there is an h and a d such that $I(h) = I_1$ and $R(h, d) = R_1$. Thus, for all R_1, the lower bound of the feasible region is given by

$$I^*(R_1) = \inf_h I(h) = \inf_{p(v|m)} I(M_0, V) \tag{5.35}$$

However, comparison of (5.35) with (5.31) reveals that (5.35) is simply a rate distortion problem with regret $R(h, d)$ used as a measure of average distortion. Thus, the techniques of rate distortion theory (Berger 1971) can be used to obtain the boundary of the feasible region in the I–R space. Since rate distortion functions are known to be convex and monotonically decreasing, we can draw the typical feasible region boundary curve shown in Figure 5.9.

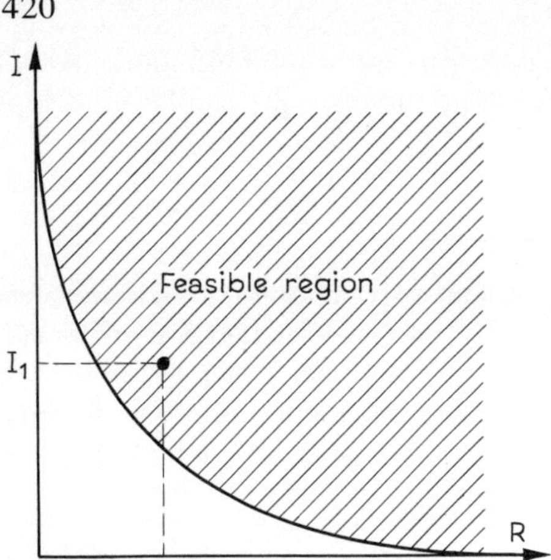

FIGURE 5.9   The I–R space.

In the above example we have related performance deterioration, as measured by regret R, to the amount of information required I. The results plotted in Figure 5.9 suggest that there is clearly a trade-off involved. These results can be used in several ways. First, when I can be related to information cost through channel capacity, a cost–regret trade-off is possible. For example, if the total cost of operating an SPDP system is $R + c(I)$ where c is a concave function representing information cost, then we can draw indifference curves (i.e., curves of $R + c(I) = $ a constant) on Figure 5.9 and locate the point of minimum total cost. Second, when there is an information constraint involved, say the agent has limited information input capacity, then Figure 5.9 can be used to relate the input constraint to performance degradation in terms of regret.

It is important to note two problems in applying this approach. First, as noted above, rate distortion theory gives only a lower bound on the rate attainable with a given regret. Shannon showed that this bound is attainable but only at the price of complex coding schemes using long blocks of data. Thus, the source-encoder must store up many messages before any are sent. While this delay may be acceptable in communication systems, it is seldom acceptable in control/decision problems. Thus, there is a need for constructive research in the area of real-time information theory, that is, information theory with a real-time coding delay constraint. Recent work on alternative measures of mutual information has led to some initial real-time results (Ziv and Zakai 1973). Second, an extension of the above approach to multiperson problems requires results in multichannel rate distortion theory. While initial results are available (Berger and Tung 1977) there are many unsolved problems. Results for multichannel real-time problems are essentially nonexistent. Some interesting relations between real-time information theory problems and dynamic information structure problems are discussed by Ho *et al.* (1978).

5.5.3. TÂTONNEMENT PROCESS

As noted in section 5.3.5, when process and goal information is only locally distributed, the function of the information structure becomes considerably more complex. If the individual agents have separate performance indices, then an information structure that simply reveals coupling information is sufficient. However, in the more interesting cases, where there is a team performance index, the agents are coupled through the team goal and additional information is needed. In this case, a sufficient information flow is one that provides a global distribution of process and goal information. However, such a solution is not informationally attractive.

Alternatives to the global distribution of process and goal information are: (a) centralization of the process and goal information in a single coordinating agent who solves the entire problem and then directs the actions of the other agents or, (b) a tâtonnement coordination process leading to optimal actions without an explicit transfer of process and goal information. Note that the former solution corresponds to a centrally planned economy while the latter corresponds to a so-called market or competitive economy.

Tâtonnement processes are planning/action processes where the planning phase consists of an iterative message exchange between the system agents and a coordinator whose sole function is the facilitation of planning. Tâtonnement processes represent an important class of solutions to the coordination problems discussed in other chapters of this book. Their properties have been studied extensively in the economics literature (see Hurwicz 1971 or Kanemitsu 1966). They have the advantage of distributing coupling information and generating optimal actions concurrently. When compared with the original solution of global distribution of the process and goal information, they appear to compress the information. However, when compared with simple centralization of the process and goal information, the answer is not so clear. An example will illustrate the point.

Example 5.15.   An iterative coordination mechanism

Consider an N-element static, deterministic system (Example 5.10; since the system is static the time index has been omitted) described by the following outcome and coupling functions

$$y_i = G_i m_i + u_i \qquad \qquad (5.36)$$

$$i = 1, \ldots, N$$

$$u_i = \sum_{j \neq i} K_{ij} m_j \qquad (5.37)$$

where m_i is an r_i-vector, y_i is a p_i-vector and G_i is a $p_i \times r_i$ matrix with $r_i \geq p_i$. Let the team objective of the system be

$$\text{minimize} \sum_{i=1}^{N} \|m_i\|^2 \tag{5.38}$$

subject to $y_i = b_i$, $i = 1, \ldots, N$, where b_i are some required values of the outputs. Process and goal information is only locally distributed. Each agent knows only his piece of (5.36) and his b_i. However, we assume that there is a coordinator (agent A_{N+1}) who knows the couplings K_{ij} for all i, j. Now consider the planning phase of the following tâtonnement process. The iterative process starts at $k = 1$ with $w_i^1 = 0$ for all i. At step k an "input prediction"

$$w_i^k = \sum_{j \neq i} K_{ij} m_j^{k-1} \tag{5.39}$$

is sent to all A_i by A_{N+1} and information on the required control

$$m_i^k = G_i^{+}(b_i - w_i^k) \tag{5.40}$$

is sent to A_{N+1} by all A_i (G_i^{+} is the pseudo-inverse of G_i). Laub and Bailey (1978) show that under reasonable assumptions regarding G_i and K_{ij}, the planning process converges to the optimum actions. That is, $m_i^k \to m_i^*$ for all i. If the system is cyclic, then the convergence is in the limit as $k \to \infty$. If the system is sequential, then the convergence is at $k = N$. When the planning phase is completed, the agents take actions according to the plans developed.

We see that the proposed tâtonnement process provides a form of information compression. The information flow is simpler than that required to distribute process and goal information globally. The iterative exchange supplies a performance-sufficient version of the missing process and goal information. In addition, the exchange leads directly to the set of optimum actions. However, the requirement of an infinite iteration (in the cyclic coupling case) is disturbing as it suggests delay and high information costs. Moreover, since there is only a finite number of parameters in the problem, one questions the need for an infinite iteration. In fact, it can easily be shown that after a finite number of appropriately chosen iterations and some additional computation, the coordinator has centralized process and goal information. To see this, note that the coordinator has access to w_i^k and m_i^k for all i and k. Since (5.39) and (5.40) can be rewritten as

$$G_i(m_i^k - m_i^{k-1}) = w_i^{k-1} - w_i^k$$

it is apparent that through a proper choice of w_i^k, the coordinator can collect enough information to know all of the G_i's. Moreover, since

$$m_i^1 = G_i^{+} b_i$$

the coordinator can also compute b_i. He then has access to all of the information required to compute optimum m_i's centrally.

Thus, when a tâtonnement process is used, the main difference between the centralized and decentralized approaches lies in the amount of data storage and computation required by the coordinator. Without consideration of this and perhaps other factors (e.g., robustness under uncertainty) a selection between the centralized process and the decentralized (tâtonnement) process is not possible. This was first noted by Marschak (1959).

While the above result was obtained for a special case, the basic argument should hold whenever the process and goal information can be described by a finite number of parameters. In such cases, it is not unreasonable to expect the coordinator to be able to estimate these parameters after a finite number of steps and select the optimum actions centrally.

5.6. COMMENTS AND CONCLUSIONS

In this chapter we have attempted to model information structures in a general class of multiperson decision/control problems. The motivating idea was the fact that information can improve decisions but information structures can also be costly. Thus the fundamental questions are who should receive information and what should they receive. We have considered two classes of problems: those where the process and goal information is globally distributed and those where the process and goal information is only locally distributed.

When the process and goal information is globally distributed, the only missing information is the actions of the chance agents $\mathscr{A}_0$. The function of the information structure is to distribute information about these actions. If these actions can be observed directly (feedforward control) the problem is solved. In general, they cannot be observed directly and some indirect (feedback) scheme is required. Here the problem is one of designing a scheme that provides the required information at minimum cost and maximum accuracy. At present, there are no systematic procedures for selecting an information structure except in the simplest cases.

When the process and goal information is only locally distributed, the agents cannot act optimally unless there is some coordination of their actions. In this case, the missing information is the process and goal information and the information structure supplies a performance-sufficient version of it. In typical hierarchical control problems, the information structures proposed are cyclic and thus noncausal and must be solved using a tâtonnement planning/action process. These processes have the dual disadvantages of being both informationally demanding and time consuming.

However, in deterministic problems they may be solved off-line and the resultant plans implemented as required in an open-loop fashion. Again, what is needed is a systematic design procedure with information costs considered.

Finally, in situations where there is both uncertainty due to the actions of a chance subsystem and only locally distributed process and goal information, the choice of an appropriate information structure is even more difficult. The use of tâtonnement planning/action processes requires a prediction of the chance actions during the planning phase. The results may be unsatisfactory if there are a number of different chance actions involved (i.e., if the disturbances change rapidly with time). At this point, we await further research on the design of information structures for realistic multiperson control problems.

REFERENCES

Berger, T., and S. Y. Tung (1977). "Multiterminal source coding." *International Symposium on Information Theory*, Cornell University, Ithaca, New York.

Berger, J. (1971). *Rate Distortion Theory*, Prentice-Hall, Englewood Cliffs, New Jersey.

Burton, R. M., and B. Obel (1977). "The multilevel approach to organizational issues of the firm—a critical review." *OMEGA: The International Journal of Management Science* 5(4): 395–414.

Davisson, L. D., and R. M. Gray (1976). *Data Compression*. Dowden, Hutchinson & Ross, Stroudsburg, Pennsylvania.

Gallager, R. G. (1968). *Information Theory and Reliable Communication*. Wiley, New York.

Hexner, G., and Y. C. Ho (1977). "Information structure: common and private." *IEEE Trans. Inf. Theory* IT-13: 390–393.

Ho, T. C., and K. C. Chu (1972a). "Team decision theory and information structure in optimal control problems—Part I." *IEEE Trans. Autom. Control* AC-17: 15–22.

Ho, Y. C., and K. C. Chu (1972b). "Team decision theory and information structure in optimal control problems—Part II." *IEEE Trans. Autom. Control* AC-17: 22–28.

Ho, Y. C., and K. C. Chu (1974). "Information structure in many-person optimization problems." *Automatica* 10: 341–351.

Ho, Y. C., M. P. Kastner, and E. Wong (1969). "Teams, signaling, and information theory." *IEEE Trans. Autom. Control* AC-23: 305–312.

Hurwicz, L. (1971). "Centralization and decentralization in economic processes," in A. Eckstein, ed., *Comparison of Economic Systems: Theoretical and Methodological Approaches.* University of California Press, Berkeley, California.

Kanemitsu, H. (1966). "Informational efficiency and decentralization in optimal resource allocation." *Economic Studies Quarterly* 16(3): 22–40.

Laub, A. J., and F. N. Bailey (1978). "An iterative coordination approach to decentralized decision problems." *IEEE Trans. Autom. Control* AC-23(6): 1031–1036.

Marschak, J., and K. Miyasawa (1968). "Economic comparability of information systems. *International Economic Review* 9: 137–173.

Marschak, J., and R. Radner (1972). *Economic Theory of Teams.* Yale University Press, New Haven, Connecticut.

Marschak, T. (1959). "Centralization and decentralization in economic organisations," *Econometrica* 27: 399–430.

Raiffa, H., and R. Schlaifer (1968). *Applied Statistical Decision Theory*. MIT Press, Cambridge, Massachusetts.

Striebel, C. (1965). "Sufficient statistics in the optimal control of stochastic systems." *J. Math Anal. Appl.* 12: 576–592.

Witsenhausen, H. S. (1968). "A counterexample in stochastic optimal control." *SIAM J. Control* 6: 131–147.

Witsenhausen, H. S. (1972). "On information structure, feedback and causality." *SIAM J. Control* 9: 149–160.

Witsenhausen, H. S. (1973). "A standard form for sequential stochastic control." *Control and Mathematical Systems Theory* 7: 5–11.

Witsenhausen, H. S. (1974). "The intrinsic model for discrete stochastic control: Some open problems." *Lecture Notes in Economic and Mathematical Systems* 107: 322–335.

Witsenhausen, H. S. (1976). "Some remarks on the concepts of state," in Y. C. Ho and S. K. Mitter, eds., *Directions in Large Scale Systems*. Plenum Press, New York.

Wyner, A. D. (1970). "Another look at the coding theorem of information theory—A tutorial." *Proc. IEEE* 58: 894–913.

Ziv, J., and M. Zakai (1973). "On functionals satisfying a data-processing theorem." *IEEE Trans. Inf. Theory* IT-9: 275–283.

APPENDIXES

A Appendix to Chapter 2

A.1. SOME PROPERTIES OF UPPER-LEVEL FUNCTIONS

Let us consider a functional $f : \mathscr{C} \times \mathscr{V} \to \mathbb{R}$, where $\mathscr{C}$ and $\mathscr{V}$ are real Hilbert spaces, and a constraint set $X \subset \mathscr{C} \times \mathscr{V}$. The properties of the following *upper-level function*

$$\hat{f}(v) \triangleq \min_{C(v)} f(\cdot, v), \quad \text{where} \tag{A.1}$$

$$C(v) \triangleq \{c \in \mathscr{C} : (c, v) \in X\}, \tag{A.2}$$

$$v \in V_0' \triangleq \{v \in \mathscr{V} : C(v) \neq \varnothing\}. \tag{A.3}$$

will be discussed in this section. The results can be applied to the penalty function method (PFM) for a fixed value of the penalty coefficient ρ, i.e., when

$$f(\cdot, \cdot) = Q_{\text{pty}}(\cdot, \cdot, \rho), \tag{A.4}$$

$$X = \{(c, v) \in \mathscr{C} \times \mathscr{V} : (c, Hv) \in CU\}. \tag{A.5}$$

This section will also be useful for mixed methods discussed in section 2.5. For the input prediction and balance method (IPBM), presented in section 2.5.3., we would have

$$f(\cdot, \cdot) = L_a(\cdot, \cdot, \lambda, \rho), \tag{A.6}$$

$$X = CU, \tag{A.7}$$

and in the case of the output prediction and balance method (OPBM) in section 2.5.4. we would have

$$f(\cdot, \cdot) = L_\psi(\cdot, \cdot, \lambda, \rho), \tag{A.8}$$

$$X = \{(c, v) \in \mathscr{C} \times \mathscr{V} : (c, Hv) \in CU\}. \tag{A.9}$$

429

The problems to be discussed are closely related to some aspects of the direct method presented in section 2.2.; in particular, if we put

$$f(\cdot, \cdot) = Q(\cdot, \cdot), \tag{A.10}$$

$$X = \{(c, v) \in \mathscr{C} \times \mathscr{V} : (c, Hv) \in CU \wedge v = F(c, Hv)\} \tag{A.11}$$

we would get precisely the case of the direct method. Therefore, we will refer to the results of section 2.2 even though we will focus on PFM, IPBM, and OPBM. These methods differ from the direct method in two ways because (A.5), (A.7), and (A.9) do not have the output *equality* constraint $v = F(c, Hv)$. First, the set X can have a nonempty interior, and second, there is an important class of problems whose control and interaction sets CU are separable. This means that

$$CU = C \times U = \mathop{\times}_{i=1}^{N} (C_i \times U_i), \tag{A.12}$$

which implies that the point-to-set mapping $C(\cdot)$ is constant. In this case, the set $C = \times_{i=1}^{N} C_i$ will be used instead of (A.2); of course V_0' can be assumed to be the whole space and is therefore unessential. In the case of (A.12), PFM, IPBM, and OPBM are especially applicable.

The following optimization problem

$$\text{for given } v \in V_0' \text{ find } \hat{c}(v) = \arg \min_{C(v)} f(\cdot, v)$$

will be referred to as the *lower-level problem* (LP).

First, the continuity of $\hat{f}(\cdot)$ should be considered.

THEOREM A.1.   *If we assume that* (a) *the set X is compact,* (b) *the point-to-set mapping $C(\cdot)$ is continuous, and* (c) *the function f is continuous, then the functional $\hat{f}$ is continuous.*

Remark.   In the separability case (A.12), assumption (b) is always fulfilled.

The differentiability properties of the functional $\hat{f}$ create a more complex question. Generally, differentiability of $\hat{f}$ cannot be ensured even if f is differentiable, and only existence conditions and formulae for subgradients can be derived. This can be done as in Theorem 2.7 for the direct method and will not be reformulated here. We concentrate on a stronger result concerning differentiability which can be achieved in the separable case (A.12).

THEOREM A.2.   *If we assume that*

1. $C(v) = C,$ *does not depend on v,*

2.   *The partial (Fréchet) derivative f'_v of the functional f exists and is continuous on $C \times N$, where N is some open subset of $\mathcal{V}$,*

3.   *For every $v \in N$ there exists a unique solution $\hat{c}(v)$ of the lower-level problem LP,*

4.   *The mapping $\hat{c}(\cdot)$ is continuous on N,*

then $\hat{f}$ is differentiable on N with Fréchet derivative

$$\hat{f}'(v) = f'_v(\hat{c}(v), v). \tag{A.13}$$

Proof.   For some point $v^0 \in N$ and element $h \in \mathcal{V}$, consider a sequence $v^n = v^0 + t_n h$ such that $t_n \to 0^+$. Denote $c^0 = \hat{c}(v^0)$ and $c^n = \hat{c}(v^n)$. Then for some $\alpha_n \in [0, 1]$

$$f(c^0, v^n) - f(c^0, v^0) = \langle f'_v(c^0, v^0 + \alpha_n t_n h), t_n h \rangle,$$

hence

$$(1) \qquad \frac{1}{t_n} (\hat{f}(v^n) - \hat{f}(v^0)) \leq \langle f'_v(c^0, v^0 + \alpha_n t_n h), h \rangle.$$

On the other hand, for some $\beta_n \in [0, 1]$

$$f(c^n, v^n) - f(c^n, v^0) = \langle f'_v(c^n, v^0 + \beta_n t_n h), t_n h \rangle,$$

hence

$$(2) \qquad \frac{1}{t_n} (\hat{f}(v^n) - \hat{f}(v^0)) \geq \langle \hat{f}'_v(c^n, v^0 + \beta_n t_n h), h \rangle.$$

When $v^n \to v^0$, $v^0 + \alpha_n t_n h \to v^0$ and $v^0 + \beta_n t_n h \to v^0$. The continuity of f'_v implies then that the right-hand sides of inequalities (1) and (2) tend to

$$(3) \qquad \langle f'_v(c^0, v^0), h \rangle,$$

i.e., there exists a directional derivative of $\hat{f}$ at v_0 and it is equal to (3). But (3) is linear and continuous with respect to h, hence there exists a Gâteaux derivative

$$\nabla \hat{f}_v(v) = \nabla f_v(\hat{c}(v), v) = f'_v(\hat{c}(v), v).$$

From the assumptions that $f'_v(\hat{c}(v), v)$ is continuous on N, it follows that $\nabla \hat{f}_v$ is continuous and is therefore a Fréchet derivative $\hat{f}'_v$. $\square$

Remark.   The continuity of mapping $\hat{c}(\cdot)$ can be derived if we assume that set C is compact; the assumption is reasonable in finite-dimensional cases.

Proof.   Taking any sequence $v^n \to v^0 \in N$, we can assume without loss of generality that $\hat{c}(v^n) \to \bar{c} \in C$. We have

$$\forall c \in C \qquad f(\hat{c}(v^n), v^n) \leq f(c, v^n),$$

hence, due to the continuity of f,

$$\forall c \in C \qquad f(\bar{c}, v^0) \le f(c, v^0),$$

which implies that $\bar{c} = \hat{c}(v^0)$ since local problem solutions are unique. Thus $\hat{c}(\cdot)$ is continuous. $\square$

Among the rather natural assumptions of Theorem A.2 concerning continuity or differentiability, the requirement of a unique minimization of the lower-level problem LP stands out. This requirement is essential and virtually determines the differentiability. When it is not satisfied at some point v, then the existence of only a subdifferential can be ensured.

Theorem A.2, together with its proofs, can be extended to the general case with a lower-level set dependent on v if this set is of the form

$$C(v) = \{c \in \mathscr{C} : G(c, v) \in S\}, \tag{A.14}$$

where S is a closed, convex cone in a Hilbert space $\mathscr{S}$. Some essential assumptions have to be made, and differentiability can be achieved only in the interior points of the set V_0'.

THEOREM A.3. *If we assume that*

1. *$C(v)$ has the form of (A.14) with G continuously differentiable on $X_N = \bigcup_{v \in N} (C(v) \times \{v\})$, where $N \subset V_0' \subseteq \mathscr{V}$ is some open set,*
2. *The partial (Fréchet) derivative f_v' exists and is continuous on X_N,*
3. *For every $v \in N$ there exists a unique solution $\hat{c}(v)$ of the lower-level problem LP,*
4. *The mapping $\hat{c}(\cdot)$ is continuous on N,*
5. *For every $v \in N$ there exists a unique Lagrange multiplier $\hat{\eta}(v)$ corresponding to constraint G and continuous on N,*
6. *For every $v \in N$ the functional $L(\cdot, v, \hat{\eta}(v))$ is pseudoconvex on $\Pi_{\mathscr{C}}(X_N)$, and $\Pi_{\mathscr{C}}(X_N)$ is convex, where*

$$L(c, v, \eta) \triangleq f(c, v) + \langle G(c, v), \eta \rangle,$$

then $\hat{f}$ is differentiable on N with (Fréchet) derivative

$$\hat{f}'(v) = L_v'(\hat{c}(v), v, \hat{\eta}(v)) = f_v'(\hat{c}(v), v) + [G_v'(\hat{c}(v), v)]^* \hat{\eta}(v).$$

Proof. The idea of the proof is identical to that of the proof of Theorem A.2, but the estimates are not so easy to obtain. For some point $v_0 \in N$ and element $h \in \mathscr{V}$, consider a sequence $v^n = v^0 + t_n h \in N$ such that $t_n \to 0^+$ and denote $c^0 = \hat{c}(v^0)$, $c^n = \hat{c}(v^n)$, $\eta^0 = \hat{\eta}(v^0)$, $\eta^n = \hat{\eta}(v^n)$. For some $\alpha_n \in [0, 1]$

$$L(c^0, v^n, \eta^0) - L(c^0, v^0, \eta^0) = \langle L_v'(c^0, v^0 + \alpha_n t_n h, \eta^0), t_n h \rangle$$
$$= L(c^0, v^n, \eta^n) - L(c^0, v^0, \eta^0) + \langle G(c^0, v^n), \eta^0 - \eta^n \rangle$$
$$\ge \hat{f}(v^n) - \hat{f}(v^0) + \langle G(c^0, v^n), \eta^0 - \eta^n \rangle,$$

since $L(\cdot, v^n, \eta^n)$ is pseudoconvex on the convex set $\Pi_{\mathscr{C}'}(X_N)$. Furthermore,

$$\langle L_v'(c^0, v^0 + \alpha_n t_n h, \eta^0), t_n h\rangle \geq \hat{f}(v^n) - \hat{f}(v^0)$$
$$+ \langle G(c^0, v^n) - G(c^0, v^0), \eta^0 - \eta^n\rangle - \langle G(c^0, v^0), \eta^n\rangle$$
$$\geq \hat{f}(v^n) - \hat{f}(v^0) + \langle G(c^0, v^n) - G(c^0, v^0), \eta^0 - \eta^n\rangle$$

since $\eta^n \in -S^*$ and $G(c^0, v^0) \in S$. Hence, for some $\beta_n \in [0, 1]$

$$\frac{1}{t_n}(\hat{f}(v^n) - \hat{f}(v^0)) \leq \langle L_v'(c^0, v^0 + \alpha_n t_n h, \eta^0), h\rangle$$

$$- \langle G_v'(c^0, v^0 + \beta_n t_n h)h, \eta^0 - \eta^n\rangle.$$

Arguing as above, one can also obtain

$$\frac{1}{t_n}(\hat{f}(v^n) - \hat{f}(v^0)) \geq \langle L_v'(c^n, v^0 + \alpha_n^0 t_n h, \eta^n), h\rangle$$

$$+ \langle G_v'(c^n, v^0 + \beta_n^0 t_n h)h, \eta^0 - \eta^n\rangle.$$

The remainder of the proof closely follows that of Theorem A.2. $\square$

It should be noted that Theorem A.3 is not a special case of Theorem 2.7 combined with Corollary 2.8 or vice versa, even if one neglects the obvious differences of the more complex form of local set $C(v)$ in the direct method. It simply creates another approach to the highly complicated problem of differentiability in the general case.

A.2. PROOF OF LEMMA 2.13

Using norm $\|\cdot\|_A$ we can obtain the lower and upper bounds for the scalar product;

$$\langle T(\lambda^{(n)} + h^{(n)}) - T(\lambda^{(n)}), h^{(n)}\rangle,$$

where $h^{(n)} = -\varepsilon_n A T(\lambda^{(n)})$ and $\lambda^{(n)}, \lambda^{(n)} + h^{(n)} \in \mathscr{P}_1$. The lower bound results from the inequality:

$$\langle T(\lambda^{(n)} + h^{(n)}) - T(\lambda^{(n)}), h^{(n)}\rangle \geq \sigma \varepsilon_n^2 \mu_1 \|T(\lambda^{(n)})\|_A^2 - \varepsilon_n^2 \mu_2 \sigma_{II}^2 \|T(\lambda^{(n)})\|_A^2.$$

On the other hand

$$\langle T(\lambda^{(n)} + h^{(n)}) - T(\lambda^{(n)}), h^{(n)}\rangle$$
$$= \langle [W(\lambda^{(n)} + h^{(n)}) + s(\lambda^{(n)} + h^{(n)})] - [W(\lambda^{(n)}) + s(\lambda^{(n)})],$$
$$- \varepsilon_n A [W(\lambda^{(n)}) + s(\lambda^{(n)})]\rangle$$
$$= \langle [W(\lambda^{(n)} + h^{(n)}) + s(\lambda^{(n)} + h^{(n)}) - [W(\lambda^{(n)}) + s(\lambda^{(n)})],$$

$$\varepsilon_n A[W(\lambda^{(n)} + h^{(n)}) + s(\lambda^{(n)} + h^{(n)})] - \varepsilon_n A[W(\lambda^{(n)}) + s(\lambda^{(n)})]\rangle$$

$$- \varepsilon_n \langle [W(\lambda^{(n)} + h^{(n)}) + s(\lambda^{(n)} + h^{(n)})] - [W(\lambda^{(n)}) + s(\lambda^{(n)})],$$

$$A[W(\lambda^{(n)} + h^{(n)}) + s(\lambda^{(n)} + h^{(n)})]\rangle \le \varepsilon_n \|W(\lambda^{(n)} + h^{(n)}) - W(\lambda^{(n)})\|_A^2$$

$$+ 2\varepsilon_n \|s(\lambda^{(n)} + h^{(n)}) - s(\lambda^{(n)})\|_A \|W(\lambda^{(n)} + h^{(n)}) - W(\lambda^{(n)})\|_A$$

$$+ \varepsilon_n \|s(\lambda^{(n)} + h^{(n)}) - s(\lambda^{(n)})\|_A^2 - \varepsilon_n \|T(\lambda^{(n)} + h^{(n)})\|_A^2$$

$$+ \varepsilon_n \|T(\lambda^{(n)} + h^{(n)})\|_A \cdot \|T(\lambda^{(n)})\|_A \le \varepsilon_n^3 \mu_2^2 \sigma_I^2 \|T(\lambda^{(n)})\|_A^2$$

$$+ 2\varepsilon_n^3 \mu_2^2 \sigma_I \sigma_{II} \|T(\lambda^{(n)})\|_A^2 + \varepsilon_n^3 \mu_2^2 \sigma_{II}^2 \|T(\lambda^{(n)})\|_A^2$$

$$- \varepsilon_n \|T(\lambda^{(n)} + h^{(n)})\|_A^2 + \varepsilon_n \|T(\lambda^{(n)} + h^{(n)})\|_A \|T(\lambda^{(n)})\|_A$$

From the above inequalities we obtain

$$\|T(\lambda^{(n)} + h^{(n)})\|_A^2 - \|T(\lambda^{(n)})\|_A \|T(\lambda^{(n)} + h^{(n)})\|_A$$
$$+ (\sigma\varepsilon_n\mu_1 - \varepsilon_n\mu_2\sigma_{II} - \varepsilon_n^2\mu_2^2(\sigma_I + \sigma_{II})^2) \|T(\lambda^{(n)})\|_A^2 \le 0.$$

We may solve this inequality for $\|T(\lambda^{(n)} + h^{(n)})\|_A$:

$$\|T(\lambda^{(n)} + h^{(n)})\|_A \le \frac{1 + \sqrt{\Delta_1}}{2} \|T(\lambda^{(n)})\|_A,$$

where

$$\Delta_1 = 1 - 4\sigma\varepsilon_n\mu_1 + 4\varepsilon_n\mu_2\sigma_{II} + 4\varepsilon_n^2\mu_2^2(\sigma_I + \sigma_{II})^2$$

is positive $\forall \varepsilon_n \ge 0$. We solve for ε_n the inequality

$$(1) \qquad \frac{1 + \sqrt{\Delta_1}}{2} \le q, \quad \text{where} \quad \tfrac{1}{2} \le q < 1.$$

After rearranging the terms, we have

$$4\mu_2^2(\sigma_I + \sigma_{II})^2 \varepsilon_n^2 - (4\sigma\mu_1 - 4\sigma_{II}\mu_2)\varepsilon_n + 1 - (2q-1)^2 \le 0.$$

According to assumption 3 and because, $\tfrac{1}{2} \le q < 1$, the above inequality has real positive solutions whenever

$$\Delta_2 = 16(\sigma\mu_1 - \sigma_{II}\mu_2)^2 - 16\mu_2^2(\sigma_I + \sigma_{II})^2 + 16\mu_2^2(\sigma_I + \sigma_{II})^2(2q-1)^2 \ge 0.$$

This condition is equivalent to the following

$$q \ge \frac{1}{2}\left(1 + \sqrt{1 - \frac{(\sigma\mu_1 - \sigma_{II}\mu_2)^2}{\mu_2^2(\sigma_I + \sigma_{II})^2}}\right) = q_0, \quad \text{where} \quad q_0 < 1.$$

For any given $q \in [q_0, 1)$, inequality (1) will be satisfied if all ε_n satisfy the condition:

$$0 \le \varepsilon_q' = \frac{4\sigma\mu_1 - 4\sigma_{II}\mu_2 - \sqrt{\Delta_2}}{8\mu_2^2(\sigma_I + \sigma_{II})^2} \le \varepsilon_n \le \frac{4\sigma\mu_1 - 4\sigma_{II}\mu_2 + \sqrt{\Delta_2}}{8\mu_2^2(\sigma_I + \sigma_{II})^2} = \varepsilon_q''$$

Note that if $q \to 1^-$, then $\varepsilon'_q \to 0^+$ and

$$\varepsilon''_q \to \left(\frac{\sigma \mu_1 - \sigma_{II} \mu_2}{\mu_2^2 (\sigma_I + \sigma_{II})^2} \right)^{-1}. \qquad \square$$

A.3.   PROOF OF THEOREM 2.18

It can be shown that $\lambda^{(1)} \in \bar{\mathscr{B}}_1$; indeed

$$\|\lambda^{(1)} - \lambda^{(0)}\| = \varepsilon_0 \|AT(\lambda^{(0)})\| \leq \varepsilon''_q \sqrt{\mu_2} \|T(\lambda^{(0)})\|_A$$

$$\leq \varepsilon''_q \mu_2 \|T(\lambda^{(0)})\| \leq \varepsilon''_q \mu_2 \frac{1}{1-q} \|T(\lambda^{(0)})\|.$$

Therefore, from Lemma 2.13 we can write

$$\|\lambda^{(2)} - \lambda^{(0)}\| \leq \|\lambda^{(2)} - \lambda^{(1)}\| + \|\lambda^{(1)} - \lambda^{(0)}\|$$

$$\leq \varepsilon''_q \sqrt{\mu_2} [\|T(\lambda^{(1)})\|_A + \|T(\lambda^{(0)})\|_A]$$

$$\leq \varepsilon''_q \sqrt{\mu_2} [\|T(\lambda^{(0)})\|_A + q\|T(\lambda^{(0)})\|_A]$$

$$\leq \varepsilon''_q \mu_2 (1+q) \|T(\lambda^{(0)})\| \leq \varepsilon''_q \mu_2 \frac{1}{1-q} \|T(\lambda^{(0)})\|.$$

Similarly, we can show that $\forall n \geq 1$

$$\|\lambda^{(n)} - \lambda^{(0)}\| \leq \varepsilon''_q \mu_2 \frac{1}{1-q} \|T(\lambda^{(0)})\|.$$

Therefore, the sequence $\{\lambda^{(n)}\}_{n=0}^{\infty}$ belongs to $\bar{\mathscr{B}}_1$, and it follows from Lemma 2.13 that

$$\|T(\lambda^{(n)})\|_A \to 0, \quad \text{hence} \quad \|T(\lambda^{(n)})\| \to 0$$

Now, $\forall m > n$ we have:

$$\|T(\lambda^{(m)})\| \geq \left\langle T(\lambda^{(m)}), \frac{\lambda^{(m)} - \lambda^{(n)}}{\|\lambda^{(m)} - \lambda^{(n)}\|} \right\rangle$$

$$\geq \left\langle T(\lambda^{(m)}) - T(\lambda^{(n)}), \frac{\lambda^{(m)} - \lambda^{(n)}}{\|\lambda^{(m)} - \lambda^{(n)}\|} \right\rangle - \|T(\lambda^{(n)})\|$$

$$\geq \frac{1}{\|\lambda^{(m)} - \lambda^{(n)}\|} \delta \|\lambda^{(m)} - \lambda^{(n)}\|^2 - \|T(\lambda^{(n)})\|,$$

where $\delta = \sigma - \sigma_{II}$ and $\delta > 0$ (since $\sigma \mu_1 > \sigma_{II} \mu_2$ and $\mu_1 \leq \mu_2$). But

$$\|T(\lambda^{(m)})\|_A < \|T(\lambda^{(n)})\|_A$$

436

and thus

$$\sqrt{\mu_1}\,\|T(\lambda^{(m)})\| \le \sqrt{\mu_2}\,\|T(\lambda^{(n)})\|.$$

Therefore

$$\|\lambda^{(m)} - \lambda^{(n)}\| \le \frac{1}{\delta}\left(1 + \sqrt{\frac{\mu_2}{\mu_1}}\right)\|T(\lambda^{(n)})\|.$$

Since $\|T(\lambda^{(n)})\| \to 0$, $\{\lambda^{(n)}\}_{n=0}^{\infty}$ is a Cauchy sequence, so $\lambda^{(n)} \to \tilde{\lambda} \in \bar{\mathcal{B}}_1$ and because $T(\cdot)$ is continuous, $T(\lambda^{(n)}) \to T(\tilde{\lambda}) = \theta$. Of course, $\tilde{\lambda}$ is a unique solution of Eq. (2.66) in $\mathcal{P}_1$. Indeed, let us assume that

$$\exists \tilde{\lambda}^1, \tilde{\lambda}^2 \in \mathcal{P}_1 \quad \text{and} \quad T(\tilde{\lambda}^1) = T(\tilde{\lambda}^2) = \theta.$$

Then

$$0 = \langle T(\tilde{\lambda}^1) - T(\tilde{\lambda}^2), \tilde{\lambda}^1 - \tilde{\lambda}^2 \rangle \ge \delta\,\|\tilde{\lambda}^1 - \tilde{\lambda}^2\|^2.$$

Since $\delta > 0$, $\tilde{\lambda}^1 = \tilde{\lambda}^2$. $\square$

A.4. PROOF OF LEMMA 2.14

We show first that condition 1(ii) of Lemma 2.13 is fulfilled. Let us take any two points $\lambda^1, \lambda^2 \in \mathcal{P}_1$. For any $w \in D$ and $\forall \rho \in [0, 1]$ the following inequalities are satisfied:

(1) $\quad \rho L(w, \lambda^1) \ge \rho L(\hat{w}(\lambda^1), \lambda^1) + \sigma_3 \rho(1-\rho)\|\hat{w}(\lambda^1) - w\|^2$
$$(1-\rho)L(w, \lambda^2) \ge (1-\rho)L(\hat{w}(\lambda^2), \lambda^2) + \sigma_3 \rho(1-\rho)\|\hat{w}(\lambda^2) - w\|^2.$$

From this we obtain

$$\rho(-\varphi(\lambda^1)) + (1-\rho)(-\varphi(\lambda^2)) + \tfrac{1}{2}\sigma_3\rho(1-\rho)\|\hat{w}(\lambda^1) - \hat{w}(\lambda^2)\|^2$$
$$\le \rho L(w, \lambda^1) + (1-\rho)L(w, \lambda^2) = L(w, \rho\lambda^1 + (1-\rho)\lambda^2)$$

where $\varphi(\lambda) = -\varphi_1(\lambda)$ and $\forall \lambda \in \mathcal{P}_1$

$$\varphi_1(\lambda) = \min_{CU} L(w, \lambda) = \min_{D} L(w, \lambda) \quad \text{[assumption 1]}.$$

The above inequality holds for any $w \in D$. Therefore, it follows from the inequality and from assumption (3) that $\forall \lambda^1, \lambda^2 \in \mathcal{P}_1$ and $\forall \rho \in [0, 1]$:

$$\rho\varphi(\lambda^1) + (1-\rho)\varphi(\lambda^2) \ge \varphi(\rho\lambda^1 + (1-\rho)\lambda^2) + \tfrac{1}{2}\sigma_1^2\sigma_3\rho(1-\rho)\|\lambda^1 - \lambda^2\|^2.$$

Since $\forall \lambda \in \mathcal{P}_0 \ \nabla\varphi(\lambda) = W(\lambda)$ we have

$$\forall \lambda, \lambda + h \in \mathcal{P}_1 \quad \text{and} \quad \forall t \in (0, 1):$$
$$t\varphi(\lambda + h) - t\varphi(\lambda) \ge \varphi(\lambda + th) - \varphi(\lambda) + \tfrac{1}{2}\sigma_1^2\sigma_3 t(1-t)\|h\|^2.$$

After dividing the above inequality by t and taking the limit at $t \to 0^+$, we obtain

$$\varphi(\lambda + h) - \varphi(\lambda) \geq \langle W(\lambda), h \rangle + \tfrac{1}{2}\sigma_1^2\sigma_3 \|h\|^2$$

and similarly,

$$\varphi(\lambda) - \varphi(\lambda + h) \geq \langle W(\lambda + h), -h \rangle + \tfrac{1}{2}\sigma_1^2\sigma_3 \|h\|^2.$$

Therefore

$$\langle W(\lambda + h) - W(\lambda), h \rangle \geq \sigma_1^2\sigma_3 \|h\|^2.$$

Thus, condition 1(ii) of Lemma 2.13 is satisfied on $\mathscr{P}_1$ with $\sigma = \sigma_1^2\sigma_3 > 0$.

Now it can be seen that condition 1(i) of Lemma 2.13 will be satisfied if $\forall \lambda^1, \lambda^2 \in \mathscr{P}_1$

$$\|\hat{w}(\lambda^1) - \hat{w}(\lambda^2)\| \leq \sigma_5 \|\lambda^1 - \lambda^2\|.$$

From assumptions 1 and 4 of Lemma 2.14, and inequality (1) above, it follows that $\forall \rho \in (0, 1)$

$$\rho L(\hat{w}(\lambda^2), \lambda^1) - \rho L(\hat{w}(\lambda^1), \lambda^1) \geq \sigma_3\rho(1 - \rho)\|\hat{w}(\lambda^1) - w(\lambda^2)\|^2$$
$$\rho L(\hat{w}(\lambda^1), \lambda^2) - \rho L(\hat{w}(\lambda^2), \lambda^2) \geq \sigma_3\rho(1 - \rho)\|\hat{w}(\lambda^1) - \hat{w}(\lambda^2)\|^2.$$

After dividing the above inequalities by ρ, adding, and taking the limit at $\rho \to 0^+$ we obtain:

$$2\sigma_3 \|\hat{w}(\lambda^1) - \hat{w}(\lambda^2)\|^2 \leq \langle V(\hat{w}(\lambda^1)) - V(\hat{w}(\lambda^2)), \lambda^2 - \lambda^1 \rangle$$
$$\leq \sigma_2 \|\hat{w}(\lambda^1) - \hat{w}(\lambda^2)\| \|\lambda^2 - \lambda^1\|$$

Therefore, $\forall \lambda, \lambda + h \in \mathscr{P}_1$ we have

$$\|W(\lambda + h) - W(\lambda)\| = \|-V(\hat{w}(\lambda + h)) + V(\hat{w}(\lambda))\|$$
$$\leq \sigma_2\|\hat{w}(\lambda + h) - \hat{w}(\lambda)\| \leq \frac{\sigma_2^2}{2\sigma_3} \|h\|.$$

Thus, condition 1(i) of Lemma 2.13 is fulfilled on $\mathscr{P}_1$ with $\sigma_I = \sigma_2^2/2\sigma_3$. $\quad\square$

A.5. PROOF OF LEMMA 2.15

From the strong convexity of Lagrangian L (assumption 4) and from assumption 3, it follows that (see section A.4)

$$\forall \lambda^1, \lambda^2 \in \mathscr{P}_1 \quad \|\hat{w}(\lambda^1) - \hat{w}(\lambda^2)\| \leq \sigma_5 \|\lambda^1 - \lambda^2\|.$$

Since $V(\cdot)$ is Lipschitz continuous on D, condition (i) in assumption 1 of Lemma 2.13 is satisfied.

438

Let us define the following functional

$$\mathcal{L}(w, \lambda, \eta) = L(w, \lambda) + \langle \eta, G(w) \rangle, \tag{A.15}$$

where $\eta \in \mathcal{D}^*$. It follows from assumption 1 that IP satisfies the Kuhn–Tucker conditions for all $\lambda \in \mathcal{P}_1$; in particular

$$\mathcal{L}'_w(\hat{w}(\lambda), \lambda, \eta_\lambda) = \theta$$

or

$$Q'_w(\hat{w}(\lambda)) + [V'_w(\hat{w}(\lambda))]^* \lambda + [G'_w(\hat{w}(\lambda))]^* \eta_\lambda = \theta, \tag{A.16}$$

where $\eta_\lambda \in -D^*$ and $\|\eta_\lambda\|$ is uniformly bounded for $\lambda \in \mathcal{P}_1$ (see assumption 5). Therefore

$$\forall \lambda \in \mathcal{P}_1 \quad \|\mathcal{L}''_{ww}(w, \lambda, \eta_\lambda)\| \le \mu_5.$$

$\forall \lambda^1, \lambda^2 \in \mathcal{P}_1$ we can write

$$Q'_w(\hat{w}(\lambda^1)) + [V'_w(\hat{w}(\lambda^1))]^* \lambda^2 + [G'_w(\hat{w}(\lambda^1))]^* \eta_{\lambda_2}$$
$$- Q'_w(\hat{w}(\lambda^2)) - [V'_w(\hat{w}(\lambda^2))]^* \lambda^2 - [G'_w(\hat{w}(\lambda^2))]^* \eta_{\lambda_2}$$
$$+ [V'_w(\hat{w}(\lambda^1))]^* \lambda^1 - [V'_w(\hat{w}(\lambda^1))]^* \lambda^2$$
$$+ [G'_w(\hat{w}(\lambda^1))]^* \eta \lambda_1 - [G'_w(\hat{w}(\lambda^1))]^* \eta_{\lambda_2} = 0$$

or

$$\mathcal{L}'_w(\hat{w}(\lambda^1), \lambda^2, \eta_{\lambda_2}) - \mathcal{L}'_w(\hat{w}(\lambda^2), \lambda^2, \eta_{\lambda_2}) +$$
$$+ [V'_w(\hat{w}(\lambda^1))]^* (\lambda^1 - \lambda^2) + [G'_w(\hat{w}(\lambda^1))]^* (\eta_{\lambda_1} - \eta_{\lambda_2}) = 0.$$

From the above equality and assumption 5 we obtain

$$\nu \|\lambda^1 - \lambda^2\| \le \|[V'_w(\hat{w}(\lambda^1))]^* (\lambda^1 - \lambda^2) + [G'_w(\hat{w}(\lambda^1))]^* (\eta_{\lambda_1} - \eta_{\lambda_2})\|$$
$$= \|\mathcal{L}'_w(\hat{w}(\lambda^1), \lambda^2, \eta_{\lambda_2}) - \mathcal{L}'_w(\hat{w}(\lambda^2), \lambda^2, \eta_{\lambda_2})\|$$
$$\le \sup_{w \in D} \|\mathcal{L}''_{ww}(w, \lambda^2, \eta_{\lambda_2})\| \cdot \|\hat{w}(\lambda^1) - \hat{w}(\lambda^2)\|$$
$$\le \mu_5 \|\hat{w}(\lambda^1) - \hat{w}(\lambda^2)\|.$$

Therefore $\forall \lambda^1, \lambda^2 \in \mathcal{P}_1$,

$$\|\hat{w}(\lambda^1) - \hat{w}(\lambda^2)\| \ge \frac{\nu}{\mu_5} \|\lambda^1 - \lambda^2\|,$$

and assumption 3 of Lemma 2.14 is satisfied on $\mathcal{P}_1$. The rest of the proof follows from Lemma 2.14. $\square$

A.6. CONSTRUCTION OF OPERATOR A FOR ALGORITHM (2.68)

Let us suppose that the assumptions of Lemma 2.15 are satisfied and make a temporary assumption that $\hat{w}(\cdot)$ and $\hat{\eta}(\cdot)$ are differentiable functions of

$\lambda \ \forall \lambda \in \mathscr{P}_1$, where $\eta(\lambda) = \eta_\lambda$; see section A.5. Then we can write the derivative of mapping $W(\cdot)$ in the form (see Eq. (2.65))

$$W'_\lambda(\lambda) = -V'_w(\hat{w}(\lambda)) \cdot \hat{w}'(\lambda) \tag{A.17}$$

After taking the derivative of Eq. (A.16), we obtain (see Eq. A.15):

$$L''_{ww}(\hat{w}(\lambda), \lambda)\hat{w}'(\lambda) + [V'_w(\hat{w}(\lambda))]^* + ([G'_w(\hat{w}(\lambda))]^*\hat{\eta}(\lambda))_w\hat{w}'(\lambda)$$
$$+ [G'_w(\hat{w}(\lambda))]^*\hat{\eta}'(\lambda) = \theta.$$

Discarding two last terms in the above equation, we can compute the approximation of $\hat{w}'(\lambda)$ in the form:

$$\hat{w}'(\lambda) \simeq -[L''_{ww}(\hat{w}(\lambda), \lambda)]^{-1}[V'_m(\hat{w}(\lambda))]^*.$$

Substituting this into (A.17), we obtain the approximation of $W'(\lambda)$. This expression can be further simplified if we fix the value of λ at, e.g., $\lambda = \lambda^{(0)}$; then

$$W'_\lambda(\lambda) \simeq A_0 = V'_w(\hat{w}(\lambda^{(0)}))[L''_{ww}(\hat{w}(\lambda^{(0)}), \lambda^{(0)})]^{-1}[V'_w(\hat{w}(\lambda^{(0)}))]^*.$$

In many cases, it is convenient to take the identity operator I or some other self-conjugated, strongly positive definite operator B instead of $[L''_{ww}(\hat{w}(\lambda^{(0)}), \lambda^{(0)})]^{-1}$. Then we obtain

$$A_0 = V'_w(\hat{w}(\lambda^{(0)}))B[V'_w(\hat{w}(\lambda^{(0)}))]^*$$

and operator A in algorithm (2.68) can be defined as

$$A = A_0^{-1}. \tag{A.18}$$

Of course, from the considerations presented in section 2.4.3, it follows that we can use any strongly positive definite self-conjugated operator A. However for practical reasons it is often convenient to use Eq. (A.18) for determining A, for example, it usually simplifies the choice of a good value of $\varepsilon = \varepsilon_n$ in (2.68).

B · Appendix to Chapter 3

B.1. PROOF OF THEOREM 3.5

Let us take any point $(c^0, u^0) \in CU_n$ and define $s \in \mathcal{U}$ as $s = HF_*(c^0, u^0) - HF(c^0, u^0)$. Then $s \in S$ and according to assumption 3 there exists $\hat{\lambda} \in \mathcal{U}$ such that $\hat{\lambda}_s$ coordinates problem (3.35) and thus $u^m(s) = \hat{u}(\hat{\lambda}_s)$. Therefore

$$u^m(s) - HF(c^m(s), u^m(s)) = s = HF_*(c^0, u^0) - HF(c^0, u^0).$$

Let us now define mapping $W: CU \to CU$ as $W = W_2 \circ W_1$, where

$$W_1: CU \ni (c, u) \to s = HF_*(c, u) - HF(c, u) \in S$$

and

$$W_2: S \ni s \to (c^m(s), u^m(s)) \in CU.$$

According to assumption 2, W_1 is weakly continuous on CU. Since W_2 is also continuous from assumption 4, we know that mapping W is continuous. Therefore, by application of the Shauder fix point theorem, for example, there is at least one point $(c^f, u^f) \in CU$ such that

$$W(c^f, u^f) = (c^f, u^f).$$

Thus

$$c^f = \hat{c}(\hat{\lambda}_{s_f}) = c^m(s^f), \qquad u^f = \hat{u}(\hat{\lambda}_{s_f}) = u^m(s^f)$$

and

$$\hat{u}(\hat{\lambda}_{s_f}) - HF(\hat{c}(\hat{\lambda}_{s_f}), \hat{u}(\hat{\lambda}_{s_f})) = s^f = HF_*(\hat{c}(\lambda_{s_f}), \hat{u}(\hat{\lambda}_{s_f})) - HF(\hat{c}(\hat{\lambda}_{s_f}), \hat{u}(\hat{\lambda}_{s_f})).$$

Therefore

$$\hat{u}(\hat{\lambda}_s) = HF_*(\hat{c}(\hat{\lambda}_{s_f}), \hat{u}(\hat{\lambda}_{s_f})),$$

and

$$\hat{\lambda}_{s_f} = \tilde{\lambda},$$

is the solution of the IBMF supremal problem. □

B.2. PROOF OF THEOREM 3.6

As in the proof of Lemma 2.10 in section 2.4, it can be shown that the dual function $\phi_s(\cdot)$ given by

$$\phi_s(\lambda) = \min_{cu} \left[Q(c, u) + \langle \lambda, u - HF(c, u) - s \rangle \right]$$

satisfies the following inequality for every $s \in S$

$$(1) \qquad \phi_s(0) \geq \phi_s(\lambda) \quad \text{if} \quad \|\lambda\| \geq r = \frac{k_0' - k_0''}{k_1 - k_2}.$$

Indeed, assumptions 1 and 2 imply that

$$\phi_s(\lambda) \leq k_0' - (k_1 - k_2)\|\lambda\|.$$

from which it follows that (1) holds. Therefore, for any $s \in S$

$$\arg\max_{u} \phi_s(\lambda) = \hat{\lambda}_s \in \Lambda.$$

But $\hat{c}(\lambda)$ and $\hat{u}(\lambda)$ are unique for given $\lambda \in \Lambda$ and it may be proved (see Theorem 2.16 of section 2.4), that $\nabla\phi_s(\lambda) = \hat{u}(\lambda) - HF(\hat{c}(\lambda), \hat{u}(\lambda))$ exists for every $\lambda \in \Lambda$. Since $\hat{\lambda}_s \in \Lambda$, we have $\nabla\phi_s(\hat{\lambda}_s) = 0$, and therefore $\hat{\lambda}_s$ is the coordinating price vector for problem (3.35) of Theorem 3.5. Thus, assumption 3 of Theorem 3.5 is satisfied. □

B.3. PROOF OF THEOREM 3.9

It follows fromn the assumptions of the theorem that $\|s^{(n)}(\lambda)\| \leq \beta_n$. From the definition of $W^{(n)}(\lambda)$ (Eq. (3.51)) and from Eq. (3.49) we obtain

$$\|W^{(n)}(\tilde{\lambda}^{(n)})\| \leq \beta_n.$$

Since $W^{(n)}(\hat{\lambda}^{(n)}) = 0$ then using a condition (ii) of assumption 1 in Lemma 2.13 (section 2.4) we obtain

$$\sigma \|\tilde{\lambda}^{(n)} - \hat{\lambda}^{(n)}\|^2 \leq \langle W^{(n)}(\tilde{\lambda}^{(n)}) - W^{(n)}(\hat{\lambda}^{(n)}), \tilde{\lambda}^{(n)} - \hat{\lambda}^{(n)} \rangle$$

$$\leq \|W^{(n)}(\tilde{\lambda}^{(n)})\| \cdot \|\tilde{\lambda}^{(n)} - \hat{\lambda}^{(n)}\| \leq \beta_n \|\tilde{\lambda}^{(n)} - \hat{\lambda}^{(n)}\|.$$

From the above inequality, it follows that

$$\|\tilde{\lambda}^{(n)} - \hat{\lambda}^{(n)}\| \le \frac{\beta_n}{\sigma}. \quad \square$$

B.4. PROOF OF LEMMA 3.5

From the assumptions, there exists $\varepsilon_1 > 0$, such that $\bar{K}(\bar{u}; \varepsilon_1) \subset \mathcal{V}(\bar{u})$. Let us take any $\varepsilon > 0$ such that $\varepsilon_1 > \varepsilon$. Thus, $K_1 = \bar{K}(\bar{u}, \varepsilon) \subset \mathcal{V}(\bar{u})$. Set $T(K_1)$ is open. Hence, there exists a positive number δ such that an open ball $K_2 = K(0, \delta)$ is contained in $T(K_1)$. Let us define mapping W_1 as follows:

$$K_2 \ni u \rightarrow W_1(u) = T^{-1}(u) \in K_1.$$

Let us assume that T_* is such that

(1) $$\forall u \in v(\bar{u}) \quad T(u) - T_*(u)\| < \delta.$$

Next let us define mapping W_2 as follows

$$\bar{K}_1 \ni u \rightarrow W_2(u) = \tilde{u} \in \bar{K}_1,$$

where $\tilde{u} \in K_1$ and satisfies the following equation

$$T(\tilde{u}) + (T_*(u) - T(u)) = 0.$$

Owing to relation (1) and assumption (3) of the theorem, mapping W_2 is well defined. Note that mapping W_2 is the composition of mappings W_1 and W_3, where $W_3(u) = T_*(u) - T(u)$. Since mappings W_1 and W_3 are continuous and the closed ball $\bar{K}_1$ is compact in $\mathcal{U}$, then, using Schauder's theorem, we conclude that there exists fixed point $\bar{u}_*$ of W_2 in $\bar{K}_1$. From the definition of W_2 we have

$$T(\bar{u}_*) + (T_*(\bar{u}_*) - T(\bar{u}_*)) = 0,$$

which implies that $T_*(\bar{u}_*) = 0$ and $\|\bar{u} - \bar{u}_*\| < \varepsilon$. Hence, the proof is completed. $\square$

B.5. PROOF OF LEMMA 3.6

Let us define the following set:

$$\tilde{U} \triangleq \{u \in U^0 : (\exists \text{ neighborhood } v(u) \text{ of } u \text{ in } U^0) \ (\forall u', u'' \in v(u))$$
$$I(\hat{c}(u', \lambda)) = I(\hat{c}(u'', \lambda)).$$

The set $\tilde{U}$ is open by definition. Let us take any $u^1 \in U^0 \backslash \tilde{U}$ and any neighborhood $v_1(u^1) \subset U^0$. It will be shown that there exists in $v_1(u^1)$ a point that belongs to $\tilde{U}$. For any $i \notin I(\hat{c}(u^1, \lambda))$ we have $h_i(\hat{c}(u^1, \lambda), u^1) < 0$. Functions h_i are continuous by assumption (2), thus

$$(1) \qquad (\exists K_1 \subseteq \mathscr{C} \times \mathscr{U})(\forall (c, u) \in K_1)(\forall i \notin I(\hat{c}(u^1, \lambda))) \quad h_i(c, u) < 0$$

where

$$K_1 = K_1((\hat{c}(u^1, \lambda), u^1); \varepsilon), \qquad \varepsilon > 0.$$

Due to assumption (1), mapping

$$U^0 \ni u \to (\hat{c}(u, \lambda), u) \triangleq W(u) \in \mathscr{C} \times \mathscr{U}$$

is continuous. Hence,

$$(\exists \text{ neighborhood } v_1'(u^1) \subset U^0) W(v_1'(u^1)) \subset K_1$$

and

$$(2) \qquad W(v_2(u^1)) \subset K_1$$

where

$$v_2(u^1) \triangleq v_1'(u^1) \cap v_1(u^1).$$

From the definition of W and relations (1) and (2) we conclude that

$$(3) \qquad \forall u \in v_2(u^1) \quad I(\hat{c}(u, \lambda)) \subseteq I(\hat{c}(u^1, \lambda)).$$

Since $u^1 \notin \tilde{U}$, from the definition of $\tilde{U}$ and relation (3) the following holds:

$$\exists u^2 \in v_2(u^1) \quad I(\hat{c}(u^2, \lambda)) \subset I(\hat{c}(u^1, \lambda)).$$

If $u^2 \in \tilde{U}$ then the proof is finished. If not, then we can repeat the procedure and obtain the following relations:

$$\forall u \in v_3(u^2) \quad I(\hat{c}(u, \lambda)) \subseteq I(\hat{c}(u^2, \lambda)),$$
$$\exists u^3 \in v_3(u^2) \quad I(\hat{c}(u^3, \lambda)) \subset I(\hat{c}(u^2, \lambda)),$$

where

$$v_3(u^2) = v_2'(u^2) \cap v_2(u^2) \subseteq v_1(u^1).$$

If $u^3 \notin \tilde{U}$ then we can repeat the procedure again. We will then obtain the following sequence of index sets:

$$(4) \qquad I(\hat{c}(u^1, \lambda)) \subset I(\hat{c}(u^2, \lambda)) \subset \cdots \subset I(\hat{c}(u^k, \lambda)) \subset I(\hat{c}(u^{k+1}, \lambda)). \ldots,$$

such that

$$\forall u \in v_{k+1}(u^k) \quad I(\hat{c}(u, \lambda)) \subseteq I(\hat{c}(u^k, \lambda))$$

and

$$v_{k+1}(u^k) \subseteq v_1(u^1) \quad \text{for} \quad k = 1, 2, \ldots,$$

Since the number of constraints is finite, we conclude from relation (4) that

444

the sequence $\{I(\hat{c}(u^k, \lambda))\}$ is finite. Hence, there exists a point $u^k \in \tilde{U}$ which is contained in $v(u^1)$. As a matter of fact, it has been shown that $\tilde{U} \neq \varnothing$ if $U^0 \neq \varnothing$ and $(\bar{\tilde{U}}) \supset U^0$. Since $(\bar{\tilde{U}}) \subset U^0$,

$$(5) \qquad\qquad (\bar{\tilde{U}}) = U^0.$$

Let us introduce in $\tilde{U}$ the following relation:

$$\forall u', u'' \in \tilde{U} \quad (u' \sim u'') \leftrightarrow I(\hat{c}(u', \lambda)) = I(\hat{c}(u'', \lambda)).$$

It is obvious that $\sim$ denotes the equivalence relation in $\tilde{U}$. Let us denote by U_i, $i \in \overline{1, L}$, the abstract classes generated by this relation. Hence

$$(6a) \qquad\qquad \forall i, j \in \overline{1, L} \quad U_i \cap U_j = \varnothing$$

and

$$(6b) \qquad\qquad \bigcup_{i=1}^{L} U_i = \tilde{U}.$$

Let us note that from the definition of set $\tilde{U}$, for any $u \in U_i$, $i \in \overline{1, L}$, there exists neighborhood $v(u) \subset U_i$ such that

$$\forall i \in \overline{1, L} \quad U_i = \bigcup_{u \in U_i} v(u),$$

which implies that U_i is open in U^0 for any $i \in \overline{1, L}$. Hence, the proof of the lemma is completed as a result of relations (5) and (6). $\quad\square$

B.6. PROOF OF LEMMA 3.7

Let $u \in U^0$ be given. From assumptions (2), (3), and (4) and the Kuhn–Tucker theorem (Luenberger 1969), there exists vector $\mu = (\mu_{j1}, \ldots, \mu_{js})$ such that

$$(1) \qquad (Q_{\text{mod}})'_c(c^i(u, \lambda), u) + h_c^{T'}(c^i(u, \lambda), u)\ \mu = 0.$$

and

$$h(c^i(u, \lambda), u) = 0.$$

Let us define operator $\Phi : \mathscr{C} \times \mathscr{U} \times \mathbb{R}^s \ni (c, u, \mu) \to \mathscr{C} \times \mathbb{R}^s$ as follows:

$$\Phi(c, u, \mu) \triangleq \begin{bmatrix} (Q_{\text{mod}})'_c(c, u) + h_c^{T'}(c, u)\mu \\ h^T(c, u) \end{bmatrix}.$$

According to (1) we have

$$(2) \qquad\qquad \forall u \in U^0 \quad \Phi(c^i(u, \lambda), u, \mu) = 0.$$

Thus, thanks to assumption (5), we can apply an implicit function theorem. By this theorem, there exists an open set $\tilde{U}^0$, $\tilde{U}^0 \supset U^0$, and continuously differentiable mappings $T^i : \tilde{U}^0 \to \mathscr{C}$, $M^i : \tilde{U}^0 \to \mathbb{R}^s$ such that

$$(\forall i \in \overline{1, L})(\forall u \in U^0) \quad T^i(u) = c^i(u, \lambda),$$

(3)

$$(\forall i \in \overline{1, L})(\forall u \in \tilde{U}^0) \quad \Phi(T^i(u), u, M^i(u)) = 0$$

and

(4)
$$(\forall i \in \overline{1, L})(\forall u \in U^0) \quad (T_u^{i\prime}(u), M_u^{i\prime}(u))$$
$$= -[\Phi_{c,\mu}'(T^i(u), u, M^i(u))]^{-1} \circ \Phi_u(T^i(u), u, M^i(u)).$$

It will now be shown that

(5)
$$(\forall i \in \overline{1, L})(\forall u \in \overline{U}_i) \quad T^i(u) = \hat{c}(u, \lambda).$$

Let us assume that $u \in U_i$. From the definition of $\hat{c}(\cdot, \lambda)$ and the Kuhn–Tucker necessary conditions we have

$$(Q_{\mathrm{mod}})_c'(\hat{c}(u, \lambda), u) + h_c^T(\hat{c}(u, \lambda), u)\mu = 0$$

and

$$h(\hat{c}(u, \lambda), u) = 0.$$

Hence, because Eq. (2) has a unique solution, we know that $\hat{c}(u, \lambda) = c^i(u, \lambda)$ and, from (3), that $\hat{c}(u, \lambda) = T^i(u)$. We next assume that $u \in \overline{U}_i$. So, there exists sequence $\{u^k\} \to u$ such that $(\forall k = 1, 2, \ldots)u^k \in U_i$. From the continuity of $\hat{c}(\cdot, \lambda)$ (see assumption (1)) and T^i, the following holds:

$$\hat{c}(u, \lambda) = \lim_{k \to \infty} \hat{c}(u^k, \lambda) = \lim_{k \to \infty} T^i(u^k) = T^i(u).$$

Hence, Eq. (5) is true and the proof of part 1 is complete. From Eq. (4) and the assumption of part 2 we can easily derive the formula on $T_u^{i\prime}(u)$, which completes the proof. $\square$

B.7. CONDITIONS FOR SATISFYING ASSUMPTION 3 OF THEOREM 3.13

Let λ be fixed in Λ_b.

LEMMA B.1. *Let us suppose that int $U^0 \neq \varnothing$ and let us take $u \in \mathrm{int}\, U^0$. We denote $J \triangleq \{i \in \overline{1, L} : u \in \overline{U}_i\}$. If we assume that*

1. *The assumptions of Lemma 3.6 are satisfied,*
2. *There exists neighborhood $v'(u)$ of u such that the assumptions of Lemma 3.7 are satisfied on $v'(u)$,*
3. *Mapping K is differentiable,*

4. *Any convex combination*

$$\alpha_{i_1}(H_{i_1})'_u + \ldots + \alpha_{i_s}(H_{i_s})'_u$$

where

$$i_n \in J \quad \text{for} \quad n = 1, \ldots, s,$$

$$\sum_{n=1}^{s} \alpha_{i_n} = 1 \quad \text{and} \quad \alpha_{i_n} \geq 0 \quad \text{for} \quad n = 1, \ldots, s,$$

$$H_{i_n}(u) \triangleq u - HK(c^{i_n}(u, \lambda)) \quad \text{for} \quad n = 1, \ldots, s,$$

is a homeomorphism,

then there exists neighborhood $v(u)$ such that $v(u) \subset v'(u)$ and mapping $I - HK(\hat{c}(\cdot, \lambda))$ is a local homeomorphism on $v(u)$.

The proof of this lemma is given in Ulanicki (1978). The lemma is an extension of the well-known local inverse theorem (Kantorovich and Akilov 1964) for nondifferentiable operators. The main assumption of this theorem is that the operator is differentiable with a reversible derivative. The theorem cannot be used here because the operator is not differentiable.

B.8. PROPERTIES OF AN ITERATIVE SCHEME

Consider a functional $f : \mathbb{R}^n \times A \to \mathbb{R}$, where $A \subset \mathbb{R}^m$. Let us assume that the following parametric optimization problem:

$$\min_{x} f(x, \alpha)$$

has a solution for any $\alpha \in A$. For given parameter sequence $\{\alpha^k\}$ such that $\alpha^k \to \tilde{\alpha}$ if $k \to \infty$, let us consider the following iterative scheme:

$$(1) \qquad x^{k+1} = x^k - \rho^k f'_x(x^k, \alpha^k)$$

where $\rho^k, k = 0, 1, \ldots$, are positive step coefficients such that

$$|\alpha^{k+1} - \alpha^k| \leq \delta^k \to 0 \quad \text{if} \quad k \to \infty$$

and

$$\rho^k \to 0, \; \sum_{k=0}^{\infty} \rho^k = \infty, \; \frac{\delta^k}{\rho^k} \to 0 \quad \text{if} \quad k \to \infty.$$

The property of sequence (1) is formulated in the following theorem (Nurminsky 1977).

THEOREM. B.1. *If we assume that*

1. *Set A is compact*
2. *Functional $f(\cdot, \alpha)$ is convex and a functional $f(x, \cdot)$ satisfies the Lipschitz condition with a constant independent of $\alpha \in A$*
3. *Sequence $\{x^k\}$ is bounded*

then the following holds:

$$\lim_{k \to \infty} \left[f(x^k, \alpha^k) - \min_x f(x, \alpha^k) \right] = 0.$$

B.9. CLARIFICATION OF THEOREM 3.21

We say that the mapping $f: X \times Y \to Z$ is convexlike on set $X(Y)$ with respect to set S if and only if the following relation holds

$$(\forall x_1, x_2 \in X)(\forall t \in [0, 1])(\exists x \in X)(\forall y \in Y) \quad tf(x_1, y)$$
$$+ (1 - t)f(x_2, y) - f(x, y) \in S,$$

or, for a concavelike function,

$$(\forall y_1, y_2 \in Y)(\forall t \in [0, 1])(\exists y \in Y)(\forall x \in X) \quad f(x, y) - tf(x, y_1) - (1 - t)f(x, y_2) \in S.$$

The definition of a convexlike function, that is, the case when $Z = \mathbb{R}$ and $S = \{z \in Z : z \geq 0\}$, is given in Sion (1958).

THEOREM B.2. (*Kneser–Fan*). *Consider a pair of topological spaces X and Y, and functional $f: X \times Y \to R$. If*

1. *Functional f is convexlike on X and concavelike on Y,*
2. *Space X is compact and for each $y \in Y$ the functional $f(\cdot, y)$ is lower semicontinuous on X.*

Then the following relation holds:

$$\sup_{y \in Y} \inf_{x \in X} f(x, y) = \inf_{x \in X} \sup_{y \in Y} f(x, y).$$

The proof is given in Sion (1958).

B.10. PROOF OF PROPOSITION 3.28

Let us assume that X^k is fixed and that there exists sequence $\{\lambda^n\}$ that converges to λ^0 such that for each n, $x^n \notin \omega$ where $x^n = \hat{x}(\lambda^n, X^k)$. We can

448

write:

(1) $\qquad Q(x^0)+\langle\lambda^n, P(x^0)\rangle = Q(x^0)+\langle\lambda^0, P(x^0)\rangle+\langle\lambda^n-\lambda^0, P(x^0)\rangle.$

The assumption implies that

(2) $\quad Q(x^0)+\langle\lambda^0, P(x^0)\rangle \le Q(x^n)+\langle\lambda^0, P(x^n)\rangle-d$
$$= Q(x^0)+\langle\lambda^n, P(x^n)\rangle+\langle\lambda^0-\lambda^n, P(x^n)\rangle-d.$$

Combining (1) and (2) we have the following

$$Q(x^0)+\langle\lambda^n, P(x^0)\rangle \le Q(x^n)+\langle\lambda^n, P(x^n)\rangle-d$$
$$+\langle\lambda^n-\lambda^0, P(x^0)\rangle+\langle\lambda^0-\lambda^n, P(x^n)\rangle.$$

Because $\{\lambda^n\}\to\lambda^0$ and P is bounded on CU_* (see assumption (A1)), we can choose $n_0\ge 0$ such that for all $n\ge n_0$

$$Q(x^0)+\langle\lambda^n, P(x^0)\rangle \le Q(x^n)+\langle\lambda^n, P(x^n)\rangle-\tfrac{1}{2}d,$$

which contradicts the assumption that x^n is the infimal problem solution for $\lambda=\lambda^n$. $\quad\square$

B.11. PROOF OF LEMMA 3.13

Let us assume that the thesis of Lemma 3.13 is not true. This means that there exist $\lambda\in\Omega(\lambda_0)$, sequence $\{\lambda^n\}$ tending to λ, and point $x=\hat{x}(\lambda, X^k)$ such that $\|x^n-x\|\ge\delta$ where $x^n=\hat{x}(\lambda^n, X^k)$ and $\delta>0$. From assumption (A4) it follows that:

$$Q(x)+\langle\lambda, P(x)\rangle \le Q(x^n)+\langle\lambda, P(x^n)\rangle-v\|x^n-x\|^2 \le Q(x^n)+\langle\lambda, P(x^n)\rangle-v\delta^2.$$

We can also write

$$\langle\lambda, P(x^n)\rangle = \langle\lambda^n, P(x^n)\rangle+\langle\lambda-\lambda^n, P(x^n)\rangle$$

and

$$\langle\lambda, P(x)\rangle = \langle\lambda^n, P(x)\rangle+\langle\lambda-\lambda^n, P(x)\rangle.$$

Combining the above relations we find that

$$Q(x)+\langle\lambda^n, P(x)\rangle \le Q(x^n)+\langle\lambda^n, P(x^n)\rangle-v\delta^2+\langle\lambda-\lambda^n, P(x^n)\rangle$$
$$+\langle\lambda^n-\lambda, P(x)\rangle.$$

Taking into account that P is bounded on CU_* and $\lambda^n\to\lambda$, we conclude that there exists n_0 such that for each $n\ge n_0$

$$Q(x)+\langle\lambda^n, P(x)\rangle < Q(x^n)+\langle\lambda^n, P(x^n)\rangle,$$

which contradicts $x^n=\hat{x}(\lambda^n, X^k)$. $\quad\square$

B.12.   PROOF OF LEMMA 3.14

Let us define, for fixed X^k, an operator Φ in the following way:

$$\Phi : X \times \mathbb{R}^l \times (t_0 - \tau, t_0 + \tau) \to X \times \mathbb{R}^l$$

$$(1) \qquad \Phi(x, \mu, t) = \begin{bmatrix} Q'_x(x) + P'^*_x(x)\lambda_t + H^{k'*}_x(\lambda_t, x)\mu \\ H^k(\lambda_t, x) \end{bmatrix}.$$

We will show that by operator equation $\Phi(x, \mu, t) = 0$ we can determine an implicit function $t \to (x(t), \mu(t))$ that is differentiable in t_0. Owing to the convexity of the infimal problem and assumption (A6), the Kuhn–Tucker necessary conditions are satisfied. Hence, there exists vector $\mu(\lambda_{t_0}, X^k)$ depending on λ_{t_0} and X^k such that

$$(2) \qquad \Phi(\hat{x}(\lambda_{t_0}, X^k), \mu(\lambda_{t_0}, X^k), t_0) = 0.$$

Let us consider an operator $C \triangleq H_x W^{-1} H^*_x$. Suppose that for certain z from $\mathbb{R}^l$, $Cz = 0$. Thus, $\langle H^*_x z, W^{-1} H^*_x z \rangle = \langle z, H_x W^{-1} H^*_x z \rangle = 0$. Because $W > 0$ (see (3.165)), $W^{-1} > 0$. Hence, $H^*_x z = 0$. It follows from assumption (A6) $(\mathscr{R}(H_x) = \mathbb{R}^l)$ that $\mathscr{N}(H^*_x) = \{0\}$. This implies that $z = 0$ and that operator C is reversible. Note that $\mathscr{R}(C)$ is a closed linear subspace in $\mathbb{R}^l$ (C is finite dimensional). Let us suppose that $z \in \mathbb{R}^l$ and $z \perp \mathscr{R}(C)$. Hence, we have $\langle z, Cz \rangle = 0$ or $\langle H^*_x z, W^{-1} H^*_x z \rangle = 0$. It has been shown above that the last equality implies that $z = 0$. This means that $\mathscr{R}(C) = \mathbb{R}^l$.

Using Banach's inverse operator theorem, we conclude that the operator $(H_x W^{-1} H^*_x)^{-1}$ exists and is bounded. Let us define an operator $T : X \times \mathbb{R}^l \to X \times \mathbb{R}^l$ as follows:

$$(3) \quad T \triangleq \begin{bmatrix} W^{-1} - W^{-1} H^*_x (H_x W^{-1} H^*_x)^{-1} H_x W^{-1} & W^{-1} H^*_x (H_x W^{-1} H^*_x)^{-1} \\ (H_x W^{-1} H^*_x)^{-1} H_x W^{-1} & -(H_x W^{-1} H^*_x)^{-1} \end{bmatrix}$$

By simple calculation we can check that

$$T\Phi'_{x,\mu}(\hat{x}(\lambda_{t_0}, X^k), \mu(\lambda_{t_0}, X^k), t_0) = \Phi'_{x,\mu}(\hat{x}(\lambda_{t_0}, X^k), \mu(\lambda_{t_0}, X^k), t_0)T = I$$

Hence

$$(4) \qquad [\Phi'_{x,\mu}]^{-1}(\hat{x}(\lambda_{t_0}, X^k), \mu(\lambda_{t_0}, X^k), t_0) = T.$$

We now compute the derivative of Φ with respect to t at $(\hat{x}(\lambda_{t_0}, X^k), \mu(\lambda_{t_0}, X^k), t_0)$. Because $I^k_0(\lambda_t)$ is constant for $t_0 - \tau < t < t_0 + \tau$, $H^k(\cdot, x)$ is also constant for these values of t and

$$(5) \qquad \Phi'_t(\hat{x}(\lambda_{t_0}, X^k), \mu(\lambda_{t_0}, X^k), t_0) = \begin{bmatrix} P'^*_x(\hat{x}(\lambda_{t_0}, X^k))(\lambda_2 - \lambda_1) \\ 0 \end{bmatrix}.$$

Using the implicit function theorem (Schwartz 1967) and taking into account relations (3), (4), and (5), we obtain part 1 of the lemma with $I^k_0(\lambda_{t_0}) \neq \varnothing$.

450

For $I_0^k(\lambda_{t_0}) = \varnothing$, the proof is trivial. Hence, the proof of part 1 is completed.

Note that for all $\lambda \in \Omega(\lambda^0)$ and $X^k \in \Xi$, $B(\lambda, X^k) = B^*(\lambda, X^k)$.

For simplicity, we denote $B = B(\lambda, X^k)$. From the definition of B (see Eq. (3.167)), it follows that $BH_x^* = 0$, which implies that

$$\text{(6)} \qquad \mathcal{R}(H_x^*) \subset \mathcal{N}(B).$$

Because operator W^{-1} is positively defined, for all $z, v \in X$ we have

$$\langle z + v, W^{-1}(z + v) \rangle \geq 0.$$

Let us take

$$\text{(7)} \qquad z = -H_x^*(H_x W^{-1} H_x^*)^{-1} H_x W^{-1} v.$$

We can write:

$$\langle z + v, W^{-1}(z + v) \rangle = \langle z, W^{-1}z \rangle + 2\langle z, W^{-1}v \rangle + \langle v, W^{-1}v \rangle$$

$$\text{(8)} \qquad \begin{aligned} &= \langle H_x^*(H_x W^{-1} H_x^*)^{-1} H_x W^{-1} v, W^{-1} H_x^*(H_x W^{-1} H_x^*)^{-1} H_x W^{-1} v \rangle \\ &\quad - 2\langle H_x^*(H_x W^{-1} H_x^*) H_x W^{-1} v, W^{-1} v \rangle + \langle v, W^{-1}v \rangle \\ &= -\langle W^{-1} H_x^*(H_x W^{-1} H_x^*)^{-1} H_x W^{-1} v, v \rangle + \langle v, W^{-1}v \rangle = \langle Bv, v \rangle \end{aligned}$$

If $v \in \mathcal{N}(B)$ then $\langle Bv, v \rangle = 0$ and from the above equality it follows that

$$\langle v - (-z), W^{-1}(v - (-z)) \rangle = 0,$$

which implies that $v = -z$ and, by (7) $v \in \mathcal{R}(H_x^*)$. Hence, $\mathcal{N}(B) \subset \mathcal{R}(H_x^*)$, which, together with (6), implies Eq. (3.168) in part 2 of the lemma. Note that it follows directly from (8) that $B \geq 0$.

To prove inequality (3.169) in part 2 of the lemma, let us observe that $H_x W^{-1'} H_x^* > 0$, which implies that $(H_x W^{-1} H_x^*)^{-1} > 0$ and $W^{-1} H_x^*(H_x W^{-1} H_x^*) H_x W^{-1} > 0$. But $B = W^{-1} - W^{-1} H_x^*(H_x W^{-1} H_x^*)^{-1} H_x W^{-1}$. Thus $B \leq W^{-1}$ and the proof of part 2 of the lemma is completed. $\qquad \square$

B.13. PROOF OF PROPOSITION 3.24

Assumption (1) yields $Q(\hat{x}_1) \geq Q(\hat{x}_2)$. Let $y \in X_1$ be a point such that

$$\|y - \hat{x}_2\| = \inf_{x \in X_1} \|x - \hat{x}_2\|.$$

Of course, such a point exists because X_1 is the closed convex subset of the Hilbert space. From the mean value theorem (Schwartz 1967), it follows that there exists $\tau \in [0, 1]$ such that

$$Q(y) = Q(\hat{x}_2) + \langle Q_x'(\tau y + (1 - \tau)\hat{x}_2), y - \hat{x}_2 \rangle.$$

Therefore, according to assumption (5)

$$Q(y) \le Q(\hat{x}_2) + M\|y - \hat{x}_2\| \le Q(\hat{x}_2) + M \text{ dist } (X_1, X_2).$$

Taking into consideration assumption (2), we obtain the following inequality:

$$(1) \qquad Q(\hat{x}_1) - Q(\hat{x}_2) \le M \text{ dist } (X_1, X_2)$$

On the other hand, for certain $\tau \in [0, 1]$,

$$(2) \quad Q(\hat{x}_1) - Q(\hat{x}_2) = \langle Q'_x(\hat{x}_2), \hat{x}_1 - \hat{x}_2 \rangle$$
$$+ \tfrac{1}{2}\langle \hat{x}_1 - \hat{x}_2, Q''_{XX}(\tau\hat{x}_1 + (1-\tau)\hat{x}_2)(\hat{x}_1 - \hat{x}_2)\rangle.$$

Set X_2 is convex and therefore the section

$$[x_2, x_1] = \{x = \hat{x}_2 + \tau(\hat{x}_1 - \hat{x}_2), 0 \le \tau \le 1\},$$

is included in X_2. Hence, assumption (3) implies that the following must be true:

$$\langle Q'_x(\hat{x}_2), \hat{x}_1 - \hat{x}_2 \rangle \ge 0,$$

Consequently, we obtain the following inequality from equation (2):

$$(3) \qquad Q(\hat{x}_1) - Q(\hat{x}_2) \ge \tfrac{1}{2}m \|\hat{x}_1 - \hat{x}_2\|^2,$$

and from inequality (1) and equation (2) above:

$$\tfrac{1}{2}m \|x_1 - x_2\| \le M \text{ dist } (X_1, X_2),$$

which was to be demonstrated. $\square$

B.14. PROPOSITION NEEDED FOR PROOF OF LEMMA 3.16

Consider a linear, bounded, self-conjugated, and strongly positive operator A defined on Hilbert space X. Then

$$\forall x \in X \quad \langle x, A^{-1}x \rangle \ge \frac{1}{\|A\|} \|x^2\|.$$

Proof. The assumptions of the lemma imply that operators A^{-1} and $(A^{-1})^{1/2}$ exist and are linear, bounded, and self-conjugated operators. Additionally, $(A^{-1})^{1/2} A = A(A^{-1})^{1/2}$. Hence, we can write

$$\|x^2\| = \langle x, x \rangle = \langle x, A^{-1}Ax \rangle = \langle x, (A^{-1})^{1/2}(A^{-1})^{1/2} Ax \rangle$$

$$= \langle (A^{-1})^{1/2}x, (A^{-1})^{1/2} Ax \rangle$$

$$= \langle (A^{-1})^{1/2}x, A(A^{-1})^{1/2}x \rangle \le \|(A^{-1})^{1/2}x\|^2 \|A\|$$

$$= \|A\|\langle (A^{-1})^{1/2}x, (A^{-1})^{1/2}x \rangle = \langle x, (A^{-1})^{1/2}(A^{-1})^{1/2}x \rangle \|A\|$$

$$= \langle x, A^{-1}x \rangle \cdot \|A\|$$

which completes the proof. $\square$

Index

Page numbers in italic type indicate figures. A *t*, *T*, *L*, or *P* following a page number indicates a table, theorem, lemma, or proposition.

456

458